THE SHARK SESSIONS

many bass, ~~were~~
~~~ more after the c~~
so, since after the c~~
me distance ~~because of~~
could get right on with
thout going through the
e boat over. I forgot about
would be swimming with sharks
t and blood pouring from the
was carrying
rks swam close beside me, especially Flora
also Martha, Cerrelline, Spark, many other
- Samaria, Medeline, cut up about 1 wk ago
Bratwurst, Madona, Cochite,
Flora went 2 ft in front of me as Madonna did last
time — circling behaviour or what passes for happiness
in a shark?
had started off worried about getting the bird in the
house, then worried about the overloaded boat
the rising wind, the anchoring problems, and
this anxiety now rose further. It was as if
me and the sharks all got excited together
— at first, just ~~Aka~~ a ripple, an imperceptible
increase of heartbeat a quicker movement,
an imperceptible speeding up of 20 sharks

~~ckily~~ all my friends — as though they'd all
realized all they had to do was count to 7
~~nd~~ we'd have a party — all who normally come over
~~e~~ whole period were there, super excited, at the
eginning

very dark little juvenile - nearly black
marks I would recognize if she comes again
soon — never saw before.
m — 1st time in ages — another male comes back
~~e~~ bag had fallen down at edge of site at the
beginning — Samaria grabbed it and got the plastic
stuck in her teeth. I was afraid I would have
a veterinary case on my hands — check last wk.

~~price~~, Apricot, small one with (Breezy?)   Charon
Amaranth

~~~ and I had ~~~
~~~ and them ~~~
died down
t I was little one with light dot mid side
at way too - both sides
~~ASAP~~ d fish as usual deep cut near corner of mouth it si
~~lling~~ is at end.   - penises of m
where   season much
                                         mor

like Venus

RAMONE   approx

♂ like   club tail
 Maaaamu   black
               edge all around
                                already
                                (lost
                                session)
                                                5:25 to 6:
                                             full moon
, 2001 Session # 154
                                             asically b
, 2002 Session # 207   4:05 t
prevented me from going the n
depression over Bora Bora
d did eventually pull out. I had period,
then I first threw all the s
wondered if she hadn't actually
Lillith
- Jesse got to eat for once
- Glammer, Shilly came. Sim

- large male -
  light brown

                    male like this at
                    Ann Bolynn   Danny

- another large female

                    - 3 or 4 shots
                    hurricane
Eclipse passed one time chasing another
giant needle fish came, start
sunset was at 6:55.
ight after that Keete sh
babies! She's big and ve
oised to photograph her
ggest white tip I've eve
hind Keete, right under
ed was just about e
d she was maybe a bi
one was with a smaller
                    No special marks. She
                                       like I was se

THE SHARK SESSIONS

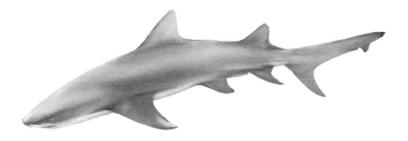

~ *My Sunset Rendezvous* ~

Third Edition

ILA FRANCE PORCHER

The Shark Sessions | *My Sunset Rendezvous*
First Edition Copyright © December 2010 Ila France Porcher
Second Edition Copyright © June, 2014
Third Edition Copyright © July 2017

All rights reserved.

No parts of this publication may be reproduced, stored in a retrieval system, or transmitted in any form or by any means, electronic, mechanical, photocopying, recording, or otherwise, without the prior written permission of the author.

This book is sold subject to the condition that it shall not, by way of trade or otherwise, be lent, resold, hired out, or otherwise circulated without the author's prior consent in any form of binding or cover other than that in which it is published and without a similar condition including this condition being imposed on the subsequent purchaser. Under no circumstances may any part of this book be photocopied for resale.

Illustrations, cover, and text design by the author.
ISBN : 9781521871232

Other books by Ila France Porcher

The True Nature of Sharks
Merlin | The Mind of a Sea Turtle

Table of Contents

Foreword: 8
Preface: 10
Chapter One: Flight to the Tropics 12
Chapter Two: A New Life 16
Chapter Three: Adventures in the Ocean 23
Chapter Four: Sharks 27
Chapter Five: Merlin 36
Chapter Six: Merlin's Recovery 43
Chapter Seven: The Search 52
Chapter Eight: In Merlin's Memory 56
Chapter Nine: The Assembling of My Sharks 60
Chapter Ten: Sorting Out the Sharks 71
Chapter Eleven: A Change of Season 79
Chapter Twelve: The Sunset Rendezvous 81
Chapter Thirteen: The Mating Season 92
Chapter Fourteen: The Season of Storms 98
Chapter Fifteen: Kimberley, Marianna, and Twilight 103
Chapter Sixteen: Puzzles 108
Chapter Seventeen: In the Dry Season 116
Chapter Eighteen: Madonna's Pups 122
Chapter Nineteen: Another Breeding Season Begins 128
Chapter Twenty: Adventures in the Storms 140
Chapter Twenty-One: Further Investigations 152
Chapter Twenty-Two: The Mirror Experiment 155
Chapter Twenty-Three: The Range Experiments 158
Chapter Twenty-Four: The Dry Season Returns 164
Chapter Twenty-Five: Meadowes 169
Chapter Twenty-Six: Poachers 179
Chapter Twenty-Seven: Shilly 185
Chapter Twenty-Eight: December 194
Chapter Twenty-Nine: Accident at the Spawning Event 201
Chapter Thirty: The Change 204

Chapter Thirty-One: Carrellina, Chevron and Sparkle Too 211
Chapter Thirty-Two: Spooky and Kim 217
Chapter Thirty-Three: The Bad Sharks 220
Chapter Thirty-Four: Thinking It Out Again 232
Chapter Thirty-Five: The Good and the Sad 240
Chapter Thirty-Six: The End of the Mating Season 244
Chapter Thirty-Seven: The New Program 253
Chapter Thirty-Eight: My Sharks 260
Chapter Thirty-Nine: Mysteries of the Sea 265
Chapter Forty: The Aftermath 276
Chapter Forty-One: The Coming of Carrellina 284
Chapter Forty-Two: Ordering My Sharks 289
Chapter Forty-Three: Thoughtful Sharks 294
Chapter Forty-Four: More on Thinking in Animals 296
Chapter Forty-Five: Christobel and Flannery 312
Chapter Forty-Six: Crisis 316
Chapter Forty-Seven: The Nurse Shark 326
Chapter Forty-Eight: Interrupted 338
Chapter Forty-Nine: The Cock Thieves 347
Chapter Fifty: Back to the Sharks 356
Chapter Fifty-One: More Trouble 362
Chapter Fifty-Two: Preparations 372
Chapter Fifty-Three: Food for the Filming 379
Chapter Fifty-Four: The Final Days 387
Chapter Fifty-Five: The British Broadcasting Corporation 395
Chapter Fifty-Six: Peace 412
Chapter Fifty-Seven: Sharks: Size Matters 419
Chapter Fifty-Eight: Life and Death Go On 423
Chapter Fifty-Nine: Jessica 435
Chapter Sixty: The Sharks Are Protected 438
Epilogue 440
Appendix 462
Bibliography 469

Dedicated to Martha

Foreword

Around the year 2000, I became aware of Ila France Porcher through an Internet list-serve called Shark-L. As a research scientist, I found the material on this list-serve to be mixed. There were abundant flame wars and an amazing amount of misinformation. But there was also a lot of interesting news and some important facts to be gleaned.

One day while perusing the submissions, I noticed one from a sender named Ila. She was reporting on some interesting and even astonishing observations she had made on her beloved blackfin reef sharks. Having been trained as an Ethologist in the European school of animal behavior I immediately recognized something of importance in her posting. It seemed that she was making the sorts of observations that I thought could only be made in aquariums and captive facilities.

So thereafter I closely followed her postings on Shark-L until one day I gathered together all Ila's information and took it to my colleague and mentor, Professor Arthur A. Myrberg. Back in the 70s Art and I had published the first purely ethological paper on sharks using a captive colony of bonnetheads in an in large oceanarium.

Art also recognized the value of her observations, findings and conclusions, as we discussed them over coffee for several hours. I clearly recall his comment on reading Ila's observations, that she was like an uncut diamond….a true naturalist and Ethologist, and that she had made the kind of careful observations that we could not even attempt. He was intrigued, and quickly struck up a conversation with Ila, instructing her in the methods needed to analyse her unique set of observations with the goal of communicating results to the scientific community. The happy collaboration between Art and Ila eventually led to several publications in scientific journals and an enduring close personal relationship.

This foreword would surely have been penned by Professor Myrberg had he not tragically passed away nearly a decade ago.

In all these years I had never actually met Ila, but I had checked out her background on line. She used to be known as Ila Maria when living in Canada and becoming a fine artist and accomplished naturalist. I learned that as a young child she was often left alone to explore in the wild forests and mountains of her native British Colombia. This fascination with nature eventually led to a blossoming of her innate artistic talent and gave her the discipline to make long and careful observations that are so eloquently presented in her book. If I may respectfully say so, as a youngster she was not only talented but a real beauty as well; I digress.

Around 2010 after the publication of the first edition of this book and her scientific papers, Ila visited Miami to give a series of lectures about her book, and the critical conservation issues surrounding the demise of her cherished shark subjects, which seemed more like personal friends than wild aquatic predators. At that time I assisted in setting up several venues for her talks and had the pleasure of attending one of her lectures. Afterwards, she kindly presented me with an autographed copy of her book. From our previous conversations, readings, and her lecture, I felt quite familiar with her study and did not think to read the book cover to cover. I basically skimmed the chapters on sharks to again re-familiarize myself about her subjects and dutifully put the book away. This was a mistake.

It was not until last month when Ila asked me to write this piece that I picked the book up again and began to read it from page one. I now confess to having missed a great joy by dismissing this jewel of a story. From the first page I could not put it down. The writing style and story were so compelling to me as a marine biologist that I can heartily recommend it to anyone that is partial to adventure, adversity and a unique and personal take on the marine realm. But there is a practical side to Ila's work.

By now, it is common knowledge that many sharks are under an enormous threat of extinction. The unbridled killing of these apex predators for Asian fin soup, is having a cascading effect on marine environments, and irreversible changes are already occurring.

Ila describes this slaughter herself in the book, which in some ways has a happy post script. A little over a year ago the government of French Polynesia placed its territorial waters, comprising over 4,700,000 square kilometres, under protection, such that no shark can be purposefully killed in the country. In other words, French Polynesia has become the world's largest shark sanctuary. Now Ila's efforts to protect sharks have come to full fruition.

In conclusion, be prepared to experience a great adventure, told in a compelling way, from the unique perspective of a truly gifted naturalist, artist and writer. I hope you enjoy it as much as I did.

Samuel H. 'doc' Gruber

Preface

I wrote this book because there was a story which had to be told—a community of sharks accepted me as a companion among them, and then were finned for the shark fin soup market. To add insult to injury, their tragedy was blocked by the media.

So I wrote down the story of my lost sharks, in hopes that the world would find out what these unusual animals were like, and what happened, when the shark finning racket arrived in these innocent islands.

For a wildlife artist and ethologist, the Tahitian lagoon was a mystical world of spellbinding beauty to explore, and I swiftly met the sharks. They were so intriguing that I launched an intensive long-term study of them. Through observing them as individuals, a window on a hitherto unknown dimension of their lives opened to me. What they were like as animals and individuals, how they behaved, how they interacted with their environment, how the other species interacted with them, when and why they roamed and with whom, their gestation period—all appeared in perfect clarity as I watched them over time.

They responded to my visits by treating me as a companion and behaving as intelligently and rationally as pets. Increasingly fascinated, I felt that a way was opening into another world, one so alien, so separate from my human daily life, that it might just as well have been on another planet. Working alone in a spirit of investigation uninfluenced by others, I had no preconceived ideas about sharks, and accepted the strange and startling things that happened as part of my journey of discovery.

After three years, when I got an Internet connection, I found that no one else had ever done such a thing, and that most people thought that sharks were vicious !

By then, I could recognize three hundred individual blackfin reef sharks on sight, and was familiar with the other species who lived and visited in the lagoon. It was evident that the sharks remembered the events in their lives and thought about them in order to make decisions in new circumstances. This was not surprising, but to my consternation, mainstream shark science had not noticed!

Professor Arthur A. Myrberg Jr. and I often discussed the indications that these wild sharks were thinking, and eventually he concluded that my observations provided the first evidence of cognition in sharks. (Cognition is the scientific word used to refer to thinking in non-human animals.) Professor Myrberg included my observations in his talk on shark cognition at the international symposium on animal cognition, at the Max Planck Institute in Germany, in 2003. He wrote a scientific paper on the subject, too, and named me as co-author. But he sadly died before his time. It was never published as planned by the journal *Animal Cognition*, so I have included it in the appendix.

For seven years I put the sharks first, no matter what else happened, so that the data would be collected with regularity, week after week, year after year. I have told each event just as it unfolded, as precisely and objectively as I could describe it. Through writing, I sincerely tried to take you with me, to see what I saw, as if you were there, looking over my shoulder, at the riveting things that happened in the splendour of the coral lagoon where I spent so much time.

Most of the story takes place underwater in the company of the sharks, but I have expanded the account of the shark sessions to include the scientific framework, and the filming of the sharks for Shark Week in the midst of their massacre for shark fin soup. A company from Singapore had begun finning sharks throughout the island nation only a few months before the film crew arrived, and the documentary was my one chance to get their story out of the country. Against the background of stormy seas, racism, and the hatred for sharks spawned by the fishermen, the effort to make the filming succeed, against all odds, is the funniest part of the book.

Since cognition had emerged as an important theme in this story, I wove in the most surprising examples of cognitive behaviour I found in the marine animals I was looking after at the time. These accounts of sea turtles and birds highlight the sharks' intelligent behaviour, and put it into a wider perspective, to help create a grand painting of the community of the sharks, the other species, and the uneasy society co-existing around it. I added a chapter on thinking in animals, to enlarge upon this controversial subject, and have since been called "the Jane Goodall of sharks" as a result of this focus.

After the book had been sent to the publisher, I worried that my readers might not know how all of these characters appear underwater, so I painted the illustrations, writer-to-reader messages in images instead of words. Following some experimentation to find the best way to paint sharks in black and white, I settled on a combination of soaked watercolour paper, ink, bleach, and lots of water, to produce these sketches, which seek to convey something of the spirit of the scene, rather than being polished works one might do to place in a gallery.

This second edition of the story of my sharks, updated in 2014, provides the latest information available about the sharks I knew, and updates all the other major themes of the book, including the plight of the sharks still left in this world. But most of the mysteries of the seas that you will discover with me are just as mysterious now as they were when I saw them.

I wrote the original book using the metric system, which is what was officially used in French Polynesia, and which had also been demanded for the scientific writings I had done. However, now I have learned that many readers are used to thinking of distances and sizes in terms of inches, feet and miles, so don't swiftly gain my meaning from the given number of kilometres, metres, or centimetres. So I have added a translation, so that all readers will gain an accurate concept of the measurement as easily as possible in the course of reading.

Since most of my size estimations for the sharks are given in terms of a metre, or a metre and a half, or two metres, and so forth, I trust that you will have gained a good idea of what these mean once the scene is set, so I no longer need to give the size in feet each time, which interrupts the flow of language. The nurse sharks in particular are often referred to in terms of their approximate sizes—the two-metre sharks and the three-metre sharks—sharks in the range of six and a half feet, or of ten feet, for example.

In other ways, I have left my original version unchanged. Blackfin sharks are more commonly called blacktip sharks, but I did not change this term. Some shark species do have different names in different regions, and blackfin was the preferred name for the species, used by the local shark scientist, Richard Johnson, when I began my study. So I have retained it, for reasons of authenticity. Scientific updates are mentioned in the text, and fully presented in the epilogue.

The launching of the study is exciting and full of discoveries. As the months turn to years, and the observations become increasingly detailed, I make sure that all the established threads of shark behaviour are pursued and described precisely, because it is important to establish the original patterns presented by the sharks. Later, their behaviour changes. At any time if you find that you want to move on to the more exciting continuation, you may skip on to the chapter "Meadowes" and from there the most intriguing part of the sharks' story begins to unfold.

Nearly a decade has passed since my study of the sharks, and it is yet to be duplicated. I give you this report not only as an observer, but as an artist, self-trained to record with high accuracy. But unlike a scientist, I am not bound to put everything into numbers, graphs and tables—I'm glad to bring you the beauty and the wonder of it all, in this strange true story of what happened in a lagoon in Tahiti.

My heartfelt thanks go out to my husband Franck, for his patience with my visits to the sharks, especially after sunset.

Amanda Rankin Barratt, my friend and a prominent shark advocate in South Africa, gave generously of her time to edit large parts of the text to help me get the story of the sharks published. Thank you so much Amanda, for your kindness and devotion, and for your work for sharks.

CHAPTER ONE

Flight to the Tropics

"Would you like to move to Tahiti?" Franck asked.

I gazed at my husband. We were huddled by the fire in the heart of another endless Canadian winter while a blizzard blasted the house.

"Sure," I said. Talking it over we realized what a perfect solution it would be. Franck was tired of my wildlife artist's life-style in the wilderness, and wanted to pursue his career as a doctor of computer science in a French territory.

I was tired of coping with eight months of ice and snow each year, especially while trying to enjoy our new hobby of scuba diving. How beautiful it would be to dive in the warm, transparent waters of south sea islands, to paint with warm hands! Step by step we made a plan for the major move, and day by day we fulfilled it.

At the end of April 1995, with snow still drifting down upon us, we drove slowly away from our mountain chalet and descended toward the airport, the first step on the long journey southward. We stayed overnight in Seattle, flew to Hawaii, then south to Polynesia, entering the night as we approached the equator. Spiralling down over Tahiti, we saw only scattered lights in the capital, and realized how small it was, a tiny town on the most remote island in the world, perfectly positioned in the centre of the South Pacific Ocean.

It was midnight when we stepped down from the plane and first breathed the air of Tahiti. It flowed over me, an exotic balm as thick and sweet as honey. All around, the brilliant stars of the southern sky hung in strange patterns in a twisting, luminous dark as we crossed the tarmac toward the bright door in the terminal building. The sense of mystery was sharpened by the drumbeat pulsating through the night, and the extraordinary fragrance of the *Tiare* flower handed to us as we entered. The drums evoked a sensation of waiting, breathless, and I felt a deep awe for the place we had landed.

After the tiresome procedure of going through customs and collecting our luggage, we were met by the owner of the *pension* (a comparatively low-cost, informal hotel), where we had reserved a room. Our host drove us around the down-town area of the capital, Papeete, showing us the main routes and other features, before we checked in.

The room was so tiny that we could scarcely walk around the bed—laying out a suitcase was impossible. Though we had reserved a room with a bathroom and shower, it was not available, so we had to use the facilities down the hall. The bathroom windows opened onto the dining room on the terrace and we were surprised by the lack of privacy and cleanliness. It cost more than the luxurious, spacious room we had enjoyed the night before in Seattle. But, it was a place to sleep, which were glad to do.

We awoke with a rush of excitement at the view of the tropical sunshine filtering through foliage outside our window, and went out to wander, amazed, through Papeete. The main boulevard along the water's edge was lined with trees and shops selling pearls, postcards and *pareus* (brightly decorated cotton wraps), interspersed with sidewalk cafés. On the other side of the boulevard was a marina where yachts were tied, and one could walk along looking straight down into the water. Beyond, the sea lay dancing with reflected lights beneath a sky of ringing blue. So close to the equator, nature's colours all seemed intensified; the sunshine was actually white. It flashed in the green canopy above and flickered hypnotically in the coconut trees.

We had a light breakfast at one of the cafés, watching the daily life of Papeete passing by, then wandered on. At a tourist information centre, shadowed by plants with enormous dark leaves and gaudy flowers, we went in to pick up some pamphlets.

The luxuriant growth of the exotic plants told me where the sun had been when it had not been shining on me in Canada. To my dawning surprise, I realised that these were not monster plants at all, but the same ones that I had been keeping as house-plants, pale and huddled by the window at the forest's edge. Here, fed by light, warmth and rain, they were a thousand times branched and extended into giant trees that lined the streets.

A huge tree by the harbour was what my rubber plant might have achieved after centuries of prolific growth in the sunshine of the tropics. An intricate network of roots hung from the branches to the ground in curtains. A rat peered out at us from this labyrinthine shelter, in a rubble of broken and unbroken dirty bottles and trash. I fantasized that somewhere there must be a planet with solar energy of a quality to empower the plants to actually move about doing things just like animals.

After the coldness of Canada, we revelled in the feel of the streets of Papeete—the cosy warmth, the people wearing bright colours, the relaxed and friendly atmosphere. Nearly all the people were Polynesian with a few Chinese and white people among them. The local French accent was quaint, quite different from the Parisian French I was used to.

As the sun climbed to its zenith, our thoughts turned toward the fabled Tahitian lagoons. There were no beaches in Papeete, but we learned of one located some distance around the island we could get to by "truck," the popular local bus. So we loaded our snorkelling gear and set off.

The elderly truck had wooden benches down the sides and one down the middle. Plexiglass windows were open along both sides. Only local people were using it—it seemed to be the popular way to get around.

The kindly driver stopped for us at the public beach access. A narrow trail down a dried stream bed opened to a white sand beach shining with the great light of sky and sea beyond. Circling a group of laughing Tahitians playing a game with silver balls, we found a shady place, sat down, and drank in our surroundings.

Coconut palms and enormous trees fringed the sands. Clusters of banana plants, palms, overgrown shrubbery, and exotic flowers rose from half-hidden gardens behind them. The lagoon lay glimmering turquoise and silver under the flawless sky, lined by distant waves on the barrier reef. Beyond, the indigo ocean spread away around the planet for thousands of kilometres. Ten miles northwest, the island of Moorea had gathered a few clouds to its summits. Its dark and spiky shape beckoned, like the fabled Enchanted Isle.

I found several small seashells and a piece of curved shell that had belonged to something enormous on the beach. But there were no sea gulls. It seemed strangely silent without them. Wandering and gazing, I kept saying to Franck, "How beautiful!" A teen-aged girl was sunning herself nearby, surrounded by boyfriends, and clad only in a g-string. When I spoke, she turned toward me, long, black hair flying, and gave me the most radiant smile. Amazed by her joy and beauty, I grinned delightedly back at her.

We sat in the water to put on our gear, then swam slowly away from shore. The water was warm as a tepid bath, salty, and crystal clear. Little white fish with one black spot on each side, were the first we saw, feeding on the luminous sand. Soon formations of coral, tiered in shades of ochre and brown, and decorated with colourful branching varieties, loomed up in the blue. It was my first look at these colonial animals, who had actually built the reef, and the atoll islands. Brilliant fish of many shapes and colours flitted around them.

As we progressed, fish of every imaginable shape and colour were everywhere, peering from holes in the coral, ranging across the sand, feeding on algae growing on old coral structures, and travelling purposefully though the blue. I recognized many of them from books and aquariums: pennant fish, angel fish, and butterfly fish with many forms.

But there were far more, that we had never seen, nor imagined could exist. The Picasso triggerfish (*Rhinecanthus aculeatus*), an extravagantly decorated creature who looked as if it had been designed by an artist, caught our attention immediately.

Any view in any direction revealed myriads of living things, each intricately coloured and patterned. Rounded corals composed of thousands of tiny branches, sheltered fish of turquoise, blue and green. They flowed together through the coral in harmony, then rose in an expanding cloud to surround it, vanishing as one into their homes as we approached. A snake-like creature half a metre (20 in) long, with brown and white stripes along its length and a head like a flower, groped in slow-motion for food in the sand.

A stingray, a member of the family *Dasyatidae,* a metre across was rummaging farther on, raising a shimmering cloud around it. We drifted closer, awed by the details of its submarine beauty. In slow-motion, it swung its two metre (6.5 ft) tail toward us, lifting its wings and rising with alien grace to fly on through the surreal surroundings. There, the lagoon was at its deepest, at about three metres (10 ft). Farther on, the inner edge of the barrier reef rose beneath us, and the coral became very thick and vivid. Gradually the water grew shallow, the sandy floor gave way to reef flats, the coral thinned and dwarfed, and *turbinaria*, a short, strong seaweed, grew profusely. The fish were different too. Many small dark blue fish flitted in mid-water, looking like tiny balloons. Others were vertically striped in black and white. The ocean pouring over the reef was crystal clear, and the sunlight slanting through the surface gilded the scene with lines of swiftly moving light.

Franck took my hand and we swam into the sparkles of breaking ocean waves. I felt like a child being led into a wonderland as the golden light broke into galaxies of stars, and the water became a thin skin of rushing molten glass. But suddenly we were flying onward, faster and faster toward an abyss of darkest blue. We turned to swim desperately away, to escape the ebbing wave tearing us from the light of safety. At first we made no headway, but just in time the torrent slowed, we drew ahead, and the next wave poured us in a shower of champagne water, back into the lagoon. We flew on, as each wave breaking over the reef brought showers of bubbles, surrounding us with stars. Spellbound, we flew with the rhythm of the underwater world—rushing shoreward, pausing, and rushing on.

The current propelled us gently through the coral garden, and back to the shore, where we dressed, and made our way out to the road again.

Back in Papeete, we stopped to eat. The streets were busy, loud with the noise of people and cars. Two starved dogs roamed past, one with a tumour the size of a fist hanging from its belly. The gutters were full of filth and litter; the street was poorly paved and old. We looked through the pamphlets we had procured at the tourist centre for an apartment that we could rent by the month, but found nothing available. We had no choice but to return to our *pension*.

The next morning was Sunday, and we found a side-walk café where we could sit, eat, and ponder our next moves. Franck bought the newspaper, *La Dépêche*, and scanned the advertisements to see if he could find an available apartment. While I drank a second cup of coffee, he made a few phone calls and found something promising. So after eating, we walked along a boulevard over-hung with giant trees, then up through a valley.

The apartment we were shown was on the third floor and very dark. Even the outside patio was barely lit, with only a narrow view out between the surrounding buildings. But through it the wild, tropical hillside half a kilometre away was visible. The apartment was spacious and there was a separate kitchen; further, it was available by the month. So we took it and went back to our *pension* to pack up and check out.

It was a long haul to carry our luggage and gear the many blocks to our new home. Along the way, Tahitians shouted greetings and encouragement, and offered to help. They seemed full of delight and natural smiles. I became delighted too, and smiled back.

We had enough things with us to set up a new home, as well as books, art supplies, and diving gear. So this first move in Polynesia was a marathon. Franck carried the two diving tanks by himself, but strained the muscles in his back on the last trip with the large suitcase. He had insisted on carrying it all by himself, while I, also fearfully burdened, trailed along behind.

Finally the big suitcase collapsed onto the ground, and Franck followed its example, saying that he could not go on. I started pulling it on its little wheels, and Franck pushed, but when we arrived at the motel, the tiny wheels had ground down to nothing, and disappeared!

The next step was the cleaning of the apartment, since every surface was filthy. It was a terrible chore without the proper tools. We had just finished and arranged our few things appropriately when the rain began. The noise gradually grew deafening, and we stood outside watching in wonder as waterfalls blew in curtains across the land. Water ran everywhere, filling the depressions and streaming down the roadway. The torrential rain fell off and on for the rest of the afternoon.

In the evening a red sun broke through in the west, and we walked through the dripping landscape to the waterfront. At the quay where the cruise ships docked were stands selling a variety of beautifully cooked food in Polynesian, French, Oriental, Italian, and American styles. We found a French cook who entertained us with local tales as he prepared speciality salads and crepes for us. The waterfront scene, the delicious food, and the exotic atmosphere felt charged with significance beneath the radiance of the western sky, and as we spent the last of our American cash, we knew we had arrived.

Back in our new apartment, we began to appreciate Polynesian life close up as we lay on the bed, reading and absorbing the details of our environment. The floor was laid with dark red tiles. The yellow walls were badly marked and stained. Across the front of the apartment, orange curtains covered the sliding doors, while on the windows the drapes were purple. Geckos of all sizes chased cockroaches across the walls, chattering in many voices. The cockroaches ran until they panicked, then took to their wings like World War Two fighter planes with drunken pilots. Televisions in apartments all around us projected a cacophony of random sound. Dogs barked across the landscape below, and roosters crowed to each other from hillside to hillside, cage to cage. These were the perpetual sounds of Polynesia, the sounds many visitors found unbearable.

Monday morning brought more rain. Lacking an umbrella, we were lucky to arrive at our sidewalk café without getting soaked. We opened a bank account, established a postal address, and Franck tackled the problem of finding a job.

When we were still in Canada he had mailed his résumé to all the companies in Polynesia involved in computer science. At my insistence the cover letter had been professionally printed on the finest, cream-coloured paper available to make the appropriate impression. We stopped at a pay phone so Franck could contact the company that had sounded most promising in its reply; it sold computers, business software applications, and services. Franck was put straight through to the owner, who was delighted to hear from him, and gave him an appointment for an interview that afternoon. He actually said, "I want you!"

We rushed back to our apartment, high up the valley, so that Franck could change. I ironed his white shirt while he prepared to put on his professional hat after so long.

When he left, I sat at the table on our hidden balcony, alone for the first time in our new country, restlessly distracting myself as I waited in suspense, by solving problems in a book on mathematics. Alternately, I gazed out at the mysterious, wild hillside, swept by veils of sun and rain.

Franck returned jubilant. The company's owner wanted him to set up a new department incorporating more high tech products and software into the company, including neural networks and paying card technology.

Excitedly discussing all of this, Franck and I went shopping for groceries and such niceties as a coffeepot, and that evening, we cooked and ate a bountiful dinner in our new home.

And that is how we established ourselves in Tahiti in just three days.

CHAPTER TWO

A New Life

Franck set up his innovative services department, and when he was free we explored widely until we became familiar with the island. By the end of the month, we had rented a fine house on our favourite beach, and once we were moved in I was freer to explore.

My first instinct as a wildlife artist was to head for the hills, but there were no trails. I found a track up the nearest valley and followed it, gazing toward the heights where cascades and flowery vines festooned the black rock walls and fell to secret places hidden by giant ferns. Awed by the beauty and beckoned by the mystery of it all, each time I rounded a bend, a new vista of human contamination appeared. There was one garbage dump after another. Between them, the narrow track was lined with piles of tires, dumped household garbage, cars, the carcasses of dogs and cats, broken appliances, and every sort of junk people might want to throw away. Some of the piles had been set on fire and were smoking and even exploding. The evil-smelling fumes spoiled my pleasure. Never had I seen such beauty and such desecration side by side.

More direct efforts to access the interior brought me into impenetrable vegetation. The jungle grew close, and the prickly tangle beneath the canopy could not be breached. The island had erupted from the ocean floor about two million years before, and there were no wild animals in the hills to study and paint anyway.

So I would go home and seek relief in the lagoon. My visits felt like forays into a mystical world singing with light and the sensation of mystery. Hundreds of species of multi-coloured fish twirled and shot in every direction. Pale predators rested unmoving, their bodies the same colour as the sand, waiting for some unsuspecting creature to stray within range. One had a wide pair of fins, followed by a long, snake-like tail, disappearing down its throat. In a pool of light was a yellow moray eel, delicately patterned with black lace. It looked up as I passed and I gazed in amazement into eyes a bright turquoise above that nightmare smile.

One of the white bottom feeders, about thirty centimetres (1 ft) in length, swam along with a slender olive-green fish of the same size, who wore a perpetual smile. The two stayed close together, responding to each other and often touching. It appeared to be a far closer association than was strictly necessary for the convenience of finding a meal. The possibility that these animals were engaging in inter-species cooperative hunting seemed unbelievable. Cooperative hunting is considered to require high intelligence, including the ability to plan for the future, think, and cooperate! All I had heard about fish thus far was that they were low animals—cold-blooded and virtually brainless. But these beautiful creatures pursuing their submarine affairs did not appear that way at all. Farther on I saw another odd couple. A tiny, transparent shrimp was cleaning the entrance to its hole in the sand while a fish stood guard beside it. These small creatures, too, appeared to be cooperating.

On a curve of coral lay something that resembled a sea anemone, but as I approached, it turned to maintain the same angle relative to me. Soon a yellow eye peeked out. I stroked the little octopus, and he raised a tiny tentacle to push my hand away. Closer to the inner edge of the barrier reef, there was a fairly clear, deeper region, where the coral formations were larger and more ornate. The bizarre head of a huge moray eel emerged from one, and swayed, as he looked out from his labyrinthine home. This was the largest local species, the Javanese moray eel (*Gymnothorax javanicus*), dark brown, thick bodied, and about one and a half metres (5 ft) in length. I gave him a wide berth. An oval-shaped fish, about sixty centimetres (2 ft) long, and intricately decorated in pastel shades, bobbed vertically along the sandy bottom, its frilly fins rippling. It swayed back and forth as it blew into the sand, raising tiny clouds as it searched for creatures hiding there. I followed and it turned to look up at me with one large eye, then dancing, it turned to look with the other, just as a horse will do when followed from behind. After that experience, I always thought of the species as the *dancing fish*, though I eventually learned it was a large species of triggerfish (*Balistidae*). Later I found another bigger one, much more dramatically coloured, tearing apart and eating a sea urchin with spines thirty centimetres (1 ft) in length. A young barracuda rested just under the surface.

Past this deeper area, the barrier reef rose. The water shallowed and the coral grew thick and very colourful there, where it was maintained in good health by clean ocean water pouring over the reef. Staghorn coral lifted delicate branches alive with tiny blue fish, and intricate clusters of pink, blue and violet corals sparkled amongst the larger golden ones. From shadowed holes beneath, the gentle face of the largest puffer fish (*Diodon hystrix*) looked curiously back with unexpected wisdom.

There were many smaller puffers, black with white spots or brilliant blue and gold. There were slender, metre-long fish, patterned with white and grey, with snake-like rippling tails, hanging motionless in the water streaming through the coral canyons. These were called flute fish (*Fistularia petimba*). Little fish striped with white, green and magenta habitually accompanied me in a small school of up to sixty.

I felt safe exploring the lagoon because I had been told by the local people that no sharks lived there. As I had lived in the mountains and observed terrestrial wildlife all my life, my knowledge of sharks was limited to the information I had gained from watching the movie *JAWS* many years before. All that remained from that brief education was that they bit and badly. Very badly. Essentially, if you met one, you died. So my vague plan was not to meet one, but I was too curious about the oceanic life in this tropical wonderland to think much about them.

One tranquil morning I was roaming upon the barrier reef, lost in a spell. The sunshine rushed in golden lines across the landscape and flashed upon the fish. It was mesmerizing. I raised my eyes to see a grey shark of about my size moving languidly toward me. Her face, the arrangement of her fins, the distinctive shape of shark, struck me to silence. The light slid over her velvet skin, and all the world seemed to be rushing by, save her, as she glided closer, graceful as a snake.

Expecting her to fly into attack mode at the sight of me, I held my breath and drifted behind a coral. But she paid me not the slightest bit of attention and passed on the other side, just a metre away. Her smug little face actually looked bored. I moved around to keep the coral between us, but when I peeked out to see her again, she was gone as if she never had been.

I decided that she must not have seen me, but now I know better. Sharks just don't care much about us.

That weekend Franck and I walked about three kilometres (two miles) along the beach to the nearest large pass at Fisherman's Point and put on our masks and fins in the shallows there. We floated away from shore over a bank of rounded stones and found ourselves high above the pale seafloor. Five eagle rays (*Aetobatus narinari*) flew in formation below. Farther on, a large shark lay resting on a patch of white sand, and we dove down for a closer look. An eagle ray accelerated with one quick sweep of its wings, holding them still on the upward beat as it shot out of sight. As we approached the shark it swam with a lazy motion into a deep cave in the coral drop-off. Franck dove again, and was able to go deep enough to look in, holding an outcrop to steady himself. The great predator turned, and Franck got a momentary look straight into its face before he had to swim back up to the lighted surface.

Farther along we saw another shark resting on the sand and as the seafloor fell away toward the oceanic abyss, the colour of blue was so intense that it appeared surreal. And there, hanging in the blue, its fins furling gracefully, was a dancing fish. I dove and was after it. A myriad of fish of different sizes, colours and shapes hung around me. Some were like large puffer fish with the faces of little dogs. Farther on was another dancing fish, and I hastened to descend. I wasn't getting much closer to the one I was following. It kept turning charmingly to look up at me first with one large eye, and then the other as it propelled itself away.

When I turned to look around, Franck was far away looking down at me from the silvery surface that formed the ceiling of what appeared as an infinite blue room. There was a sense of disorientation, surrounded by the blue abyss with fish in all directions, at different distances and levels, the water filled with their chittering voices. So I swam back to Franck who told me not to go off diving like that into the blue.

A Napoleon wrasse (*Cheilinus undulatus*), a heavy, slow moving animal about a metre and a half (5 ft) long, searched lazily below us, then entered a coral valley where it descended and disappeared from view.

On the way back, another dancing fish appeared and I swam away after it. But when a shark shot in front of it, it apparently decided that it was more scary than I was, so it turned around and swam back up toward me, still turning its head to focus first one eye and then the other, upon me. Back at the surface, we were surrounded by a loose school of transparent fish catching the sunlight as they turned—they appeared and disappeared in flashes of blue as if from another dimension. As we swam on, everything else was viewed through this wondrous curtain—the Napoleon fish, the coral, the dog-faced puffer fish, and the sharks.

Walking home, we found an eel, a little over half a metre (20 in) long, lying on its back on the sand. I thought it was dead at first, but Franck noticed that its gills were moving and put it back in the water. The animal unhesitatingly undulated back onto the beach. Franck put it in the water and kept stroking the sand off it, talking to it as if it could understand French. It returned to the beach. Each time we gently replaced it in the sea, the eel swam back out again. Finally, it oriented itself, curled its tail into a spiral, screwed itself deep into the sand, and vanished!

July 13 was the fourth anniversary of our meeting, and Franck rented a car for the long weekend. We went out to dinner at the most luxurious hotel on the island. A Tahitian man dressed in a native costume made mostly of leaves drove us from the lobby down to the restaurant in a little, open car. It consisted of five circular areas, each with a thatched roof, built right over the lagoon. The serving area was in the centre, with the tables around the perimeter, each overlooking the coral garden, which was lit by hidden spot lights.

The service was perfect, and an exotic drink arrived as we perused the menu and absorbed the tropical paradise created around us. Franck, saying all the most romantic things like the true gentleman he is, gave me a beautiful wedding ring; we had married just before coming to Polynesia. By that time, the alcohol had begun to smooth the ruffled edges of my consciousness, and I felt as though I, Franck, the surroundings, and the lagoon were glowing softly with light and happiness. As Franck guided me through the decision making regarding the many menu choices, a small dollop of bird liver fat artistically displayed on a bed of alfalfa sprouts was placed in front of me. As I watched with interest, an ant broke from the cover of the sprouts and raced, clearly panic-stricken, around the table.

The whole meal was a magical experience. Two more ants escaped from my salad—I wanted to help them but there was nowhere to put them. (I had been a vegetarian since having had to protect my pet raccoon from being made into soup as a child.) As we finished, I noticed that a crab had crawled out of the water onto the rock wall, and pointed it out to Franck. He couldn't see it, so I made more of an effort to explain where it was, and for the next few minutes tried to help him to locate it. Then the elegant woman at the next table arose. She had been listening to our conversation, and for some reason her group cared to see this crab. Finally, everyone got up from their tables and went to the railing to view it. For many minutes, this ordinary crab received so much attention that I wondered why these hotel guests didn't have a look at what you could find in the lagoon!

The next morning we took advantage of having the car to take our diving gear to the pass at Fisherman's Point and investigate it in greater comfort. As we slowly moved along the base of the coral wall we had already seen from above, we found ourselves outside the cave of the sharks. I paused and stared in wonder. Two lay together there. Suddenly the large one arose and swam out past me. She was dark grey and about two metres (6.5 ft) long, with shining white tips on her fins—she had large eyes and a short nose like a cat. She made a wide circle around me and returned to her place beside the other shark. Slowly, I drifted away, since I had obviously disturbed them. Franck was at the side of the cave, watching too. I swam above him and caught onto the edge of one of the huge boulders forming the cave, to support myself in the slight but relentless current of the pass. There was a hole under my face, and there, within, was the tail of the male shark, lit up by the sunshine pouring in from above. My eyes ran over every detail and paused upon the animal's male organs. Yes, he had two! They are called claspers. So close, I could see every detail and was amazed that each white sex organ was finely decorated with lacy grey shapes that resembled the flowers that grew on the island.

Each outing revealed some new wonder. The thin skin of the lagoon, a mere sliver between the earth and sky, was glowing with life on every scale. Compared to terrestrial forests, where I could wander all day and would be lucky to see a squirrel, it was surreal. My first love was the sea-shells. I learned where they lived, what the animals in them looked like, and the appearance of their shells at different ages, drawing and painting them repeatedly just for the pleasure of running my brush over their graceful contours.

I drew the fish I had seen after each excursion, and developed paintings of them in the views that had impressed me. But it was difficult to realistically paint an environment of such intricate detail and ephemeral lighting, full of creatures coloured in precise patterns that were unfamiliar to me. I could paint the forest with a few rapid strokes of a brush, but the coral habitat required a different approach with new tools and brush strokes. In the end I developed a new technique altogether to capture the submarine beauty.

One day I was having trouble completing the details of a painting which included a school of the iridescent blue fish that I loved, who formed clouds around their coral homes. After many days of alternately observing and painting them, I became frustrated and decided to catch one in a jar and place it in front of me while perfecting the details of the many fish in the painting. A camera would have been a perfect solution, but it would be years before I acquired that luxury. So I took a large jar, located a coral structure which was home to a large cloud of the shining blue fish, and gently slid the jar into one of its many holes. The fish all swam straight over to it. In a loose figure of eight pattern, they milled around in front of it until they had all had a good look, then they pursued their affairs while giving it a wide berth. They were smarter than I thought and I was obliged to complete the painting from memory.

Sometimes Franck and I would laze in the shallows on the beach and watch the baby fish. They were tiny versions of the adults living in deeper water. One of our favourites was the Picasso trigger fish. These babies lived in little holes in the old coral substrate that emerged from the beach in many places, and we were able to stretch a finger inside the holes to stroke them. With each stroke, they squeaked! It was at the same time funny and touching to hear that squeal of protest with every touch.

As the lagoon grew more familiar, I began venturing across to the barrier reef at sunset to see who emerged with the darkness, while Franck had his nightly swim. One evening I was waiting and watching on the reef when a movement caught my eye. Through the shadows in the distance, an eagle ray with a two-metre (6.5 ft) wing-span was flying. The shape of its wings and its spots indicated its identity, but those that I had seen until then were just a third of that size. I waited without moving, hoping for a better view as it circled me at the limit of visibility, its soaring form little more than a movement against the background. With the next turn, it spiralled close enough for me to see that it swam with its mouth wide open! Its shape was bizarre. With its lower jaw hanging down thirty centimetres (1 ft) at a ninety degree angle, it resembled a large shark with wings, and I was awed at the sight of its gaping black mouth when it turned, just three metres (10 ft) away, and swam straight toward me! But just before colliding, it veered enough to pass on my left, brushing me with its wing before accelerating away. Shaken, I headed homeward in the twilight, and it was then that I met my second shark while alone in the lagoon—she passed me from behind. Breathless at the sight of such fluid beauty and understated power, I followed. But she quickly drew ahead, became a moving shadow, and vanished in the darkness.

Sometimes I went later and took my dive light. At night there were very strange looking species of fish that didn't appear in daylight, and bizarre crabs that scuttled away from my light. Peering into the holes in the coral was like looking through a transparent apartment building, with layer upon layer of inhabitants moving within. Often, a frilly vision of dark red-and-white stripes hovered—a lion fish (*Pterois radiata*). With each extravagantly patterned spine glowing in the darkness, it presented a flower-like, lacy beauty.

One night as I neared the shore, a small octopus shot through my light beam. Zooming along with its tentacles streaming behind, its shape was so fish-like that it took a moment to recognize. It settled near me on the sand, its little skirts spread around it. I walked my hand toward it across the sand, and it wrapped its tentacles around my forearm briefly, its light touches like those of a playful kitten. For a while we played a little game of chase, and unexpectedly, this small wild invertebrate responded in play, enchanting me by repeatedly encircling my hands with those light, brief touches, after initially evading them. But I was tired and very cold by then so had to leave.

There were a variety of micro-habitats in the lagoon, and curious places to explore. Directly across from our house was a tiny pass, a dip in the reef of less than half a metre (20 in). I visited it often to drift just within range of its ebb and surge. The sparkling ocean waters poured through, and oceanic fish visited or dwelt in the open area just inside the reef from this tiny passageway. If conditions permitted, I would go to it and gaze out, as though through a window into the mysterious region beyond. The dark shadow of the ocean hung on the other side of the veiling light. It beckoned, and occasionally I spontaneously let go, to fly out upon the out-going current through a sinuous, narrow channel onto the outer slope of the barrier reef.

The outside of this ancient coral wall was worn smooth by aeons of resisting the ocean's relentless onslaught. Narrow channels less than half a metre (20 in) wide cut deeply into its roots. They appeared, in some cases, bottomless. A large eel undulated far down in the sinuous, green recesses of one of them. Virtually spherical green shellfish, the size of a human head, hid in deep crannies. Sometimes I had seen their badly worn and broken shells after they had been rolled over the reef by tumultuous waves during storms. Other than these, there was little life in the mysterious canyons. I was often able to peer in or dive into the wider parts, but had to be alert to the torrents tearing along the narrow passages, which threatened to scrape me painfully along the sides, or catch me in the darkness beneath an overhang.

The outer slope was barren to a depth of about seven metres (22 ft), where the water was still enough to allow small corals to grow. Though the coral habitat was an attraction on ocean scuba diving expeditions, the coral in the lagoon was bigger and more varied, and supported a richer ecosystem of fish, molluscs, and invertebrates than the reef's outer slope. But the mystery of the ocean depths, the oceanic species, and especially the shoals of flashing blue fish that appeared and disappeared, enticed me. I never ventured out for long, though, for fear of not returning safely.

Many people had drowned in that seemingly innocent passage through the reef. The height and power of the ocean waves could rise with unexpected speed, unnoticed while swimming through the long, gentle undulations of the ocean.

The highest part of the barrier reef was worn smooth by waves, and the edge inside the lagoon fell off in a steep cliff, to form a channel. This channel formed a natural highway which was used by countless unexpected fish and other creatures that I saw nowhere else, especially visiting oceanic species, such as barracuda and large jack fish (*Carangidae*). I explored it on calm days in both directions, discovering its inlets and secret passageways into the reef.

Sometimes, I found the enormous viridian snails, first seen in the twisting channels on the reef's outer slope, and lifted them to view the animal living inside, but there was only a glimpse of deepest green before the moon-like operculum slammed shut. I always hid these shellfish so that no one would find and eat the occupants. Since they were the grandest of possible snails, I named them *racehorse snails* and drew them repeatedly.

I loved to hold onto the edge of the reef, peering across it under the thin layer of pouring water. Within this rushing film streamed fish of many shapes and colours to where the curve of the reef cut off my view. Sometimes a shark came wriggling through the shallow water, surfing over the reef to arrive in a cascade of champagne water. When the bubbles vanished, if I watched quietly, the shark often approached to turn a circle around me, its eye fixed on mine. It was the first wild animal I had met who came to me instead of fleeing.

These sharks were usually smaller than I was, and it seemed that each was coloured differently. Some were grey, with or without stripes on their fins, and others were bronze. Some were slender and others were

heavy.

The local library was small, costly, and had almost no scientific information, so at the time, I was unable to learn anything about them, and thought that these creatures were the same species, but came in different colours, like horses. When we finally bought a reference book on local fish I learned that they were actually three different species.

The shark lacking markings which I had first seen, was the grey reef shark (*Carcharhinus amblyrhynchos*). The sharks we had found at Fisherman's Point were whitetip reef sharks (*Triaenodon obesus*), and the one I most often saw in the lagoon was the blackfin reef shark (*Carcharhinus melanopterus*). Its colour ranged from gold to dark grey, and it had black tips on its fins and a white band across its dorsal fin. It was also called the blacktip shark, I read, but that led to confusion with the blacktip oceanic shark, a different species.

One morning while roaming through the coral, I encountered a heavily built shark about my size, and swam after her. I was able to keep her in sight as she cruised parallel to the reef, circled, and returned to form a large figure of eight. She followed this path twice, then pursued a fairly straight course as she wound through the coral in the opposite direction. I swam about three metres (10 ft) behind. Sometimes she would get so far ahead that I could scarcely see her, but then she slowed down and I was able to catch up. I marvelled at the way the highlights slid along her satin skin, while every detail of her behaviour etched in my memory. Only once did her relaxed motions change, when she suddenly turned at a right angle and glided across my path close enough to touch. I stopped moving and she went on.

An enormous school of yellow convict tang (*Acanthurus triostegus*) appeared in front of us, a dazzling, shifting wall of moving gold. The shark ignored them and the little fish cleared a path for her, scattering light in all directions. After thirty-five minutes, she slowly drew away from me and I lost her. She had travelled over a kilometre along the lagoon.

CHAPTER THREE

Adventures in the Ocean

Franck and I bought ocean kayaks so we could dive outside the reef—the ocean was inaccessible without a boat. The ocean kayak was designed to be used as a diving platform. Its hollow, plastic hull was wide and flattened so that one could sit on it sideways to get into the water. It was moulded for comfort with a well in the back for the dive gear. Holes through the hull allowed waves washing over the gunwale to join the sea again, without accumulating in the bottom. So on Sunday morning, we set off for Fisherman's Point for a practice run.

The Pacific islands were originally formed by the movement of the earth's crust over a hot spot, where super-heated rock pushed through a thin place from deeper inside the earth. The resulting volcano eventually broke through the ocean's surface and formed an island. With time, the crust moved the volcanic mountain off the hot spot, and eventually another belch from the earth spewed forth another island farther to the west. Thus chains of islands formed across the Pacific Ocean.

The weight of each island pushed down on the crust and caused it gradually to sink, while erosion from rain and wind wore down the volcanic peak, creating soil and river valleys, so that a limited terrestrial ecosystem could form. Coral grew along the shores, but since it cannot live in fresh water, the places where the rivers flowed into the sea remained coral free. Meanwhile, the sinking island shrank in size, and a lagoon formed between the shore and the coral, which continuously grew to stay at the surface of the sea, and formed a protective barrier reef. When the island finally sank beneath the surface, only the coral barrier reef remained where the shores of the original volcano had been. Such structures are called atolls.

The gaps, or passes, in the barrier reef, where the rivers have always emptied into the sea, have a profound effect on the dynamics of the flows of water through the lagoon. The ocean's waves pour over the barrier reef, into the lagoon, and out again through the passes, which become extremely dangerous when the waves are high, due to the funnelling effect of these narrower channels. Unknown to me, we were approaching one of the most dangerous passes on the island. Fisherman's Point was at the mouth of one of the island's major rivers, and the pass there was very wide, allowing the ocean's breakers to strike the shore.

The lagoon was a vision of tropical beauty as we paddled happily along, gazing around and watching the fish through the glass-like surface. As we neared the pass, Franck turned to talk to me, and we drifted briefly, discussing how we should proceed. We didn't realize that while we talked, we were being carried more and more swiftly—the current can accelerate unexpectedly as the water becomes shallow.

Franck turned away and I started paddling again, realizing immediately that we had a problem. Waves colliding from different directions flung themselves upward into peaks in the image of agitation, a pattern I had never seen before. The roar of the breakers in the pass took on a threatening tone. Franck was trying to turn around, shouting at me to go back. I struggled to follow his example, but the current had swept me so far that no matter how hard I fought the water, it carried me onward, into the breakers crashing in the pass.

Being unaccustomed to kayaking, I was already tired from keeping up with Franck during the three-kilometre (two mile) trip down the lagoon and knew there was no hope of regaining control. Swept broadside, and accelerating toward immense breaking waves, it was obvious that the sea was taking me, and I became terrified. The bottom was shooting past just below the boat, and without further thought I swung my legs over the gunwale and planted my two feet firmly on solid ground. The current swept around my ankles, and

I held tight to my boat while arranging my paddle and putting on my thongs for protection. Then I put one foot firmly and carefully in front of the other, with the intention of walking to safety.

For a while this worked. Franck had managed to reach the shore and was trying to walk out to meet me. It wasn't until he turned back in chest-deep rushing water that I realized I had stepped out in the only shallow place, where the barrier reef curved in toward shore at the pass. Between the reef and the beach was a deep channel, which was now a white-water torrent. Breakers entering the pass were rolling up this tumultuous river, toppling white foam in all directions. I wondered how I had managed to forget, or conveniently ignore, this detail when I had snorkelled there so many times.

I kept walking slowly, trailing the kayak, and managed to get much closer to the shore, but the water was getting deeper with every step, and my footing felt more and more uncertain. Finally, I couldn't keep my balance enough to take another step or retreat to shallower water. The powerful current sweeping around my chest threatened to topple me at any moment, while the kayak twisted restlessly and yanked me in unexpected directions. Franck gesticulated frantically on the shore, which was now lined with Tahitians, apparently waiting to see what would happen to the stupid white woman who had gotten out of her boat in the current. I stared out to sea, telling myself, *Think! Think!* Then I looked at Franck, who was making paddling motions, with the silent, watching crowd spreading out on both sides of him. I gazed back out to sea again, all the time fighting with the boat, which kept jerking at me like a living thing.

Why couldn't I figure out what to do? The white water and crashing waves beyond, filled me with terror. I kept looking at it, at the crowd of Tahitians lining the shore, then back across the wide, tranquil sea, desperately keeping my balance. After an interminable period of indecision, the answer came as a voice speaking inside my head. It said, "Well, any minute now you are going to be going through that white water. You can either stay here until you lose your balance and get dragged into it by this plunging boat, or you can get back into the boat and go through it with some dignity."

The problem was getting back into the boat, which was performing its wild dance at nearly shoulder height. Could I get in? At that moment I slipped. Using my last contact with solid ground, I leaped, threw my weight on my hands on the boat, hauled myself in, jammed myself into the seat, and started paddling with all my force. The kayak shot across the channel and nosed the beach just at the point of land beyond which I would have been dashed into the surf. Franck was grabbing the front of the kayak, and the Tahitians were all clapping their hands. Two women came over to sympathize with big smiles of relief for me.

On the way back up the lagoon with Franck, who had received me on the shore with a big hug and many kisses, unnatural lights appeared in my vision, and I felt foggy and unwell for the rest of the morning. The experience wasn't the best introduction to my kayak and instilled in me a profound distrust of the sea.

Nevertheless, one quiet morning after Franck left for work, I saw the tail of a humpback whale disappear into the ocean outside the reef. After pouring my half-drunk cup of coffee into a jar, I dragged my kayak into the water, loaded it with fins, mask, snorkel, coffee, and a few bananas, and headed for the passageway. Two whales surfaced in front of me as I flew through on the outgoing wave and, for the next half hour, I chased after them as they roamed the area. They were surfacing every two to three minutes, and though once I was only about thirty metres (100 ft) away, it was not close enough.

It was the first time I had seen great whales up close. The sight of animals so large and so close, and the sound of their breathing was awe inspiring. Finally they stayed down for seven minutes, and when they surfaced again, they were heading straight out to sea. But soon they turned parallel to the reef, apparently following the drop-off. One was smaller than the other, suggesting that the pair were a mother and calf. Each time they surfaced, they left an expanding circle of glassy water in the restless ocean. At first they were two and a half minutes ahead of me, but I gradually narrowed the distance between us, until I was reaching the place they had breathed just a minute later. If they slowed to look at something or stopped to play, I would be there with them. After five kilometres (three miles), we approached a bay, and they turned into it, but at that moment, a third whale surfaced on the far side. I guessed that my two would go to join this one, so I headed toward the third whale. As expected, the mother and calf turned, crossed the bay, and started back out into the ocean. Slowly our paths began to converge, and when I reached the other side of the bay, I was directly in front of them. When they surfaced about thirty metres (100 ft) away, I put on my mask and snorkel and slid underwater. My paddle and everything else was tied to the kayak in case they overturned it. At first there was nothing but blue, and I waited, counting the seconds and estimating the moment they would appear, so excited I could hardly breathe, but making a little squeaking sound to attract their attention. And out of the blue swam two humpback whales. Right in front of me, the smaller one paused, hanging in the water motionless, looking at me. We stared at each other. He was about five metres (15 ft) long, so he must have been a yearling. His mother had vanished, and when I tore my eyes away to look for her, she was descending straight down beneath me, apparently spiralling from the way her tail twisted as she disappeared into the deep. The baby had surfaced beside my boat before going on—there was one of those circles of glassy water there. Beyond, dolphins were cavorting along, having joined the whales.

Inspired by this experience, Franck and I went looking for whales together one weekend. We found a mother and calf and took turns going underwater to watch them—one person had to look after the kayaks. Far below, I could see the baby whale outlined in white, swimming above the dark back of his mother. I was able to follow them briefly until darkness obscured them. When again they surfaced, Franck, drifting quietly, watched for many minutes while they rested. The mother was quite undisturbed by our proximity.

When they surfaced again, we approached quietly, and I slid underwater. Far off, the baby whale was diving down with great, slow sweeps of his tail. He twisted and spiralled as he dove with a wonderful feeling of freedom and joy. His mother was already deep, and the baby swam under her nose, where she rested stationary in the indigo shadows of the sea. The great whale wrapped her fins around her baby. They were cuddling, unmoving, hidden in the deep. Touched with awe, I started to sing, spellbound at the sight of this secret of the ocean. Diving down, I gained a clearer view of the great whale, the lighter line of her mouth encircling the upper surface of her head, the baby moving beneath her, clasped in her fins. I went on singing, the music emerging unplanned from the heights of reverence and love.

The baby, who looked very small so far below, wriggled from his mother's grasp and swam straight up toward me, growing surprisingly larger as he came. How beautiful was the motion of his sculpted form, his prominent eyes fixed on mine. His fins were held out horizontally from his body, and I held mine out too, as we came closer and closer together. As he turned past me, his fin was just centimetres from my hand. I reached out just that much to touch it, and he moved it back just far enough so that I could not. His large, soft eye looked into mine as he passed, and it was one of those long moments, suspended in the blue abyss, gazing at each other. Then his silver side that had seemed so small, expanded into a moving wall in front of my face. It was already quite marked, as though he had often scraped against things. I put out my hand and felt his sleek skin. His body was hard beneath. My hand trailed along his side until his tail caught me under the arm and flipped me onto my back. When I turned, he was diving back down to his mother.

When I found them again, the great whale was raising her wings and undulating vertically in slow-motion to ascend. I savoured every detail from about five metres (15 ft) away as she surfaced; her baby was on her other side. She was very dark, and her face was hard to make out. She was surprisingly rotund and had small white spots decorating her back. At about three times the size of her baby, I estimated her length to be about fifteen metres (50 ft). She passed like a train in slow-motion, resting on the surface, breathing. Where her black colouring met her silver underside, she looked as if she had been painted with careless strokes of a wide brush. She had a five-centimetre (2 in) deep V on the trailing edge of her right fluke, and the tip of the left one was missing. They drifted on, and were only shadows in the blue as they dove again into the deep.

Christmas in Tahiti was a joy for me. Instead of the psychological chill of snow-fields and black and white mountains, icy cold, and perilous roads, it was delightful to awaken each day to cosy, warm sunshine, and new red flowers opening—nature expressing itself in the gentlest way. To celebrate, we donned our diving gear and swam across the lagoon to the tiny pass. The sea was strangely quiet for December, and we slipped through the reef's window and swam through the lighted channel full of champagne water and fish.

As we descended the gentle slope past the surf zone, coral appeared and grew larger. The island's downward slope grew steeper almost gradually enough that one scarcely noticed at first, but the low hills and shallow valleys all slanted away more and more steeply from the island. They resembled rolling hills covered with sage brush and tumble weeds, like some dry foothills on earth, all dimly lit by a cloudy sky.

Franck led, always just a little too far ahead for me to reach him, and increasingly, I seemed to be drifting down rather than swimming. I was watching for sharks and large ocean fish, and each time I checked my depth gauge, I was surprised by how far we had descended. At twenty-five metres (80 ft), the slope was comparable to that of a steep mountain, which one could, nevertheless, hike up. At a depth of forty metres (132 ft), the slope of Tahiti into the depths had become very steep. The bright ocean blue deepened and became a mysterious, violet twilight, while the coral landscape flowered with layer upon layer of enormous purple roses. Still I drifted down, Franck ahead and beneath me. The stillness, the weightlessness of drifting ever on, ever down, and the alien beauty of the dark roses, became dreamlike. At fifty metres (165 ft) in depth, we signalled "okay" to each other, and I pointed to a place below, where Tahiti appeared to end.

Franck swam down while I continued my slow-motion free-fall, gazing down at him, the coral precipice, and the blackness below the drop-off. It looked like any crumbling mountainside, where some places fall off more steeply than others, but at the drop-off, it ended. The great shadow of the ocean hung over the scene, while the utter stillness of the supernatural roses felt dreamlike. Franck alighted there, peering straight down into the abyss, as I slowly fell toward him, nothing but water and the abyssal plain below.

A sharptooth lemon shark (*Negaprion acutidens*) appeared from beneath the drop-off and swam slowly, inexorably up past us, and glided on over the next hill. My depth gauge now signalled that we were sixty-five metres (213 ft) beneath the surface. I paused, finning to keep myself in place in front of Franck, who still stared downward, clutching the coral as if he could fall. Then he took my hand, and as we finned back up the slope toward the light, I kept looking back at the mystical place where Tahiti ended.

We ascended slowly, looking around. A large, green sea turtle (*Chelonia mydas*) was nibbling at something under a coral branch. Though we watched from less than two metres (6.5 ft) away, he paid no attention to us. Far above, a barracuda hung between us and the surface. We met an enormous school of yellow striped snappers, each about thirty centimetres (1 ft) long. I became fascinated by them when they began swimming around me, blocking my view of anything else and lay motionless in the water, just looking at them. Incredibly, they responded to this by looking back at me!

As I watched, hundreds of fish turned toward me, and one after another, they positioned themselves about thirty centimetres (1 ft) in front of my face, each one taking a deep look into my eyes for a few seconds before moving on. The precise position they had chosen for this action was then taken by another fish. Their eyes were large and four centimetres (1.5 in) apart, and their look was serious and intent. Fishes' faces filled my view as they continuously moved toward me to take, one by one, the position in which they could look directly into my eyes. This went on for a very long time and would have continued indefinitely, it seems, had Franck not come into the cloud of fish to see what we were doing.

Eye contact is significant to us and seems to create or enhance a communication with not only humans but animals as well, as everyone who has a pet knows. This experience with the fish changed my thoughts of them, for it was the first clue they gave me of their inner lives. On other occasions I found that if I stopped moving and relaxed on the bottom, fish would come over to look at me. Then, on finding me looking so intently at them, they would approach close to my mask to look into my eyes. I never had an opaque school around me again, but every time I tried, some fish came. One small, brown grouper with purple spots was so interested that it kept coming to look into my eyes even after I was swimming away.

Soon we were back in the little pass, shooting through it with the waves and into the sunny lagoon. That night, the sunset was one of the most spectacular we had seen, the sun sending rays of red light to the zenith from below the horizon, the sky behind nearly green above the embers on the ocean. Above, the deep blue faded to luminous violet, then purple, as night rushed over us from out of the east. And all was reflected, shimmering in the lagoon.

CHAPTER FOUR

Sharks

One night I dreamed I was walking at the edge of a supernaturally brilliant lagoon brimming with life. Fascinated, I leaned down to look deeply in, and saw two large sharks, one black and one white, gliding towards me. Without a pause, the black one swam straight into my arms. As I held him, he spoke into my ear, and I realized that all his fins had been cut off. At that moment I awoke in a panic, trying to remember what it was that the shark had said. It seemed terribly important to remember, but I could not. It was one of those striking dreams one can't forget, and it troubled me.

Two days later *La Dépêche* announced that a ship full of shark fins had come into the harbour at Papeete. A whole page was devoted to the damage wrought by shark finning. The shark's fins and tail were sliced off and the living animals were thrown back into the sea alive and shocked, suffocating, as they sank slowly into the abyss. It was an indefensible waste to kill a large animal and throw it all away except for its fins. The article claimed that oceanic sharks had been seriously depleted in the South Pacific to keep Asians supplied with shark fin soup, a luxury for the rich. It was estimated that many thousands of sharks had been killed just to fill this ship with fins.

Franck and I took the truck down to the harbour to see the ship and spoke to some of the crew. They knew only enough English to keep repeating, "No problem! No problem!" The harbour authorities had nothing to say about it.

The dream snapped into perspective. Sharks were being finned nearby while I had been dreaming!

In Canada, I had focused on the problem of hunting, which emptied the forests of life, and this concern came out in my art. In Polynesia, shark finning was responsible for emptying the seas of the top predator, one of the most majestic of marine animals. I had been searching for a focus since arriving in Polynesia, and now I had found it. Considering the dream a call for help, I decided to begin painting sharks, and see what I could contribute to their protection.

So I immersed myself in shark images, painting them full time, searching for them in the lagoon, and drawing them at the local lagonarium, a fenced section of the lagoon where one could watch the enclosed animals as one does in an aquarium.

Franck took me on a shark feeding dive to help me, and I found myself all dressed up in scuba gear, sitting on the side of a dive boat wildly rocking upon a disturbed ocean. I had never entered the water from a dive boat before, and when told to kick up my legs, I decided that that was something I could not do. So I began re-adjusting my mask. However, the dive leader persisted in nagging me about it, and finally I was obliged to flip over backward. Warm water broke around me and I was looking into the blue. Far away, I could see the bottom, grey in the cloudy light. It was easy to descend holding the anchor line and to settle myself beside Franck with the other divers on the bottom. Someone pointed upward, and there, coming down the anchor line, was our dive leader, a little round man who was carrying a plastic bag of fish scraps, the left-overs from a fish store. A few sharks were going in and out of the cloud of fish that surrounded him like balloons around a clown. He arrived among us, and, still accompanied by his entourage of excited fish, he led us to a place on the side of one of the gently sloping hills that disappeared into the blue. We settled there watching, while sharks gathered, appearing out of the veiling light, and circling majestically. As the dive leader held up a fish spine for them, once in a while one darted in and took a swipe at it, circled tightly, and approached again. The density of sharks grew thick close to the dive leader and decreased proportionally with distance out to the limits of visibility. Nothing had prepared me for the drama of the scene—there were sharks as far as the eye could see in all directions high and low.

Each time I counted them, there were about twenty visible. They seemed to ignore us, their audience, and swam among us as if we weren't there. I studied the expressions on their little faces, but could discern nothing, and wondered if they sensed their environment mostly through other senses, rather than their eyes, to look so interiorized. But I had heard that they could see as well as us. It was daunting to watch them and know that they had stopped evolving surprisingly early in the history of the earth. That meant that they had been tried and found perfect; no further mutation could improve them. And how beautiful they all were, flying through their blue world with such grace that it took my breath away to watch them.

After the feeding, we explored the area in two groups. Once, as Franck and I topped a rise, a large shark swam close by me. I reached out my hand to stroke it, and it changed direction. The sharks accompanied us all the way, sometimes swimming parallel to the group in visual range and sometimes leaving and returning. They even continued to swim around the boat after we had climbed in to leave. It was as if just being with us was more interesting than what they usually did. Otherwise, why would they swim along with us like a pack of wild dogs?

The next time we went diving, I waited to see a shark and was disappointed when none appeared. We were with a dive club, and I was trailing behind, blissfully looking at things, my usual artistic trance enhanced at thirty metres (100 ft) by the effects of nitrogen narcosis, also termed *rapture of the deep*.

When a shark appeared in the distance, I stopped and watched him. From the limits of visibility he shot toward me, straight as an arrow like a lost dog called from the forest, and by the time he was very close, I was back in my dream and ready to take him into my arms. But at the last minute, he turned just enough to pass on my right.

Franck hastened back to get me moving again, and the shark came. I was delighted that for the rest of the dive he roamed the vicinity, returning often to circle near me. This unexpected rapport with the shark seemed to put an official stamp on my dream and decision to paint them.

When I had completed five large paintings, I found a gallery to work with. But since sharks were not popular, the gallery owner asked that I paint other subjects, too, suggesting Polynesian women with the sharks. He would not hang the shark paintings he said, if I did not.

This was an artistic puzzle I found insoluble. Each time I had seen a shark, no Polynesian woman had been anywhere near. The two subjects seemed mutually exclusive.

So I did a series of chalk drawings of Polynesian dancers to show I could draw people too, then designed a woman reclining on a coral reef, caressing the nose of a passing dolphin. With the scene a third above the water level and two thirds below, it had a visual fascination. I roamed through the coral in the mornings, and painted in the latest beautiful fish or coral formation I had seen on my return, working with an airbrush from memory. I tried to have a Polynesian woman pose for the figure, but couldn't seem to get the right position, so rendered the woman by alternately observing myself in the black glass doors of the stereo cabinet, then getting back in my chair and painting the part I had just studied.

When the gallery owner saw it, he said he would have it printed, and I was sure that my art career in Polynesia had been launched. So I airbrushed another under-and-over the surface painting of a woman in an outrigger canoe being followed by a large hammerhead shark, and called it *The Guardian*. The owner of the gallery loved it and said he would print that one too.

But the art market seemed virtually non-existent. The Chinese and Polynesians had their own styles. Tourists favoured small, inexpensive watercolours of Polynesian huts beneath palm trees by the lagoon, and semi-draped Polynesian women if they were interested in art at all. The French, who made up less than ten percent of the population, liked impressionism and the style of Paul Gauguin, rather than my North American wildlife art style, reflecting wild nature. Further, they were rarely interested in large paintings, since they usually returned to France, and a large painting would not be convenient to take to the other side of the

planet.

While I could always sell enough to have a reliable income in Canada, in Polynesia, for the first time in my life, my paintings were not selling. They sold so slowly that the gallery owner changed his mind about printing my work. Finally I changed galleries when I was invited to hang an exhibition in a new one that was opening.

By then we had been on Tahiti for two years, and when the opportunity came, we moved to a quieter island and rented a house at the lagoon's edge. This lagoon was similar to the one we had left, though the island was older, so the reef was farther away, and a deep, wide channel separated the shore from the lagoon.

To enter the water, we had to step down a low wall into the shallows and sit there to put on our gear. One evening, a little black and white-ringed form went snaking under the sand just where we put our feet prior to stepping up out of the water. A sea snake!

I watched for it until darkness fell and it appeared again. This time, I clearly saw its triangular, serpentine head. It was about forty-five centimetres (18 in) long, and evenly ringed with black-and-white alternating bands from head to tail, each ring about two centimetres (0.75 in) wide. The appearance of this rare and poisonous beast quite impressed us, living as it was on our door-step. It was a reminder that nothing could be taken for granted.

It was many months before we were able to find a reference to it and learn that it was a snake eel. The eel looks more like a snake than sea snakes do, since the marine reptile has a vertically flattened tail, while the eel's is snake-like. It was a remarkable example of convergent evolution. Only once did I see another—a large snake eel swimming across the sand on the other side of the channel. It was a rare one.

But at the time, I had no doubt that an offspring of the poisonous marine reptile itself was living beneath our sea-steps, so I always entered the water with due care and attention. There were dinner-plate-sized flatfish too, so well camouflaged that they were nearly invisible and prone to be stepped on. The water was very shallow from our sea wall out to the edge of the fringe reef lining the channel ten metres (33 ft) from shore, and soon I began to put on my gear at the channel's edge, where I could sit on a smooth, pillow-shaped formation of dead coral far from the hidden sea snake.

One evening when I arrived and reached out my hand to steady myself, I nearly put it on a lion fish who was reposing there. Beautiful as it is, plumed and veiled in lacy patterns of red and white, it carries poison in every decorated spine. I put my face into the water, centimetres from the little monster, to admire it. It glided slowly across the coral and out of sight on the other side, where I found it when I came back from my expedition. After that, I often saw it on my coral pillow on my evening outings. And if I went early, before it emerged, I could look into its arched cathedral beneath the coral and see its spines and laces hanging down and gently waving as the little creature slowly danced, perhaps to attract its meal of crustaceans. My reference book said that it had evolved its remarkable appearance in order to attract them. What did that tell us about crustaceans? Each time I saw it, I wondered.

My book on fish was useful to an extent for identification and for the bit of information it provided. But it was written for fishermen. The fish described were often pictured dead, or held up dead, by proudly smiling men, instead of alive in the sea. Many fish, including moray eels, sharks and needlefish were characterized as being dangerous, which caused me to be fearful when I first encountered them. I had to learn what each fish was really like for myself. The photos were too poor for a definite identification of many species, and I remained ignorant of their English names.

So I made up my own names for them. With no idea what the lion fish (*P. radiata*) was called, I thought of it as *the fish that looks like a flower* or *flower fish*. The bright fish that swam with me during my travels on the reef were *plastic fish* named for their psychedelic colours. It wasn't until years later that I learned they were called Hardwicke's wrasse (*Thalassoma hardwicke*). I called the white bottom-feeders goat fish from the appearance of their faces. It turned out that they really were called goat fish (*Mulloïdichthys flavolineatus*)! Then there was the *dancing fish*, a large triggerfish (*Balistoides viridescens*), and my *racehorse snail* was actually called the Great Green Turban (*Turbo marmoratus*). I gave most of the creatures I was getting to know descriptive names in this way, for convenience.

While this is an unimportant detail, it illustrates the degree to which I was isolated from much needed information, due to the lack of scientific resources and other written information in Polynesia. I was on my own in my efforts to learn about the natural world surrounding me.

I didn't always have time to cross the channel to the lagoon on my evening swims, so often I just explored along the fringe reef. It fell, a dead coral wall, to a floor covered with garbage. There was not only the usual fishing trash of discarded nets and fish traps, but refrigerators, tires, car batteries, car parts, twisted pieces of wire mesh, kitchen utensils and appliances, bottles, and plastic bags. Not only were these plastic bags snagged among trash on the bottom, but they were floating in mid-water. Myriads of unidentifiable objects presented straight lines and right angles in proof that they were the trash of man, drifted over with silt.

The visibility was bad. It was like swimming in a dirty fog, and I finally learned the reason. The sewage from the houses lining the shore, in many cases, emptied straight into the lagoon. It was hard to comprehend such a practise in a culture dependent on fishing. The rule is self-evident: don't mess where you eat.

The channel was thirty metres (100 ft) deep, and carried the effluent away, so the lagoon beyond was cleaner, rinsed by sea water coming in over the reef.

Two hurricanes passed close by that season. The surface of the lagoon fluttered like a flock of nervous birds at their approach, and darkness fell as the shadow of each storm brushed by, trailing tumultuous waves expelled by a tormented ocean. We were an hour away from being evacuated on the approach of the second one, then it suddenly turned away after laying waste to Bora Bora and a string of other islands. There was no doubt among the inhabitants that it was their prayers that had saved us.

In its wake, I found a large spider conch (*Lambis lambis*) out in the open on the fringe reef. This shellfish constructed a virtual palace for himself, as shown in the illustration, so elegantly coloured and shaped along its twenty-five-centimetre (10 in) length, that it was prized by the Polynesians for cash sales to the tourists. They extracted the secretive builder by hanging the shell in a tree until he became so tired that he fell out. These were among the animals I hid when I found one, and I took the wanderer back to the lagoon. (The only shells I collected myself were ones I found empty.)

As we moved out into the channel, the eyes of the creature emerged to lie against its pearly shell, gazing ahead through the torrents. In spite of the depth, in the wake of such a storm, the water was on the move, and I swam slightly upstream to compensate for the current. The visibility was poor, but in the cloudy light a whirlpool of spinning *turbinaria* torn from the reef, and plastic bags in pastel shades, appeared ahead. It spun from the surface all the way into the depths. When I got close, the canopy of a tree became visible far below, slowly turning as it travelled erect down the channel, branches and leaves extended in place as if it were still growing on land. But the sea had taken it. For a long time I watched its progress toward the ocean, while the eyes of the spider conch gazed calmly on. When I returned, dozens of young eagle rays were visible below, flying in place in the current, much as we had seen them off the lagoon at Fisherman's Point.

Sometimes, Franck and I went to a nearby jetty to dive off into the channel, climbing back up the jumbled rocks after each dive. A variety of surgeon fish lived in the crevices among these boulders, and they would rush out and bite us on the behind as we left the water. When we went snorkelling, they watched us from the safety of their crevices, usually in a defensive manner as if they were territorial. They would raise their hackles as we passed but did not attack. They attacked only when our heads were above water and preferably high above the surface. Further, they were fully aware of which direction we were looking and always came to bite from behind us. Once, only my feet were still in the water when one of them rushed toward me. I crouched down over it, waving my finger above it in a gesture of negation. It backed into its hole just far enough to feel protected, making itself small with its fins at its sides, and its hackles half raised, looking up through the surface at me. As soon as I stood up, it shot out of its hole and gave my foot an extra hard bite that bled. It had clearly been aware of my attention as I bent down over it above the surface, and the extra hard bite seemed to be the result of its alarm at my threatening gesture.

It seemed that fish differentiate between a scuba diver and a snorkeller at the surface. The same ones that regard a snorkeller with suspicion are willing to crowd around a diver and even eat from his or her hand. They evidently seem to understand, as well, that a swimmer, head out of water, is unable to see them. Often I was bitten by fish while swimming for pleasure, but never when I was wearing a mask and looking back at them underwater. They are sensitive to eye gaze.

While exploring our new lagoon, I searched for sharks, hoping to learn something of what they were doing. Here, there was no channel along the inner edge of the barrier reef where they travelled—they were usually just roaming around randomly in the outer third of the lagoon, though occasionally I found one cruising along the edge of the channel. Sometimes I swam for hours, unwilling to go home until I had seen one. When, exhausted and disappointed, I began to drift homeward, that was often the moment a shark would appear.

Going regularly on long expeditions, I soon had a detailed map of the submarine terrain showing the different ecological features, including an underwater river where the current concentrated, which I quickly learned to avoid. The reef was much farther away, so it took longer to swim out that far. It seemed a wilder place, and the water always felt colder.

Our new location was close to a bay to the eastward, and there the reef curved toward shore and ended in a wide shallow region. For a while, I found baby sharks, or shark pups, there. They were a bit less than half a metre (20 in) long, and varied in colour and size. They were often curious—perfect tiny sharks with avid eyes coming for a swift circle before vanishing in the waves. Though the shallow area was protected from the main force of the ocean swell, it was continually flushed by torrents rushing across the reef, and seemed a strange place for babies of any species to live. But many of the other species who lived in the region were babies, too! It was a micro-ecosystem of the juveniles of the lagoon in waters too shallow for large predators to come. Conditions were rarely calm enough for me to explore underwater there.

Around the shark nursery, reef flats alternated with strands of sand where there were many stingrays, and the occasional large, solitary eagle ray sometimes appeared, flying slowly on unknown business. This region gave way to a coral lagoon habitat very similar to the one I had known on Tahiti.

Once I was motionless in the water, watching the activities of a small fish swimming around its coral home, when I had the strong impression that someone was watching me. It was as though a shadow had fallen on me. A survey underwater revealed no one, so fearing that a fisherman in an outrigger canoe was close by watching me, I looked above the surface, turning quickly to see in all directions. But the surface shimmered calm and empty off into the distance.

When I sank underwater again, I found myself face to face with a large shark. She had come knowing I could not see her, since my head was not underwater, but once we were eye to eye, she turned at right angles and disappeared. In this new lagoon, the sharks seemed much less accustomed to encountering people. They seemed skittish and would not approach, often accelerating swiftly out of sight. However, I found that when I drifted slowly, motionless in an artistic trance, sharks would sometimes appear at the limits of visibility then come very close. But the slightest movement, even a switch to left-brained alertness on my part, would cause them to instantaneously depart. For them to feel comfortable, it was necessary to stay very quiet mentally as well as physically.

So I tried to remain as meditative as possible as I drifted, watching the surroundings, and found that the quieter I remained mentally, the more confident the shark seemed about approaching me. Were they aware of human brain-waves? An Internet connection to the rest of the world was still far in the future, so I had no one to ask about what sharks might be able to sense with their electric sense, or the distance at which it was effective. What sharks were actually aware of was completely mysterious to me.

On one occasion I had been exploring along the reef for so long that I was stupid with cold. When I finally started back, I found that my snorkel was leaking and I had a sharp pain in my foot where my fin was rubbing. So, flying along in the current like a bit of wood in a river, I took off the fin and unwrapped the bandage I used to protect my feet on these marathon forays. However, with my numb hands, my slate, and the seashell I was carrying, I couldn't seem to get it re-wrapped while intermittently drowning in my snorkel water. Finally, realizing that I was in a hopeless mess, I put on the fin without the bandage, and when I turned in the torrent, found a large, pregnant shark right in my path. She was less than half a metre (20 in) away, and as I braced for the collision with the apparently stationary animal, the slate clattered against the seashell, and she accelerated away. She seemed not to have been aware of me before that moment, so once more, I rethought my theory of me and sharks being hyper aware of each other!

One morning when I went outside at 5:00 a.m., a single white feather fell from the empty sky to lie upon the surface of the lagoon, and as if in a spell, two shark pups appeared in the shallow water in front of the house. They were scouting the area, their circles, figure of eights, and cloverleaf pathways creating mesmerizing moving patterns. Sometimes they swam separately, sometimes they followed each other, and sometimes they swam after each other in a perfect circle, each with its tail fifteen centimetres (6 in) from the tail of the other, only to return to their complex ballet of graceful curves. The school of baby tangs that lived in the shallows formed a little yellow cloud, pointing toward them as they passed. A mynah bird swooped down to look. Sometimes they followed other fish; sometimes other fish followed them. The rhythmic side-to-side motion, like a heartbeat, which typifies this most graceful creature, rang like a beat in their music. Though I had formed a theory about being able to estimate the size of fish from the ripples generated on the surface, the passage of these little sharks showed not a sign.

At 7:00 a.m. the French neighbours started trying to shoot them, and bragged all weekend of their fruitless efforts, as though they had taken on visiting monsters.

Not long after that, I was busy in the house when I heard children shouting as they passed along the shore and glanced up just in time to see a baby shark flying head over tail through the air. I went out and found a group of teenaged Polynesians playing with two shark pups, so began trying to bring the suffocating animals back to life, pointing out to the children that the sharks were so stressed that they would die if not left alone at once. Franck joined me, and we launched one of the kayaks to take them back to the shark nursery. But one of the little pups was already dying, and when the end came I left it in the current that returns to the sea. Franck paddled while I sat in the back, holding the remaining shark so that water flowed over its gills. It was on the edge of survival; sometimes it undulated as sharks do and sometimes it stopped briefly, only to begin again, its movement in my hand back and forth, back and forth, steady. But again it would stop, only to start again, regular as a pendulum's swing, determined as a beating heart. Then the baby shark arched its back, its nictitating membrane slid across its eye, and its body relaxed forever.

It was not the first time we had tried to save animals being openly tortured by Polynesian children. Since we had moved there, it had been a frequent necessity.

I continued to paint from nine to five, taking a lagoon break at times, but as the economy gradually slowed down, my paintings began to accumulate. Art stopped selling in Papeete, and finally my gallery closed its doors. I found another to display my paintings on the island on which I was living, and planned to make prints. They would be inexpensive enough to appeal to tourists. Unfortunately the cost of printing turned out to be exorbitant, so I waited, trying to save my francs, while producing small paintings for the tourist market and giving lessons in drawing and watercolour painting. I continued to paint, though, in my newly evolved style of sprayed layers of pure colour, trying, time after time, to capture the mystique of the Polynesian islands and the beauty beneath the sea.

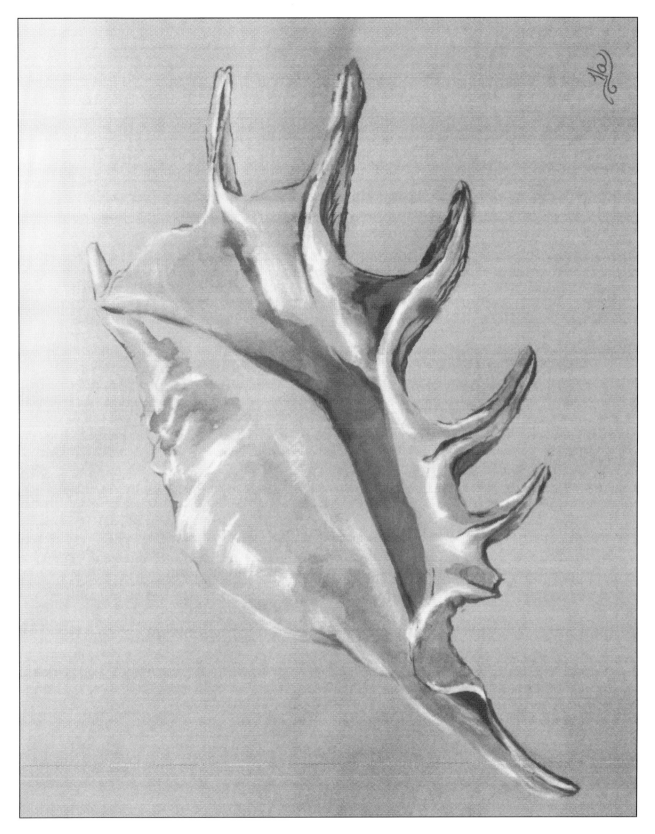

CHAPTER FIVE

Merlin

The humpback whales passed through the islands during our cool season, and I always watched for them when I knew they were in the vicinity. One morning I set off in my kayak in hopes of intercepting some outside the reef. As I came out of the pass, two Polynesian fishermen called out to show me where a sea turtle was floating close to the reef. They asked if I would take him home and look after him since he was very weak. I had no experience with sea turtles, so it was with some doubt that I approached the creature. When I got close to him, it was obvious that he was floating high in the water and was helpless as the waves swept him rapidly toward the surf. His shell was only about thirty centimetres (1 ft) wide, so though I had no idea how much he might overbalance the kayak, I decided to take a chance. I accelerated past him, and with a hand under each side of his shell, managed to heave him onto my lap seconds before we would have been caught in the surf and smashed upon the reef.

At home, I installed him in a large, blue basin in the kitchen and settled beside him. The book identified him as a Hawksbill turtle (*Eretmochelys imbricata*). The movements of his wings, weak as he was, were indescribably graceful, and the intricate patterning of his scales was fascinating. I gently cleaned the algae from his scales, and when he took a breath, wrote down the time. His breaths were spaced about fifteen minutes apart. I believed he was near death, but phoned one vet after another for guidance on his care. No one had any suggestions. When Franck came home, he called every governmental department that might employ an expert who could advise us, but there seemed to be no help available and, indeed, no interest.

Since he floated so high in the water, it seemed safe to conclude that he had too much air inside. A lung infection was an obvious possibility, combined with exhaustion from floating too long and fighting the waves while being heated by the sun. We fed him, gave him vitamins, and let him rest. Finally I contacted the vet employed by a local hotel for the dolphins kept as a tourist attraction. For a hundred dollars, you could get in the pool with the dolphins and touch and caress them as you wished. Naturally these wild captives did not live long, and a vet was required to keep them alive as long as possible. This vet assured me that even if the turtle had a respiratory infection, I should not give him antibiotics.

The next day, the turtle's movements were stronger. I sat beside him with my sketch-book, drawing his scales and noting his breaths. When I returned after an absence, he would slowly move his head toward me. Realizing he was conscious of me, I caressed his head and talked to him. He began breathing more often, at times every minute for ten minutes, then slowing to half that rate, depending on whether he was sleeping. I didn't know if this meant difficulty in getting enough oxygen from each breath, or that he was recovering.

A week passed. Daily I changed his water and took him for a swim in the shallows in front of the house, but he did not seem to enjoy it. He was too weak and was repelled by the light. Doubtless sea turtles feel best in the still, blue depths, and the motion of the sea's surface seemed to distress him.

On the eighth day, I designed a large painting of my exotic visitor around the detailed drawing I had already made of him, in which each scale had been carefully defined. Sitting on the floor beside his basin, I drew and glanced at him less and less. By mid-afternoon, he was sleeping, and I had become so engrossed in shaping the purple rose coral he was swimming over that I stopped keeping track of his breathing. Later when I bent to stroke his head, I found it rock hard. Dead! I couldn't remember the last time I had heard him breathe and realized that he probably hadn't for some time. It was a special loss because the Hawksbill turtle is critically endangered and the species is the only one in its genus.

Small bubbles of foam appeared around each nostril. When Franck came home, we removed his ventral shell and found that his stomach was crawling with brilliant, scarlet flat-worms, each one much more lively than its host had been. The following day I carefully examined each organ. As I suspected, his lungs were full of foam, and bruised looking, which explained why he had been floating, but the main problem appeared to be cancer riddling his small intestines. I must have found him at the end of a long decline, exhausted and overcome by parasites and bacteria.

Just two weeks later on November 16, 1998, a sea turtle appeared in front of our house, carried in from the sea in the waves. I installed the stricken reptile in the blue basin, where he steadily tried to dive down. Though very high in the water—only the edge of his shell was submerged—he seemed in better over-all condition than the Hawksbill turtle had been. His skin was the texture of satin, and his limbs felt alive and vital. He was strong and alarmed by his inability to dive. He breathed about once a minute.

This was a green sea turtle. He was a rich amber colour with intricately patterned wings, head, and hind fins. His beautiful, nearly circular shell was thirty-nine centimetres (15.5 in) in length. Having learned something from my former sea turtle patient, I administered worming medicine and asked Franck to bring home antibiotics for the respiratory infection. We gave him an injection that evening.

The next morning, he had calmed but still spent much of his time trying to dive. Though he took lettuce from my hand that first day, he didn't seem able to chew and eat it. I had to push his food down his throat. When I stroked his wing, he turned his head to press his cheek against my hand. Franck and I found his responsiveness very touching. Whenever we put a hand into the water beside him, he would move his head to press against it. We named him Merlin.

Merlin's condition declined day by day until he lay immobile on the surface. He didn't swallow the small fish and vegetables I put in his mouth, and the ones I failed to get down his throat rotted overnight, stuck to the sides of his mouth. His flesh lost its vitality, and his respiratory rate slowed to a breath every fifteen or twenty minutes. When taken daily to swim in the lagoon, there was no muscle response. His head and fins hung motionless, and his mouth was slack. When his course of antibiotics was finished, mucous seeped from him, filling the water in strands, and his breath reeked as if he were rotting inside.

The weeks dragged on as Merlin waited at death's door. Franck and I gave up hope that he would live. It seemed unbelievable that he could go on so long in such a lifeless state. Two or three times a day, I washed his mouth and supported him on rolled towels while changing his water. During a visit to the lagoon, a sea turtle like him was flying through the coral like a bird, and I wished it could have been Merlin flying there.

At the beginning of December 1998, we moved into a larger house on the edge of a deep bay. There was a beach in front of it, and beyond that about fifteen metres (50 ft) of coral habitat—the fringe reef. Then a coral wall fell away about ten metres (32 ft) down to the sloping, sandy floor of the bay, where the depth increased rapidly. The visibility was so bad from the pollution pouring into it from the agricultural region at its head that we never did explore further than that.

There came a day when Merlin began to move weakly. As he slowly returned to the world of the living, we noticed him holding his head up just enough that his eyes were above water level—he was watching us moving around the house. That December was stormy, but whenever the water was not too rough, I took him out in the shallow water on the beach. He had no strength and had lost a lot of weight. It seemed vital to get him moving again in order to help clear his lungs and rebuild his muscles. It was important, too, for him to see that the world outside his little basin was still there.

When he began to make progress, I took him for a longer swim out over the drop-off, so that he could see the deep water. As we drifted slowly along, I opened his mouth to check that he had no rotting food inside. Not getting a very good look, I opened it again a few minutes later to see. Merlin, looking right at me, opened his mouth so that I could see in all the way to his stomach, and powerfully expelled a cloud of rotting fish into my face. I took it as a statement and never gave him those little fish again.

My sister, Kerry, was visiting at the time, and she picked a handful of snails off the stones along the shore, crushed them with a hammer, and offered them to Merlin. He fed willingly for the first time.

In his basin, he watched more frequently above the surface and responded more quickly if approached. He seemed to enjoy being stroked. But if someone bothered him, he folded his wings over the edges of his shell, shortened his neck so that he was looking down, and stuck his little hind fins straight out at the sides. It seemed to be a protective attitude.

We bought an inflatable pool, installed it in the shade on the deck, and put him in. He drifted around in a slow circle, then stayed beside us to be stroked. I held some of his snails under the surface and waited to see what he would do. Extending his neck, he slowly waved one wing back and forth to propel himself over, intent on what was in my hand. When his nose was just thirty centimetres (1 ft) away, he paused. Water entered his nose as he sniffed. Then he lunged onto the food. He had trouble getting it into his mouth since he was floating and could not brace himself. With difficulty, he would get a tiny snail between his jaws but was unable to manipulate it into his mouth. Even when I steadied him he couldn't eat, in spite of all sorts of manoeuvres and thrashing about, and was so frustrated with his difficulties that I put the food into his mouth, as I had done when he was very sick. He learned to relax his jaw for me, and never did he bite. This seemed amazing considering that sea turtles can crush large seashells with their powerful jaws.

Now that he was feeling better, Merlin required a surprising amount of food. Keeping him supplied with shellfish was increasingly difficult, and daily I searched the fringe reef for plants, sponges and other possible foods, putting them in his pool to see if he liked them. He was still unable to dive, but he was riding lower in the water with his shell now half submerged. He wanted continuous attention and swam to me whenever he saw me. Floating in his pool, he gazed up at the trees arching over him, the mountains and sky, and he watched anyone in view while scenting the air. It was a strange phenomenon, a marine reptile so exquisitely adapted to life in the ocean whose interests suddenly lay in the world above the surface.

I spent hours with him, drawing the patterns of his scales, and for that, I needed a side view. The problem was that he always faced me. He would swim toward me and rest his chin on the edge of the inflatable pool as if he wanted to swim out. I would stroke him and talk to him, give him a shellfish, and try to sneak around for the side view. Each time, he turned with me. He seemed to want as much attention as he could get, and day after day, I spent more time fussing with my model than I did drawing him. Usually while he was occupied with his bit of food, I could draw him uninterrupted for a few moments, but not for long.

Once, my patience was wearing thin, and instead of getting him more shellfish, I gave him some lettuce that happened to be handy. He shook his head and spat it out, then violently smacked his wings on the surface several times and turned his back on me. Now, try as I might to get the side view, he kept his tail end turned exactly in my direction! It took him fifteen minutes to forgive me and drift back toward me again.

One day near the end of January, a stranger came walking in, asking to borrow our wheel-barrow. I was not pleased when the man noticed Merlin and went uninvited to look at him. I briefly explained that the turtle had been seriously ill and walked him back to the road.

Two days later, he returned with a serious looking Chinese man, who demanded in an authoritarian way to see the sea turtle I had in custody. I took him to see Merlin, trying to lighten the conversation with friendly talk and a description of how ill Merlin had been, but failed. He stated bluntly that he was an authority concerned with sea turtles, and that I had no right to have him, as he snatched Merlin from the water and roughly examined him. I froze with anxiety and began to describe how my husband, who spoke French without an accent, I mentioned, had called every branch of the government to ask for advice and help but failed to find anyone who could advise us. We had been obliged to treat the stricken reptile on our own.

"This animal is seriously emaciated!" he accused me.

I agreed and described how I had found him floating, and too ill to swallow food for over two months. But he was now eating well, gaining weight, and swimming daily in the sea. I described Merlin's respiratory infection and asked what a healthy turtle's respiratory rate should be. This was the information I had wanted ever since my first sea turtle had come with his erratic breathing. The man ignored the question. Anxious to learn, since he was a sea turtle expert, I asked again. He said it depended.

"But when a healthy animal is at rest," I asked, "how often does it breathe? What is the normal respiratory rate?" Again he replied vaguely, and I realized that he did not know! He told me brusquely that the turtle would be transferred to a government holding facility at once and said that someone would come by later that day to pick him up. With no further comment, he left with the wheel-barrow borrower.

Sick with worry, I paced around, reviewing the situation. It would probably take some time to get the appropriate person to come from Papeete, which would mean taking the boat. The need to transport the turtle would eliminate the possibility of flying. No one could come before late afternoon. . . Unless there was someone on my island who could come! I flew to Merlin and carried him into the house, where I laid him on cushions while closing and locking the front doors. Then I set his old blue basin in the back bedroom where it could not be seen from outside, filled it as fast as I could, and placed Merlin in it. Then I closed the curtains and sat down beside him with my latest drawing propped in front of me. The weather had become brooding and dark, and I could scarcely see well enough to draw.

Merlin threw a violent fit to find himself back in his basin in the house, and splashed so much sea water around that it took the next twenty minutes to clean it up. Finally he calmed as I sat stroking him, poised and listening for someone approaching. I was terrified. Not being French, I was uncertain if I had any human rights in the country, and was afraid that if I was caught breaking any laws, such as the one governing the touching of sea turtles, I could be deported.

On the other hand, while the rest of the world was working to save sea turtles from extinction, Polynesia tolerated people eating just as many as they wanted. They were very religious and believed that God had put sea turtles in the sea for them to eat—even to the very last one! But it was wrong to try to save one from death. I sat, trembling with anxiety, systematically drawing. The day darkened, and eventually I just sat looking out across the grey sea and comforting Merlin. When I was beginning to think that I had overreacted, a vehicle drove in onto the lawn. I could hear it but dared not look. Instead, I threw a cover over Merlin, and retreated to the bathroom, where I waited, arms crossed, leaning back against the counter.

A man began yelling, "*Allô!*" Listening intently, I followed his progress around the silent house. Hiding and clutching myself, I imagined myself making a plea to a judge on Merlin's behalf. Only I knew his needs and it would endanger him, after all he had been through, to put him into a strange facility. I began to gain a little confidence. Surely, I was in the right. . . There came the sound of footsteps mounting the deck and approaching. Merlin's pool, decorated with seaweeds, sponges, and the seashell he liked to clutch, was empty, and the person paused there a long time. Then the footsteps retreated, the car door slammed, and the sound of the motor faded. Franck was using our car, so it must have appeared as if no one was home.

When Franck returned, I told him what had happened, and he reached for the phone. He had a long conversation with his contact at the Department of the Sea, who claimed not to know of someone having been dispatched to take the sea turtle. The holding facility the man referred to was on another island, so it was illogical that someone from there could have come to take the animal. Further, this facility had serious problems so was not a solution for Merlin. The woman informed Franck that there was now a veterinarian charged with overseeing sea turtle cases, and she put Franck in touch with him.

Weeks later, an article in the newspaper described how, due to poor care, all the sea turtles in that facility had died. I began to wonder if the wheel-barrow borrower and his Chinese "expert," who was ignorant of the respiratory rate of a sea turtle, had not hatched a plan together to steal him to roast for dinner.

Merlin was doing so well that we built an enclosure for him on the fringe reef, about five metres (16 ft) in diameter, in water one and a half metres (5 ft) deep. We put him in it whenever conditions permitted. But instead of exploring, he waited for us in the corner nearest the house in a cloud of fish. Days passed before he began moving around in the area he had. I think that he was lonely after so much attention close to us on the deck. He waited especially anxiously in the evening to come back into his pool.

At sunset, I lifted him out of his enclosure, carried him to the beach, and scrubbed him with a little brush to keep algae from growing on his skin and shell. He had been sick for so long that in the tropical warmth, algae had proliferated. He soon became used to this, and followed my movements with delicate touches of his wings. Working by feel as I gently rubbed his underside, he would clutch my hand for support with his round hind fins. His delicate, curved bones shaped his fins like fingers with soft webbing in between. Each night, it felt as if a small, human-like creature grasped my hand between his two, with cool, gentle fingers.

By mid-February, Merlin had become more comfortable in his enclosure and spent less time waiting, looking up at the house. He roamed around the perimeter and drifted in the middle, facing into the ocean waves and finning to hold his position. Once, when high waters lifted him over his fence, he circled all around it then swam to the beach. With exercise and good food, his limbs were filling with muscle. Daily, there were new signs that he was becoming stronger and more alert. The bones of his wings extended like those of a bird, as he subtly altered their shape to push against the water in the most effective way. His movements were graceful as the flow of water itself. But, while some days he was lower in the water than others, in general, there had been little change in his buoyancy over the three months he had spent with us.

When the water was too high, the waves too exhausting, or the sun too hot on his exposed shell, Merlin stayed in his pool on the deck where he was comfortable. He spent the nights in his blue basin inside for protection from marauding dogs and Polynesians. He wanted to eat almost continuously, and keeping him well provided with good food was increasingly problematic. Having always been a vegetarian myself, I didn't like to kill the snails for him. He was eating a variety of common shellfish, including a large one favoured by the Polynesians, but his diet needed to be supplemented with a variety of other types of seafood to provide balanced nutrition. Each day, I explored the fringe reef and the wall of the drop-off, studying the plants and animals growing there for possible foods, and collecting his shellfish. His pool was always freshly stocked with the latest variety of seaweeds and sponges I had found, in hopes there would be some that he liked. Unfortunately, his favourite seaweed was rare on our fringe reef, but I found a bank where it grew thickly on the border of the lagoon at the end of the bay. Every three days, no matter what the conditions were, I had to replenish Merlin's supply. It was a long way to paddle the kayak, and a long way to dive down, time after time. Fighting the current while picking as much seaweed as I could carry and getting back to the surface in time for the next breath was an exhausting struggle. But I could see no other solution.

Since Merlin had begun to eat again, his digestive system was upset. His copious diarrhoea smelled unbelievably foul, worse than a rotting whale carcass. It was a worrying sign. I had hoped that this would improve once he was eating natural foods. He was very picky about eating vegetables, though green sea turtles are supposed to be vegetarians. Unable to locate information on green sea turtle requirements, I had only my instincts to depend on, and added potato to his diet, hoping it would soothe his digestion. But it didn't.

Once when I arrived at the sea bank to collect his seaweed, an enormous sea turtle surfaced near my kayak. Then a second one appeared, took a breath, and eyed me, brilliant in the morning sunlight. When he dove, he descended right beneath me, while the other faded into the depths beyond. They were green sea turtles more than a metre in length, and their shells were very elongated compared with Merlin's, which was nearly as round as a pie. He was little more than a baby. No wonder he didn't want to eat his vegetables.

One day while looking for seaweeds, closely studying the textures and colours below me, I was struck by a curl of emerald shining in the sunlight. It was a green so vivid and rich that I wanted to touch it. Considering this, I glimpsed something oddly familiar. It looked like part of Merlin's wing. It had to be part of an animal, but the area around it was identical to the rest of the substrate, and I just couldn't understand this visual puzzle. Then the wing flicked, and I saw a second wing shape on the other side of what had to be a thirty-centimetre (1 ft) long, ancient stonefish (*Synanceja verrucosa*). His one startling shine of green blended perfectly into a scattering of tiny green plants, in a jumble of ambers, rusts, and purples. I found a reed and gently poked the animal to see if it was really there. The only difference between it and the seafloor, was that its skin moved flexibly. I could see where the mouth and eyes should be but couldn't make them out.

I started back, swimming awkwardly along the drop-off, so fully loaded with shellfish and plants that I was barely able to keep afloat. A little bit farther along, I found a chair. You can find anything you want in the waters off the islands of Polynesia. It struck me that this chair was just what I needed to place on Merlin's underwater hill for him to hold onto when he wanted to rest. So I loaded it up and went on.

In the floodlit, blue waters below was a moray eel of a species I'd never seen before. Over a metre long and slender, he was flowing through the water very lazily, curling here and there, as though he had nothing much to do. He was rust-coloured, and shining, sky blue rings wound around him every five centimetres (2 in). Mesmerised by this new wonder, I slowly sank, chair and all, as the eel glided into a hole in the coral and proceeded to entwine himself in the open maze below in an incomparable display of serpentine beauty. Finally, the need to breathe overcame the spell as I sank past the sparkling eel's retreat, and barely made it back to the surface in time to get some more air. I was frantically trying to manipulate the chair into a position that didn't produce an intolerable amount of drag when I saw a purple and white, lacy-patterned stonefish on a shelf three metres (10 ft) below, stopped finning, and sank. As I drew near, he shot at the speed of light into a hole, startling me badly. With difficulty, I regained the surface, and struggled home, thinking about the numbers of lethally poisonous stonefish living invisibly in our vicinity. Why had I never noticed them before? The great invisible world is what you don't see because you don't look.

By then I was suffering seriously from spaghettification, the condition in which one's muscles turn to spaghetti after a prolonged submarine excursion. I quickly tied the chair upside down in Merlin's enclosure, and fitted a large seashell over one of the legs for him to hold onto, then attached a coconut frond to another. Merlin soon came to investigate the floating frond, upon which he crawled to rest.

Now that I was looking so closely at the intricately detailed reef, I saw many stonefish that I would otherwise never have noticed. Another time, I was picking clumps of a lichen-like plant and putting them in my net bag, when I saw a tiny flower fish, centimetres under my fingers in the maze of micro-caverns in the dead coral beneath. It was hard to believe that this fantastic fluff of ruby lace glowing above the green could be a close relative of the hideous stonefish and carry the same cobra-like poison.

March had come. Since Merlin was so much stronger, I took him with me to look for food, and when I saw something promising, carried him down to see if it interested him. Though we didn't find anything new this way, the outings stimulated him and gave him a chance to see his environment. Though he was gaining strength, his buoyancy was no better than it had been when he was ill. In spite of all the exercise he was now getting, his respiratory rate was too fast. His abdomen itched. He scratched at it and rubbed himself against the racehorse snail shell in his pool—the crevices between his ventral scales seemed infected.

So twice a day I dried him off, soaked the areas with betadine (an iodine solution) to kill infection, and let them dry while he waited, propped on rolled towels on the deck. Then I applied a healing, antibacterial ointment in a Vaseline base, which would not come off in the water. Very slowly, the condition improved. One of his eyes seemed infected as well, and I treated that with an ophthalmic ointment. With no one to advise me on his care, I was always unsure, analysing every facet of his behaviour for signs of trouble.

Each evening I cleaned his pool and changed his water so it would be the same temperature as the sea he had just left. The air was always colder than the sea, so Merlin's pool and basin gradually cooled to the temperature of the air. His reptilian body acquired the temperature of his surroundings, and I felt he should not be subjected to a sudden temperature change, but I didn't know what the best temperature was. Filling the pool required heavy lifting to carry the sea-water in pails, and one night my back was so painful that Franck said he would do it when he got home. But when he arrived, Merlin had waited an extra two hours, and it was dark. He was very upset and threw a tantrum, rapidly striking the surface of his pool with his wings as violently as he could and spinning himself wildly in the water. But Franck had brought a bonito (an oceanic fish) for him, and when I held a piece of the fish meat in front of him, he extended his neck and grabbed it. It calmed him right down. Suddenly he was thrilled and ate an astonishing number of pieces.

After that, I prepared a large salad serving bowl of food for Merlin each morning, mixing together lots of cubed bonito, several crushed, boiled potatoes, a package of spinach, and some lettuce. I had a second salad bowl heaped with the seaweeds he liked, and he consumed the contents of both in one day.

The deterioration between the scales on his ventral shell cleared up, and I stopped treating it. Now that he was passing the days in natural sea water and he had recovered from his illness, I didn't think it would return. But over the following weeks, the outer layer of his scales began peeling off, starting in the crevices and moving outward. As the condition progressed, irregular, parallel red lines appeared, about two millimetres (less than one tenth of an inch) wide, following the lines of the crevices between his scales. The red lines appeared first, but the problem changed so slowly that I didn't realize what they meant. By the time it became really noticeable, the big crevice down his mid-line had seriously deteriorated.

So again Merlin had to be brought from the sea and propped on towels to have his treatment. This time, I soaked his ventral shell in potassium permanganate, which kills fungus. Then an application of betadine and an anti-fungal ointment completed the ordeal. Merlin was furious on finding himself tortured anew in such a manner. Back in his pool, he beat on the water faster than I had ever seen him move, spun himself around, drove himself into the side of his pool, and writhed and thrashed about, all the while scratching frantically.

Two weeks passed before his shell showed signs of healing. Imagining Merlin's ventral shell giving way to release his innards into the environment, I tried to figure out an effective treatment even in my dreams.

One Saturday morning, I took him out for a swim on the fringe reef and pushed him down to the bottom to see if I could interest him in any of the plants growing there. For the first time, he flew along the seafloor, easily following its contours for several metres before slowly rising to the surface. I summoned Franck, but Merlin was too breathless to repeat the performance. It was the sign we had been awaiting so long—he had briefly achieved neutral buoyancy. Hours later, the wind began to rage wildly down the bay, whipping white waves ahead of it. The atmosphere screamed all night long. I awoke late the next morning since no birds sang. Monstrous brown waves tore down the bay, wreckage covered the shore, and the sea had deposited a thick layer of sand far up under the house. Merlin was hiding on the bottom of his pool. His enclosure was gone, and the bay was far too wild for swimming, so he had to stay in his pool for a few days, as cold, torrential rain fell and the wind howled. Merlin returned to the surface and remained afloat.

Since it had taken me a week of exhausting work to erect Merlin's enclosure underwater, I couldn't face rebuilding it. Franck was in France, so was unavailable to help. I took Merlin swimming often, and in between I left him on the beach. I was afraid to take my eyes off him, though, in case he strayed, so I set up my easel and air-brushing equipment on the deck at a location from which I could see him.

I had been anxious to finish the painting of my first sea turtle, the Hawksbill who had died while I was designing it. All I wanted to do was immerse myself in the vision of the exotic creature flying over the convoluted coral rose. But Merlin was there, so it was him I should paint. Therefore, I incorporated the sketches I had done when he was ill, to format a large painting of him flying the ocean depths. Since I had already prepared the board for the other painting, I just took it for Merlin's and started right then and there so I wouldn't have to leave the deck from which I could observe my precious sea turtle.

But to my surprise, Merlin stayed reliably in the water in front of the house, so rebuilding the enclosure was unnecessary. Each morning I put him in the sea and left him there, keeping an eye on him as I worked. When he wanted to eat, he came to the shallow waters in front of the house, looking up toward me, and if I did not come immediately, he came clambering out of the water to find me. On our swims, he was sometimes able to remain submerged longer than usual. And then he was a different animal. For the first time, I glimpsed the alert, watchful being he truly was. He could submerge himself by stroking down vertically to investigate a crevice in the coral, then when he began to rise, he stroked with his wings again, moving on to look into another crevice. But it was never long before he returned to the surface.

Once I dove with him beside the drop-off, thinking that he would like the deep, dark water. But he panicked and swam up over the drop-off and back to the beach by himself. Often he played in the waves on the beach, surfing up and down. At first I rushed down to rescue him, gently putting him back in the shallow water beyond the small breaking waves, only to see him come surfing up again a few waves later. It seemed that he preferred playing where he could touch the bottom. After all, it was not natural for a turtle to be forced to float all the time. He began playing in the waves on the beach each evening while I prepared his pool for the night, sometimes crawling up toward me. His gesture was proof of his wish to return to his pool in the evening.

CHAPTER SIX

Merlin's Recovery

Looking after Merlin took up all my time. Only once in the four months since we had moved was I able to get away to explore for a while in the lagoon, which was now half an hour away by kayak.

It was the western end of the same lagoon I had been exploring from our former house. It was about four kilometres (2.5 miles) long and one kilometre (0.6 miles) from shore to reef, bounded on each end by a deep bay. Our new home was halfway down the bay on its western border, while previously we had lived on its eastern edge. The lagoon was on the northern side of the island, which put it in the lee of the prevailing oceanic waves approaching from the southwest, and helped to make it an especially rich coral habitat.

On my one visit to the western end, it had been like a witch's cauldron, the surface leaping into peaks as waves clashed together from three directions, and the current had been too strong to swim through. I had looked around while holding onto the kayak, but even so, I had been circled by a whitetip shark. So I was anxious to go again, especially since Kerry had given me a disposable underwater camera for Christmas.

The camera was an unprecedented luxury. I couldn't wait to start snapping photos. With the shark feeding dive in mind, I wanted to use the remains of Merlin's bonito to attract them. So on my first morning alone after Franck left for France, I put the scraps from Merlin's last fish into the kayak and set off.

The morning was still as I paddled down the bay, with winds rising as I approached the reef. Moderate waves were tumbling into the lagoon, but the current was tolerable. I crossed over the border region, anchored, slid into the water, and looked around. The area was about two metres (6.5 ft) deep, with large coral formations separated by open strands of white sand, perfect for seeing sharks approaching in the distance. I attached the bonito spine to the side of the kayak on a string, so that it would hang in mid-water, poured in the rest of the blood and scraps, and drifted away to await the mêlée that was sure to ensue.

But no sharks came. Presently, I decided that the fish spine was floating too high up in the water and needed to be lowered. So I returned to the boat, and untied the knot with difficulty, increasingly anxious as I clumsily fumbled. Waves splashed against the boat and into my face, while the current unbalanced me.

Checking underwater, I saw a large shark approaching and redoubled my efforts. Finally, I got the knot undone and retied. The shark was circling some distance away, and I was satisfied that the spine now hung at the correct level. A few pieces of skin and the bonito's head lay on the sandy floor. I held up the camera and gripped a dead coral projection for stability as another shark lazily undulated toward me. It circled and passed close by as I snapped its picture. They were investigating this strange situation but seemed very shy.

Still, in the presence of food, they came closer and looked at me more intently than usual. They passed me, circled the boat, circled the area, and returned in their rhythmic way, while I waited, poised, and watched the situation develop. A third shark arrived and joined the relentless circling around. The current was so strong that I could scarcely maintain my position, and the waves washed over my snorkel so that I had to lift my head frequently to clear it. Suddenly, one of the sharks accelerated and grabbed a bit of skin off the bottom, shaking it violently as it shot away. A second picked up the excitement and passed by me very fast, but all this happened too suddenly to photograph. They ignored the spine hung in mid-water.

The two sharks that remained continued to circle. All was calm again. They circled more and more widely until they disappeared for five or ten minutes at a time and ceased to come close. After another half hour, I was very cold, so I detached the bonito spine, laid it on the sand, and climbed back into my kayak. I was taking off my gear when a great, dark shadow passed below the boat, but though I watched for several more minutes, no other came. Eventually I went back down to retrieve the anchor, and saw that the bonito's spine had gone. They had been waiting for me to go.

It was a long time before the government's sea turtle veterinarian could come. His practice was on Tahiti, and he rarely came to our island. However, on a Saturday morning near the end of February, he announced that he would be arriving to examine our sea turtle. Merlin was resting on his coconut frond when he came, and I extricated him from it with difficulty and carried him to see the vet, who stood watching on the shore. He was very gentle with Merlin, and his examination was brief. When he finished, he put Merlin down in the water, where we were standing knee-deep, and said he was the healthiest looking turtle he had seen in custody. Considering that he had been so ill, it was the opposite of what he had expected to find. We stood in the shallow waters of the beach while talking, and Merlin stayed nearby, often coming close to touch our legs, wanting attention. The veterinarian was very impressed with his freedom, his trusting behaviour, and the way he used the toys we had tied up for him. He said he was glad to know that there were people able to care for these specialized marine reptiles and that he would keep us in mind in the future if a temporary home for a sea turtle was needed, prior to release. We were very reassured by his words. The vet left no directions about when Merlin should be released, leaving it to our judgement.

My forays in search of food for Merlin continued to reveal surprising numbers of stonefish in the vicinity. I found three lying low in the coral right in front of the house and two together partway down the coral wall. One was very big, and the other was smaller. They were cuddled together as if in love, and a broad pale stripe crossed both fish. I had wondered if the fish camouflaged themselves so well because some inner design caused the skin to imitate the pattern on the fish's retina. But that didn't explain this mutual stripe.

In the evenings, many flower fish of all sizes appeared. The tiniest were less than five centimetres (2 in) in length, and their colouring had yet to develop the white patterning. Wine-coloured ghosts in the deep green, more frills than substance, they drifted tranquilly, their spines and fins so exaggerated that they looked wraith-like in the gloom.

Merlin began to explore more widely, and discovered a thick anchor rope floating on the surface at the drop-off about fifteen metres (50 ft) up the shore. He began making a slow, wide circle in front of our beach, stopping there to play. One day, a fisherman in an outrigger canoe cruised slowly past, watching him. After that, whenever I saw him playing there, I took him looking for food so he couldn't make it a habit.

But Merlin did not like to accompany me when I searched for his seaweed, and he became increasingly adamant that I respect his wishes. Though he came along passively in the beginning, he experimented with various tricks to avoid coming on these expeditions, which were sometimes quite long. He would swim passively along while I had my arm around him, but once he was free, he would whip his wing out of my reach and take evasive action with surprising rapidity. When he was especially irritated, he would smack my hand with his wing when I reached out to guide him in the proper direction. Whenever I dove down to pick some seaweed, he would take off for home as fast as he could, and I would have to speed after him to catch him and turn the protesting little sea turtle around to continue our search. He could be very hard to see since he was often able to swim just underwater. The reflections hid him from me, when I scanned above the surface for him.

On one occasion, after I had taken him to see a group of small anemones about three metres (10 ft) below, I left him and dove down to get one for him to try later. When I resurfaced, he had vanished. I searched for him underwater and above the surface, but several minutes passed before I saw him climbing up the beach in the distance. I swam quietly nearby to watch. There was no doubt about his intentions; every time a wave touched those little hind fins, he hoisted himself farther up the beach. Eventually he struggled, like a little bulldozer, right to the top, where he was partially hidden in shrubbery.

After that, I procured his seaweed by myself. But there were times when I was with him, and he got tired of being with me, or wasn't in the mood—he flew home and climbed out of the water to get away from me, just vanishing from the underwater world. No matter where we were, Merlin always knew where home was.

One night I was giving him a last feeding in his pool. He was a barely perceptible dark shape, and I fed him by feel, enjoying the delicate touch of his jaws against my fingers. He was eating a piece of fish when I had the impression that he had spun around and snapped something off the side of the pool. But there had been nothing there. Since he was still in the same place, I wondered if I had suffered an illusion. However, he began thrashing his head back and forth, and through gentle touches, I learned that he was wildly shaking a leaf that he had grabbed in a flash from the other side of the pool. He had a remarkable sense of vibration. Yet during the day, leaves frequently fell into his pool, and he paid no attention to them.

There were other occasions when his behaviour was different at night. Sometimes when approached, he would tuck his wings over his shell tightly, arch his neck so his head was pointing downward, and extend his hind fins out from his body, fingers spread. This position was automatically assumed whenever he was alarmed by someone's approach, or if a light came on after dark. Stroking him gently didn't relax him.

I tried to learn more about sea turtles and particularly about their senses, but the only books I was able to find that mentioned them, focused on their reproductive cycle and their use by humans.

Franck was away, and I worked on my painting of Merlin during the day, while watching his activities in the sea, and spent the evenings looking into the pool of light cast by a lamp I affixed to a bamboo pole. The fish seemed to avoid exposing themselves to the light. Passing schools separated to go around it and individuals accelerated if they found themselves alone in the light. A puffer fish drifted through, as it moved among the rocks, but never did a great, dark shape come near. In time, a swarm of insects gathered around the light. Sometimes they fell into the water, and this, in turn, attracted baby fish who came to eat them. Many of the fish on the fringe reef were the young of species who lived in the lagoon as adults. One species was a metre in length and as thin as a broom handle on maturity; the adults lived on the inner flats of the barrier reef. Their young appearing in my light were just fifteen to twenty centimetres (6 to 8 in) in length and much thinner than a pencil. They looked impossibly delicate, gliding on the surface and snapping up insects with amazing grace. It was obvious that they could communicate with each other, and I wondered how they lived and how they managed with such delicate bodies.

The translucent, dark sea lapped around me, luminous with starlight, rising and receding by powers unseen, as I sat upon the rock, taking it all in. Sitting thrilled at the edge of the waters full of life, gazing across the starry sea, while the galaxy of diamonds broke over me like a wave across the black abyss of the night, I would hug myself and feel right at the centre of a universe of wonders.

On the night Franck came home, before I had a chance to tell him about my experiment and the amazing things that appeared in the illuminated water, an outrigger canoe darkened the pool of light, and a man started throwing out a net. I ran down and turned it off. The next day, some Polynesian children came to our rocks to kill crabs, and knocked my light into the sea. That was the end of it.

At night, Merlin stayed in his pool on the deck, and I put him in the sea each morning. He roamed around and spent hours playing with the floating objects I tied in front of the house, coming often to the beach for food. He stopped going out to the floating rope. Most of the afternoon, he rested on his coconut frond.

One evening, I was bringing him in for the night, holding him close in my arms as I swam. As we swept up the beach with a wave, I saw the extraordinary face of a stonefish pass a few centimetres under me on the last clump of coral just where we entered and exited the water. It was a large one, with seaweed growing on it. The next day it was still there, and it remained in the vicinity for several weeks. We became used to it and enjoyed keeping track of it. It rarely moved at all, even when we went close to watch it. I had assumed that though these poisonous fish may be around, we didn't need to worry about them close to the beach where the waves broke. As long as we put on our gear there and didn't touch anything farther out, we could safely avoid being poisoned by them. But here was one within a metre of the water's edge.

I always fed Merlin in the shallows of the beach, and his expanding cloud of fish ate the titbits scattering around us. I counted about a dozen species of reef fish among them, but the most mobile and alert were small silver jack fish. When I went to check on him and feed him, they appeared around me in the lapping waves as I sat down to put on my mask, snorkel, and fins. When I glided out, they surrounded me, swimming in formation, the leaders a metre in front, escorting me to Merlin—I never had any trouble finding him as long as he was nearby. While I fed him, they formed a cloud around us, and though they took the particles of food drifting away, there was no greedy rush forward in spite of the hundreds of individuals. In deeper water, the multi-species school filled the volume of a room. When I left Merlin to return to the beach, some of the jack fish always came with me, a few taking the lead, while others swam companionably around me. When Merlin and I swam together, they came too, but if he roamed too far away, they would not follow. When looking for him, scanning the water from the shore, I would often see them coming back, which told me the direction he had taken.

These animals together seemed to form a companionable community, one in which I was accepted and welcomed, something I had never experienced with wild animals before in spite of all the time I had spent observing wildlife in the Canadian wilderness. The presence of the stonefish so close to our feeding area was unlikely to be a coincidence; it was part of our little community.

One morning in early March when I put Merlin in the sea, he finned slowly along the beach. I fed him, and arranged his set of toys. He came over while I worked on it, and floated beside me. When I walked back to shore, he spun around and swam quickly, almost anxiously, after me, looking up at me as he did. But when I went inside, he didn't wait for me to return—he swam away.

The next time I looked out, he was gone. Walking along the shore, I soon caught sight of his pale hind fins catching the light in the turquoise water as he investigated the bottom, so I went underwater to see what he was doing. His entourage of fish announced his presence up ahead, and I drifted closer trying not to alarm him. Poised on the sand on his wing tips, he was poking his head into holes in the coral, apparently looking for something to eat. Suddenly he saw me and froze, alert with wide eyes, and he crept out of sight behind the coral outcrop. I drifted away, not wanting to disturb him. He spent the rest of the day on the bottom in a metre or two of water, not going far, languidly looking around and surfacing every four or five minutes to breathe. But by late afternoon he had tired, and lay resting, clutching one of his floating toys.

The next morning he dove and remained submerged most of the time, but by evening he was having trouble staying down, and after that he resumed his habitual surface swimming. Another month passed before he was able to dive as he wished. When he was frightened, though, he found the strength.

One evening, Franck was playing with the stonefish and offered the lazy creature a chunk of bonito. There was not the slightest response, until, moving too fast for the eye to follow, it shot from its shelter and engulfed the food. Merlin and his entourage of fish were also there eating, but when the stonefish moved, they all disappeared. Merlin dove to the bottom, though he had been floating for days, and didn't reappear at the surface again until he reached the drop-off. We speculated that the sudden lunge of the stonefish had startled him and the fish badly. Given their sensitivity to vibrations, the sudden movement had doubtless been a shock. When I looked back, the stonefish had disappeared.

Once, when a stranger came to talk to Franck on the beach and petted Merlin, he flew far past the drop-off, and three motorboats roared over him. Another time, three of our neighbour's friends came over to meet him. I went to get some food for him, and on finding himself alone with strangers, he vanished and appeared later past the drop-off. Merlin dove and swam to the drop-off in fear, most often when it was getting dark, apparently due to his enhanced sense of danger at that time. Often we could not discern the reason.

One early morning, a whitetip shark of about two metres (6.5 ft) in length came up from the depths behind me as I dove to pick some seaweed. Perhaps she had smelled the fish meat I had just fed the little turtle, for she circled me, looking closely, and glided into the shallows in front of our house. I followed to see that she didn't bother Merlin, playing on the surface not far ahead. However, she completed her turn and swam back over the drop-off, down in the direction from which she had come. Two days later, she came up the drop-off when Merlin was with me, passed less than a metre away, and slowly descended again.

Franck and I took the scraps from Merlin's bonito, and put them on a shelf on the coral wall as night fell.

Many fish flocked around us, excitedly eating the bits of meat scattered in the water. When it was nearly dark, the shark appeared, a dark purple torpedo in the indigo water, her white tips glowing in the dark. As she came along the drop-off below us, there was a slightly startled reaction as she became aware of Franck and I suspended from the surface, though she did not change her trajectory. Neither did she return to eat.

I tried to feed her again a few days later in the morning, and saw a blackfin shark far off near the surface of the bay. But no shark came to eat while I was there.

After our experiences at Fisherman's Point, I was curious to find the home of our resident whitetip shark, sure that it wasn't far away. But scuba diving explorations along the bottom of the wall revealed no holes where sharks were sleeping and not even any places where sharks might want to sleep. The steep, sandy floor that fell away from the fringe reef's wall about ten metres (32 ft) down was littered with trash, layered with mud, and appeared lifeless.

I took the scraps from Merlin's next bonito to the lagoon late one morning after turtle chores. This time there was less current, and the weather was calm. I put the sharks' food near a formation of dead coral from which I could watch quietly without disturbing the water. Soon, a grey blackfin shark appeared and circled at the limit of visibility. Slowly he came closer and finally passed me, looking intently. He glided over to the food and returned to me several times, moving in ever smaller circles, and each time he passed, I took a photo. The rhythm of the shark's increasingly rapid circling, accented by the click of the camera and my flowering excitement, brought a sensation of being together in tune. When the shark was satisfied that there was no trick, the crescendo came. He shot to the food, grabbed it, and accelerated away.

Months later when the photographs were developed, I found that this shark was the male resident I later named Meadowes. He was one I was to know very well. As I took my collection bag and began foraging for Merlin's seaweed, sharks hurried by, going in the direction from which I had come.

We bought a bonito for Merlin each week, and Franck came with me the next time, to give the scraps to the sharks. We put them near a dead coral formation we could hold onto when necessary, and Franck shredded bits of meat into the water for the fish while we waited. A multi-species cloud of fish formed, darting about to snap up the pieces. A green and brown spotted grouper swam off with a chunk the same size as he was.

Then the sharks came. There were four, a large one, of over a metre and a half (5 ft) in length, two slightly smaller, and one about a metre long. They were interested in the food but not pleased to see us—they turned and glided away the moment we looked at them. They circled around in the distance until I became so cold that I began to shake. Occasionally, one of the sharks would swim up to the scraps of food then swim away again. Once, the three larger ones approached in a triangular formation but never came close enough for a photograph.

We began discussing the situation above the surface. I took a quick check underwater to see what the sharks were doing and saw that they had taken advantage of our momentary distraction to dive onto the food, and gobble it. As soon as they saw us watching, they scattered.

Moved by their shyness and inspired to reassure them, I dove down, chose a good scrap, and swam to meet one of the medium sized sharks, holding the food outstretched toward her. She didn't turn away, but she didn't take the food either, and glided under me, practically brushing me with her dorsal fin. That was enough for Franck, who announced that we were going.

We swam to the boats, and began talking.

"Did you see how they began to eat the moment we started talking?" I asked him.

"Yes," he said. We both had the same idea at once and looked underwater. The sharks had followed us to the boats and were circling us, much closer than they had come when we had been underwater watching them. As soon as we looked at them, they accelerated away. When we got home, Merlin had clambered out of his pool and was lying on the deck, one hind fin still stuck on the rim. Luckily, he had not fallen off and broken his neck. It was another sign of his increasing strength that he was able to climb out.

He was also roaming farther away. One morning, he disappeared when we were at breakfast and a frantic search up and down the shore failed to locate him. So I launched my kayak, and, travelling down the bay, caught sight of a fisherman in an outrigger canoe farther along the shore. Since fishermen posed the greatest danger to Merlin, I glided near. The huge man was gazing intently at something underwater and angling his boat closer, so I approached too. When he saw me, I explained that I was looking for my sick sea turtle, hoping that he would get the message that if it was sick, it wasn't good to eat. He pointed, and as my boat glided on toward the place, I saw Merlin in the shallow water, trying to wedge himself into a hole. I was greatly relieved when the man didn't object as I stepped into the stonefish-infested coral and hoisted the straying turtle into the boat. He could easily have stuck a harpoon into him right in front of me.

The practice of eating sea turtles, in spite of it being illegal, is especially unfortunate since it was the French who convinced the Polynesians to use them for food. Before they arrived, the islanders held them sacred and did not eat them. These exquisitely evolved reptiles take twenty-five to fifty years to mature and breed, a long time to survive in islands full of sea turtle hunters. Protecting Merlin from fishermen became my greatest worry since I never knew when an outrigger might cruise past the oblivious sea turtle playing on the surface.

When I brought Merlin back to the beach, his entourage of jack fish was still there, apparently anxiously waiting. Like cascading silver coins, they flashed around us, more than a hundred of them, taking up their positions around Merlin. As many as could fit underneath him clustered against his ventral shell as if for protection.

There was another beach on our property about twenty-four metres (80 ft) away. As Merlin extended his range, sometimes he clambered out on it, and I would go and carry him back. His fish companions would be at our beach waiting by the time I arrived with him on foot. On another occasion, Franck went to fetch him in his kayak when he strayed too far. The fish were swirling around my feet as I waited on the beach, but when Franck was still ten metres (32 ft) away, they all went streaming out to meet the kayak!

However, the most striking incident illustrating the faculties of Merlin's fish occurred when I was walking along on the black rocks that lined the shore between the beaches, looking for him. His fish came streaming toward me along the shore, so I knew I was looking in the right direction. Though I was not touching the water, they saw me and milled around in the water below. Obviously they recognized me from under the surface, though they had never seen me there before. I must have looked very different standing up on the rocks above, than I did when underwater. They were remarkably aware.

When Merlin disappeared from the region in front of the house, one of us would go looking for him underwater. We were puzzled that he reacted as if he were terrified when we found him, when he was so familiar and comfortable with us on the shore. He could hear us coming long before we could see him and flew away, making it hard to find him once he could dive and swim swiftly. I speculated that it was because we looked so different when approaching in the distance underwater, than we did when we were with him on land. Our size and poised, alert attitude as we approached, plus the turbulence of our movements to which he was sensitive, possibly warned him that he was in the presence of predators about to attack.

His body language showed that he was truly frightened of us when he saw us coming. So I taught him to recognize me underwater by holding him so that he could see my face and hands only, as he had originally, then feeding him. Gradually he learned to recognize me in spite of his alarm when I first appeared. He would relax and come to me.

I continued to treat the recurrent deterioration of Merlin's ventral shell. It seemed to heal, then the problem started again, slowly and subtly, elsewhere. Waiting on the deck for me to prepare his treatment, with his weight supported on towels, he would sigh, roll his eyes toward me, and settle visibly, as if to say, "Ho hum, not this again." He had grown strong enough to support his own weight on land, and sometimes he would begin to lumber around the deck, slobbering as he went. I had to rush out to make sure he didn't tumble off. He had become used to his treatments and didn't fuss very much over them any more. I would hold him vertically with my left arm around his shell, his weight resting on my lap, and pat the solution onto his ventral shell with my right hand, while talking to him in an encouraging tone.

Courtesy of Franck Porcher

In this position, his face was centimetres from mine, and he would gaze into my face, his head inclined toward me, following my movements with his.

His inflatable pool was no longer large enough nor strong enough for him. He needed a safe place to spend the night and an alternative to being in the sea. I rarely went out, due to the need to watch him, but at times it was unavoidable. And during storms, he required a sheltered refuge.

So I deepened a natural depression behind some of the huge black rocks lining the shore, to form a pool. When it was finished, it was six metres (20 ft) long, shaded by trees, sheltered from the waves, and floored with white sand. Merlin could neither climb out nor injure himself there. Best of all, wave action provided continuous flows of fresh sea water between the rocks, keeping it perfectly clean. That evening after he ate, I carried him to his new pool, which he minutely explored. It seemed to please him, and he stayed there contentedly.

When I put him in the sea the next morning, he swam so far out into the bay that he vanished from view, so I took the kayak to search. When a careful scan of the bay didn't reveal him, I assumed he had done a large circle and returned to the shallows where he seemed to feel safest. Searching along the shore, I was lucky to catch sight of him as the kayak glided above the place he was resting, poised on his fins on the coral bottom a metre below. But while I was manoeuvring to get out and grab him, he disappeared again. So I had to wait for him to surface for air. When his little head emerged in deep water some distance off the drop-off, I managed to catch up and hoist him into the boat. It was worrying that he was so easy to catch.

He was rapidly becoming stronger, and was able to swim so fast underwater when he wished, that it was hard for me to keep up with him. Often I lost him, and had to wait for him to surface to relocate him. He came up to breathe about once every five minutes.

When I swam with him, he would often take me on a wide circle far beyond the drop-off, then return to our beach, where he hoped, it seemed, to ditch me. There, he would behave as if he wanted to eat, but when I ran to get his food, he would fly off again, and I would have to run to get back in the water fast enough to stay with him. But either through exhaustion or affection, he would soon cease his evasive techniques and come to me. I would pick seaweed for him and we would roam lazily, then off he would go again, a dramatic spectacle flying fast with strong strokes of his wings, his school of silver jack fish filling the water around him in fighter-plane formation.

A tiny octopus lived outside Merlin's pool. When I first caught sight of him, he was sitting straight up on some coral, his skin patterned and prickled for camouflage, while a green rock cod peeked at him over the edge of his hillock and circled around him like the hand of a clock. The cod stayed just far enough away to be out of reach of those little tentacles. Several other fish were also interested in the octopus' activities as he busied himself working on excavating and cleaning his home. He carried arm-loads of sand out and tossed them down the bank. When a fish came too close, a tiny tentacle shot out toward it, and the fish darted away. When he sank into a camouflaged heap, the fish lost interest. The next day I found him in the sun, clinging to the opposite wall of his alcove. He saw me and turned crimson, his golden eye glaring. The colour change suggested fear or aggression or both, so I sneaked away, not wanting to disturb him.

Since Merlin turned out to be such an interesting little character who had survived a severe illness, I wrote down his story and submitted it to the Tahiti Beach Press, a magazine for tourists and English speaking residents. It was accepted for publication.

While we had lunch, Merlin disappeared, so I hurried out in the kayak to look for him. After watching the glassy surface for a long time, I saw his little head appear, and by the time I caught up to him, he was swimming into a beach a few hundred metres along the shore. I glided up beside him and snatched him from the water before he got there. There was a house beside the beach where a Polynesian family was watching from their deck. I waved and smiled, explaining that the turtle had been sick and was recovering. The woman called to me, "You'd better keep that turtle at home. The people here eat sea turtles."

We had discussed the possibility of penning the beach area with many fishing nets, but even if we could afford the extremely high cost of such an endeavour, we didn't see how we could make it work. It would be almost impossible to attach a net close enough to the bottom, all along its length, so that Merlin couldn't escape, and he would probably be able to clamber over the top. Further, Franck was sure that it was illegal to fence off part of the sea, and if a boat ran afoul of it, we would be liable for any damages.

My goal was for him to resume the normal life of a wild turtle as soon as he was able, but he was not yet strong enough. He was spending far too much time on the surface and in shallow water, and he had not resumed diving deep outside the drop-off. He needed more time.

We considered attaching a little orange float to him so his location would be immediately apparent. However, there were many problems with that idea. The line connecting him to the float could catch on a coral and hold him underwater. His shell was alive, so the line would have to be glued on. Most glues wouldn't hold underwater, and I didn't like the idea of messing with him at all. Nevertheless, I went shopping for a glue we thought would work, painted a small soda pop bottle bright orange, and set it in the sun to dry.

Often, Merlin still spent the entire day in front of the house, exploring underwater between the two beaches, playing with his toys, and lying around on the beach waiting for attention.

He was eating more food than ever. Daily he consumed eight potatoes, mashed and mixed with a package of spinach, a head of lettuce, a kilogram of raw bonito, a salad serving bowl of seaweed, and three or four eggs. When he swam in to the beach and lay looking up at the house, I ran down and sat with him, feeding and caressing him, while his fish flashed about us. Yet he would frantically protest that he was starving to death when I carried him down the stone pathway in the evenings to his new pool. So he would get a few extra pieces of fish meat before I left him for the night. He had grown several centimetres in length since he had come, and put on a lot of weight. His vitality, the satiny quality of his skin, and his increasing strength indicated that his diet was fine.

Merlin knew what my white net collection bag was, and loved to sink his teeth into it whenever he found it unguarded. He also developed a fascination with biting toes. I wasn't sure if it was the colour he was attracted to, or our reaction that he enjoyed.

It started when I was dangling my feet in the water while sitting on a rock, watching the activities of the resident octopus. Without warning, Merlin swept up and tried, with surprising determination, to bite my toes! After that, any time that he saw our white toes in the water, he just couldn't resist trying to get a bite. While being fed, when with us on the beach, or any other time he saw someone's toes dangling in the water, he would come flying from the depths with gleaming eyes and try to bite them. The sight of toes was irresistible to him, and between bites of food, suddenly, his graceful wings would practically meet above his head, and down he would plunge for a bite.

One early morning, a woman with three boys came to fish on the rocks in front of our house. She lived along the shore, and I had often greeted her in a neighbourly way when out walking. Merlin was in his pool behind a clump of trees some distance to their left, and it wasn't long before the children's shrieks of discovery announced that they had found the sea turtle. By the time we had arrived, they were pelting him with rocks, which, luckily, hadn't done any damage so far. Franck spoke with the boys, while I went to talk to the woman, who had permitted the children's attack. I explained to her that the turtle was recovering from a serious illness, and asked that she not let the children hurt him. She said that she would not, but it was obvious that had I not been there, he would have been badly hurt or killed. I realized that our sea turtle could not ever be left alone.

Five months had passed since Merlin came, and April was drawing to a close. I worked on the painting of him whenever I had a spare moment. The water around him was finished, and the intricate patterning on his shell, which was a finicky business using the air-brush, was almost completed.

Two days after the woman and boys had found Merlin in his pool, I had everything required to stick the orange-coloured float to his shell and planned to do it before putting him out again. That morning, I had to paddle the heavy kayak out to the lagoon for a load of seaweed for him and left him in his pool until I returned, watched over by Franck, who was working at home on his computer. By that time he owned his own company, named *Smart Technologies*.

I planted some of the seaweed, to live and grow in Merlin's pool for his pleasure, and fed him some. But then I received some alarming news by telephone, and distracted, I put Merlin in the sea, and fed him his potato, fish, and vegetable mixture. He stayed around the area while I put away the boat, looked after the house, and made lunch. Noticing him waiting on the beach for something to eat, I fed him again, and planned to attach his float when he came to the beach after lunch. Then we ate, and I made coffee while assembling the things I would need to attach Merlin's orange floater. I served Franck his coffee but never drank mine.

Merlin was gone.

Chapter Seven

The Search

I searched from one end of the bay to the other for the rest of the day, but Merlin had vanished. As night came, the sea was still as glass. Anything on the surface was visible far away, and I made one last search from the end of the bay to the lagoon, with no sight of his little head breaking the surface. Above the mountain, the full moon arose. When the water toward shore became too dark to see, I turned the boat. To the west, the glassy bay perfectly reflected the sunset sky, and as I paused, looking across the glowing expanse, an enormous head and back broke the surface just in front of my boat. I thought it was a person swimming at first, but it was a sea turtle of enormous size. I heard him breathe, then he submerged, leaving only a shining circle of turbulence. I paddled home, amazed in the moonlight, wondering if Merlin had left for the ocean.

The next morning, Franck and I went out at dawn to search all the way to the ocean before the wind arose to ruffle the surface and obscure the view of a surfacing turtle. I was awed by the waves on the reef, thinking that there was no way that Merlin could cope with life in the wild yet. And why had he not come home to eat? That was the question that bothered me the most. Having some experience in rehabilitating wild animals, I knew that they didn't change their patterns suddenly; their behaviour evolved slowly over time as they grew more capable and expanded their horizons.

I had a bad feeling about Merlin's sudden disappearance.

In the afternoon, I walked along the road, talking to the neighbours. I asked that they call me if they saw him and stressed his long illness, trying to subtly convey the message that eating such a sick turtle wasn't a good idea. Since he tended to avoid deep water and swam mostly on the surface along the shallow fringe reef, even coming to the beach, he would be visible to anyone looking out from their seaside home. All the residents were fishers, any of whom might fancy a sea turtle so easy to catch.

Some of the people were friendly and talked to me, but some just laughed and made signs of eating. Some were hostile. Never had the racism in the society in which I lived been thrown into such a stark light. In particular, the woman who had told me to keep him 'at home' said she knew nothing and would not talk further. Her attitude had changed, another ominous sign.

At the neighbourhood store farther on, the owner told me that the day before, people he knew had caught the sea turtle and eaten him. I nearly collapsed. However, I asked if I could see the shell to confirm whether or not the turtle had been Merlin. He told me that he would try to get the shell, and I should come back the next day. Further enquiry informed me that his throat had been slashed, and he had been delicious.

On the way back, I made my way along the shore for a distance, looking around in case his shell had been thrown out around any of the shoddy little shacks in the area, but I couldn't find anything. I was sick with shock, and unable to eat, sleep, or do anything else until the next day when I went back to the store.

The man told me dismissively that there was no shell; it had been thrown out with the garbage. It was gone, so was the turtle, and he obviously expected that I would be gone too.

"Well," I said, "the vet thinks that the turtle's medication could be dangerous to someone eating him. . . His heart could just. . . aaahhhhh. . . *stop*." The man, who carried well over a hundred kilos—a good three-hundred pounder—stopped grinning. His eyes grew round, and so did his mouth. "So anyone who ate that turtle should go to the doctor right away," I continued. After a long look, the man changed the subject and turned to other things. I left.

I believed that Merlin had been caught, killed, and devoured. Otherwise, he would have come home to eat. Whenever I thought about him, which was most of the time, I felt nauseated, and I could not bear to work on the painting of him. Unable to let it go, I continued to search from one end of the bay to the other because it was the only way to feel hope that I might find him, anyway.

My story about him appeared in the Tahiti Beach Press, and on the same day, a whole page of *La Dépêche* was devoted to Merlin's fate. *Merlin, Disappeared by Enchantment*, the headlines read. The *gendarme*s were at the beaches checking boats to make sure that no fishermen were bringing sea turtles back with them. For once the law was being enforced!

Since I was referred to in the article as a '*naturiste*,' which means 'nudist,' instead of a '*naturaliste*,' men were phoning up wanting to meet me. Others called me to say that they had seen Merlin in different locations near and far. The last thing I wanted was to be seen out in public, but since our car had broken down, I was obliged to spend much of the time out on my bicycle. With mask, snorkel, and fins in my basket, I pedalled as fast as I could to investigate all the reports. When I came home, in an anguish that could not be quenched, unable to rest, I set off in the kayak to search for Merlin.

A friend had sent me a newspaper article about cruelty to sea turtles in Canada, where they were displayed alive on ice. When someone wanted a piece of meat, the store keeper simply sliced a steak off the living animal. Finally the humane society had intervened, and a law had been passed requiring that the turtle be killed before it was cut up. The head had to be crushed since the head of the sea turtle would otherwise stay alive and suffer for two days! I couldn't get the image out of my mind of Merlin's bodiless head thinking and suffering in an overheated garbage can somewhere.

One morning during another frantic search for Merlin, I put the remains of his last bonito in the boat and dumped them in the lagoon when I got there. It seemed logical to search a bit for Merlin in the lagoon, too, since he favoured shallow water.

Sharks passed far off in the coral, but it was nearly half an hour before any circled closer. Then a new one arrived. She was grey, the same colour as the male who had come alone. She seemed more confident than he had been. She glided by, looking hard at me, then swooped down to the fish scraps lying on the sand. With no further fuss, she ate them one by one, while the other sharks continued to pass in the distance. Dark clouds had obscured the light, and a storm threatened, so I climbed out, and sat on the side of the kayak, dangling my feet in the sea. While watching the coming storm and scanning the surface for Merlin, I suddenly realized that a shark was nosing my ankle.

Glancing down, I saw two more beside her at the surface. The three had come to investigate the creature who had brought food, now that it was no longer underwater watching them.

When I went back to pick up the anchor, the shark who had eaten the scraps swam up to me, turning away at the last minute. I snapped a picture of her so was able to confirm her identity later. For the second time, I had encountered the shark I would later name Bratworst, the worst brat. But then, I did not know we had already met—my sad thoughts were on Merlin.

I put the anchor in the boat, climbed in, and drifted away, paddling intermittently as the rain began falling. I was crossing over the border where Merlin's favourite seaweed grew when a shark appeared half a metre (20 in) away at the surface at right angles to the boat, swimming toward it. Her dorsal fin was actually out of the water. At that moment, the shark became aware of my paddle in the air above her, startled, and vanished. I recognized her dorsal fin—it was Bratworst!

It seemed an astonishing display of curiosity for a fish, particularly after she had eaten. Intrigued, I asked at the local grocery store if they had any fish scraps and was given a few.

Anyone who saw a sea turtle at that time was still calling me to tell me they had seen Merlin, and three days later, I was assured that he was alive and well and living across the bay. So I decided to search the perimeter one more time, and took the fish scraps with me. As I passed the lagoon, I took a detour.

Bratworst

Laying the food on the sand in the same region I had stopped before, I waited in suspense. After a few minutes, a large shark appeared at the limits of visibility then disappeared again. Moments later, three dark grey, female sharks glided toward me in triangular formation. Instead of going to the food, they came to circle me closely, then circled the kayak. I speculated that these were the large sharks who had watched from the periphery so many times, finally approaching together for a closer look at the situation. One was Bratworst; the other two were bigger. Cautiously, they circled the area for many minutes, passing often in front of me, before descending to swim over the food. One grabbed the spine and accelerated away, shaking it vigorously to extract a bite, while the others flew after her. One of them snatched the spine as it fell, to shake out a bite for herself. The three sharks soared through the coral, each shaking and dropping the scrap to be caught up by the next shark, until what remained fell into a crevice. I retrieved it and put it with the rest of the pieces. When the sharks came around again, they looked for their morsel where it had fallen and could not find it, turning repeatedly around the coral where it had been. Apparently at close range they depended on their eyes, not their noses, to find their food.

One by one they inhaled the few remaining pieces until all that was left was the big head of the fish. The largest shark had a black mark above her left pectoral fin. She tried to get her teeth around the smooth head but succeeded only in catching it by its fin, which she rejected. When they circled away, a large moray eel who had been watching, came undulating out of his shelter, grabbed the head by its side, and blasted back under the coral. Presenting the image of frantic haste, he tried to find a place on the smooth surface of the fish's dead face, where he could grip and turn it to access the interior from the safety of his retreat. But when the sharks passed again, they ignored him and continued to patiently, languidly circle through their brilliant coral garden.

I was trying to imitate the rhythm of their slow motions so as not to disturb their peace, but they were passing me as if I weren't there, and I had difficulty keeping myself balanced in the river-like current. When I was watching a shark passing at arm's reach in front of me, there was also one equally close behind and one at my side. I was afraid of kicking one of them in the face, and trying very hard to keep balanced while watching all around me, without moving quickly or suddenly, and exciting or alarming them. They were evidently not afraid of me. While thrilled that I had won their trust, I began to wonder if they were circling around me the way they were circling around the bonito head. Would they bite? I did not know.

Finally they drifted away, disappearing into the coral for increasingly long periods, so I climbed back into the kayak. No shark approached it this time, but there was a great commotion at the surface some ten metres (30 ft) away, and for a moment a shark's fin broke through. Maybe one of them had taken the fish head away from the eel when I left, and they had gone on another wild chase through the coral.

I continued the long, long search around the periphery of the bay. Noon had come, and the merciless sun burned down from straight above. Once the sharks were left behind, grief over Merlin's fate came surging back as no little turtle's head ever appeared at the surface, as far as I could see in any direction, no matter how long I looked. But still, I continued to restlessly search for him, as though to stop would be to give up on ever seeing him again.

One day, about two weeks after Merlin had vanished, I walked along the shore in the early morning and sat down in a wild place from which I could survey the entire head of the bay. I still imagined that he would surface and I would see him when he breathed.

On the rock sea-wall, rats were pursuing their affairs. One, I idly noticed, had skin problems. Another was very young. About half an hour after I first noticed it, the rat with the skin problems emerged from the forest. He had actually found a plastic package half full of cookies. As he dragged the cumbersome thing along the rocks above the sea water, he suddenly noticed me. Startled, he jerked his head up, letting go of his prize. The bag of cookies tumbled down the rock wall and into the sea. He watched it sink and took a few steps toward the water. I could see the moment he realized the futility of trying to rescue them. He went back the way he had come and disappeared.

CHAPTER EIGHT

In Merlin's Memory

A month after Merlin vanished, a woman phoned to say that he had been caught in a net and was at her house. In a frenzy of anticipation, I leaped in the car and sped around the bay to her home on the other side. She met me at the door and led me into a darkened room. Soon she emerged from the back of the house, carrying a green sea turtle. It was exactly Merlin's size, but it was not him. As soon as I saw the turtle's face I knew it was a female, and that Merlin must have been a male. Her hind fins were covered with scrapes where the scales were missing, and the tips of both front wings were broken. Her right wing was broken in a second place halfway from the tip to the wrist joint. There was a deep hole in the centre of one wing, the flesh around it red and badly swollen, obviously a harpoon injury. She had many other lacerations and a rounded chunk was missing from the back edge of her left hind fin, including some of the bone.

The woman handed her to me. She struggled strongly, beating those broken wings with surprising force, more strongly than Merlin ever could have. The woman's husband approached, and I began to feel uncomfortable. Pointing out that the animal needed veterinary care, I thanked them and left. Putting the turtle on towels in the back of the car, I sped home while the poor animal lumbered around, trying to find a way out.

Safe in Merlin's pool, she refused to eat anything, and when I tried to feed her, she clamped her jaws shut and nearly took off my finger. She flinched at the sight of my hand, and behaved as if she were being beaten on the head, repeatedly blinking her eyes. Her eyes were much more mobile than Merlin's and she blinked them often. My approach so alarmed her that I feared she would injure herself further in her frantic efforts to find a way out, bending the raw breaks in her wings. She had had nasty experiences with people.

Her shell was forty-three centimetres (17 in) in length, was more elongated than Merlin's, and heart-shaped. Her wings were shorter and smaller, and the pattern on her shell was different, but otherwise, they shared the same look. Like him, she must have been very young. Later I found her on the bottom of the deepest part of the pool, her head in one of the seaweeds I had planted. Once in a while she roamed around it. I went to the lagoon's border to gather seaweed for her, the kind that Merlin loved.

While I was gone, the couple's son arrived and asked Franck for money, which he refused. I had been so overwhelmed by the idea that they had found Merlin, that I hadn't thought it through. They had hoped for a big reward in cash, and instead I had simply taken the turtle that they would otherwise have made into a meal, leaving them nothing. What if they had deliberately caught the turtle for a reward? Nevertheless, I drove across to the store, and bought a six pack of beer. The young man was sitting idly in their driveway next door. He was delighted with the beer and I thanked him and his family for their kindness to the turtle.

The next morning the turtle seemed calmer and lazed on the bottom of her pool, nibbling her seaweed.

We called the veterinarian, who arrived from Tahiti the following Saturday. I held the powerful little animal tight by her upper arms while he examined her, sewed up her cuts and breaks, poured a syrupy disinfectant over each of her injuries, and gave her shots of antibiotics and vitamins. Then he said she could be released. Her injuries, would heal much better if she were free. If she remained confined, she would keep re-breaking her wings in her efforts to escape, and her condition would deteriorate. I was deeply relieved.

As Franck and I paddled out to the pass, she lay quietly on my lap, looking down into the water and resting one wing on the side of the boat and the other on my leg. Amazed that for the first time in my arms she wasn't struggling, I wondered what she was thinking, seeing the sea water so close. Sometimes I felt her little wings tense as though she was planning to attempt a leap, but luckily she did not try; I had to use both hands to paddle. When we approached the ocean, she must have heard the powerful reverberations of the waves exploding on the reef and became excited. I had to hold the struggling creature with one arm around her, and manoeuvre the kayak with one hand. Franck got into the water to watch her response to sudden freedom, and as soon as I could bring the boat around, I put her in the sea beside him and slid in after her. All I saw was her distant silhouette flying into the blue so fast that in seconds, she had vanished. It was a dramatic display of how fast a sea turtle could move. Such action had been far beyond Merlin's ability.

The family from whom I had taken her lived beside the store across the bay, and from time to time, I saw them. About a year later, the adult daughter of the family, Clementine, came to talk to me as I waited on the beach there, hoping to buy fish for a seabird. After an exchange of pleasantries, she told me they had another sea turtle. Curious, I asked her more about it, and she offered to show it to me.

I followed her into the dark house. The family room was full, and a hush fell over everyone when I appeared. I smiled and greeted Clementine's parents and the children, but Clementine was already at the other side of the room, urgently beckoning me to follow. She led me into a dim back room. There, a sea turtle lay in a tub with a circumference barely larger than her shell, so that her wings were forced up vertically, cramped in between. Her neck was twisted, and her head was also forced up, pressed against the confining tub. Her hind fins were crumpled between it and her shell, and the water was thick with excrement.

"The tub is awfully small," I said, not wanting to criticize, but shocked at the turtle's circumstances.

"Yes," she said.

"Are you going to let her go?"

"No." Clementine told me that the turtle had been speared through the wing, but I saw no sign of that. Her mother, who came in from the next room, said they had had the turtle for a year. Their son told me as I was leaving that the turtle had been caught in a net recently.

I thanked them all and left, but couldn't get the plight of the sea turtle out of my mind. The next morning, Franck and I went to talk to the family and encourage them to set her free. We tried to buy her. But Clementine's parents were adamant. They would not sell her, and they would not let her go. Neither arguments about compliance with the law, the sea turtles' decline toward extinction, nor humane treatment could sway them. They were Polynesians, they reminded us, so the white man's law did not apply to them. Sea turtles belonged to them, and they would eat them if they wanted. We pointed out that the ancient Polynesians believed that sea turtles were sacred, honoured them, and did not eat them. It was the French who convinced them that sea turtles were good food. The man was angry most of the time and treated us to a long rambling monologue about the evils of white people and his own special rights and powers. He was especially angry that I had seen the turtle and threatened to charge me with trespassing.

When we left, we asked that he think about releasing the animal so that she could grow up and reproduce. It would set a good example for the younger generation. He said finally that he would consider it, and we left them with the government's publication about the need to protect sea turtles, written in Tahitian.

In the meantime, Franck talked to the veterinarian and the governmental department for the Protection of the Sea about the case. The French vet dissuaded us from interfering, saying that for the people, the turtle was no different than a pig. They wanted to eat it, and that was that. The government representative, however, a Tahitian herself, was insistent that the family be persuaded to obey the law.

A week later, Franck went back. The turtle was still there, and though some members of the family thought that she should be set free, the parents were adamantly against it. So we contacted the government's representative again. This time, she advised us to call the *gendarmerie*, and ask them to recover the turtle and bring it to our house. When the veterinarian had seen the turtle and filed a report, she could be released.

I put a pillow in the old blue basin and covered it with a layer of plastic and thick wet towels, to support the turtle during transport. When the *gendarmes* arrived at our house, I took it out with me, thanking them for acting on behalf of this turtle, who had suffered for such a long time. But they did not smile and informed us that they had known about this sea turtle for six months. The man who was keeping her had told them that we were trying to get her away from him to eat. They talked of a turtle we used to have in possession with a missing leg, which had disappeared. Since it couldn't have been released, we must have eaten it.

"That wasn't us," I told them. "We never had a turtle with a missing leg."

Franck repeated that it was the department for the Protection of the Sea who had asked for their intervention in the case, and suggested they call them. The leading *gendarme* said they had never been in contact with that department. Trying to lighten the atmosphere, I offered to show them Merlin's pool so that they could see how well the turtle would be housed until its release after the veterinarian came to inspect it. They looked it over but said that they wouldn't bring us the turtle, since they suspected that we wanted to eat it.

"Then let it go!" I said. "Just put it in the sea!"

The leader looked at me closely, then said that they would investigate the circumstances of this turtle and make a decision based on its welfare. While Franck talked to them, I spoke to an officer who was watching from the background, and who had seemed receptive to my points. I told him that sea turtles drink almost continuously and that when I had seen the turtle, its water had been thick with faeces. I expected them to find the turtle in similar circumstances and hoped that they would remember this nauseating point.

They did not take the basin with them when they left. I sat on the deck and watched as their vehicle parked at the house on the other side of the bay. After fifteen minutes, it left, and soon turned into our property. An officer got out alone and walked over to the blue basin. He said that they were bringing the turtle.

When the police van returned, Clementine was with the *gendarmes,* and stood alone to the side of the group as the turtle was placed in the pool. A *gendarme* told me that she had wanted to see its new home.

I waited until everyone left, to check over and feed the turtle. She was trying frantically to escape and it was a relief to see how strong she was after her ordeal. Soon she settled down, explored, and eventually rested on the bottom of the deepest part, with her nose pushed into a flow of fresh sea water. She looked like the other female turtle, but she was in better shape. Examination revealed superficial cuts and old scars on her right wing, two growths on the underside of her hind fins, an old scar on the back of her neck, and a badly damaged ventral surface from lying in the tub. I began treating it with a healing ointment.

We had to wait until Saturday for the government vet to come and check her before we could release her.

That afternoon, the turtle's ex-family arrived, and Franck went out to talk to them. Shouting soon broke out, and when I started hearing threats of violence, including death threats against me, I sneaked around the periphery, took the turtle from the pool, and brought her into the house where she would be safe. Franck managed to calm everybody down eventually, but I continued to feel extremely nervous about the situation.

Clementine's brother was an athletic, handsome young man with a sharp gaze. I speculated that if anyone was going to make a move to get the turtle back, it would be him and had visions of him sliding in his outrigger canoe across the bay in the moonlight and lifting the turtle from her pool. So I made a little bed on the grass beside it, tied a soft string loosely around her shell, attached the other end to my wrist, and slept there at night, awakening regularly to check on her. Each time, she was reposing in the moonlight, fins gently waving, her nose in the flow of water from the sea. No canoe was visible slinking over the water.

The vet came on Saturday morning, checked her, filled out a form, and said she could be released. But the turtle's ex-family suddenly arrived. The man told us that he and his friend wanted to release the turtle on Sunday after church, taking the children to witness the event. Franck agreed, but I was very disturbed about the plan, seeing no reason to trust this man after all that had happened. Still, Franck took the turtle and went with them, and luckily she was successfully freed while the man performed a ritual over her. She disappeared so fast that none of the waiting children so much as glimpsed her. I was deeply relieved to see Franck return, having imagined him being tossed overboard, and the turtle kept to be cooked for dinner.

It was seven o'clock on Monday morning when the young man came over the bay, but he was piloting a broken surfboard instead of an outrigger, and greeted me wreathed in smiles. He had come to thank us for getting the turtle released, and told me that his sister, Clementine, had shown her to me because she knew that I would find a way to set her free. A great weight lifted from my heart to know this.

I slowly gave up my search for Merlin, but it was a year before I could bear to work on his painting.

CHAPTER NINE

The Assembling of My Sharks

I went on collecting fish scraps at the store while doing the grocery shopping, and taking them to the lagoon for the sharks. After years of seeing them pass in the distance or approach briefly, I was thrilled to be interacting with them through this simple gesture of benevolence.

I wanted to find out what they were like, not only as animals, but as individuals. I wanted to know them. Until then, I had remained the suspicious alien in their world. But this was a way to become part of it and see what I had been unable to observe before—what they were doing when no one was watching. Because now that they were used to my presence, they were behaving as if no one was watching!

There was always something new to see. At first, sharks passed sporadically in the distance just within visual range for twenty minutes or half an hour. When one came near, it was one I recognized from earlier visits. Each was different in colour, size, and build. Their colours ranged from gold to deep purple-grey; one was a beautiful golden-bronze colour with exaggerated black tips on his fins. By noting the characteristics of each, including any distinguishing markings, I was able to keep track of most of the sharks I encountered, but always there were new ones. Increasingly fascinated, I had the feeling of a window opening for me into another world, one so alien, so separate from human daily life, that it might just as well have been on another planet.

Besides the reef sharks, I was making the acquaintance of the wide variety of other species sharing their community. One day, after I had waited while a large, golden female shark passed in the veiling light and a grey one cautiously circled the area, a group of six arrived together from the direction of the bay. I had the impression they had travelled together from the ocean. The big golden leader had a remora (*Remora remora*) stuck on her chin, and every so often, she tried to shake it off.

They swept down to the food I had brought without a moment's hesitation. One grabbed up a scrap and soared off shaking it, while two others followed in close pursuit. They dropped a piece on a coral, where they all converged, trying to get hold of it. They played with it for a while, then the excitement faded. Their scrap had dropped onto the sand, so I picked it up and put it back on the coral where it had been before. But when the sharks came around again, they couldn't find it, having apparently remembered exactly where it had dropped on the sand. On their next circle, they located it on the coral.

Meanwhile, the shark with the remora had successfully dislodged it, and the displaced creature flew first after one shark then another. It was about thirty-five centimetres (14 in) in length, a streamlined silver fish with a black racing stripe and widened head.

Remoras have evolved a clever strategy to survive. They accompany a shark or other marine predator to feed on the scraps generated when it eats. They are not parasites, but they do use the shark for transportation. The upper surface of the remora's head is roughened, and the remora can use this patch to attach itself to a shark when it is feeling too lazy to swim, as it often does after eating, and for convenience in staying with a travelling shark. Its transparent fins and tail, in combination with its exaggerated horizontal undulations while swimming, give it a showy look.

I returned to the place where the rest of the scraps still lay. Soon the remora came flitting back too, and waited there, moving around in apparent agitation. He swam a few metres in one direction, returned, and after a while, took a similar foray the opposite way. Time passed while we both waited for the sharks to return, but they did not, and eventually, the sleek little fish began throwing speculative glances at me. A shark did pass finally, so far away that I could hardly distinguish the movement in the blue, but the remora saw it and went winging after it, its fins a blur.

But it did not catch up to the shark, and soon returned to roam around in my vicinity, sometimes going away in one direction or another to look more widely. Eventually, it came wriggling tentatively over to me. It flowed up my leg and moved around my body, trying to stay out of sight as I bent to stroke it. It didn't like to be stroked but I stroked it anyway, and when I got one of the bits of fish meat to feed it, it eagerly ate until it was full. Then it settled comfortably on my leg and attached itself. I was cold by then and wanted to go, but the remora posed a new problem. Never could I have imagined that remoras who had lost their sharks were things I needed to worry about. I roamed around the area feeding different fish, including a Javanese moray eel. However, instead of taking the treat, he reared up toward me and opened his jaws.

The remora's ceaseless undulations up and down my legs and around my body were a unique sensation, and I couldn't help caressing it softly. I tried to feed it again, but it wasn't interested. It rested against my thigh, opening and closing its triangular mouth, its flat, lizard-like head conveying an oddly cold and alien look. I didn't want to leave it, but had no choice so tore apart the last fish head, and as the cloud of blood and fluids drifted down-current, I climbed into the boat, hoping a shark would return for the remora's sake.

I went to the lagoon very early to avoid the wind that blew up as the sun rose and warmed the atmosphere. The large, steel-grey female blackfin, with the small black dot five centimetres (2 in) above her left pectoral fin, usually met the kayak. She was one who had first approached in triangular formation with Bratworst, and when I slid underwater, she would be gliding nearby, placidly circling and watching me.

One morning when I put the few fish scraps on the sand, she drifted slowly above, scenting the different pieces. She was very calm. She picked up a piece with a scarlet fin still attached, gave it a vigorous shake, and glided on, swallowing the mouthful she had bitten out. I picked up the piece she had dropped and put it back with the others, admiring her bite mark—about fifteen centimetres (6 in) across. In this casual fashion, she ate three pieces, moving around me confidently and even swimming underneath me. I swam down to hold onto a piece of dead coral as she approached the food, and she glided over to look at me. I snapped a close-up of her face filling the viewfinder. She must have been only centimetres away. More sharks were moving around the periphery as I circled around for another dive to meet the big grey shark. All the while, a small golden juvenile intermittently approached me, circling to take a closer look.

Focused on the sharks, it was a surprise to catch sight of a great, dark shape coming along on the surface in a swirl of bubbles, and it took me a few seconds to recognize a large male member of my own species. When I turned back, the sharks had vanished. The man was moving from one coral to another, hacking shellfish from them with a large knife, an illegal activity. I put the remains of the sharks' food in the kayak, and pretended to be a tourist photographing the coral. The man, surprised to see me so far away from the beach, began warning me about all the dangers out there. He pointed at one of the Javanese moray eels, told me that it was dangerous and said that he would shoot it.

"Now?" I asked in horrified surprise, thinking of the sharks lurking just out of sight and what might happen if he saw those.

"Oh yes," he assured me and set off for his boat to get his spear-gun. Since his boat was quite far away, though, and he didn't have fins to make swimming easier, I decided that it was unlikely that he would return, and put the sharks' food back in the water.

Soon two dark females came, both so like the one with the grey circle above her pectoral fin that I could scarcely believe they were different individuals. A smaller female came, and I recognized Bratworst. More sharks cruised the periphery. They took the food piece by piece during periods of excitement when one flew off shaking a scrap, and the others soared after her. Eventually they disappeared into the surroundings, leaving the fish spine on the sand. The small, golden female was still circling and looking at me. I drifted with her away from the food, and when I returned, the eel had glided from his hole and taken the spine. He was twisting his head at every angle as he tried to get his awkward prize through the narrow entrance to his hole.

He finally succeeded, even as I was thinking of stealing it back for the little shark.

At home, I began writing a list of the sharks I had seen, numbering them from one to nine. Number one was the big grey female with the black dot five centimetres (2 in) above her left pectoral fin; she was usually the first to appear. The second was the curious shark, Bratworst, and the third was the one with the light mark on the right side of her dorsal fin and the rather maternal build. I filled in the descriptions for the others and wrote an account of the session, as I always did, from notes taken on my slate. I was anxious to go back and sort out the identities of several others, grey torpedoes only glimpsed so far, flying with the others and showing no obvious distinguishing marks.

The grocery store had some scraps that week from cutting up its fish, but I was unable to get away for several days, so I saved them in the freezer.

One morning I left early, gliding swiftly along the bay, out from beneath the shadow of the mountain, and into the flood of light beyond. The sea lay like a mirror all the way to the reef, reflecting a pale, glowing sky, while behind me the emerald island flickered green fire in the rising light. It was one of those magical mornings. Crossing the lagoon, the water was so still that it was like gliding over the surface of a huge aquarium. Every detail of the coral and the fish pursuing their affairs was clear to see, the shadows luminous with teals and purples, contrasting with the golds of the coral.

But as I got ready, the wind began to blow. The first light touches of air sent feathers of disturbance over the waters, obscuring the scene below. As the ripples spread and rose into tiny waves, the view underwater became ever more distorted as wave patterns of sunlight rushed over the fantastic landscape with the wind.

The big shark with the dark circle above her left pectoral fin was waiting for me when I slid underwater. The fish scraps were still frozen, and knowing that they would swiftly thaw in the warm water, I tried to separate them while she circled me. There were two small, plastic bags of scraps, one of white meat, one of red. The scraps of white meat came apart, but not the bag of bloody red scraps. Naturally it was this chunk of frozen meat that interested the shark, and I decided on the spot to drop it as she came over.

She investigated, and I swam down to take her photograph. She turned to look into the camera, and I snapped her portrait. Then she took hold of the frozen chunk and gave it a shake. Failing to extract the expected bite from something smelling so good, she circled again, chose a piece of white meat, and ate it. Then the curious shark, Bratworst, swept in, and the grey one sped up. The two shot around the area, and on returning, the grey shark grabbed up the bloody mass of frozen fish and whipped it back and forth, her head and tail nearly touching with each swing. She charged upward and through the surface as blood filled the water, and swarms of excited fish snapped up the scattering particles of food. What power! Stunned into silence, I drifted, wondering what this tropical shark was feeling with her many teeth sunk into ice.

At that moment, a movement caught my eye, and I turned. Bratworst was zooming in behind me. I was looking at her, and could see her looking back, centimetres from my face, yet with the next beat of her tail, she failed to change direction! Instinctively, I leaned back, brought up my knees between us, and finned water into her nose. She startled, turned, and shot away.

Again, the big grey shark picked up the heavy chunk of frozen fish and shook it. Agitated, she accelerated, soared in a tight circle, then grabbed and whipped it violently back and forth at lightning speed. Bratworst flew after her, grabbed the food, and arced away with it as several more sharks soared out of the blue and joined them. After passing from shark to shark a number of times, the chunk of frozen red scraps fell in the coral, and in a moment of calm, I replaced it on the sand. Never had I seen anything so tortuously mangled in my life. More sharks were arriving, all of them thrilled with the situation. In consternation, I drifted away from the food, watching in growing confusion and trying to keep track of everyone.

Again, I sensed motion behind me and turning, found Bratworst shooting up my body. By the time I turned she was already passing my face. Another larger, agitated shark was beside her, also centimetres away from me. But the two changed their trajectory the moment I faced them. The newcomer had a mass of scars on her right side, thirty centimetres (1 ft) from her tail.

Badly frightened, I retreated farther from the food, and stared around at the endlessly appearing, disappearing, circling sharks. I was frozen in place, afraid to move, unable to get back in the boat, stunned by the circus I had precipitated. Sharks were passing in the coral everywhere I looked to the limits of visibility. I tried to note dorsal fins and distinctive markings, turning continuously to keep track of the confusing scene and make sure no one was coming in from behind me. Still worried that the shark with the dark circle above her left pectoral fin might be suffering from a bad reaction to ice, I watched her worriedly as she came toward me, though she paid me no heed as she passed. But I saw, with a pang of guilt and pain, that her mouth was cut. (To my relief, the photograph I had taken of her on my arrival, eventually showed that her mouth was already injured before she bit into the frozen food that morning).

Bratworst roamed around with the rest, often passing close by me, too. Finally, mesmerized by the continuously moving scene, I gave up and just stared round and round and round. . . and that was when it came to me: all the sharks were females!

Gradually the excitement faded, and the sharks slowly dispersed. Finally, there was no one but the fish going about their business. I got in the kayak and paddled home, swept on my way by the trade winds.

Comfortably installed on the deck with a fortifying coffee, I sat with pen in hand, and wrote down what had happened. Thinking it through, I realized that when I had moved away from the food to let the sharks feed, I had drifted with the current; this same current was carrying the scent. I had been in the scent flow when Bratworst had zoomed at me. That had been a mistake I shouldn't have made, but at the time I was so concerned about the predicament of the frozen shark food that I didn't think about it. I resolved to be much more careful in the future about making careless mistakes while alone with a pack of sharks. The behaviour of Bratworst and her companion was the most troubling, since it had felt like a threat. What did it mean?

It was much later that I learned in conversation with spear fishermen that they habitually trail the fish they have caught behind them. This attracts sharks to come darting in and try to take them. Perhaps Bratworst, fearless and possessed of a curious nature, had learned this through experience, and was looking for the food she could smell behind me, with no intention of harming me. Since neither she nor any other shark ever soared up behind me in such a way again, it seemed possible that they had learned a new routine in the new situation: the food was on the sand and not trailing behind me.

Another factor involved in their excitement must have been the bloody food. If you invite a roomful of children to help themselves to a pile of candy on a table, you'll get a different reaction than if you put some pieces of bread on the table, I speculated. The trouble with this theory was that this was not the first time I had taken red scraps to the lagoon, and there had been no such reaction to them before. The sharks had appeared to prefer the white pieces, permitting the fish to carry the red pieces away into their coral retreats.

Further, Bratworst had badly frightened me. I had never been frightened with sharks before. Perhaps this had made them more confident and overbearing.

There were other factors for which I could find no explanation. I had visited the lagoon many times, and each time, the same few sharks had come within the first half hour. They had not seemed to consider eating very urgent, and had not been excited, except briefly while carrying the food away at times. Considering the way the situation had developed, it seemed that the shark with the dark circle above her left fin had become agitated over her difficulties with the heavy, frozen chunk of food, and that this had communicated instantly and dramatically to Bratworst. Had this general mood then communicated to all the new arrivals? I was not sure about that. This time there had been many times the usual number of sharks all present very quickly, and all of them seemed excited and bold. Something had been different, but I did not know what it was.

It became one of my unsolved mysteries that events like this unfolded sometimes for no apparent reason.

Then there was the factor that nearly all of the sharks were female. This had to be an indicator of a facet of the species' social life. Given that there was a place at the far end of the lagoon where baby sharks came into the world, I considered the possibility that the female sharks lived in the lagoon around the nursery areas. I had seen on our dives outside the reef, that the much smaller male sharks dominated there. The fact that the males and females occupied different habitats, revealed an unsuspected level of organisation in their society. Since quite a few were juveniles, too, it appeared logical to assume that they grew up in the lagoon after leaving the nursery. I recalled the shark pups I had seen playing in front of the house when we lived across from the shark nursery at the opposite end of the reef. The tiny sharks apparently left it while they were still very young, and they moved deeper into the lagoon.

But my reference book stated that the males grew to two metres (6.5 ft) in length—this came from the official records from fishing sources—while the males I had seen were not much longer than one metre (3.3 ft). They were sleek and muscular, whereas the females were rounder and much bigger—often more than one and a half metres (5 ft) in length. Where were these two-metre males?

My list of identified sharks had grown much longer by the time I finished updating it. But referring to them in my notes by their descriptions had become unworkable. I was tired of writing, 'the large grey female with the dark spot above her left pectoral fin,' every time I referred to her. The obvious solution was to give them names. Using a number to refer to a familiar animal seemed ludicrous. Plus, it would invite mistakes; already, I kept confusing numbers fourteen and nineteen.

But I hesitated. There was something funny about naming fish. And what does one call a shark? I pondered this, sitting exhausted on the deck, watching the wind whip the waves down the bay. Finally I wrote "Madonna" beside the description of the number one shark. Next came the curious shark, who had sniffed my foot as I sat in my boat, then followed me out of the lagoon—the one who had frightened me so badly. She became Bratworst due to being the worst brat. The next shark on the list was the large female who had initially approached me in triangle formation with Madonna and Bratworst. She was such a placid, maternal shark. I named her Martha. The male who had come to eat by himself at my second session became Meadowes, the name of a character in a book I was reading. The small golden female who had been so curious about me, coming often to circle and look at me and who had a diamond shape of light marks on her right side, became Diamond. Since the shark with the scars on her right side had so frightened me, I decided to give her a very nice name, and called her Gwendolyn, which brought to mind the gentle undulations with which sharks move. In this fashion, I continued down the list.

The next day, I swam out from shore to the place I visited the sharks, and explored it thoroughly. Though occasionally a shark passed in the distance, the beautiful coral garden felt safe and friendly with its complement of fish pursuing their affairs and looking out from their coral refuges. There was no sign of the gathering of thrilled sharks that had assembled the day before, and I returned home feeling reassured.

But for many nights after, I lay half asleep clutching Franck, while sharks soared and turned through the swirling darkness, and I got up before dawn.

That week I re-read the few books I had on sharks, which yielded only superficial information. But they helped me to review the factors at play when I was with this new and unfamiliar class of animals.

Sound and vibration are very important to sharks. Sound travels a long way in water, spreading in a uniform spherical pattern, and sharks hear well. They are particularly sensitive to low-frequency vibrations, such as those caused by movement in the water and crashing waves. They can also detect pressure waves with the sense organ called the *lateral line*.

The lateral line is found in fish, sharks, and some amphibians and is made up of a series of receptors in a line along the length of the animal. The receptors consist of a sensor within a cupola of jelly, which is directly affected by pressure in the water, just as the hair cells in our inner ears, which keep us balanced, are directly affected by movement. It is thought that the lateral line and the inner ear have a common origin far back in evolutionary time when life was selecting the basics. Via a complex nervous system, which analyses the incoming information in each tiny receptor, sharks can perceive events beyond visual range. They are aware of a person or large animal moving in the vicinity while remaining unseen, beyond their blue curtain.

I was familiar with this phenomenon from my underwater activities with Merlin, and it explained much about shark behaviour. Sharks in the area would be aware of an uproar caused by excited sharks feeding. Could this have precipitated the sudden gathering of sharks once Madonna had begun in earnest to try to shake a bite out of the huge chunk of frozen food?

I learned, too, that sharks can try to enforce dominance and wondered whether such motives had played

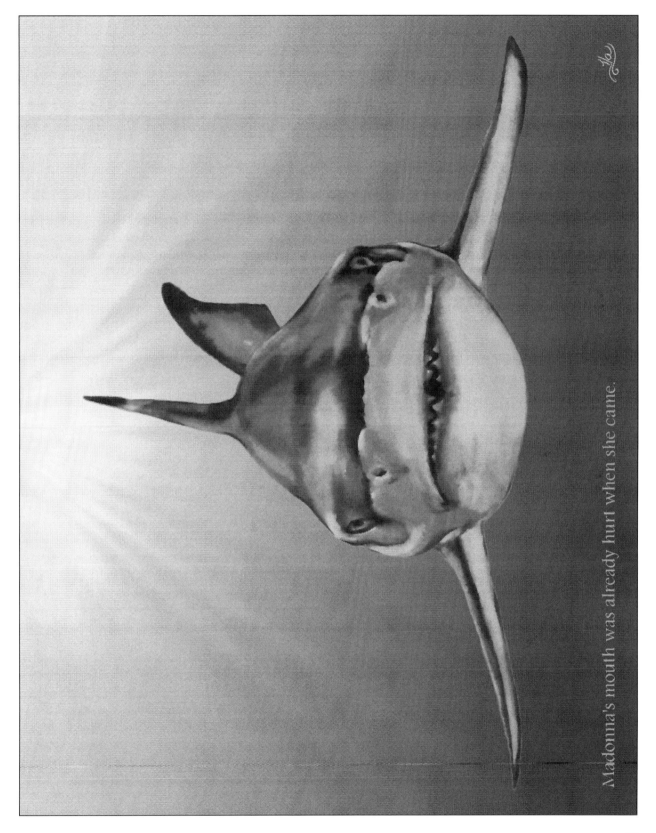

Madonna's mouth was already hurt when she came.

a part in my alarming experience. Once I was frightened, it was hard to objectively analyse what followed.

Then there are the electro-receptive organs, the *ampullae of Lorenzini*. They detect voltage—electrical potential across a barrier. This is not the same as the ability to detect current flowing through a conductor, such as along a nerve. They couldn't detect brainwaves, but they could hear the beating of my heart. They were picking up other signals about my subjective state. It was possible that they could sense whether I was stressed or frightened, or completely relaxed, which could indicate to them that I was not in attack mode. Perhaps that was the reason that sharks tended to approach me closely when I was in a meditative state, and not when I was purposely finning and looking for them—they sensed I was quiet and safer to investigate.

The ampullae of Lorenzini can pick up the voltage emitted by working muscle tissue. Sharks easily detect such voltage down to the microvolt range, which is the range emitted by most sharks' prey. Since sea water is saltier than blood, the difference in ionic concentration produces an electrical potential between the inside and the outside of fish. The animal's skin shields most of it, but there are places, such as the gills, that emit a faint electric field that sharks can detect. I read that it had been calculated that in a perfect ocean, a typical shark could detect a one-and-a-half-volt AA battery from a distance of hundreds of kilometres! But the ocean is full of background noise that limits the range to about a metre. So sharks and rays use their electro-sense to detect living prey at close range. This works even when the prey is hidden in murky water or, as in the case of other species such as rays, when it is buried in sand.

Lastly, a shark can taste and feel its prey with its sensitive mouth, which is the only part of its anatomy designed by nature for contact with the solid environment.

All of these senses are used in different combinations depending on species and circumstance, just as we at times use hearing more than sight and at other times are focused on an odour or touch. The approximately five hundred species of sharks inhabit a wide variety of environmental niches, so their senses are adapted accordingly and doubtless vary widely, just as birds adapted to different habitats are very different.

From my reading I learned that even if one of these reef sharks bit me, it was unlikely to remove flesh, but rather leave a circle of tooth marks. That was so reassuring. Madonna might be small compared to a great white shark, but she was still as big around as my waist. And she had a big mouth. I worried about how it would go for me if I were bitten. Would the others bite too? Even if I could get back into the boat, how would I get the anchor up? How would I paddle the kilometres home?

Exactly a week after my fright, I prepared to return to the lagoon. I cut the sharks' food into bite-sized pieces to make it as easy as possible for them to eat, hoping to keep them orderly. Over my bathing suit, I wore my skins, a clingy suit of light material, designed to provide protection from the sun and scrapes against the reef. Though I usually never bothered with the outfit, except during the middle of the day when sunburn threatened, wearing it with the sharks seemed prudent. I remembered Merlin's fascination with our white toes, and it seemed plausible that the sunlight flashing off my pale skin could stimulate a shark in the same way. The skins were black with a green and a blue stripe down the sides of my arms and body, which was a more appropriate colouring for a sea creature.

I carefully chose an open, sandy area with a dead coral formation up-current, where I could hold on for stability while watching the sharks coming up the scent flow, and anchored the kayak so that it would steady behind me. If at any time I felt uncomfortable, I could turn away from the sharks and climb to safety.

I placed the food, stationed myself, and waited. After five minutes, Gwendolyn made a cautious pass. She seemed calm as she cruised slowly through the shafts of sunlight and, after a few circles, glided over the food, scenting it. Madonna arrived with a small remora clinging to her head. She was cautious of me at first, perhaps because of my changed appearance in the skins. But after circling majestically and coming over to look at me more closely, she descended to the sand and began nosing over the fish scraps. Her cut mouth had healed. Bratworst came and joined them. The sharks soared around as they took the food and swallowed it, swimming with and after each other, and everyone seemed content. I noted the distinguishing marks of some new sharks who came, but there were not many there, and when they disappeared, the two fish heads were still lying on the sand. I scattered the meat from them to feed the fish. The needlefish

(*Platybelone argalus*) especially enjoyed that. An expanding cloud of fish fluids and particles drifted away, filled with fish excitedly feeding. But no shark returned even then.

The following week, I had been given more scraps than usual, and a lot of it was bloody. I expected large numbers of excited sharks to come quickly, a repeat of the former dramatic session or worse. So I located a place with a dead coral formation to hold onto, up-current from an open region, moved the anchor until the kayak stabilized correctly in the wind and current, and put the food in the water.

By then, a pall of rain had obscured the sun, so it was dark underwater as I waited, breathless, for the circus to begin. But ten minutes passed, and no one came. Just as bright sunlight broke through the surface, a large stingray flew toward me along the sand and I manoeuvred for a photograph. That was when Bratworst appeared, gliding suddenly over the food. I forgot the ray and photographed my special shark as she calmly circled and ate a few pieces of food, ignoring me. After eating she left, and half an hour passed before another shark came undulating lazily up the scent flow toward me. It was Martha. Having by this time scrutinized the photographs from the first session, in which Martha, Madonna and Bratworst had approached me in triangular formation, I had found that almost every photograph I had taken had been of her. Strangely, it had always been Martha who had come closest to me.

I swam down to the sandy floor where she was scanning the odours from the scraps to see what I had brought, and took a close-up of her face. More sharks came up the scent flow, and I watched each one for identifying marks on both sides, noting them on my slate so that I would be able to recognize them in the future. I considered it impossible to draw a shark's dorsal fin in detail during a session, because of the need to watch my back, but memorized and noted the main features of each one. Ten minutes later, all the food was gone, and the sharks vanished into the surrounding coral. There had been no frenzy, no startling incidents of being charged from behind, just tranquil sharks coming, eating a scrap or two, and leaving—bloody red fish meat in the scraps had not been the cause of the congregation of excited sharks.

A week later, I returned with some good fish scraps properly cut up for the sharks, and some animal scraps too, neatly rolled into meatballs. However, these floated away on the surface, accompanied by happily snapping needlefish. It took fifteen minutes for Madonna to come, acting uncharacteristically skittish. She ate and left, and more sharks cruised in. One was an exceptionally large and heavily pregnant grey female with a companion I had not seen before. A male also came this time, but it was not Meadowes.

Every time I raised my eyes, little needlefish surrounded my head, appearing and vanishing against the reflections under the surface. Having read that these fish could bite, and that one had once caused a brain injury, it took time to get used to them, but they looked so trusting, curvetting by beneath the waves, their large solemn eyes fixed upon me, that it was hard to believe they could be dangerous, even with the seven-centimetre (3 in) spike that preceded the elongated silver creatures.

Madonna returned twice during the forty minutes in which the sharks were eating. Being able to distinguish one individual from another added a whole new dimension of understanding to my observations.

One day I went to see if there was a shark nursery within the curve of the reef where it ended in our bay, as there was at the other end, opposite to where we had lived before. There was a shallow, quiet region there, though much smaller and somewhat deeper than the one at the other end of the reef. Though there were no shark pups, it was populated by myriads of baby fish of countless varieties. It was truly a nursery!

Franck and I went diving outside the reef to a popular location where the sharks were often fed. Two female lemon sharks, about three metres (10 ft) long and probably eight times the volume of my little sharks, were languidly roaming around together and swam close by me when I drifted over to see them. The water column was strung with cruising blackfin and grey reef sharks, and I watched closely to discover the proportion of grey blackfins to brown and golden ones. To my surprise, they were *all* golden! I had wondered if the grey colour of my population of females was genetic or due to other factors. Now it seemed that the dark colour of the females in the lagoon could be related to sun-tanning in the shallower waters, though genetic factors probably played a role, accounting for the variety of shades of grey and brown that they displayed. According to Richard H. Johnson in *The Sharks of Polynesia*, sharks did get suntanned.

I confirmed that nearly all of the sharks were adult males. They were slender and more muscular than my matronly females and few were more than a meter long. Once again, I wondered where the two-metre males were. It seemed important for my understanding of the population dynamics.

When Franck left for France, I went to the lagoon before dawn each morning. The setting full moon lit the darkness as my boat slid over luminous waters, out from under the mountain's shadow, past the island and across the gulf to the lagoon. The glowing moon in the west and the flowering light of the approaching sun in the east cast a supernatural aura over the lagoon lying still and dreaming. Floating on the gleaming disk of ocean between the sun and moon, I felt at one with a wider universe, set apart from the little daily human-life affairs. Above me was the perfect vault of a sky overflowing with light, beneath me, the unknown dark, the abyss, the mystery, as I paddled, enchanted, toward the sharks.

The pale turquoise water of the lagoon revealed once more the coral garden brimming with life awakening to the new day. A shark swam toward the boat, turning three metres away. I anchored and sat watching, waiting for the sun to clear the line of clouds at the horizon and provide a bit of light. The wind began to ruffle the surface with the lightest touch. The ragged peaks of the island glowed as if with their own light. Tangled jungles swept to the shore, where the plumage of coconut palms waved slowly in the stirring air.

Each morning was so beautiful that it was nearly unbelievable.

One day, as I waited, watching the sun clear the horizon, I turned to look out to sea and felt a brilliant pain in my toe. A bee had somehow crossed the lagoon, found my little boat in all that shiny water, and crash-landed in the bottom. He had just been climbing up onto my toe when I turned and threatened to crush him against the side. I set him on a safe, dry surface of the kayak, where he cleaned himself and rested, swiped the stinger out of my foot, and tried to ignore the pain. Slowly the sun and wind rose. The multi-coloured, empty plastic pop bottles that roll forever across the lagoon's surface increased in speed. When the sun emerged, casting long wings of light across the water, and illuminating the magic garden below, I slid into the water.

I put the food in the chosen place, and a fog of blurred, bloody scent drifted up and slowly moved away, obliterating the visibility down-current. Soon after, Madonna and Martha came swimming up the scent flow side by side and lazily cruised the area, sometimes in single file, sometimes abreast. Martha looked pregnant, and Madonna did not, but otherwise, they seemed identical. Enchanted by the grace of their movements as they soared through waters flooded by the first slanting sunbeams, my attention was taken by a swift movement coming from the direction of the reef. A pale shark shot straight to the food, though it had approached from up-current and so couldn't have scented it, and swirled around it momentarily with Martha and Madonna. It was a juvenile lemon shark, just a little larger than the blackfin females, and its oversized, wing-like fins and pale colour gave it an ethereal appearance. In spite of it being of a different species, the sharks moved together in a relaxed and non-competitive way, often touching—no one gave way to another.

Finally the lemon shark snatched up the fish head and carried it away toward the reef. It didn't return. Bratworst came, and my three favourites ate calmly together, accompanied by a large remora. A tiny nurse shark pup (*Ginglymostoma cirratum*), all frills and gracefully flowing, floated on top of the pile of fish scraps and settled himself for a munch.

Many more sharks soon arrived, and I was kept busy noting identities and sketching unusual dorsal fin patterns. When over an hour had passed, I sat in the sun on the kayak and watched them gliding under me. It was like looking into a magic window on a dramatic and closely held secret, and I was gripped by a longing to know them that drew me back whenever I could get away.

Time after time, Bratworst appeared within moments of my arrival and I began to wonder how she could always be there so fast, as though she was just beyond visual range whether I came at noon or seven-thirty in the morning. Perhaps she recognized the sound of the boat. It had taken Merlin's fish a very short time to come swarming around our feet when we walked into the water on the beach in front of our house, so there was no reason why Bratworst couldn't have learned to associate the sound of the boat with my arrival.

Martha, Madonna, and the lemon shark.

Every time my paddle struck the plastic boat, it made a unique sound, quite different than the sound of the local outrigger canoes, the only other boats used on the lagoon. The sound of my paddling and of the waves lapping against the hollow hull, could all contribute to a unique sonic signature made by my approach across the lagoon.

I took a special trip out to investigate, and anchored at the same place and time I normally held the feeding sessions. Sliding underwater, I drifted down the path that the scent flow would take, watching.

Within two minutes, a shark appeared on a course that intercepted my own. It was Madonna. She looked at me curiously, watching, as if to see what I was going to do next, and circled, spiralling closer and closer till she was within arm's length. I watched every move she made, every movement of her eyes. Finally she began to glide away. I called to her as I sometimes did on finding her underwater, and drifted after her. She turned and swam back to me to circle again a few times. Then she swam slowly on, and I followed.

She pursued a sinuous path through the coral in the general direction of the border of the lagoon and the bay. Sometimes, she sped up a little, and I barely kept her in sight. Then she would slow, and I was able to follow more closely. It seemed that she was more troubled by my presence than the shark I had followed in the Tahitian lagoon. After ten minutes, we entered an open space where Martha and Bratworst were gliding toward us. There was no sign of communication between them, but I was certain that it was no accident that we had all met there. Martha passed by me just at arm's length, looking. Bratworst watched us pass in front of her without altering her direction. I followed Madonna as she crossed the open space into shallower reef flats close to the bay. Just a few minutes later, she sped up enough to leave me behind. I returned to where Martha and Bratworst had been, but they were not there, so just roamed through the area. A juvenile passed too far off to identify, and a little later Martha came swimming from the direction of the reef. I tried to follow her, but she accelerated away as soon as she realized what I was doing. I continued to search, hoping for another meeting, especially with Bratworst, but didn't see any of them again, only another distant juvenile as I returned to the kayak.

The sharks must have been there because they had recognized the sound of the boat. It was intriguing that Madonna had led me to Martha and Bratworst. I was sure that had I gone there alone, I would not have found the two of them approaching me with interest like that. Never had it happened before in all the time I had been roaming the area alone. But what communication, if any, had passed between them had been impossible to discern.

Franck returned from France with his three children, Julie, Romain, and Marie, to spend July with us. During their visit, I took them one at a time to see the sharks. It was intriguing that the sharks noticed that I had a child close beside me, and several came over to look at us as soon as they saw us. Martha especially paid attention to my small companions. Once she came so close that she scared herself and accelerated past my ear, scaring me too! Then she returned, and approached us repeatedly throughout the time we were there.

But I could not stay long with the children, since being so small, they got cold very quickly.

CHAPTER TEN

Sorting Out the Sharks

As the weeks passed, Madonna, Martha, and Bratworst came regularly to the sound of my kayak. Gwendolyn was a common visitor, Diamond appeared frequently, and there was a very dark female without the slightest imperfection, whom I named Isis. Meadowes came with another male whom I named Ruffles, since he seemed ruffled not just the first time I met him, but quite often. The pair was in the area regularly but infrequently. Though other males passed through, Meadowes and Ruffles seemed to be the only permanent male residents of that part of the lagoon, and they were nearly constant companions. Visiting females who were enormous compared to Martha, Madonna, and Bratworst sometimes appeared. I watched them pass in awe and avidly collected photographs of them. It was a challenge since they were very shy.

At a given feeding session, I considered that I would see all the sharks who passed through the scent flow during the hour in which I watched, as well as those close enough to hear the vibrations created by those already feeding. Thus, I would have a record of which sharks were in the area at that time. I assumed that quite soon I would have met all the sharks using the area, which I estimated to number about thirty.

Since the little bag of scraps I had been bringing was sufficient only to provide Martha, Madonna, Bratworst, and a few other locals with a treat, there was nothing for the sharks who came later. So I expanded my search for fish scraps in order to bring enough for sharks who arrived half an hour into the session.

Slowly, I understood that each individual had a preferred region, his or her *home range*. Usually, the shark could be found there, but sometimes he or she wandered. Since the scent flow from the fish scraps drifted westward toward the border of the lagoon, those whose ranges lay in an easterly direction rarely crossed it, and it took longer for me to meet them. On the other hand, the sharks who ranged between the feeding site and the border were alerted by the sound of my kayak going over, so were waiting when I arrived. Rare visitors also passed through, some from very far away, so sorting them all out was a challenge.

It soon became obvious that a description of size, colour, gender, and distinguishing marks was not enough to identify an individual. If I saw it soon again I knew which shark it was. But if weeks passed in which it did not reappear, while in the meantime I accumulated many more descriptions of other grey torpedoes that had gone shooting by, I could no longer be positive of its identity. Some sharks were easy to recognize due to unusual distinguishing marks, but most individuals had little to distinguish them from the others. Further, if a distinguishing mark was on one side only, I would not recognize the shark if I only saw its other side when it reappeared. Many sharks showed only one side, and it wasn't easy to see the other.

Visitors tended to keep the eye with which they had first seen me, fixed on me as they circled. My manoeuvres in hopes of seeing the shark's other side were parried by the animal, who appeared reluctant to take that eye off me—I was lucky to gain a brief glimpse of its second side when it turned and accelerated away.

So I began drawing the shape of the white band on the sharks' dorsal fins and didn't consider that I had enough information for a definitive identification of a new individual until I had both sides drawn and verified, in addition to a physical description of the shark. It was easy to verify whether a new sighting was a shark I had seen before simply by redrawing its dorsal fin. Further, I found that the simple act of drawing etched it into my mind, so that I far more easily recognized sharks whose fins I had drawn before. In order to be efficient in recognizing the sharks, I memorized the appearance of each one on both sides so that I would know it the instant I saw it.

It took time to learn which scars were permanent and which would heal and vanish. But with a drawing of both sides of a shark's dorsal fin, I could be sure of its identity, even if its marks, colour, or size changed. I still thought I would soon know them all, and had the drawings of their dorsal fins with me on my slate.

At one session, I had many more scraps than usual. On the sand, they oozed a greenish drifting plume like a smouldering fire. A large grey shark passed in the coral as I was preparing, but Madonna did not arrive for ten minutes. When she came, she passed at a distance, circled widely, and disappeared again. She did this for so long, that I wondered if she was simply not hungry. She finally approached, then accelerated as she passed my side. The speed at which a shark could suddenly move was one of the startling things about them. A few minutes later she came and looked slowly over the food, left again, came back, and chose a piece, taking it away with her without shaking it or accelerating.

I had expected much more action. Once again it was clear that the behaviour of the sharks was not predictable, and that many unknown factors must influence their appearance or lack of it. Sometimes the food interested them, and sometimes they ignored it.

Meadowes arrived. He and Madonna greeted each other with a close, undulating pass in which they may have touched. He circled and perused what I had brought for them as Bratworst glided up the scent flow. She behaved as Madonna had, cruising the area without approaching. Finally, she left for half an hour without eating. When she returned, she circled slowly over the food, then chose and ate three large pieces, one of which she swallowed down whole with repeated gulps. So the reason she hadn't eaten on her first visit couldn't have been that she was already full—Johnston in *The Sharks of Polynesia* had stated that sharks of her size normally eat the equivalent of one thirty-centimetre (12 in) fish weekly.

Martha glided slowly into view, and a large stingray flew into the sandy circle and settled in front of the food. Braced on its wings, it lifted its nose and arched its body. Martha swam languidly up behind, brushing it as she swam over it. There was no aggression between the sharks and other species. As with Merlin's fish, I had the impression of a multi-species community in which there was a sense of comradeship. Martha ate in her languid way, and left. Diamond took advantage of the lull to eat with Nightwind, a juvenile of the same size. Two shark pups appeared who moved about over the food at the speed of light in all directions. They were so excited that I wondered if the thrilled little sharks could bite just for fun. Here was further evidence that newborn sharks left the nursery to grow up in the lagoon. Already I had seen sharks of all sizes, whereas outside the reef were adults only, nearly all males.

Gwendolyn arrived and cruised the area for many minutes before she came to eat. Her face was quite unique; one could probably learn to recognize sharks by their faces. Finally, she too vanished, and the blue surrounded me, still and silent with only the flashing fish going about their business in the coral, and a few perch busily nibbling the food. How strange, that with all that food lying on the sand, no one was eating, no one was there. I swam around the area, enjoying the beauty of the morning light in the lagoon, and was thinking of leaving when, quite suddenly, sharks materialized in the coral in all directions.

Returning to the site, I found Isis in a state of excitement with many sharks I did not know. Others that were passing in the veiling light came flitting forward one by one to enter the site in single file, accelerate, and swirl around the food. Eventually one grabbed a scrap and shot away with it, triggering a wave of excited feeding. With so many sharks eating, the scene was clouded by particles raised from the sand. Gliding in from the coral appeared three large grey females in triangular formation. It turned out to be Martha with two strangers. Once again a shark had left and returned with companions.

One of the newcomers had a curved white band on her left side and perfect dark grey skin. She was a magnificent shark, with a black streak between her eye and mouth, giving her face a daemonic look. While Martha steadily circled the site, she whipped around the food, swept down to it, and ate. I named her Madeline.

The other stranger had blurred areas on the right side of her dorsal fin's white band, and her skin was not the uniform dark velvet of the others. It was a paler brown, and she looked as if she had suffered many wounds and healed up many times. There was a hardening of the lines defining her colours, which was soft in the others. She seemed older. She glided slowly, appearing to be careful of her swim-way as though she wished to avoid sudden turnings and collisions. She circled for a long time before she came to eat. I called her Samaria. In time, I learned that Madeline and Samaria spent most of their time about half a kilometre (a third of a mile) to the east and often travelled together.

An enormous, pregnant shark began cruising at the limits of visibility. She vanished, then a few minutes later, returned and drifted behind me. Gradually she spiralled closer. She was the biggest blackfin I had ever seen. As the sharks shot away in another wave of excited chasing through the coral, she joined in. Losing her shyness, she passed close by me with the others. Eight of the very large sharks were now feeding, soaring and turning with the many medium-sized ones and juveniles, their satin skin haloed with sunbeams. They swirled over the food a metre in front of me, filling my view, and passing me as though I wasn't there. One zoomed through the triangular opening formed by my arm when I held up my camera.

And Martha, my placid shark, left the food and all the excitement, and swam directly to me, then fell away to the left just as she reached my face. Scarcely a session passed in which she failed to do that at least one time. She was the only one who continued to give me such special attention. Madonna never really had, and even Bratworst was taking me for granted by then.

One of the big sharks suddenly appeared to threaten a little one with a bite, coming up behind it so that it shot ahead, but they were behind a small coral outcrop at the crucial moment, moving almost too fast for the eye to follow. For a split second, I was able to view the small shark and see that it was uninjured before it disappeared in the throng.

A tiny shark swam around behind me while this fiasco was occurring. She had a distinctive black crescent shape extending down from the black tip of her dorsal fin on the right side, and on the other side she didn't have a proper black tip at all, just a little dot at the top of her fin. She seemed to be swimming lop-sided, though I wasn't able to watch her for very long. She hid in the background and did not come close.

Waiting poised with my camera, I was finally able to photograph the biggest shark and called her Shiloh. She had come once several weeks before; she had been less swollen with pregnancy then. I watched for her in the weeks, months, and years that followed but never saw her again. When, one after the other, two sharks materialized centimetres from my side, eating, I swam to the boat and got in, too cold even to register what I was seeing or cope further. Sitting in the sunlight, I watched the sharks cruising slowly or shooting excitedly under me. Through the glassy surface, every detail in their liquid world was floodlit with supernatural clarity, while beyond, the glowing green mountains lifted to the sky whenever I looked up.

Unable to leave, I sat and wondered what their lives were like, and what they thought of my descent through the silver surface into their world. The behaviour of the different individuals was so intriguing and unanticipated. How strange it was that Martha continued to leave the food and the other feeding sharks to swim to me, right up to my face! It brought into question the idea of feeding competition. There was something else going on in her shark mind. Through the lucid water below, they flowed with infinite ease, their secrets safely hidden from me. Finally, I pulled up the anchor and rode the winds and the waves across the lagoon, the wide sparkling gulf, down the bay, and home.

One morning, I came to the lagoon in the early morning after a prolonged battle against the wind. Though the bay was protected by the surrounding volcanic rim, beyond, for thousands of miles, no land impeded the winds that curvetted out from the equator as the planet turned. That year, they flowed with deceptive gentility over my island, often waiting until I was crossing the lagoon before frolicking down upon me to dance with the sea. At first only a ruffling showed on the shining surface as, here and there, the wind caressed it. But as it lowered, the ruffles spread in fanning patterns, grew to wavelets, then powerful flows, slamming the kayak and forcing it back. Sometimes, it was all I could do to hold the kayak into the wind. Lacking a keel, it wanted to fly off with the rushing air, and sometimes all my force put into each stroke was barely enough to move the boat ahead. Though I knew the region well, at these times, it was impossible to see anything beneath the tossing surface. I had to find an open place down-current from a dead coral structure tall enough to hold onto, then quickly judge wind and current in order to anchor appropriately. It was important, no matter what else was demanding my attention, that I made sure when I threw the anchor, that the unravelling rope didn't flip the camera, mask, or anything else out with it. Just moving in the overloaded boat was treacherous, and that morning I realized that my plan to throw the sharks' food overboard, and jump in afterwards, wasn't going to work. Having carefully estimated the current and wind before anchoring, still I had misjudged; the boat steadied over living coral.

After sitting uncertainly, blasted by the wind and unable to see anything, I slid through the surface to look around, and gave the soft call I made on entering the water, so that the sharks would know it was me. Swarms of rainbow wrasses and needlefish shot around me, and even before a shark appeared on the periphery, it was obvious that I was expected. So I hoisted the first pot of scraps out of the boat and pulled it, filling with water, to a nearby open area of sand, dumping it out with difficulty. The second followed. Beyond the rising cloud of blood and fluids, two sharks were visible. I went back for a sack of fish spines. They were hard to shake out, requiring a lot of agitation as I grew increasingly impatient. I was still trying to extricate them from the bags, when out of the curtain of blood obscuring my view, a shark flew in, grabbed a piece, and swallowed it down. Isis. I managed to finish serving the sharks' breakfast, but when I returned with the fish head, they had taken over the area and were excitedly eating.

Martha, Madonna, Bratworst and Isis seemed very hungry. I hovered on the periphery, head in hand, and offered it to the little Bratworst as she glided by, but she ignored me. It was several minutes before I could push it toward the rest of the food without getting in their way. By then, Gwendolyn and Diamond had joined in, and more sharks arrived. This was the first time that they had not waited politely, with surprising patience, for me to place the food. This time, they seemed ravenous!

Sharks kept arriving, including Nightwind and Ruffles. Madeline, huge and pregnant, entered the circle. Two slender young females glided in, bronze and glowing with vitality. One was flawless, with not a mark to identify her so that for a long time, I thought of her as the perfect shark, so beautifully coloured, proportioned, and radiant did she seem in this period of her life before she matured. I gave her the name of Christobel. Her companion had a bright, light circle in the middle of her right side. She circled lazily, looking over the situation. I called her Flora. The sharks left after eating, and more came drifting in to investigate the situation while I happily sketched and watched them. After fifty minutes, Gwendolyn was the only circling shark. She always stayed long and circled widely, passing me every two to five minutes.

Eventually, the pup with the crescent-shaped tip on her dorsal fin came to eat in the lull, and for the first time, I was able to watch her. Her dorsal fin seemed to be injured, and she had something stuck in her gill on the right side. It looked like a crayfish—a parasite? She was very small; she must have been born that year. She roamed about, trying to extricate various morsels that she found, dropping them and foraging on. I laughed to see the little shark burrow into the sand when she tried to shake a bite out of a scrap. All the other sharks rose up when they did that. Small ones even shot through the surface, while the larger ones bounced off it. She was cute and funny, and I named her Trinket, but wondered if she would succeed in growing up. She appeared to have inherited a poor set of genes.

More sharks began to pass, gradually circling closer, then gliding in over the food. One was a pale, creamy brown adult male, the first large male to come to a session. Yet, still, he was smaller than the females. Since he appeared to be an aged animal, I wondered again about the two-metre males.

Suddenly another very large female drifted by after a cautious pass in the distance. She too seemed old, and was the colour of charcoal with a black smudge on the left side of her head above her eye. I named her Lillith, wondering why, in five months, I had never seen this striking creature before—she too, lived far to the east. So many sharks drifted in and joined the feeding that collisions became a problem. Clinging to a dead coral up-current, I was in the swim-way of sharks coming up the scent flow and gliding over the food. Once, as Lillith crossed, nosing over the scraps to find a bite, she accelerated to avoid a collision and rose suddenly. Swiping sidelong against another shark, she brushed by me as she left the site. I watched until the sharks drifted away and only the fish remained, then returned to the airy world, and sailed home.

By this time, I knew nearly all the sharks who arrived in the first half hour, except for the juveniles.

Martha

They seemed to roam far and wide without having a home range, and their general pattern was to come a few times then disappear. Unfortunately, I could not know if they had moved on or died, because some, such as Diamond, did stay around most of the time. The large group of unfamiliar sharks that sometimes arrived later remained a puzzle. I had no idea where they came from, or who they were, or if they were generally the same group each time, though I thought not. I could deduce only that the species liked to travel in groups. It took months to identify everyone, so rarely did each individual come, and years to learn where they were coming from. Many died before I could find out.

Getting an accurate identification required being able to watch the shark closely while it passed, drawing its fin, noting down any distinctive features on that side, then keeping track of it until it passed closely again, showing its other side. Only in this way could I have both sides accurately drawn with no doubt about the drawings being of the same shark. The effort was most confusing, as grey torpedoes soared by, flashing light marks and strange dorsal fins and disappearing. When a rare shark that I might not see again for months appeared, sometimes I tried to compel it to turn and show its hidden side by moving subtly to block its path. But, time after time, on arriving home and going over my notes, I was thankful that there was always another day. Though having an accurate drawing of one side of the dorsal fin and a description of the shark was better than nothing, it could, and often did, lead to two entries in my records for one shark, one for each side. In the case of rare visitors, years were sometimes necessary to sort out the mistake.

The food was vital to attract the strange sharks close enough to see them in detail. Once they began feeding, they circled over and around the food, showing both sides while I drew their dorsal fins, and double-checked the drawings for accuracy. When drawing quickly while being tossed by choppy waves and pushed and pulled by surge, a chance to compare the drawing with the shark, and perfect it, was essential.

Initially, I thought that it would be easy to keep the sharks sorted out with descriptions and rough sketches, but since no end to the flow of unknown sharks was in sight, it was soon clear that to differentiate lookalikes, each drawing had to include the subtle details of the sharks' dorsal fins, not just the general appearance. Initially, I had often described the dorsal fin tip line as "a shallow, rounded W" or "a plateau joined at trailing and leading edges by an angled line," and so forth. Having gained the confidence to draw more intently in the presence of excited sharks, and with months more experience, this now seemed absurdly simple. My photographs of the sharks were not of high enough quality to use for identification unless I had been lucky enough to capture both sides of the shark at close range. This was virtually impossible in the case of new visitors. The most interesting were the mature, visiting females, and they were very shy.

Frequently, I drifted through the region, in hopes of seeing some of the visiting sharks again. Often, I was lucky, but other times the effort to go all the way out there revealed no new information. But it was always a pleasure to spend time with the residents, watching and moving with them, outside of the feeding sessions.

One day when I was able to collect a few scraps, I took them to the lagoon in the middle of the day. Gwendolyn was there, and Trinket, Nightwind, Diamond, and a few other juveniles soon joined her. Trinket had lost the odd cylindrical object that had been stuck in her gill. One of the new youngsters, a little brown female, circled around me, looking curiously, and ignoring the food at first. I drew her dorsal fin, and named her Sybyl. Our intimate meeting and the resulting drawings permitted me to recognize her at once when she came again, more than a year later! As always, Gwendolyn widely circled the area, including me and the boat, while the little ones ate. Trinket gobbled down a surprising number of big chunks of food that had been too old for the fish shop to sell. She gulped repeatedly as each long fillet went down whole.

After twenty-five minutes, a blur appeared in the distance, streaking forward, and I watched closely to see who could be coming in such a state of shark agitation. It was Bratworst, apparently frantic to think that she might have missed something. Arriving in the clearing, she grabbed one scrap after another, giving each a half-hearted shake as she desperately sought something worth eating. Isis followed, and seeing her and Bratworst, heads together, nosing among the remains of the food, the colour difference between them was clear—Bratworst was a warmer, browner colour while Isis was steel grey. There was another grey female there with her, with no distinguishing marks, but an inverted heart on the line of her black tip on the left side. Naturally, I named her Valentine. She was like a twin to Isis, who had begun arriving with Martha, Madonna and Bratworst at the beginning of the sessions.

While the sharks circled and fed, a Javanese moray eel of the two-metre sort slid from his hole and insinuated himself subtly around the periphery. An octopus scuttled swiftly away from the fish scraps with a little piece of food clutched under its body and all its tentacles curled inward beneath it. It disappeared into a hole in the coral outcrop beside me. Though blackfin reef sharks are said to eat octopi, not one took any notice of the courageous little creature.

I was cold and decided to go when the sharks dispersed, so picked up the anchor, but let it go again when a big shark glided in. I was surprised that she didn't wince at the sound of it hitting the reef-flats. It was Martha. She slowly circled and Madonna joined her—the two had been travelling together. I called to her and she swam over to me, passed closely, and circled away. They saw that there was no food and left. Tiny Trinket had stuffed down all the good fillets I had brought for my favourites. This time, the usual three had not been close-by, perhaps because I had come so much later. Mostly only juveniles had come at first.

I left, drifting and rinsing the crumbs from the boat for the needlefish who followed.

On arrival at the site, there were always needlefish down-current, facing the boat. I brought the food in two containers that could not possibly leak, but still they were waiting. Either they could pick up scents from the air, which must fall in microscopic amounts on the surface, or they, too, came to the sound of the kayak. Underwater, needlefish and swarms of rainbow wrasses greeted me in increasing numbers. Martha, Madonna, Bratworst and Isis were usually circling near, while more sharks, just shadows in the distance, appeared and vanished. While placing the food, I scattered crumbs for the fish. The delicate needlefish snapped up particles in the silvery lights of the surface, their huge eyes on me, while below, a multi-species cloud of excited fish darted—butterfly fish, triggerfish, angel fish, pennant fish, trumpet fish, cod, groupers, and other brightly coloured species, in waters so alive that even its molecules seemed to vibrate and sparkle.

At one session, after Martha glided over the scraps, chose a piece, and gulped it down, Bratworst came. After a couple of passes, she tried to pick up a scrap, and came up with a piece of broken coral instead. She dropped it with a little shake of her head. Lacking a limb designed for contact with the solid environment, the sharks used their sensitive mouths to manipulate things, but they were not always accurate.

Suddenly, two large, confident, dark grey males swept in. Since Meadowes and Ruffles were light-coloured, I didn't recognize them until Meadowes turned and showed the dark mark on his back. He had injured his mouth. Somehow the two had turned dark grey since I had met them in April, when I was still looking for Merlin. Ruffles had been light brown—his colour was clear in an early photograph I had taken of him. Apparently, the sharks' colour differences and changes were not only genetic and due to variations in their exposure to sunlight, but the result of other factors as well.

Two sharks shot towards me from deeper in the lagoon. The first turned away just before the site; the second was Madonna. More and more sharks cruised in. Their mouths deformed as they detached their jaws to engulf the food. Sometimes it seemed to take a bit of manipulation to get their jaws back into place afterwards and several gulps to swallow the larger pieces whole—I could see the food going down into their stomachs. Only the large pieces were shaken, which thrilled the little fish, as they shot through the scattering particles. It excited the other sharks to fly after, and catch the falling food when the leading shark extracted a bite. Great excitement ensued as the sharks took turns catching the food in mid-water and shaking it wildly, eight or nine together soaring through the coral with breathtaking gracefulness, almost too fast for the eye to follow. For a moment, four sharks had hold of a fish head and were all shaking it at once.

During a quiet moment while the sharks were occupied, a Javanese moray eel, who had approached around the periphery from one coral refuge to another, undulated into the open to take a fish head and retreat into a hole. As the sharks came shooting back, Gwendolyn was behind Madonna. She was uncharacteristically agitated, and there was a wound in the middle of her right side in front of her mass of scars, a hole about three centimetres (1.25 in) long surrounded by redness. It looked as if it had been made by a harpoon.

Martha's pregnancy had progressed so dramatically that I started manoeuvring for a photo. I had a good position on a clump of dead coral overlooking the area, from which the light was perfect. However, I had become distracted by the moray eel. He had put the fish head in a hole under a coral structure, and glided in his serpentine fashion back out into the open to take one of the biggest and best pieces of meat, which I had been lucky to get; they had become too old to sell. I wondered if the sharks would do anything to protect their food. Little Pontiac zoomed up to the eel, but turned away before coming close. The other sharks ignored him. He was bigger than most of them. He retreated with his prize under the coral formation next to mine. I kept an eye on his activities.

Martha passed and passed again, but not close enough to photograph. Another shark tried to pick up a scrap and came up with a piece of dead seaweed instead—she swallowed the seaweed. By then, the food was nearly gone, and the sharks were in search mode, locating every last little piece. Gwendolyn nose-dived under a clump of coral to get a piece that had fallen there, going vertical so that her tail brushed the surface. She badly scraped her nose.

While the sharks were busy, the eel slid from his hole, and targeted another choice morsel that the sharks had failed to find. Outraged by his wanton thievery of the food I had gone to such trouble to procure, I swam over and waved my camera at him, indicating that he was to back off. The eel undulated forward, gazing up at me gesticulating at him on the surface, and lifted his front half toward me. He opened his mouth, with a venomous look one normally sees only in nightmares, and thrust his gaping jaws repeatedly in my direction, while waving slowly back and forth in the turquoise waters. I glared at him and he at me as sharks shot around us. I was hoping that one of them would do something. But no one did, and when I returned to my perch, the huge eel undulated into my coral formation and repeatedly thrust his head out at me. I had to retreat to another coral structure.

I continued to take notes and draw, but the lighting was not right for photographs from the only other formation available to hold onto. When a new, larger shark appeared, I returned to my original one, and waited, camera poised, for her to come close. But I had underestimated the emotional state of the eel, who now appeared under my nose, energetically trying to push himself through a hole just a bit too small to permit him to reach my face. I had to change locations again. When the big shark finally did approach me, just the movement of my finger when I took the picture startled her. She accelerated away at a right angle, and didn't come close again. But I managed to draw the white band of her dorsal fin on both sides and eventually named her Princess.

Several pups, all a cool, grey-yellow, arrived as the others dispersed. They flew around the site like lightning. One was covered with little raised light marks like warts. Another had many light marks all over it, including some which formed a straight line along most of its back. One shot straight toward me, and at the last minute, I raised a hand. It turned a right angle and flew off. I never saw them again.

The sharks ate and slowly dispersed until only Princess still passed in the distance. She was so pregnant that she looked as if she had swallowed a barrel. A few juveniles approached but turned and sped away when they saw me; they must not have attended before. Whenever I looked at the moray eel, he menacingly thrust his head toward me. Did eels experience subjective states? I was convinced that they did!

I left, then, so that Princess could eat without further disturbance, hoping she would be less shy the next time we met. As I went to the kayak, Madeline soared though like a jet plane and disappeared into the blue.

When I got home, I found the fresh remains of a large white spotted puffer fish *(Arothron hispidus)*, washed up on the beach. A shark had neatly bitten away its body from head to tail, removing all but some entrails and the skin of the ventral surface. The force of the bite had turned the fish's tail inside out. Its huge eyes still gazed with a shocked expression from its big head. The shark's mouth was more than forty centimetres (16 in) across. It seemed that very large sharks cruised past our beach, biting things.

CHAPTER ELEVEN

A Change of Season

As the weeks of September drifted by that first year of my study in 1999, there was a subtle change in the sharks' patterns. Strange, pregnant females began passing through, and the residents appeared less predictably. I had been checking the shallows in the curve of the reef for shark pups, yet had not found any.

One day when I had some scraps, I put them in the well of the kayak, and plugged the holes so that fluids would accumulate. Inspired, I added a chicken I had found freshly killed on the road, and two of Franck's steaks. Once in the lagoon, I took out the plastic plugs, and paddled in a wide circle, trailing scent to attract all the sharks in the region to the site. But though I saw sharks from the boat, when I slid underwater none were visible for the first time in months. Those who eventually began passing into visual range were strangers, who would not approach. I scouted the area for a good site, avoiding the one where the angry moray eel lived, and put in their food. The change in routine had produced a change in results—the first shark to cruise in was Samaria, who usually didn't appear until the end of the session. The elderly shark had lost her nervousness around me; I noted again how carefully and slowly she moved. Bratworst, Martha, Isis, Diamond, and Pontiac soared in and began excitedly choosing morsels and gulping them whole. One swallowed a piece of seaweed with her fish. But the sharks rejected the chicken and steaks.

The slender, bronze females, Flora and Christobel arrived. They were radiant with health, graceful young beauties without a flaw on their satin skins. I hadn't seen them for weeks and wondered if they were travelling companions, as Martha and Madonna seemed to be. The sharks were excited, and the entire group shot around in the coral shaking and eating their scraps, while the little fish swarmed through the disturbed waters dining on the particles. Two sharks flew off, wildly shaking a morsel together and causing a dramatic commotion on the surface. While I was taking a photograph of Martha, a little shark zoomed under my arm.

Soon all but the chicken and steaks were gone, and finally, Meadowes scooped up the chicken and gave it a shake, only to drop it again. Maybe he didn't like the feathers up his sensitive nose, or the taste. He did not return to it. The steaks were ignored. The sharks showed no taste for the flesh of birds and mammals.

Madonna and Bratworst still circled when the others dispersed, and when Madonna tried to extract some meat from a fish head, I looked in the boat, found a last scrap, and waited for her to circle back. She didn't come so I went looking for her. In a sunny glade, a large grey female approached me, wrapped in golden light. She swam unhesitatingly forward, and I held out the piece of fish to her. She passed by within a metre without the slightest reaction to my movement, and it was only then that I realized that this was a new and unknown shark. A remora decorated her throat. Treating me as if we were old buddies, she came with me to the feeding site, though all the other large pregnant females I had met had been so shy! As she ate the last scrap, I drew her dorsal fin and noted her general appearance. I never forgot this intimate meeting, because when she returned to the study area several months later, she was so fearful it took her months to come close enough to identify! I named her Annaloo, the mystery shark.

Some little juveniles gathered, including one with a narrow white band in the shape of a V—I called her Chevron. I drew the fins of the others, but none of them ever came again, and Chevron, so easy to notice, did not reappear until the year of her maturity. Lillith passed in the distance. She didn't approach the site, and I was lucky to have seen her.

When I left there were whales visible in the ocean, so I paddled over the barrier reef to see them.

Franck had bought a motor boat with his business partner, enabling us to go diving outside the reef more often. On one of these trips, I threw in some fish scraps before jumping in. Though we were far from any location used for commercial shark feedings, as we descended, a pale gold male blackfin shark appeared. Franck and I went on, exploring deeper in the ocean, accompanied by a school of yellow perch. After a few minutes, he soared over us with a companion, like two little planes, then swooped down and did a tight turn around us. I turned with him to see what he did. The other stayed a little farther away, but they were together, and they came with us. Like dogs on a walk with their human companions, these sharks accompanied us, scouting about in the vicinity but always coming back to circle and swim with us. I couldn't imagine Madonna or Bratworst leaving the food to accompany me, but that's what these males did. The yellow perch came, and one of the sharks often swam through them, as if it excited him to scatter them.

Back at the scraps, several more sharks were gliding around. The school of yellow perch found a good morsel and fell on it all together, so that it was hidden by their shimmering, shifting forms. Their collective nibbles lifted it up off the seafloor, and our shark noticed, swooped in and snapped it up. The other sharks were eating too. When one grabbed up a scrap and flew off with it, the entire school of yellow perch streamed after the shark, a flowing river of living gold. They were just beautiful.

These male sharks were even smaller than Ruffles and Meadowes, and once again, I wondered where the two-metre males could be, if they weren't outside the reef. Even on shark feeding dives, I had never seen a shark the size of my females. We left reluctantly to do our safety stop at a depth of five metres (16 ft) where black fish with very long pointed tails hovered around us in the blue, like birds in the sky, and the sharks rose, too, to circle us. I noted again how differently the fish treated us when we were scuba diving.

One morning at the end of September, I went looking for Madonna, taking a nice fillet for her. No scent leaked. As I crossed the lagoon, a shark encircled the boat. It looked like Bratworst, but a breeze started to ripple the glassy surface just then, so I couldn't tell if she followed me or not. I anchored in the usual place and sat alternately admiring the view and watching for the shark. The surface became transparent as glass, though the rest of the lagoon was ruffled, as sometimes occurred along the area just within the reef.

Then Diamond glided in and passed beneath the boat, turning on the other side to circle. Martha passed. Gwendolyn glided by, her mass of scars shining, and Bratworst curvetted around me. I wondered to whom I would give the piece of food—there was no doubt that this time they had come to the sound of the kayak.

But why had they never come to the kayak at the surface before, when I had a load of scraps for them leaking copious fluids? I decided to give the fillet to whoever came first, transferred it from its wrappers into a water-tight pouch, and slid underwater.

Bratworst came and I unwrapped the piece of food. She watched but passed by. Then Martha glided to me and I offered it to her. She circled, looking, but continued on without coming close. Ruffles approached next. I offered him the piece, but he swam by and did not take it. I drifted slowly with the sharks, offering the piece to one after the other. Gwendolyn passed but did not come close, and many sharks circled farther away. Martha returned to orbit me three times, but would not come to take the food. I offered it a second time to Ruffles, and to Bratworst again. Gradually, the sharks dispersed. As another approached, I held out the piece of fish, thinking it was Martha. But instead of Martha's dorsal fin, the pointed black tip of Anna-loo appeared—the shark I had recently tried to hand-feed. She drifted away into the coral. Madonna did not come. I got back in the boat and began absently planning how I would cook the fillet for Franck for dinner.

Fifteen minutes later, Martha surfaced beside the kayak, and I tossed it into the water over a light area of sand. Her dark shape swept down then accelerated away. I was glad she got it.

It was good to know that my sessions were not encouraging the sharks to pester people for food, something that had worried me, given the hatred toward sharks often expressed by fishermen. The sharks' natural caution and reluctance to approach me closely, even though they were familiar with my presence, seemed to confirm that they would not closely approach others either. They were capable of distinguishing between different situations and recognized my sessions as being different from someone spear fishing. So I was reassured that their behaviour toward fishermen would remain unaffected by their experiences with me.

Strange, pregnant females began passing.

When I returned again, it was a windy, sunny day. I had needed to hitch-hike to pick up the scraps from two fish shops, one several kilometres in one direction and the other a similar distance in the other, so by the time I returned with the shark food, I was ready to collapse from the heat.

A southern breeze blew me down the bay as clouds gathered and darkened, and there was rain in the strengthening wind as I prepared. I couldn't tell if there were sharks in the wind-lashed water from above. It was already 2:30 p.m., later than I had ever visited the sharks, and I wondered if I would see the same ones.

Flora passed as I placed the food and Madonna, Isis, Samaria, Bratworst, Diamond, Martha, and Gwendolyn, whose wound had healed to a dark line, swept in and began to eat. Sharks kept arriving quickly, as though they had been waiting for me to finish, and for the next while, I was busy writing down their names. Every time I looked up, I saw another one, as they calmly ate and swirled at arm's length in front of me.

Samaria had scraped her head on the coral, leaving a square white mark, and she seemed darker. She moved slowly, checking methodically for scraps, eating some, and searching for a scrap that the curious juvenile, Tiddlywink, had taken from the place she had dropped it. The sharks left, while she still foraged alone, passing me with no sign of nervousness. Trinket appeared after another long absence. I looked out to see if the sunshine that had unexpectedly flickered into the gloom would hold, and back underwater, found an enormous stranger passing, swollen with pregnancy. In spite of her size, she had perfect, velvety bronze skin, darker on her sides than on her back, which dramatically emphasized her lines. Two tiny, bronze sharks came into the site just then, and I named them Pixie and Fawn. Like Trinket, Pixie had an unusual dorsal fin pattern, so she would be easy to recognize again, and Fawn was unique, too. Both these little pups disappeared after this sighting, and though Pixie never reappeared, Fawn returned months later, and stayed.

The bronze shark drifted through again, truly the healthiest looking of the very big sharks I had seen. I named her Carmelle. Another large female passed briefly into view. She resembled Madeline, but Madeline was smaller and not so shy! Whenever I looked up, I would see her looking, then at my glance she would disappear, only to slowly pass again minutes later.

The wind had risen, whipping the surface into a foaming froth of spray and waves, and since functioning above the surface was impossible, I swam to the shelter of the island before getting back in the kayak.

The shark who resembled Madeline came each time to cruise back and forth at the limits of visibility. Only once, I happened to turn and saw her passing behind me, to take a closer look, without being seen. Five sessions passed before she approached and circled in a more relaxed way, showing her second side. I named her Keilah.

As September gave way to October, increasing numbers of unknown, pregnant females passed through the area. They would usually remain tantalizingly far off in the blue, and I was lucky if they eventually approached. If they did, it was often with a familiar shark, such as Martha or Madonna. Though they often seemed to travel alone, they felt safer with companions in unfamiliar situations. Often, these pregnant strangers would approach after a long period of passing far off in the blue in groups of two or three.

I continued to check the local nursery for pups, but so far none appeared.

As the wet season approached, the atmosphere grew turbulent. A procession of storms blew over the island, driving the ocean before them. Waves exploded over the reef, poured into the lagoon, and flowed toward the border. The strengthening torrent bore particles raised from the seafloor which obscured the visibility. Through the troubled surface, it was rarely possible to see the sharks who awaited me when I arrived.

Once, I took a magnet and buried it under the sand to see if the sharks would react to it. They didn't, and as I left, the boat overturned when I raised the anchor in the whipping wind. My belongings scattered fast in the careening waves, and I was lucky to recover my camera and most of my things. But the magnet vanished, as well as my slate with drawings of twenty-one shark fins on it.

So I bought a blank book and began to keep a proper record of the dorsal fins of the sharks. With four sharks to a page they were about half life-size. Page after page was soon filled with drawings of dorsal fin patterns, the two sides facing each other, with the shark's name written in calligraphy between them.

I kept my new slate clean for drawing, and a steady parade of unknown sharks continued to appear.

CHAPTER TWELVE

The Sunset Rendezvous

I had wanted to hold a session at sunset and never had, due to the inconvenience of the time. Increasingly, I found myself rehabilitating wild birds, who required treatments and care as night came, and then it was time to prepare dinner. After hours of meticulous painting, I didn't feel like a marathon at the end of the day. There was also the problem of the strong winds that rose in the morning, and didn't die away until the sun sank. Several times when I had planned to go at the end of the afternoon, the wind had been too strong.

But one day in late October, I went anyway, counting on the calming of the atmosphere as evening came.

In fact, though it took me ten minutes longer than usual to get out to the site, the wind was tolerable. Concerned by the height of the waves on the reef, I chose a place farther into the lagoon than usual, where the water was deeper and the current weaker. Dropping the anchor over a region of sand, I peered into the choppy water as the boat steadied. As if by magic, an oval window of glass opened the view for me. There were no sharks, but fish were streaming forward in a radial pattern from all directions.

The sun was still fifteen degrees above the horizon, but Franck had told me he wanted me out of the water by 5:00 p.m., and it was already 4:45, so I slid underwater to look around. I arrived in a beautiful place—a wide-open area with clear, deep water, studded with tall, widely spaced coral formations. With the sun plunging toward the horizon, I pulled the boat into position and tossed in the big fish heads that were on top, then struggled to dump out a package of spiky scraps. Two large sharks flew together into the falling shark food. Martha and Madonna, swimming alertly and close together, swerved over to affirm that it was me there, then circled tightly, snatching the morsels from around me. The assertive way they had come was stunning, as if for the first time since I had met them, I had found them wide awake!

I pulled the other package of scraps in, and tried to dump it out. My level of anxiety increased as Valentine and Keilah, the big shark who had watched from the periphery so many times, soared in and were swiftly joined by many more. They shot around me inhaling the food as it fell through the bloody water, apparently unconcerned by me and my struggles. Finally, I slung the wet plastic and cardboard back in the boat and floated, looking down. The sharks were all big females, delightedly grabbing pieces, shooting away together, and making a wild commotion on the surface when their energetic thrashing brought them through it. The sharks seemed to feel much more free and confident in the light of evening. There was such a dramatic difference compared with their behaviour in the morning that I wondered if they even normally ate during the day. Nearly all the regulars from the morning sessions were there, but no juveniles, and Meadowes was the only male I saw. Since I had lost my slate and used up my disposable camera, I had to memorize the distinguishing features and dorsal fins of newcomers, and there were many!

It was gratifying after so long to see Keilah closely, now passing me like a jet plane with no sign of nervousness at all. The evolution of her behaviour over time was interesting. Lillith arrived, and as the light faded to twilight, more and more unknown, larger females congregated. I concentrated on each one, trying to note something distinctive that would allow me to recognize each again, awed by the presence of so many huge sharks. Many looked old; you could see it in their faces, and it was etched in the quality of their skin. I tried to take a mental snapshot of each dorsal fin flashing by to draw when I got home.

Since it was getting dark, it was possible that a tiger shark could come up the scent flow from the border, so I kept watching down-current, and soon caught sight of a shark so big that I needed a second look to be sure that she was a blackfin. She had a wild dorsal fin pattern, all semi-circles fitted together into abstract W patterns, and her violet skin was mottled with old scarring. She flew into the site, snatched up a scrap, flew out, and circled beyond visual range. Minutes later she executed another hit-and-run swipe at the food. She returned at long intervals until I left, and I named her Amaranth. Keilah had left after eating, but she returned to join this large congregation of older female sharks.

Some male sharks arrived, but by then, it was too dark to see enough details to recognize them. One nearly swam into my face while trying to swallow a big bite and get his jaws back in place at the same time.

A small nurse shark appeared, attached itself to a scrap, and rested there, its pale body the colour of the sand, its wide fins stirring, wing-like, as it adjusted its position to feed. Soon, a second nurse shark came, drifting in uneven circles around the site, while an eel undulated out of the coral formation I held, and dragged away a fish spine. The countless fish were whipping around the nurse sharks in an agitated cloud, eating the scattering particles of food. Then a third, much larger nurse shark came. He began trying to extract some meat from a fish head, pushing it forward with his efforts, until he got his head inside and reared upward, wearing the fish head like a mask. More nurse sharks gathered, and there was a moment in which three of them had pushed their faces into fish heads and were wearing them like masks, vertical in the water column, slowly swaying back and forth in their efforts to scrape out a snack. Sharks orbited in and out of the drifting clouds of sand, appearing and vanishing eerily, and fish flashed dimly around them like little stars. I wondered how many planets you would have to visit to see anything as perfectly bizarre as this.

The reason for the grey-brown-violet colouring of the reef sharks was suddenly obvious—they were the colour of the twilight water, and as night gathered, they became virtually invisible. I watched until they were only movements in the dark, and had to go through them carefully when I went to the kayak.

There, I looked above the surface to check the conditions and plan my leap back on-board, and when I glanced underwater again, one of the nurse sharks was swimming up to my face, apparently to have a look while my head was above the surface. It re-descended, and I hoisted myself into the boat.

Air and sea lay entranced in a spell of silence. Wings of light from beyond the horizon fanned out to illuminate the zenith with shades of deepest violet, while the layers of cloud and sky between shone with the colours of an opal, all reflected on the shimmering sea. The intermingling of subtle yet contrasting shades in intricate patterns on the surface created a visual puzzle of endless fascination which, no matter how long I watched, I could never unravel or imagine how to paint. It could not be captured, but only enjoyed.

October passed. Franck and my sister, Kerry had bought me a proper underwater camera for my birthday, and it finally arrived. I visited the sharks in the evening as often as I could, sure that I would soon have a complete collection of photographs of them.

Isis, Flora, and Valentine now waited for me with Martha, Madonna, and Bratworst, when I slid into the water, and the six of them swirled around me as, with difficulty, I brought the awkward containers of scraps and dumped them. Sometimes I tossed them a treat, but only Martha ever caught on, targeted it, and took it.

Meadowes and Ruffles were the only males to come regularly, and apparently lived in the lagoon, though most males lived on the outer slope of the reef in the ocean. It seemed another expression of the flexibility and individuality of sharks, that some would differ so much from the norm. Another who lived in the lagoon appeared at this time. His dorsal fin tip formed a memorable, pointed W on one side and a V on the other, and I was sure that I had not seen him before, yet my early photos revealed that he had come the very first time I had taken the scraps of Merlin's bonito to the lagoon. Meadowes and Ruffles often came together, but he was never with them. I called him Avogadro. More strange males began appearing, usually when it was almost too dark to see. One large one who visited during this period had come just once months earlier. I called him Kilmeny, after the poem "*Kilmeny*" by Alfred Noyes, since "nobody knew where Kilmeny had been." And the beautiful creature I called Flannery, passed through for the first time.

Keilah attended the sessions for a few weeks before disappearing, while Amaranth came regularly. Once, she soared in preceded by a seven-centimetre (3 in) yellow fish, which swam perpetually in front of her

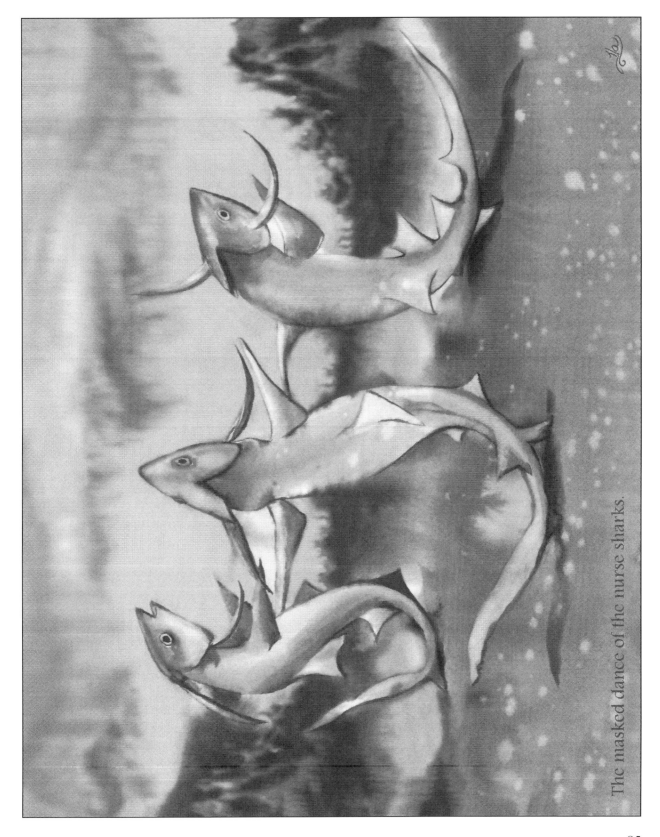

The masked dance of the nurse sharks.

nose in the dusk, a bright, fluttering flag. An odd coincidence occurred the following week, when I went sneaking underwater along the other side of the bay to a submarine enclosure where I had been told that sea turtles were being held. It was thick with swiftly circling fish, but I saw no sea turtles. While I swam back down the bay, the same species of yellow fish flickered in front of my nose too! In all the years I watched the sharks and travelled underwater, that was the only time one accompanied me, though occasionally sharks appeared with these tiny flags.

Once, Carmelle came up to me shaking something, the great muscles of her jaws prominent, her head a third of a metre (over a foot) across. After that she spent many minutes circling around me and coming for a close look, suddenly curious about me. Another female visitor soared through and alarmed me by passing close at a very high speed, though she'd never seen me before. Because of her behaviour and appearance, I called her Samurai. She had a five-centimetre (2 in) jag in her colour line behind her gills on her left side. She would be easy to recognize in the future, at least if I saw that side of her when she passed.

A pregnant visitor who appeared at this time was light grey with a black freckle on her back near her dorsal fin. She was very calm, eating and circling with the others, and she came close enough that I was able to get a good photograph of her. I named her Merrilee. At the same session, I photographed an adult male, also with a black dot near the base of his dorsal fin, and named him Bree. These photographs were unusually clear, and in the following months I often looked at them, wondering about these sharks and why I had never seen them since. Each new visitor was checked to be sure he or she could not be one of them. Then one evening two years later, I was feeding the fish two kilometres away when Merrilee cruised by. She was so calm that I swam around her and drew the other side of her dorsal fin. Remarkably, no sooner had I finished than Bree glided in, providing more evidence that sharks travel together.

Another visitor had a black circle on her white band just off an elegantly curved black tip. I called her Venus, since the configuration was reminiscent of the planet Venus low over a hill at sunset.

Once a tiny shark pup came, possibly one born the previous year. She passed cautiously at a distance until the adults left, then came alone to eat by herself. One of the big females circled back, and when the baby took a fish scrap, she zoomed after her until her nose was even with the baby's side. I watched to see if the little one would drop her morsel, but instead, she shook it like sputtering lightning with such force that she shot straight out through the surface and flew off at the speed of light. I called her Lightning and the big one Storm, impressed at this new evidence that the sharks treated each other the same, irrespective of their size. Storm could have bitten Lightning in half rather than follow her in case she dropped a scrap.

Though sharks give birth to live young, they do not mother them, so a motherly attitude on the part of the big shark could not be responsible for her treating the pup the same way she would treat a shark of any other size. The lack of intra-specific aggression amongst the sharks seemed to hold true amongst all sizes and ages, and there was no sign of a hierarchy based on size playing a part in their society.

Diamond eventually reappeared at the evening sessions, the only shark to have failed to discover my change of schedule quickly. I wondered if it had something to do with her being a juvenile. She had only missed two of the daytime sessions, not consecutively, since she discovered them. Perhaps the smaller sharks withdrew to sheltered regions at nightfall, which could explain why few juveniles came at sunset.

Occasionally a whitetip shark floated through, its colour of deep purple-grey making it nearly invisible in the darkened waters, while its white tips glowed. It gazed back curiously and did not eat.

More and more nurse sharks came to the sessions, undulating lazily around the area in submarine displays of bizarre beauty. They were heavily made sharks, with large, graceful fins, an improbably long tail, and small eyes. They could lie still on the sand without needing to swim forward to breathe as the blackfins did. Plunking themselves down on top of the food, they slowly munched, and sometimes gyrated while the blackfins turned patiently over them, showing no aggression toward these slothful sharks, even when they blocked access to their food. When a few nurse sharks accumulated, their contortions as they tried to extricate the meat from the scraps filled the water with particles, spoiling the visibility.

I would stay watching the alien ballet of the sharks until it was so dark that they faded into the gloaming.

Only Martha learned to target a treat I threw.

They swirled and glided around me, the colour of the twilight water, becoming ghosts, then just movements, and when I couldn't see them at all, I would drift away through the silent evening, darkened rainbow colours still shimmering across the sea. It was a magical time, as detached from my human life as a dream.

After several weeks had passed, I went in the morning, to see who was around. The session was a strange one. I waited for forty-five minutes, and none of the residents, Martha, Madonna, Bratworst, Gwendolyn, Diamond, Flora, Isis, Valentine, Meadowes, Ruffles, or Pontiac, came. Once in a while, a shark passed at the visual limit but didn't approach because there was no group of sharks to join; he or she didn't know me.

Finally, a group of visiting females came gliding in with Martha, who apparently gave them the courage to approach. One of these newcomers circled me curiously while I drew her dorsal fin—it was my first meeting with Marianna. There seemed to be no other sharks in that part of the lagoon.

Another morning session the following week revealed the same mystery. Eventually Ruffles and Isis appeared, moving so languidly that I thought I had never seen such sleepy fish, and after half an hour, tiny Tiddlywink passed by, but did not eat. When an hour had passed, Bratworst came from deeper in the lagoon and passed over the food one time before disappearing, and much later, an enormous female passed, and eventually came close enough for me to recognize Princess. She was with a pregnant shark I didn't know, who joined her after a few cautious passes in the distance, to languidly peruse the food. A few other sharks appeared then and Madeline swooped in, moving fast. I recognized her right away by her no-nonsense manner. But after a brief investigation of the situation, they all drifted away, and finally I did too.

These sessions suggested that there were days when the resident sharks were there, and days when almost no one was around. The reasons remained obscure. At the beginning, I could count on finding Martha, Madonna and Bratworst in the region on nearly every morning visit. Apparently something had changed.

On November 12, 1999, Franck came to see my amazing gathering of sharks. I placed the food while he prepared, and as the excited animals came soaring into the site, he paddled overhead and unceremoniously threw several fish heads in through the surface, badly frightening everyone. The sharks were eating when he swam up to a dead coral structure a metre to my left to watch. They retreated when he first came, but soon returned to swirl around the food, and I pointed out my favourites.

Madonna had been missing for a few sessions. She had never disappeared before, but the population had undergone a seasonal transformation of some sort with so many pregnant females passing through and less juveniles around, so I had speculated that her disappearance was connected with that. With so many strangers on the move, it was becoming a mystery that the local residents remained there so faithfully.

But my pleasure at her return changed to shock when she glided in front of me. Black lacerations covered her from head to tail in a fine latticework, and there were several deep cuts. One, low on her side in front of her pelvic fin, was gaping open. Her pectoral fins were shredded to the leading edge.

She must have had a bad accident! My thoughts flashed to the turtle who had suffered so many lacerations in a net, and it followed that something similar had happened to her. But how could she have escaped? I watched in awe as she circled. She moved abnormally slowly and didn't stay long, but she did eat a bit.

Flora glided slowly over the food, scenting each piece. She found a large fillet, discarded because it had become too old—when fresh it would have sold for about ten dollars to provide a meal for a family. She opened her mouth and inhaled it. Then she picked up a backbone and shook a bite out.

Martha glided into view, so obese with her pregnancy that she appeared huge. Franck picked her out as the biggest shark, though some months before, she had been similar to Madonna. She swam by him very closely, and the two looked at each other. She was the only shark to make a point of going over to see him.

It was a beautiful session—many sharks were excitedly feeding, graceful figures soaring and turning across a stage of sunbeams. As the sun sank and the nurse sharks gathered, Franck was fascinated by the way they moved and fed. The sharks circled, the nurse sharks thrashed and writhed as they nibbled, and a fleet of stingrays flew in and covered the food as they settled on it to eat. They stayed for fifteen minutes. I was happy to have such a good show for Franck. Never before had stingrays come in a group and eaten.

Even Keilah came after a long absence. One two-metre (6.5 ft) nurse shark took a bite of food and languidly cruised around prior to making a decision, it seemed, as to what to do with it. Madeline, her nose caressing the tip of its exaggerated tail, followed at the same languid pace all around the area.

Finally the light grew dim, the excitement faded, and we decided to leave. I was about to take the leap into my boat when I saw that a large moray eel was curled out of its hole in the place I would have taken off from, so I moved the boat and tried again, checking all the time for sharks.

Very worried about Madonna, I returned a few days later, spilling a jar of blood collected from the scraps into the lagoon as I crossed it, to make sure that she was aware I had arrived. The sky was dark with the promise of rain, but the storm moved slowly away to the east, and I got only an occasional rain-drop. The windy conditions paused just long enough to let me locate the place I had used for the previous session.

It was a large, oval, open space in the deepest part of the lagoon, just before the inner edge of the barrier reef arose. There was a dead coral structure with actual hand-holds for stability, in a long, intricate formation forming a wall along the up-current, northerly side. This structure functioned as a barrier that separated the sharks' region from mine.

Looking out across the site from there, to my right was a wide open area for the sharks to pass and turn in. Bordering the western end was a mountain of coral, an ancient structure topped in *turbinaria*, which waved just under the surface. Beyond it, across from me and to the right, was an open region where the scent from the food naturally flowed away—that was where the sharks entered. Directly across from me, about five metres (16 ft) away, was another massive coral formation with a wide canyon between it and the great coral structure that defined the site's eastern edge. It, too, reached the surface and was topped with *turbinaria*, matching the structure on the western edge. These two enormous formations, about ten metres (30 ft) apart, affected the water flow at the surface, so I could see them, spaced just so, as I approached across the lagoon, and thus locate the site. I couldn't mark it for fear of attracting attention to it.

Between the coral wall on the north side, and the eastern formation, was a wide passage through which the sharks passed as they left the site, and to the left of my lookout was a dead coral shelf on which I could stand to brace myself while pulling the shark food into the water. It was much appreciated in bad weather.

It was the best place for the shark sessions that I had discovered, and I began using it each time. Quickly, I thought of it as "the site," and later, when I held sessions in other places, it became "Site One."

Several large sharks were circling as I prepared, that time, in the kayak. The space in front of me was heaped with two garbage bags of fish scraps, and putting on my gear in the cramped space, on a boat that threatened to overturn with every motion, was challenging. Two fish shops were beginning to count on me to take away their scraps, and while there was often none, other times I was given more than I could handle.

Strands of blue sky appeared in the west, and a few beams of sunlight broke through as I put in the food. Though a nurse shark ambled over before I finished, the blackfins waited until I was ready. When they came there were many strange sharks, some of them large and all of them agitated. I was out of breath and unbalanced and just watched for a while. The wind whipped up the waves again which bounced me around so powerfully that it was difficult to draw.

Gwendolyn came for the first time in the evening, suggesting that each shark had his or her own schedule and did not stay in the same area all the time, though Martha, Madonna, and Bratworst seemed to do so. Pontiac, who was a juvenile, rediscovered the changed feeding schedule that evening. The beautiful, bronze Carmelle came and circled until the end, swimming past and around me repeatedly at arm's length. I had seen her only three times, and it was a joy to watch her curious investigations. A new shark called Droplet was there, named for an exaggerated loop at the front of her fin-tip line, that resembled a drop of liquid. A young female with two round bite marks on her back and side circled in the site. It was the first time I had seen a juvenile shark with bite marks and I named her Sparkle. But Madonna did not come.

I was worried about her and had hoped to get some extra food to her. An animal injured over such a large fraction of her body could well be in danger. So I returned as soon as possible with more food and found her waiting for me. Her skin had healed dramatically. Instead of cuts everywhere side by side, there was one about every two or three centimetres (1 in). Her pectoral fins had healed almost completely, with just two V-shaped cuts left on the trailing edges, each about two centimetres (less than 1 in) long, instead of all the way to the leading edge. I wondered how much worse she had been right after her accident, if she could have healed so well in just a week. Madonna swam slowly, but otherwise appeared to have her normal strength, and she ate all she wanted.

I had discussed her injuries with our gardener, my main source of information about marine life at the time. He had told me that sharks chew their way out when caught in a net. It was one of the reasons why people did not like them—nets were costly. I was relieved that her mouth seemed all right—she had not hurt it while chewing her way to freedom.

The usual crowd soared in and ate. Gwendolyn grabbed a scrap in mid-water, dropped by another shark. It disappeared down her throat, then she brought it back up and spat it out. Storm drifted in and stayed, patrolling around the area. She no longer seemed the enormous shark that had followed Lightning when the little one had shot out through the surface. Had I not had an accurate drawing of her dorsal fin on both sides, I would not have recognized her. Puzzling it over and wondering how I could have been so mistaken about her, I realized that she must have had her babies!

A young, golden female swam in with Droplet. She flew around, poking athletically into every nook and cranny where a scrap might have fallen. Such an energetic and radiantly alive animal was a sheer joy to watch. For her vivacity and association with Droplet, I named her Splash.

Venus came with another large female—Emerald appeared for the first time.

At the end, all the remaining pieces of food had been pushed under my lookout, and a big nurse shark was beneath it, eating and thrashing leisurely around. Because of his ability to lie still on the sand, he could get under the overhang and munch on the pieces there, though the blackfins could not. Gwendolyn swiped a piece of backbone away from him, as they all flew around below me, trying to get a bite. As I leaned out to peer over the edge of my coral lookout, Venus passed centimetres below my nose. Finally I dove down to the nurse shark, shooed it away by clapping my hands in front of its nose, and put the food out in the open. But the blackfins did not return.

The next day I was given some chunks of tuna that had become too old to sell, and took them gleefully in search of Madonna. So many sharks were aware of my sessions by then that I knew I wouldn't have enough for everyone, but since I had fed them the night before, I didn't expect many to come. They often seemed to go roaming after a feeding session, and when I went on consecutive nights, fewer sharks came. Those who attended were generally not the ones who had fed the night before and this occasion was no different. Flora, who had come late the previous night after the food was gone, carefully hunted through the chunks of food. She inhaled one of the big ones, then methodically sniffed out more without leaving the site to circle. I had saved a few pieces to make sure that Madonna got one and waited impatiently for her to arrive.

Finally, Madonna came gliding through the gloom with Madeline. I tried to hand her the chunk of food that I had saved for her, but she didn't see it. As she circled around, I put it on the sand, but it was Madeline who found it. She was at least thirty centimetres (12 in) away from it when she saw or smelled it, and then she just opened her mouth, and in it went! Madonna cruised by one more time and then left. No one came back, though I still had a nice chunk of fish for Madonna.

It was interesting to see how poorly they actually located the food. They seemed to know it was around but had trouble finding it. If they located it by sight, and couldn't see well enough to see it, why didn't they use their noses? I was sad that Madonna didn't get anything, but later decided that she may not have been hungry. She hadn't searched intently the way Flora had.

Still worried about Madonna, I went again soon. The wind was so strong on the way out that I was blown backward. With all my strength, I could scarcely hold the kayak into the wind, and repeatedly thought I would capsize in the unpredictable rushes of the gale. Advancing about a centimetre per minute, focused totally on keeping control, I didn't know how long I would be able to go on, and vowed that, in future, I must be more careful and cancel trips in strong winds before I got in trouble.

As I crossed the lagoon, a shark approached the boat at an angle, turned, and left at the same angle in a V-shaped path. It did this twice. It was the right size and colour to be Madonna, but in the turbulent water, I couldn't be sure. Scarcely able to control the boat in the merciless winds, and washed over by waves, I found the site with difficulty and felt ill and badly stressed by the time I slid underwater.

Then, in a twinkling, I was in that golden place, all shadowed in purple and turquoise. As I reached up to start pulling the food into the water, Madonna glided by. It was like a dream to be in such a terrible battle alone on the surface then in a split second in another world of perfect calm and silence with Madonna beside me and fish flitting forward from all directions. The needlefish surrounded me on the surface; rainbow wrasses, butterfly, and angel fish were rising near, the school of yellow perch was swiftly filing over the sand, and two large, long-nosed grey fish waited patiently in the circle of the site.

As the food fell, Madonna swam under me with Bratworst and Martha, as though I weren't there. Flora, too, glided around to see what I brought, but she wasn't so comfortable and accelerated away when I moved strongly. By the time I had finished placing the food, dark torpedoes were shooting into the site from the lagoon, all my familiar sharks. Madonna stayed, often passing close by me, for the first half of the session. She had only a scattering of black lines and marks on the middle part of her body by then, and her pectoral fins were nearly healed except for nicks at the edges.

Suddenly, I looked up and my heart stopped. A creature a metre high was coming through the coral canyons, straight toward me from the other side of the site. Never had I seen such a thing or imagined one could exist and I waited, breathless, as it entered the site. It seemed to be some sort of giant jack fish, like the little fighter planes who used to accompany Merlin, but two metres long, with a huge, extensible mouth. This was one that was definitely not in the reference book!

The creature went straight to the food and began to inhale the fish scraps. When a shark approached, it accelerated after it, whipping around in a circle, whereon the shark turned and chased the giant in a circle in the opposite direction. This went on for several minutes, like a bizarre scene from a science fiction film, though no one was bitten. The huge fish ate everything it wanted, since the sharks were not ready to defend their food from such a monstrous animal. Finally, it vanished into the coral. There seemed to be no end to the secrets held in the shadows of the sea. If you waited, eventually something came!

Madonna cruised by again as I was leaving. It was the first time she had stayed for so long, and returned repeatedly throughout the session. Each time she passed, I called to her. Carmelle and Emerald drifted through, just as it was becoming too dark to see. Carmelle had stayed in the area for a long time on this visit, and I wondered again where she came from.

The little juvenile named Sparkle came late, and poked about in the scraps that were left until darkness obscured the site, one more time.

CHAPTER THIRTEEN

The Mating Season

November passed, and the sessions followed the same general pattern. Droplet, who had come infrequently, disappeared for weeks, then suddenly joined the group of sharks who were waiting when I slid underwater. Madonna's injuries continued to fade, and she became, if anything, more faithful than ever, almost never missing a session. Since I continued to use the same location, more and more fish gathered there, and each time I slid into the water, I was stunned anew by their numbers.

On one occasion I caught sight of a new species of silvery, frilly shark disappearing into the coral. But then it reappeared—it was a tiny blackfin accompanied by four remoras nearly as big as it was. A smaller remora came with me to the kayak when I went to get the food, briefly treated the hull as if it were a shark, then eagerly came to eat from my hands as I crumbled some food for it. The big remoras paid no attention to me but decorated one shark after another as they swept around the site, adding a new dimension of underwater beauty to the grace of the sharks.

When the excitement died down, Avogadro still roamed the site with a fat remora reposing on his back. This remora, his fins folded on his chest, was the picture of indolence with his bloated stomach, like a person flopped on his back on a couch after a big meal. The little fish stuck on well; Avogadro's energetic shaking of his food as he prowled around feeding, didn't bother him one bit. The remoras apparently left their sharks to eat, and when they were full and wanted to relax and digest, they would stick fast to a shark and ride slothfully away.

In early December, Bratworst appeared with bites in several places, small parallel lines, some on her back behind her dorsal fin, and some on her side posterior to that.

As well as the visiting females, increasing numbers of strange males passed through. One had a tiny triangle on the tip of his dorsal fin, very similar to Trinket, the only other shark without a full-sized black tip. I named him Mordred and wondered if this could be Trinket's father.

I became ill in mid-December, but knowing that the sharks were waiting for me and that they would experience a shark version of disappointment if I did not come, I finally set off to see them with a few scraps.

It was the wet season, the season of storms, and while I had been semi-consciously trying to fight off the tropical fever known as *la dengue*, the ocean swell had changed direction, and was coming straight in over the reef from the north, instead of from the southwest. I found a huge surge charging into the lagoon like terrified white horses, their manes flying in the wind. At the same time the east wind swept powerful waves, capped with furls of white, across this current, turning the lagoon into a witch's cauldron. I had trouble finding the site and manoeuvring to anchor. The underwater torrents were so strong that at times I couldn't swim against them, and the visibility was terrible.

Madonna, Bratworst, and Isis were circling, and several others quickly appeared as I struggled to hold the plunging boat and push the shark food out of the well in the back with my other hand. Flora turned closely to see what I had brought, and Isis joined her, but they paid little attention to me, going instead to the scraps pouring away from me. By the time I got all the food in, thirteen sharks were eating excitedly, but I could scarcely see them through the particles raised by the current and their whirling. In the submarine torrents I could hardly keep a grip on my coral lookout, and letting go of it to take notes was impossible.

Isis had several bite marks, parallel cuts, similar to those that Bratworst had recently presented.

When the excitement died down, I tried to gather up the scraps that had spread in the current, and moved to the massive coral structure on the opposite side of the site, from which I could face the current while watching over the sharks. Though down-current, I was out of the scent-flow, and high up. Madonna passed closely one time after the clouds of sand settled. Her terrible wounds had healed into a few black lines, but the gaping slash low on her side was still a heavy black mark. Her abdomen was nice and round when she slowly swam away into the murk, so she had fed well.

Everyone left, except for some little sharks, including a cute one I had newly identified and called Rumcake. They ate excitedly together while needlefish watched me hopefully, waiting for me to feed them. There was one quite close whose delicate silver form had been badly mashed and bitten near its tail. I began shredding some fish meat into the water. The thrilled clouds of fish darted through, and the needlefish picked out the tiny pieces that floated near the surface, while the delicately shaded perch, the colour of white sand with yellow accents and purple trimming, fed on the scraps on the sand.

A pale, almost white spangled emperor (*Lethrinidae nebulosis*), came flitting in. She was forty centimetres (16 in) long and twenty centimetres (8 in) tall, with lacy frills giving her a feminine look. I broke off larger chunks of food, floating them toward her. She took them delicately one at a time. An adult male shark cruised in from the left, raised his nose, inhaled a chunk intended for the frilly fish, and exited right.

Simultaneously, five more male sharks swept in from three directions, and I was riveted by the sight of an extraordinary face shooting toward me at eye level. As he swept by, I could see that a large concave bite had removed the front of his face! After his close inspection, he zoomed around looking for food, but once more came to look at me, granting me a second view of his injury. The front of his face was white and featureless where it had healed over. I called him Mephistopheles, a striking name for a striking shark.

The other males swiftly scouted the site, and Gwendolyn came to undulate along with one of them. The males seemed in a macho mood, moving fast and unpredictably. Then Isis and Valentine returned. I was back at my usual lookout, trying to remain motionless, while the power of the waves moving over from behind me actually pushed the air out of my lungs, and it took all my force to stay balanced in position. I floated the piece of food I had been using to feed the fish toward them, and Isis picked it up.

What a beautiful dark velvet shark she was. I watched her, noting again the bites she had, and for the first time put it together—the bites, the visiting males, the undulations of the females beside the males, and the air of excitement. Could the bite marks be the result of mating? Never before had a gang of strange males appeared like this, all together, and usually the females did not return at the end of a session when they had fed well.

A reproductive season could explain the change in the pattern of the sharks' movements. With the males and females living in different regions, a change in their movements would be required for them to mate.

At the beginning of the following session, Princess swam into the site very fast, with two yellow fish flitting at her nose in perfect formation. It took me some time to confirm her identity, since I remembered her as being enormous, and here, though still a big shark with a heavy build, she looked much smaller. She must have birthed. Apparently, pregnancy caused the whole shark to swell up rather than showing just as a bulging abdomen.

She arrived with a large group of sharks who remained hidden until I had the food in place. But when everything was still, they came from the west, parallel to the reef rather than up the scent flow. They came in a long, staggered formation with an inter-individual distance of about two thirds of a metre (2.5 ft). Apart from Princess, they were the residents, and apparently they had just been waiting, all nineteen of them.

Princess was very flighty, and didn't come close for a long time. Then she passed slowly, very close behind me, as if wanting to reassure herself without me seeing her. Watching her performance out of the corner of my mask, I turned with infinite care and snapped a photograph of her. As the flash went off, she shot away like a bullet, her fish still in place at her nose! She continued to cruise the site but did not approach me again. None of the other sharks had reacted to my new flash. With them was a male with a white crescent through his black tip, named Gabriel. I had only seen him once before, at the beginning of my study, and had wondered where he had been. There were a lot of travellers!

Another big jack fish a metre long darted in, and all the little fish streamed away from it. The fish who came to eat with the sharks were frightened of the big jack fish, whereas they never behaved as if they were frightened of the sharks. They just casually stayed out of range of their mouths. Carmelle was present too, but there was no aggression between the big shark and the big fish.

Amaranth, Storm and Mordred joined the sharks. As Mordred passed for a close look at me, I took his picture. Though he had only passed through once before, his reaction to the flash was to turn straight toward me for a close approach! Each shark really was different.

Madonna did not come that time, but when I returned three days later, she was waiting for me underwater. Her skin was perfect, except for the black line on her belly where the gaping cut had been. It was just under six weeks since she had been injured. Bratworst's cuts were also healed. However, the rest of the resident female sharks were absent. Martha had been missing for three weeks.

Amaranth swept into the site early and stayed for half an hour, long after the others had left. It was the first time I had been with her alone. She was very pale violet, and there was slight damage to the top of her dorsal fin. She, too, had birthed.

Finally, only the nurse sharks remained, languidly writhing around the site amid the flitting fish. There were three nurse sharks about one and a half metres (5 ft) in length who usually came, and one of about two metres (6.5 ft). After the blackfins left, they would scrape and suck out the contents of the fish heads, wriggling all over the site in clouds of sand, wrasses and yellow perch. One of them lay under a coral formation, close beside an enormous Javanese moray eel. The two of them were touching all along their sides, the nurse shark eating, the eel looking calmly out at me. For two species renowned for their aggression and even for being dangerous, the sight was unexpected, enhancing the feeling of being in a community in which a certain camaraderie existed, one whose true qualities no human mind could conceive.

The following week, enormous waves were still pouring over the reef, generating a series of rolling swells that moved slowly across the lagoon. Mercifully, there was no wind. I had trouble locating the site in the turbulence, and when I arrived, I anchored farther away than usual—on a former occasion the waves had washed me over the big coral structure on the western end of the site, nearly overturning the boat.

Once in the water, I mistook the eastern coral structure for the western one, and due to the clouded visibility, I couldn't find the site. Thinking I had mistaken the location, I went looking for another suitable place to use. The fish had come to meet me in their hundreds, including the silvery spangled emperor. They surrounded me in a gigantic multi-species cloud that reached many metres in all directions. No doubt wondering what the matter was, the sharks, including a nurse shark, soon emerged from behind their curtain of blue to join us, as I roamed around like a lost soul, wondering what to do. The spangled emperor swam at my side like a little dog while the twenty-three blackfins spread out through the coral in formation around me. Thus, I cruised through the area until I finally found an appropriate place, noted its direction from the boat, and returned to bring the kayak over, still in my cloud of fish and sharks. Then I realized my mistake, and found the site just seven metres west of the boat.

This time, I had put the food on a plastic bag in the well in the back of the kayak, so was able to flip it quickly into the site to make up for the delay. By the time I gained my lookout, the sharks were all eating, and Martha was sailing into view. She was her old familiar self, having had her babies! Her whole body had been affected by the pregnancy; even her head had lost the massive appearance it had gained. She looked young and slender again.

My search for newborn sharks had shown that the nursery at the western end of the reef in my study area, was not much in use. I had found a group of four little sharks there, which hardly corresponded with the numbers of pregnant females passing through. Yet the wider, better protected, and shallower nursery on the lagoon's eastern end was alive with shark pups. If the sharks travelled three and a half kilometres (two miles) to go there to birth, that would help to explain Martha's three-week-long absence. It seemed impossible to know where the different sharks went to birth or why they chose the sites they did. It was even

Mephistopheles

possible that they went much farther than that, to another part of the island. A shark could circle the island completely in three weeks.

Two new juveniles were there—they had begun to come regularly. Both had wavelike dorsal fin tip lines, and they were the same length and colour. They seemed to be associated with each other, though one was male and the other female. I called them Cochita and Cochito and wondered if they were brother and sister and had been together since they had left the nursery.

Drifting later across the shining waters in the afterglow of the sunset, looking into the enchanted garden below, I heard a distant sound of splashing near the edge of the lagoon. The sound came and went, but finally I could see that there were fish jumping there—black surgeon fish (*Ctenochaetus striatus*). Wondering if a shark was frightening them, I paddled and drifted closer, making a minimum of noise in the water, to where the fish were leaping in scattered places. Soon, under the boat, the water became solid and black with those fish, which normally live distributed evenly through the lagoon. They coiled and milled, and filled the waters from the surface to the floor.

Suddenly, the dorsal fins of blackfin reef sharks appeared, slicing cleanly through the silken surface, three together in front of me, two at the two o'clock position, and another two at ten o'clock, moving in a circular pattern relative to each other. The line connecting them formed an arc about six metres (20 ft) across. They seemed to be herding the fish, and working together to do it, from the deeper water outside the border into the shallows at the edge of the lagoon. Since they rarely swam high enough to show their dorsal fins, it was probable that there were more sharks below. The black fish seemed to take shelter under my drifting boat. Sharks' dorsal fins broke the surface in that circling pattern a few more times, and a few fish jumped, but the kayak passed over the area, and all was quiet. I saw only the blue surface darkening into night with no sign of what was passing beneath. Who can say what sharks are doing when no one is watching? When you watch, they stop.

How interesting to see that my sharks had gone hunting together to supplement their supper. I had recently read about cooperative hunting in dolphins so knew that it was considered a sign of intelligence since it cannot take place without planning for the future, a social network that includes communication among the members, and a spirit of cooperation. However, I cautioned myself that I could not be certain until I had observed the sharks underwater, which I set out to do the following night.

I went to the site first. Madonna, Martha and Bratworst were waiting and for the first time in months, for a few moments I had my three favourite sharks alone. Then more joined them to eat the few scraps I had brought, and when they all dispersed, I drifted silently to the place where I had seen the sharks seeming to herd the surgeon fish the night before.

Again I found the fish jumping, the turbulence in the water, and the passing sharks, their fins slicing the surface as they circled the leaping fish. Sometimes one entered in among them, then lunged ahead. The water streaming out of the lagoon under my boat was opaque with black fish. Two whitetip reef sharks passed like ghosts in the deeper waters downstream. I threw in the anchor, slid into the water, and quietly swam forward. Ahead, a darkness filled the waters from sand to surface, and when I got closer, I found the area filled with black surgeon fish in a natural congregation. To the limits of visibility, they streamed in a dense black cloud. Here and there, a few bunched together and shot to the surface like fireworks. They were spawning! Clouds of spawn flowed away with the current. When it was nearly dark the activity died down, and the surgeon fish disappeared. I saw no sharks attack them.

I had to forget about my theory that the sharks had herded them in communicative cooperation. The importance of observing events underwater rather than making assumptions about them on the surface could not be overestimated.

The next night I returned before sunset, anchored in the deep water outside the border of the lagoon, and finned toward the shallows, where I found the black surgeon fish swimming amid the coral as they usually did. But they seemed restless. The current was very strong in the shallows, and I found a deeper area where it slackened, and it was possible to maintain myself against it. After exploring it, I held onto a dead coral formation to rest and look around. Strangely, the fish collected around me. Approximately two hundred remained under my body and in the surrounding coral, while there were none anywhere else in sight. They were so skittish that, if I moved, they all jumped fifteen centimetres (6 in). Drifting farther, I found that they were filing into the area, one after another, through every opening in the coral from the direction of the lagoon. When I stopped moving, the black fish again collected around me, now in an opaque cloud that obscured the surroundings.

Cochita passed, looking very bored and paying no apparent attention to the fish gathering in greater and greater numbers. Two juvenile blackfins appeared, and a larger pregnant female shark came over to look at me. I was sure I'd never seen her before. Madeline drifted by, gazing at me very curiously. She circled me and came back several times as if she couldn't understand what I was doing there. She had a nasty cut on her left shoulder. Martha also came and circled me but not closer than two metres away. Bratworst approached. Samaria glided infinitely slowly through dense fish, paying no attention to them, and the fish showed no reaction to her either. Two whitetip sharks passed together, down-current of the congregating fish, at the place where the lagoon ended and the sand sloped away into the depths of the bay.

I drifted with the fish as their numbers increased; my environment was opaque and in motion with them. Their solemn faces looked intent as they flitted by, following their intricate ancestral patterns. They had turned a pearly violet colour with rose highlights, still outlined with black on fins and tail. Streams of them flowed away and returned; they would rise slowly toward the surface and re-descend, and all the while a sense of tension grew among them. Eventually, their tendency to collect around me caused the whole event to crystallise right there. They grew increasingly energetic, rising and rushing past each other in twos and in larger groups, only to calm again and sink in a swirling mass until some unseen signal triggered the beginning of their dance.

Suddenly, my environment of rushing black fish erupted in groups that shot to the surface together, then descended. At the moment they flicked the surface, they released tiny pale clouds. I turned and turned, mesmerized. Wherever I looked, fish were streaming and shooting to the surface. I tended to drift, since there was nothing to hold onto, and ended up at the downstream edge of the school, unable to see anything since the water had become opaque with their spawn. Swimming to the side, I found a clearer area from which to watch the spectacle. Behind me the border fell away into the depths of the bay and darkness.

A whitetip shark swam slowly under me and on along the edge of the cloud of spawning fish. He flipped on his back to wriggle for a few seconds in a patch of sand in the reef-flats. Then he languidly returned and cruised under me again.

With the gathering darkness, the fish dispersed. Some half-heartedly swam in little groups toward the surface, but without their former zeal. Another whitetip, a large male, approached to look at me. When he came too close, I raised my camera, and he curved away. I realized it was nearly dark, that I was on the edge of the bay, and that the sharks, including surprise visitors who could have been attracted to the scent and sound of the spawning, could be excited.

So I finned back to my kayak, climbed in, and paddled slowly home as the light of the last day of the second millennium left the sky. Once again came the feeling that the intangible powers of the universe had opened a magic window for me. I drifted, watching the first stars flower in the sky and shine, strewn across the sea, still seeing those pearly fish dancing in my mind. A star fell from the zenith, like a cosmic exclamation upon this mystery of the sea.

The next night, I returned with Franck, and the spawning event replayed once more. Inclined to mathematical computations as he is, Franck estimated that ten to fifteen thousand fish had congregated there.

CHAPTER FOURTEEN

The Season of Storms

To celebrate the new year of 2000, Franck and I went diving at a popular dive site outside the reef, known for the varieties of dolphins, fish and sharks that congregated there. After a long tour in the spectacular coral, we explored for a few minutes in the vicinity of the boat before ascending, looking at the fish and watching the circling blackfin and grey reef sharks. The blackfins were all paler than the lagoon sharks, pale bronze to light gold. It seemed reasonable to conclude that pale gold males passing through the lagoon were visiting from the ocean. Living deeper, they were pale in the same way that a person who stays in the shade is paler than one who spends a lot of time in the sunshine.

The grey reef sharks we saw there were, as always, female! I never learned where the males might be.

While we were roaming there, a sharptooth lemon shark glided up from deeper water. We both paused as he swam slowly and steadily straight toward us. His head moved lazily from side to side with the languid undulations of his three-meter (10 ft) form, and as he came near, we saw that his mouth was open. We were able to see all his teeth with perfect clarity. I was impressed to realize that he had visited a joke shop and replaced his normal teeth with five centimetre (2 in) spikes just to scare us. He was a heavy-set shark, many times larger than the biggest of my familiar sharks, and why was that cavernous mouth open? This seemed a most suspicious sign, particularly by the time he was two metres (6.5 ft) away. Gazing at that huge face with its wide, alien smile, moving back and forth, back and forth, and coming inexorably closer, I felt my body prepare for a fight. . . Then he turned, and just as slowly, he left on a course at a corresponding angle to his approach, back down toward the depths from which he had come. He had been looking to see if we had brought food, we guessed. Both of us were stunned into silence by the unexpected attention of this majestic shark. With such dentition, he could easily have munched on us rather than looked to us for a scrap.

I was awed by the contrast between the way sharks actually behaved, and their awful reputation. It seemed that one could forget everything he or she had ever heard about sharks, start learning about them all over again, and be better off.

Back with my sharks, I found more of them bitten. Bratworst had many bites on her back and sides, Madeline had a few cuts, and Christobel was badly bitten all over, very deeply in places. She had always been one of the most perfectly proportioned and graceful of sharks. Possibly some male sharks thought so too. Her skin was no longer flawless—she was a mess. It seemed a strange way to make love—biting the female to bits in the process. Perhaps Madonna had not been caught in a net after all but had mated!

I had waited late each evening hoping that the shark with the bite out of his nose, Mephistopheles, would return; I was determined to photograph him. The night he had come, the batteries in my camera had suddenly died. A whole series of camera problems had developed, in which the apparatus failed to work the moment there was something incomparable to photograph, as if it were a law of nature.

One evening in early January, I waited for him until it was nearly dark, and afterwards sat resting in the boat, reluctant to leave. I drifted for a while. The breezes flowed down the island as its heights cooled the air, so I was gradually pushed toward the barrier reef. Eventually I paddled back over the site for one last look before going home. But I couldn't find it, and twenty minutes had passed when I finally glided above it and saw dark sharks passing below, silhouetted against the pale sand. One had the concave face of

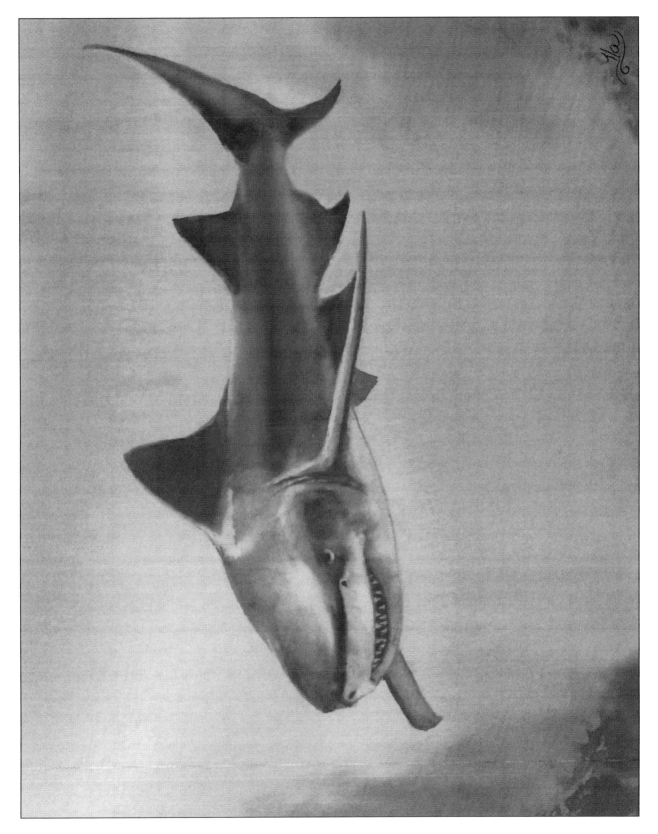

Mephistopheles. He was as skittish and fast moving as I remembered him. He was still in the area!

I returned as soon as I could get more scraps, equipped with a new roll of film and new batteries in my camera, determined to photograph the awesome face of Mephistopheles. The waves were high, sending a powerful surge through the lagoon. Underwater, the waiting fish flowed toward me, and Mordred was passing by, though he did not come back to eat. I placed the food as usual, but no sharks came. Madonna, Bratworst, Martha, Isis, Flora, and Valentine, the six who reliably waited, weren't there.

Finally, Droplet swam in alone, coming from the east rather than up the scent flow, which moved southwest away from me. Christobel appeared, and while I was watching her, a very badly slashed, dark grey female shark swam into the site. She looked unfamiliar, but I wrote down that she was Flora's size. Her injuries were similar to Madonna's—from head to tail she was covered in criss-crossing cuts, many of them deep. She grabbed a scrap and disappeared into the gloom without approaching, so I was unable to draw or even discern her dorsal fin pattern. Ruffles glided into view, soon followed by Meadowes. I had not seen them for more than a month and had assumed they were roaming as other males were doing. They had returned together, so perhaps they had been roaming together. In the middle of the site, Meadowes suddenly turned over on his back and wriggled, rubbing the back of his head in the sand.

Martha, Gwendolyn, Madeline, Storm, Rumcake, and Pontiac arrived during the next half hour and picked through the food. Though there were many chunks of meat, Martha didn't take any, just gave a fish spine a shake. She didn't seem hungry, though she stayed such a long time, moving slowly, disappearing and repeatedly returning. I wondered if she and my other favourites had been busy at the fish spawning site. It made sense that an important source of food in the vicinity would affect the sharks' behaviour at the site.

A few small juveniles were roaming through the region, coming sometimes to pick at the food. One was Muffin, who had tiny budding claspers. Another was Teardrop, a little, golden shark with a pale mark on her back that resembled a water droplet. After everyone left, when I was alone with the nurse sharks, and hoping to see Mephistopheles, a small female shark came swimming into the site. She had come before, arriving late each evening after the rest of the sharks had left. Each time, she scouted energetically, sniffing everything, looking everywhere, staying long, and eating nothing. She had a prominent rectangle protruding from her tip line, and I named her Carrellina, after the French word *carré*, meaning square.

As darkness fell, a massive, pale form appeared off in the coral, weaving in and out of view as it floated cloud-like through the alien landscape, waving an unbelievable tail in slow-motion. It waltzed into the site, fins spread wide as it pressed the water left, then right, as if to an unheard rhapsody. It was the biggest nurse shark I had seen, at least three metres (10 ft) in length, with a body massive as a draft horse. As it floated in slow-motion across the site, a female blackfin appeared suddenly from the left, nearly collided with it, and shot away. The huge shark startled, moved on, and undulated with its beautiful, lazy ballet through the area until I left. Mephistopheles did not come.

No fish scraps were available for the next two weeks due to fishing-boats being in dry-dock or broken down. Each evening I thought of my sharks circling hopefully at the site, and being disappointed as they waited for me to come to our sunset rendezvous. The weather was poor and I could not go to swim with them either.

When I paddled out in late January, squalls were blowing over the ocean to the north, and the wind was frightening in its force and unpredictability. I was obliged to go along the island far past the site, then turn at right angles and paddle frantically to close the distance to the reef while being blown back, broadside. Guiding my trajectory, I managed to be blown over the site and threw the anchor in as accurately as I could so that the boat would stabilize in the right place between the current and the wind.

While I prepared, one shark after another, sometimes in pairs, approached the boat at the surface to within a few centimetres. This was unprecedented. I was afraid that they would have forgotten me, but instead, they seemed much more anxious than usual to reassure themselves that I had come. Many fish came to the kayak at the surface from their usual realm on the sand, which they had never done before. Their behaviour indicated an awareness of time passing.

In the water, two spangled emperors approached me rapidly, both bigger than the white feminine one. I always did a three-hundred-sixty-degree, underwater survey on arrival. Halfway around, I found fourteen sharks gliding slowly by. It had taken Bratworst months to get over her startled reaction when I slid into the water as she passed, and I was incredulous that the sharks, passing together, didn't react at all to my splashy entry. I had the impression that nothing frightened a sea creature more than a splash on the surface. The sharks had learned! The fish were pouring toward me out of the coral as usual, the yellow perch streaming along the bottom, the rest in clouds in mid-water. I wondered what it was like for fish to have to wait.

The first shark who came close was the same badly cut up dark grey female that I had noticed the previous time. It seemed odd that a stranger would come to me like that, but I didn't have time to think about it then—my sharks were hungry. I grabbed the back of the boat, pulled it to the site, and flipped the scraps off the plastic sheet into the water.

The sharks began to eat, soon joined by the nurse sharks, while I focused on the injured, dark female. Even her dorsal fin was slashed. When she turned, a cut and smudged white mark in the middle of her right side became visible—it was Flora, inexplicably turned from light bronze to dark grey!

She had changed so much in colour that I hadn't recognized her. Her dorsal fin looked very different with the soft, pale flares in her white band turned to hard, dark lines, and her mating wounds had initially obscured the prominent, white mark on her side. How remarkable that a shark could change colour so fast. This time the colour change appeared to have a hormonal basis.

Cochito, originally the colour of Cochita, had become a deep purple-grey, while Cochita was the colour of dark chocolate. Most of the juveniles were a rich bronze or rust colour, but since some were grey and some shark pups were born grey, a genetic factor could not be denied.

Repeatedly, Madonna passed me. She had become quite brown, whereas when I first met her she had been dark steel-grey. Once as she passed through a beam of late sunlight, she actually looked golden! Her skin looked mottled in places. Samaria and Sparkle cruised in, both badly cut up, though Sparkle was only a metre long! Even tiny Diamond had a bite on the diamond-shape of pale dots beneath her dorsal fin!

All the food was gone in minutes, but the sharks still circled. Every time I counted, there were thirteen to fifteen. Though they habitually left after eating, this time they were like children who had been made to wait too long to have a party, and couldn't believe it was already over. When there was a quiet moment, I looked around in the coral for scraps that had fallen out of reach but couldn't find any.

For the first time a whitetip shark approached me closely and swam beneath me. Unfortunately the batteries in my camera were again dead, having taken only half the number of photos I got from the first batteries I used. My camera failed to work when there was something exceptional to photograph. The slender shark, with his large eyes and flattened face, gazed wonderingly at me as he drifted by again.

I took one of the fish heads to my lookout and started feeding the fish, who filled the site in a multi-coloured cloud in lively motion. The sharks surrounded me too, maybe hoping there would be something more for them, and when I finally dropped the fish head onto the sand, they took turns trying to get a bite out of it. But they could not extract the meat from the fish heads. With the whitetip shark, they stayed close to me, circling. I was sorry I had brought so little food when they had waited for so long.

As shadows enveloped the scene, the huge nurse shark floated through. The others lay on the floor of the site, each munching on a fish head, wriggling and undulating as they scraped the meat from inside. Once, they were all writhing around the same one, the waters above them alive with feeding fish and orbiting blackfins. At times their pale forms rose vertically until their pennant tails waved at the surface.

When everyone had gone, the tiny shark Carrellina came and poked all around the site by herself.

Three days later, I returned and slid underwater to find myself face to face with the two-metre (6.5 ft) nurse shark. A smaller one rose beside him, and in spite of my sudden plummeting into their realm, neither shark reacted at all, but continued tranquilly floating up toward me, looking. They must have started their approach when my feet passed below the surface, and did not startle when I appeared suddenly so close. Normally, they seemed to pay no attention to me, but they, too, could be curious. They slowly turned and began to swim away. Gwendolyn passed just beyond them, where Samaria, too, was gliding through the silent, slow-motion world. As I turned to put my hand out to pull the kayak to the site, Droplet passed on my other side. Many more sharks were beyond, but this time I didn't count them.

By the time I placed their food, grabbed the camera from the boat, and returned to the site, it was in an uproar. Six nurse sharks were thrashing about in the sand amidst clouds of rushing fish, including the spangled emperors. Madonna, Valentine, Flora, Isis, Cochita, Droplet, Gwendolyn, Samaria, Sparkle, Lightning, Teardrop, Meadowes, Cochito, Amaranth, Splash, Bratworst, now badly bitten, and many visiting males and juveniles were orbiting, snatching scraps from around the nurse sharks and flying around with them. The sneaky new visitor, Carrellina, was there. Suddenly she had arrived at the beginning of the session. Yet, she had never before come in time to eat, always appearing about five minutes after sunset.

Meadowes and Ruffles arrived together, both looking paler. Perhaps they had been spending more time in the ocean during this period when so many sharks were roaming, and had lost some of their suntan.

Lillith came, and the resident sharks remained circling in the area. The usual, immense pregnant visitors passed on the periphery while I waited, poised, for a view of their dorsal fins. Once a big shark approached me head on; Madonna was returning, yet again. She was not interested in eating, and swam very slowly. Gazing at her lovingly, I called to her as she passed, and badly frightened Bratworst, who shot off at the speed of light—the speed too fast to see.

As the sharks searched out pieces of scraps fallen in the coral, the tension built, and they would all go after each piece they found. Once, Ruffles, Meadowes, and some others zoomed off together, passing above and under my body. The scraps were always washed under my coral formation by their swirling as they ate, so later they would come to try to get them out, passing underneath me all around, close enough to touch. This was trust, since they couldn't see me right above them. Mordred arrived as I was hoisting myself into the boat, and swam up to me at the last minute, but it was too dark to stay.

As January ended and February passed into March, the sessions were similar—most of the same sharks attended off and on. The continuing parade of strangers, female and male, did not abate. I recorded the identity of each one, drew their dorsal fins, and copied them into my book. The sharks mated, one by one. In Lillith's case, I saw her in perfect condition, and five days later, she was badly sliced all over, so she had mated in that short time. Her wounds had begun to heal, forming black lines, so she must have mated right after I had last seen her; mating wounds didn't become edged in black until four to five days had passed.

The gorgeous spangled emperors were waiting each time in the site, moving languidly in mid-water. The very pretty one I had first noticed had not come again, and I wondered uncomfortably about fishermen, who would love to kill such large and magnificent creatures.

A striking visitor who came through for the first time during this period, was a pale grey male covered with dark blotches, the pinto of sharks. I called him Dapples. He was the only shark with such markings, so was instantly recognizable. A young female, whose dorsal fin tip formed a rounded W, came to a session at the beginning of February, and attended regularly after that. She was the colour of a doe with satiny, perfect skin. I named her Windy to remind me of the W shape in the line of her fin tip. After that, I named other sharks whose fin tips formed W shapes after winds.

A male shark I named Marco came for the first time. He passed, and I sketched his fin, then looked above the surface to check the sun relative to the time. It was rolling like a red ball upon a dark horizon.

Raphael and Shimmy both came for the first time during this period, along with many others whom I saw more or less frequently in the following years but who have little relevance to this story.

Finding names for the sharks was a challenge, and I kept a list of words I liked and names from books with which to label new sharks when required. I could never have imagined that I would meet so many, and had I not named them, I would have had no way to think about them—they would all have been a haze of differently patterned dorsal fins, pale circles, light dots, black markings, and flying torpedoes with waving tails in my mind. But each name acted like a psychological handle, permitting me to pull the shark and its history out of my memory on sight.

CHAPTER FIFTEEN

Kimberley, Marianna, and Twilight

One evening as night fell, a large, charcoal-coloured female blackfin began passing off in the shadows. After many minutes, she glided out over the pale sand. She had a prominent white spot on each side of her head, glowing in the gloom, and moving back and forth as she swam. In time, she came close enough for me to sketch her dorsal fin. I named her Kimberley.

Kimberley appeared again at dusk soon afterwards. She looked very old. She wasn't too shy, and I was able to see that the white marks on her head were natural. They were about two centimetres (0.8 in) across, irregularly but cleanly outlined, and perfectly white, as if someone had pasted twin snowflakes on her. Her dorsal fin pattern was a complex tangle of circular shapes, and her skin was mottled as though it had been badly hurt and healed many times. The boundaries of her colour scheme were frayed looking but sharply contrasting, as if the curlicues making up her colour lines had been darkened with a felt pen above, and lightened below, with shiny white paint. She was the most remarkable looking individual I had yet met with her white decorations; she circulated through the shadows socializing.

Another large female shark, pearl grey, was with her that time. She had mated in the past day or two and was covered with nasty bites and slashes. She had a tendency to circle always in the same direction, so I waited for her to show her other side. It was nearly too dark to see anything. The blackfins had congregated around some nurse sharks who were tearing up a fish head beyond the site, so I drifted over. A new female, Jessica was there, and she appeared frequently at dusk from then on. Emerald passed me, and more big females were passing beyond. One appeared at my side, and I searched for an identifying feature. She was pale and brown. . . My eye fell on the black dot above her left pectoral fin. Madonna! I called to her, and she circled very slowly, looking at me. Bratworst and Martha appeared and circled me too, before moving on. All three had come back together to join this late evening gathering! Several other residents had also returned, apparently for social reasons. Finally, the shark who had just mated passed again showing her other side. I had seen her some weeks before, roaming in the lagoon with some visitors—it was Marianna.

After an hour and a half underwater, I had to battle a gale, and white-capped waves from the south, for fifteen minutes as I struggled homeward. Then utter calm fell. Three hours later, nearly hurricane-force winds blasted the house where Franck and I were relaxing after dinner. It was another uncomfortable reminder of the fickle unpredictability of the sea.

Soon after, the fish shop cleaned out its freezer and gave me a big load of scraps. So as the sun lunged toward the sea, I prepared to take it to the sharks, thrilled at the possibility of seeing Kimberley again.

The day had been beautiful with just a bit of wind at midday, but as I started to get ready, high winds began to blow. They subsided before I left, but beyond the shelter of the bay I could see a dark storm lowering beyond the island's central mountain, and raging toward me across the sea. I could scarcely keep the heavily laden boat from being blown broadside and paddled hard to maintain control. All my strength was barely enough to stay in one place and progress seemed impossible. I seemed to be interminably paddling into frothy grey waves as time passed and the day darkened, frequently coming close to losing my grip on the paddle as powerful gusts threatened to blow it out of my hands. I planned to anchor the moment I reached shallow water to rest, then proceed by swimming, though trailing such a heavy boat against the current could be worse. But, when I finally gained the pale waters of the lagoon, the strength of the wind had abated just enough to go on. Knowing that I could throw in the anchor if I had to, helped my state of mind, and before the winds increased again, I was able to pull ahead far enough to turn broadside to the wind, and paddle on the western side only, to keep the kayak straight. Paddling fast toward the reef while being blown westward, I arrived at the site and threw down the anchor.

But sharks were circling in the agitated water, and even the nurse sharks approached. When I slid through the surface from the overwhelming storm into that silent world, I was struck again with the unreality of the underwater scene. The second it took to pass from one world to the other enhanced its dreamlike quality. One moment I was alone, fighting the natural elements, the next, surrounded by large and familiar friends, the whole environment alive with multi-coloured creatures flying toward me through the peaceful coral landscape. Martha and Madonna glided by together, and I called to them as I pulled the boat into place. As far as I could see in the cloudy light, sharks were moving toward me in slow-motion. The spangled emperors hovered in the site where Bratworst slowly circled, waiting. I swept the scraps from the back of the boat as they swirled around me and swam to my lookout, passing above several strange sharks.

Gwendolyn arrived, badly bitten all over in the past two or three days, with a yellow fish, vertically striped in black, swimming beside her dorsal fin. The fish didn't change its location as she circled and fed, and was still in place when she departed much later. Emerald glided in, and swept around the food with the others. A remora with a full stomach who had been sticking to one of the large nurse sharks, switched to her as she passed, and left with her, as the group that had come at the beginning dispersed. A few visiting males roamed in with some juveniles, and a tiny pup who had hovered on the periphery came to eat in the lull.

Marianna came. Her countless mating wounds were healing nicely. When Venus arrived, the fat remora who had left with Emerald fifteen minutes before, was reclining on her head, looking just as stupefied as it had when it left! Apparently, Venus and Emerald had passed close enough to each other during that time for the remora to have changed sharks! Was this how remoras changed direction, like a person changing planes? Why had it aroused itself and switched? Venus had mated, too; she had been absent for two months. Kimberley came as darkness fell, and didn't stay long. She picked up some of the pieces off in the coral that the other sharks hadn't taken, and left.

A whitetip reef shark suddenly materialized, blending into the dark. After roaming the area, he swam into a hole under the big coral structure to the west of the site where the nurse sharks often retired to eat. A few minutes later he emerged and glided away. I had never seen a visiting whitetip eat.

All was quiet, and the rainbow wrasses began agitating, whipping around in the water just in front of my eyes. They were concentrated in an oval volume there, with countless multitudes around and beyond them, their density decreasing with distance. I watched them curiously, looking for a piece of fish that they were eating. But there was nothing there, and I finally realized that they were waiting for me to feed them. The surface was glimmering with the constantly circling needlefish too, looking at me gravely with their big eyes. Hundreds flashed through the silvery surface lights while some circled solemnly up-current, watching.

So I found a fish head, and began shredding the meat into the water. Marianna came circling, gradually approaching. The next time I looked up, there were three large sharks level with me at the surface. One passed centimetres in front of my eyes. Her dorsal fin pattern was very rough with the edge formed of circles and dots, like Kimberley's. She didn't come again, but her appearance was unforgettable. It was my first meeting with Twilight. Jessica glided up the scent flow through the swarming, feeding fish. I handed her the fish head, but she didn't take it. The females were with some visiting males, who were passing off in the darkness. I was watching one shark after another, my peripheral vision blocked by the mask, so I didn't see exactly what happened, but it one of these females had a near collision with a male who had grabbed something to eat and both of them passed too close and too fast; the male brushed against me as he flew by.

The pale, three-metre nurse shark came ghostlike through the coral and swirled around the site in his slow-motion drift. For the first time, I saw tiny claspers behind his ventral fins. They were not easy to see on nurse sharks. The big one was a male, not a female.

I swam to the boat in my cloud of needlefish and wrasses, put in the slate and camera, then went to pick

Marianna

up the anchor, still surrounded by fish. Drifting away in the kayak, I could see them following briefly, and was touched by their behaviour. I had had no such experience with terrestrial animals—I had fed the birds all my life, but they did not fly down around my shoulders when I went outside. Yet that was what these exquisite birds of the sea were doing. No matter how long I lived, I knew I would never forget their trust, their complicity, and our secret rendezvous at sunset under the silver surface. The clouds blew apart and left the night perfectly still, a quarter moon and stars brightening, as the glow faded from the western sky.

It had troubled me that people passing might be met by my enormous congregation of creatures at the site and set out to kill them. So the following morning I returned, left the boat inside the lagoon's border, and swam there. A few rainbow wrasses, and the odd individual of other species, flitted toward me when I arrived, and the wrasses accompanied me as I looked around. A shark circled at the limits of visibility.

Then I paddled the kayak over, anchored as I would for a feeding session, slid underwater, and looked around again. A nurse shark was sleeping under the big coral formation to the west, but it was unlikely that anyone casually passing would notice it. No sharks came. Drifting and watching, I was increasingly satisfied that a hunter could pass there and see nothing unusual, and that anchoring boats would not be greeted by twenty-one circling sharks and clouds of fish. There was no sign of the wild parties held there at night. Looking around through the empty landscape, it was hard to believe myself that so many huge predators could materialize there. The small wrasses who came toward me in the site itself would not attract attention or strike anyone as being odd. Wrasses often follow people, and perhaps these fish would not respond to a different person. Merlin's fish had seemed to specifically distinguish me as an individual.

Even the remains of the sharks' food had disappeared. There were no backbones or remains at the site, only the odd vertebra lying in the coral around it. I picked them up and put them back in the kayak.

Kimberley and Marianna attended the next sessions, always as darkness fell, and not together. They didn't come every time, but often enough that I knew they were staying in the area. Emerald, who had also mated two or three weeks before, was still present too. Arcangela, pale and majestic, came, and I named her after Galileo's daughter. She was waiting with the others on one occasion, but soon left the area. A tiny male appeared in the region and stayed. He was a rich reddish brown colour, and I named him Tamarack.

Late in February 2000, the little Trinket glided in. She had been absent for five months, and her last visit had been after an absence of two months. She passed closely, then startled to see how intently I was watching her. She was the only very small shark who appeared so sporadically. The juvenile whose curious circling around me a year before, had resulted in an accurate recording of her unusual dorsal fin pattern, reappeared too. She had grown and was beginning to acquire the rounded look of a female shark.

A large stingray with a broken tail often roamed the area, rummaging up clouds of sand, landing briefly in the site, then flying gracefully off in another direction. Sometimes he nibbled on a scrap, and sometimes he carried something away with him. Often I could see him in the distance, poking around.

There came a session when Marianna was waiting for me when I slid underwater, and for once she fed with the others. It was good to see such a badly injured animal healing up. Since she had been in the area, she had come regularly, but always at dusk when there was virtually no food left. But the next time, she didn't come until everyone had gone, and I was feeding the fish. Suddenly, she appeared from out of their clouds, and swam straight up to me. I put the food down and drifted for a while. She came again from the shadows, passing me at arm's length with that momentous ease of the big sharks. I had no food left for her. Her skin was a network of scars, but only her deepest mating wounds remained unhealed.

Five days later I was able to return, and saw something odd on the surface as I paddled across the gulf between the bay and the lagoon. It was flat and white, yet looked substantial, with small, lapping waves breaking around it. Though anxious to get to the sharks as soon as possible, with the sun sinking past twenty degrees above the horizon, I went to investigate. The whiteness turned out to be the throat and upper abdomen of a pale grey shark, dead. Her head was thrown back, her fins extended behind her, and her tail pointed straight into the depths. Gas expanding in her abdominal cavity had brought her stomach area to the surface, which was badly eroded by the waves lapping over it. Her gills were coming apart, and I could see traces of damage on her skin—the familiar latticework of old mating wounds—but no sign of the cause of her death. Her mouth was closed and perfect. I circled around and around the body looking at it from different angles. It was badly bloated and deteriorating. There was a five-centimetre (2 in) long cut through the body wall, but I couldn't be sure that it hadn't happened after death. On the other hand, something must have killed her. My mind flashed back to the stab-like injury Gwendolyn had suffered. I had to pull up her dorsal fin to be able to see it, which was very hard. She was incredibly heavy, threatening to capsize the kayak. But it was Marianna. Her tip line was unmistakeable. She must have died in the lagoon near where I had last seen her, and been carried out into the bay by the current, right into my path.

Nothing had bitten her or fed on her. The sharks, so delighted to devour as much fish as they could get, had not tried to eat one of their own. Her body was so decayed that bringing her home was out of the question. I didn't even want to touch her again because my sharks would smell her on me. Judging from the rate of breakdown of a small dead whale I had once found, I believed she had died approximately forty-eight hours before. Cadavers rot rapidly in tropical seas.

I got to the site, got ready, got in, and apart from the fish, there was no one! I had a sudden horrible thought that there had been a great massacre of sharks, of which I had found only a fragment of evidence. Soon Madonna appeared, cruised over the food, and passed on into the coral; she was not interested in eating. Later, she came back to nibble. She ate slowly, moved slowly, and didn't seem hungry at all. Minutes later, Martha arrived, and picked disinterestedly over the food. One at a time, over the next minutes, five more sharks, including Carrellina, glided in, looked around, and left. No one seemed interested, and for the first time in months, I was alone.

I was sure that this peculiar behaviour was connected with Marianna's death, far fetched though it seemed. Odd stories I had heard about sharks fleeing when one of them was fished, haunted my thoughts. Perhaps the hole in Marianna's side had been made by a spear fisherman, and her distress had communicated to the other sharks before she died. Logic argued against this, since Marianna was one of those big, shy females who simply would never show herself to a spear fisherman, never mind come close enough to get stabbed or shot. Her death, and the strange feeding session that followed it, troubled me. Marianna had seemed perfectly well five days before, and couldn't have become ill and died in such a short time, I was sure. But her death remained a mystery. It had been such an improbable coincidence that I had found her body that it served as a reminder that some of the sharks who disappeared may not just have moved on. Marianna had been a visitor in my area, and had she simply stopped coming to my feeding sessions, I would have assumed that she had returned to her own home range. At least in her case, I knew she had died.

I asked around and was told that no one would have speared a shark. Nor were the sharks fished by the Polynesians, who had never wanted their sharks either fished or disturbed. But I recalled our French neighbours at our former house and their drive to kill anything and everything that they could. Other whites, too, were voracious shark killers. Our landlords were renowned for it and had left many old shark hooks lying around the property when they had rented it to us. So even if the Polynesian fishermen didn't kill sharks, there were others who did. Certainly, no one else had developed my attitude to them.

Kimberley left when Marianna died, but Twilight appeared again at dusk. She passed just once, then off in the distance, I could dimly glimpse her prowling. Patience and a liberal scattering of fish meat were rewarded, and she glided closer. It was impressive how much she resembled Kimberley. Her black tip line was full of circles, some of them floating free in her white band like Kimberley's. She turned and showed a left side nearly identical to Kimberley's right, and when I glimpsed a faint white snowflake on her head, I wondered if it was Kimberley after all, her snowflakes being injuries of some sort which were healing.

At home, my drawings of Kimberley's fin confirmed that the two sharks had virtually the same tip shapes on the opposite sides of their dorsal fins, so she was, indeed, a different shark. However, the resemblance between these unusual sharks was startling and I wondered if they were sisters. Though their visits to the area had overlapped, I had never seen them together. Yet they had been in the area during the same period. I was convinced that there was a connection between them.

Puzzles

As March ended, the wonderful new camera began to leak and had to be returned to the manufacturer in America. Eventually, I bought a disposable underwater one, but it lacked a flash, so was useless after the sun lost the altitude to pierce the surface, and most of my interesting visitors came at sunset or afterwards.

The lagoon was agitated all that month, and the current was worse than it had ever been. The visibility was often so bad that I couldn't see across the site, particularly when there were thick clouds in the west. I had the impression that the sharks did not like to exert the extra energy necessary to manoeuvre through the coral in the underwater torrents, and retreated to deeper water when conditions were very bad. Sometimes, it was all I could do to cling to my perch—letting go to take notes or draw was impossible.

Many of the sharks were growing darker in colour. Sparkle became nearly charcoal. Perhaps their colour changes had a seasonal component too. Yet, Chimera, who had visited the year before, reappeared, and she was still an unusually light shade of grey. Grace passed for the second time, also very pale.

Among the sharks, I caught the odd glimpse of a dorsal fin with a protrusion at its tip, and finally singled out a young female of a rich, rusty colour. Her tip lines were matching convex arcs. She had an easily identifiable configuration of three light dots on her right side under her notched dorsal fin. I called her Vixen for her foxy colour. A young female who resembled Sparkle became a regular attendant, and I called her Sparkle Too. The juvenile, Chevron, named for the shape of her white band, also came again, and one day, Fawn flitted in. He had visited on only three occasions seven months before. That some of the juveniles disappeared for long periods, then reappeared, reinforced the impression that they roamed widely while young—indeed, it was later discovered that some were travelling between islands! Amaranth reappeared at the beginning of April; she had mated two to three weeks before. Gabriel cruised through again.

Cochito arrived after an absence with a long, vertical cut from near his dorsal ridge to the curve of his ventral surface, just behind his gills. It healed quickly to a black line and finally to a white scar that gradually faded. He and Cochita were maturing and were not together as much. Cochita had acquired the look of a rounded, female shark, while Cochito was a slender, fast, and muscular male. The two had come to the area at the time they would be due to settle in a home range, during the year prior to maturity.

The stingray who had been coming sporadically still often appeared, floating in with increasing confidence like an improbable flying saucer, feeding, and taking occasional pieces of food away.

I was working at the time as an assistant for Dr. Michael Poole. As a marine biologist studying the local spinner dolphins and humpback whales, he conducted dolphin and whale watching tours around the island. On the days that I did two tours, I was late to see the sharks. One evening, strong winds slowed me down so much that the sun was already sinking below the horizon when, with difficulty, I found the site. Failing to pay proper attention, the rope flipped my snorkel into the lagoon when I threw in the anchor, so I had to put on my gear in the water.

The fish had gone to sleep, and Madeline glided in with a companion I didn't know. The two swam within arm's length many times while I sketched her dorsal fin. She was the lovely, pale violet colour of Amaranth, and I called her Marilyn. Samurai arrived with Madonna. She had grown so much since our meeting months before, that if it hadn't been for the inverted V on her colour line, I wouldn't have known her.

Many familiar sharks came to eat, roam together socializing, and evaporate into the dark, while the nurse sharks writhed lazily, pulling meat off the scraps. A small one inhaled a chunk of food with the volume and general size of a stack of magazines three centimetres (1.25 in) thick. I was awed by the volumes of food they could inhale. No wonder their bellies were so fat! The three-metre nurse shark was there, rolling around on the floor of the site, and his magnitude was stunning. But night fell swiftly, and I had to leave.

Morning excursions had revealed few sharks in the vicinity, and sometimes, I could find no one. One morning when I took some food, I had to wait ten minutes to see Shimmy pass on the far side of the site. He returned many times before he snatched up something to eat. Once, a larger shark followed him close to the site but didn't approach; there was no group, and each cautious individual stayed back. Later, Ruffles and Mordred came and ate avidly, circling and returning several times. Shimmy passed far off, then came to take a piece of food while the other two males were there. A strange male cruised through, and eventually Pontiac came briefly. All males!

After a long wait, I got back in the kayak to warm up in the sun. But after fifteen minutes a shark passed under the boat and I went underwater. Chevron had come to look around. She was the only female I saw that morning, but she was a juvenile who was not a resident. After another hour, I went home.

This was the first time I had found only males in the area. Where were the females? Could they have all gone roaming together? Nearly all the resident males had appeared. I was reminded too, that though sharks were often in the same place at the same time from one day to the next, this was by no means predictable. Their schedules were flexible. But since I was working and caring for sick, injured, and orphaned wild birds, I was not free to investigate the activities of the sharks as intensively as I would have liked.

Two days later, I returned at sunset. The sea was the colour of a black pearl, reflecting swirling storms to the north, and the kayak moved restlessly to the surge of a darkly troubled ocean. My landmarks had disappeared among the waves. Roaming around looking for the site, and peering into the dimness below, something appeared on the surface and I focused on Martha's dorsal fin! She was swimming with the boat, no doubt wondering what was the matter with me. We were just past the site, and as I manoeuvred to anchor, the fish were so excited that they were splashing through the surface. Underwater, Martha appeared at my side. She was covered with mating bites. I had been wondering if she would mate this season—she had recently birthed. She did so, but very late. She had been absent from her range for two weeks to mate.

One of the spangled emperors was in the site, and the water was opaque with fish. Sharks and fish swam over as I pulled the boat into place, and flipped in their food. Martha cruised around me as the food rained down from the surface, taking a drifting piece centimetres from my body. She was such a quiet shark, calm and so perfectly at ease, that I rarely saw her accelerate. Madonna, Bratworst, Isis, Gwendolyn, Carrellina, Avogadro, Tamarack, Pontiac, Meadowes, Windy, Cochita, Shimmy, and Rumcake were circling.

Droplet, Storm, Amaranth, Christobel, Cochito, Diamond, and Chevron soon joined them with Arcangela, a pale beauty in the gathering gloom. She circled briefly before disappearing with another female, who had stayed beyond the coral on the far side of the site, too shy to approach to eat. Trinket appeared among the circling sharks and began trying to grab a fish head nearby. She got her teeth in the chunky part, shook it, then orbited it in tiny circles. She left, returned, and ate one of the big chunks of meat overlooked by the others. I watched it go down to create a large bulge in her abdomen. She looked no different than she had when I had met her, nearly a year before.

The stingray came again. It had become used to me and even swam under me as I drifted, sketching.

I fed the fish as Meadowes circled. He seemed hungry and irritated by the fish. As he spiralled in toward the food, his nose followed the tail of one fish after another as though thinking of snapping one up. Then he tried to grasp a fish head, got the fin in his mouth, shook it, circled, and tried again.

There were nine nurse sharks, three of them about two metres long, forming a churning carpet that almost covered the floor of the site. Two of the huge Javanese moray eels were in my dead coral structure, one looking out of a hole beneath my body. By then I was cold, there were black clouds over the setting sun, and the view was becoming impenetrable. Jessica began to prowl around as she did each night when darkness thickened. Nightwind and Sybyl roamed in, and were still passing in the darkness with Jessica when I left.

As I paddled away, the three-metre nurse shark was flying through the coral toward the site with energetic horizontal undulations; I had no idea he could move so fast!

Two young females moved into the area at this time. One was chocolate coloured, named Filoh, and the other was golden with a W-type tip line, who became Mara'amu, the Tahitian word for the south wind. Two light-coloured males appeared then too. One was Peri, who stayed in the region for long periods, interspersed by long absences. The other was Jem, who passed through regularly, like Avogadro. The males didn't seem to be as settled as the females and were always on the go.

Beautiful Arcangela remained in the area for a few weeks.

Lightning suddenly developed tiny, budding claspers. I hadn't realized that the males weren't born with them, and grew them later, and had wondered why all the small pups were female!

I had been keeping track of every shark, since each was a part of the shark life of the lagoon I was trying to understand. So when Valentine, Splash and Venus, who had been missing for many weeks, all reappeared at a session two weeks later, in mid-April, I was pleased. This turned out to correspond to the end of the reproductive season. In my first year of observation, I had identified a hundred and fifty sharks, most of them visitors, since I first wrote down the names of Madonna, Bratworst, and Martha.

At the beginning of May, I glided out on a supernaturally silent morning, across a glassy sea. Since I no longer fed the sharks at that time, those I found would be there through habit, rather than because they were waiting for me, and I wanted to see if the females were around after their odd disappearance on the previous occasion. Some were wandering through the coral near the site, and when I slid into the water, Bratworst and young Flora were passing. Excited fish were gathering, and as she moved across the site, Flora, with a lightning gesture, suddenly snapped one up! It was my first observation of predation by one of my sharks.

As I placed the food, Madonna, Windy, Valentine, Isis, Samaria, Martha, Meadowes, Sparkle Too, and Ruffles languidly entered the site from different directions. Soon Madeline swept in. She and Valentine chose a scrap together, nosing around and over each other and the food. They moved so slowly, compared to their performance at night, that it was possible to analyse their motion: they turned by curving their bodies in the direction they wanted to go, and pushing against the water with their tails. Pontiac sailed in, and Trinket came, hurrying up the scent flow. They fed and glided through the waving shafts of sunlight.

Trinket ate well, and the tiny shark had a very swollen abdomen when she left. All the long-term regulars came. The session had been earlier than the previous one, when no females had been present, so perhaps they went elsewhere in the middle of the day. They lived in such a different world, and it was so hard to venture out to see them, that learning what they did seemed at times to be an insurmountable problem. But it was intriguing that their lives were so much more complicated than I had speculated, that they seemed to decide to roam sometimes and other times stayed home, that they had social lives and ways to stay in touch with each other. There was always something new, something curious, funny, or startling.

Some weeks later, I arrived at the site an hour earlier than usual for the evening feeding, and the sharks who came were the same males who had attended on the morning that no females had come. They seemed to have stayed in the area while the females had left for the day and not yet returned. Gwendolyn and Bratworst came later, and eventually two large sharks approached, swimming close together, side by side. It was Martha and Madonna. Later, Amaranth, who had been absent for more than a month, appeared.

The parade of visitors ended in May. Though a handful of older juveniles such as Filoh and Mara'amu, had joined us, only the residents now attended the sessions. None of the visitors had stayed due to the food. Carrellina, the little shark who had come in December and attended every session since, had vanished.

There was a curious session in early June. It began as usual, except for the presence of an unknown female waiting amongst the residents when I slid underwater. After the first group of sharks left, I retrieved the scraps that had drifted under the coral, and placed them back on the sand. Droplet swept down to them as I rose to return to my lookout. She took a long strip and swallowed it, bit by bit, swimming around with

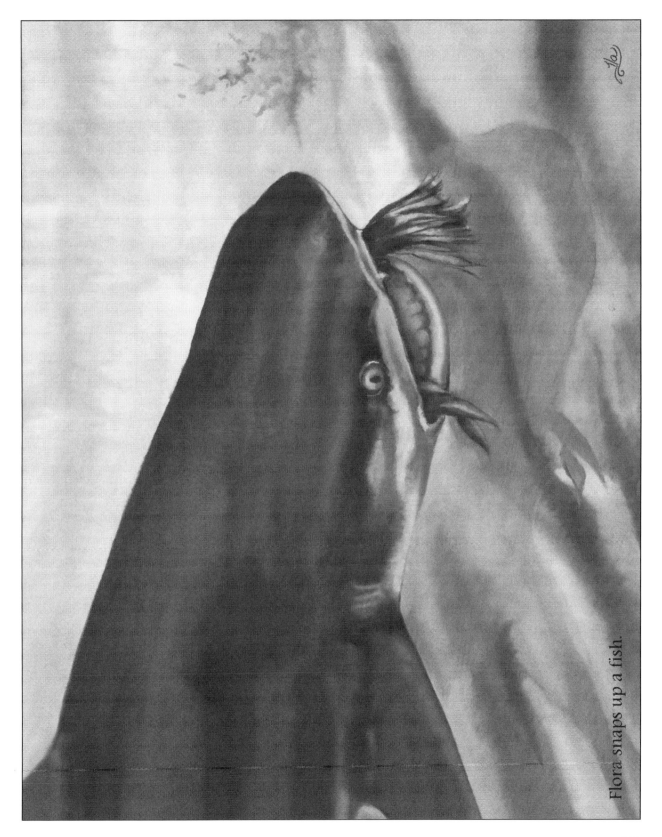

Flora snaps up a fish.

it trailing from her mouth. Her usual companion, Splash, glided in after her, followed a few minutes later by Venus. The three had again been absent since mid-April.

The little shark pups, Trinket, Sybyl, Nightwind, Fawn, and Rumcake, were taking advantage of the lull to devour the scraps, shaking bites from the backbones, and gulping down the other bits. The core of their food had been frozen and was still lying in a chunk. When they found it after an amazing delay, there was wild excitement amongst these little ones, who took turns taking bites out of it until it was gone.

Martha, Valentine, and a few more residents drifted through, leisurely picked at the few remaining fish spines, and circled the region as night gathered. I went around picking up the scattered scraps, accompanied by my entourage of wrasses, perch, goat fish, long nosed butterfly fish, and a striking, yellow trumpet fish who had been staying close beside me at my lookout during recent sessions. Suddenly, one of my pretty rainbow wrasses was grabbed by a lizard fish while I was swimming with them above the sand! The others flitted rapidly around their stricken comrade like birds do, as he was slowly swallowed.

Then, a shadow with twin snowflakes glowing in the gloom, drifted toward me from across the site, passed, circled, and floated above the scraps. Kimberley! Twilight followed. They were together! They cruised the area with the rest of the mature females, their wide circling patterns keeping them in the vicinity and bringing them back into contact time after time. I could see how they could move slowly through an area, staying in loose contact like that, as endless orbiting brought them repeatedly across each other's scent trail. Their swimming patterns resembled those of the two shark pups who had played in front of our house, as though demonstrating how sharks stay in touch.

Marco, a wanderer like his namesake Marco Polo, came as darkness fell. He approached precisely as he had when I had met him, triggering the memory of how the sun had rested like a glowing ball on the horizon. I raised my head to look at it above the surface, and it lay again, a fierce hot red, upon the horizon!

A new female soared through the site just before I left. She was very large, very dark, and on her right side, her white band was almost completely filled in with dark lines. She returned one time to pick something from amongst the remaining bones and shot away. I named her Eclipse.

It was curious that Droplet and Splash, as well as Twilight and Kimberley, two pairs of companions, had suddenly reappeared after a long absence, with the rare visitors Marco, Venus, and Eclipse. Thrilled, too, that my speculations about an association between Twilight and Kimberley had been so unexpectedly confirmed, I scrounged up more scraps and returned the following evening.

Putting the food among the thrilled fish, I waited, but eight minutes passed before a shark came, though I was there at the same time as the previous night. It was the first time that this had happened when I had come on time. Valentine, Martha, and the others who had appeared at dusk had not eaten the night before; their gathering had been social, yet except for Martha, none of the sharks who usually met me—Madonna, Isis, Bratworst, Flora, Valentine, Gwendolyn or Samaria—ever came, though it was dark when I left. Again, the impression that these females travelled together was reinforced. There had been other times that residents had left after a session, but never was it so clear as in this instance.

Could the resident females have accompanied the visitors when they departed?

Sparkle Too drifted in but didn't eat. Avogadro picked up a scrap as he passed through. Finally, the usual juveniles came in together and fed. I began scattering crumbs for the expectant fish. There were so many needlefish that I couldn't see anything else, and were so excited that one of them got caught in my hair.

While I was occupied trying to free it, a large, dark female glided very slowly through the site. I barely glimpsed the line of her dorsal fin tip—straight with an angle upward at the leading edge. She exited right stage into the darkening waters, and did not return. No one did. I was there alone with the fish and the pile of scraps, still drifting its plume of blood and fluids, lying on the sand.

There had been chunks of fish meat among the spines and bits of skin that time, and I was about to collect them when Madeline swept in and cruised slowly above the food, sniffing. She chose a piece of skin! The sharks seemed poor at recognizing the chunks of flesh, which actually contained some sustenance. I found a chunk of fish meat and threw it to her when she returned, but she did not notice. It fell to the sand

and was approached by one of the nurse sharks. I had already been anguished to see them inhale several of these precious pieces, so dove down, shooed it away, got the chunk, and threw it again to Madeline. But once more, she failed to see it. It fell to the sand, and after she passed, the nurse shark returned, went unerringly to it, and ate it. Maybe I was underestimating these blimp-like animals.

Martha had begun passing without approaching. After a very long time, she glided to the food, located and took a fine chunk of meat, and swallowed it in several gulps as she exited the site. She didn't come back. The sharks dispersed in a bored fashion, and I recovered the chunks of fish meat and put them back in the boat, stirring up scent as I did so. No one returned. It was dark, and the first stars were already appearing and glimmering in the lagoon as I drifted away homeward.

But the following week, everyone was back, and the session passed as usual. Off in the coral, a shark moved with exaggerated, horizontal undulations. When she drew near, I saw that she was deformed! Her tail waved high above her back, and her head was lowered. Her spine was vertically curved in a mild S-shape, whereas a normal shark's body is as straight as an arrow. Yet she was the size of Madonna, so she had grown successfully and matured in spite of the extra energy she had to expend to travel and hunt.

Droplet soared in and took a swipe at little Filoh as she left the site with her scrap; Splash zoomed behind her. Splash never appeared in the absence of Droplet, though Droplet often came unaccompanied. Splash acrobatically soared around in the coral, looking for leftovers, and passed close by me for the first time. The slender, golden shark was the image of grace, and a delight to watch. I checked Filoh later, and she was unhurt, so I assumed that Droplet had just tried to snatch away her trailing scrap of food.

At last Kimberley emerged from the deepening blue beyond the site, and I reached for one of the fish chunks in the kayak. She was on the far side of the site when I tossed it close enough for her to see, but not so close that the splash would frighten her. When she came abreast of the falling food, a metre to her left, she turned and took it by the corner, pulled it to her, and gulped it down. I saw it pass into her belly. Amaranth and Lillith, also elderly sharks, came after her and were each rewarded with a chunk of food.

Mordred, Pontiac, Jem, Meadowes, and Tamarack (all males), came up the scent flow, and excitedly devoured the last of the scraps. I was watching them chase each other through the coral to my right, when I turned my head and saw the same dark female passing, from left to right, as she had the previous week. She was moving so slowly that it was almost unbelievable. Just the tip of her tail propelled her. I swiftly sketched that side of her fin tip. She passed through the middle of the site high in the water and maintained the same altitude, about forty centimetres (16 in) below the surface, until she disappeared into the darkness. At the same time, a stranger appeared on the periphery. She had very light patches on her sides and tail that glimmered in the dark. It was Glammer's first appearance.

Clouds had darkened the island on the day in late June when I had planned a session, though there was some sunshine on the sea. So I considered the southern wind, the *mara'amu*, a blessing, and surfed out to the lagoon upon its waves. Just as I arrived, the last of the sunlight faded; clouds filled the western sky to the horizon. Sharks were swimming to the boat in the rioting waters, turning underneath and leaving at the reflective angle. The large eagle ray flew by below, just before I slid over the side.

Underwater, I was met as usual by the sharks and the expectant, watching fish, converging to escort me to the site. They soared around excitedly as I placed the food, swimming under me and grabbing pieces. I gained my perch as the grey torpedoes fell on the food, and the water blurred with sand as six nurse sharks undulated into place. For several minutes, the site was alive with flying sharks as the thrilled creatures grabbed bits and pieces, soaring around and chasing each other with the scraps before dispersing in the coral as more came swiftly up the scent flow.

Madeline swept in fast and aggressively, as usual. She was very dark by then. She grabbed a bite and left immediately, while Sparkle Too steadily glided through the area, and was still there when I left. Trinket had come again. She always seemed skittish and slightly off balance, sometimes startling and changing direction for no apparent reason. She snatched up a scrap and started to shake it, but instead of rising in the water as normal sharks do, she again burrowed down into the sand, raising a cloud around her. Later as she made her way single-mindedly around the food, she bumped straight into Rumcake.

After watching the vain efforts of the sharks to pick through the scraps that had landed in a pile, I spread them out and collected up the ones that had fallen in the coral. Animals who possess hands just have no idea of the difficulties experienced by those who lack them. Bratworst paid no attention, passing me within centimetres as she circled. Another female drifted slowly over me as I worked. It was Droplet arriving. Her companion, Splash, passed as I returned to my lookout, and Valentine drifted in with Venus. She passed rhythmically time after time, then rose vertically with her white front facing me and her fins arced behind her, to touch a piece of seaweed on the surface. As she curved back down, her dorsal fin pulled a shining stream of bubbles behind it. Martha came as darkness fell, without Madonna this time.

Flora arrived at the same time relative to the sun, as she had at the last several sessions. Methodically she searched for a scrap. The experienced sharks seemed more able to find something they wanted to eat, and searched the food more closely. They learned. Flora found a chunk of meat that I had put in the open for them, and swallowed half of it, dropping the rest. Martha flew after her and inhaled the falling morsel.

I was watching this off to the right, and when I turned back, the large dark female, tip line angled up at the leading edge, was passing. It was like watching a rerun of a movie as she glided by, just as she had before—the same speed, same path, same altitude, same time, precisely the same image. She glided infinitely slowly, barely moving her tail, straight across the site. Again, she had come while I was engrossed in watching something in the opposite direction. Later, when I happened to turn my head, I caught sight of her passing behind me, but had I not turned at that moment, I would have been unaware of her sneaky passage.

Storm arrived, circled the site a few times, and left. She never stayed long. It was as if she just stopped in to see what was happening as she passed by on other business. Sharks living eastward in the lagoon could come to the border to hunt in the evening, become aware of the session, and come to see, while on other occasions, they didn't stop. Sometimes I saw them pass by down-current. This appeared to be the case with Jessica, and possibly other infrequent, but regular, visitors. Peri took the last belly strip, the others flying behind to get some, and the sharks began to disperse, though some of the food remained. Martha, Flora, Splash, Droplet, Fawn, and Sparkle Too still cruised the area with a few juveniles.

Waves were tossing me so powerfully by then that I could no longer draw, so I looked above the surface. The wind had increased and changed direction. It was lashing the water into white caps and whipping off the spray to form a haze over the sea. The northern horizon was a wall of blackness, while overhead clouds roiled uneasily. It was not very late, but darkness was coming anyway, falling like a curtain by the moment. As I fed the fish, the needlefish, who had been lost in the waves, were no longer at the surface; they formed into rivers of silver moving in synchrony half a metre (20 in) underwater. They streamed away, and another group, smaller and higher in the water, streamed by. Never had I seen them do that before.

Marco appeared just then, and as he meandered closer, it began to rain so hard that the underside of the surface looked like an animated sculpture in bubbling glass. Each heavy raindrop pushed a thick cylinder of water five centimetres (2 in) into the lagoon. Marco rhythmically patrolled through the site, back-dropped by the rivers of silver fish, the deep green darkness, and the menace of the storm. I checked my watch. Behind the clouds, the sun was on the horizon, as it had been the last times he had come.

I picked up the anchor, slung it in the kayak, and climbed in. Pelted by sheets of cold water, I took off my gear and when I looked up, all I saw was the rain, falling so thickly that it was like sitting in an endless waterfall. Briefly, the reef remained visible through a space behind me, so I began to paddle away from it. I had noted minutes before that the wind was coming from the east, so I moved broadside to the driven waves and paddled forward hard. After what seemed like an endless battle in the storm, while searching through the darkening cascades for any view of land, I glimpsed the faint silhouette of a hill I did not recognize. I fought onward, trying to make sense of it, thinking I had crossed the bay. But when I neared shore, I was far up the lagoon! The wind had shifted again and was coming from the north, so my course had been off by ninety degrees. Mercifully, the atmosphere cleared as I paddled all the way home along the shore.

Her spine was curved...

Chapter Seventeen

In the Dry Season

I had fewer surprises with the sharks during the dry season, which is the cool time of year in the southern hemisphere. It was mostly the resident sharks who cruised the study area. But in mid-July there was another riveting session. A group of visitors contributed to the excitement, and I had a good load of food—I had been given a whole tuna that could not be sold, which I had cut into pieces.

The period of unsettled weather had passed. The ocean was tranquil, and the waves coming over the reef were gentle ones, generating little current in the lagoon. It was glassy, and I located the site as I dreamily paddled in the right direction by seeing the school of yellow perch below quickening in recognition as I glided over them. They swam with me, hundreds swimming to join the hundreds who waited at the site.

No sharks were visible from the boat. When I slid into the water I found Fawn drifting there as if in meditation. I put in the food, and the perch streamed over the bottom, expertly choosing morsels to carry to shelter and eat in peace. When a group of them carried away one of my few chunks of dark meat, I took it from them and replaced it on the sand. They went back and took it again, so it was a deliberate choice, specifically chosen among all the sorts of scraps available.

Madonna came first, and drifted over the food. Meadowes shot in, gulped a chunk of meat, turned, and took another. As he passed me, his mouth and throat were bulging, and the bulge quickly passed into his stomach! Madeline zoomed in. Madonna had taken a piece of white meat, but she dropped it, and Madeline snatched and swallowed it. I found another good piece for Madonna and tried to pass it to her, but she didn't turn toward it. Madeline zeroed in on it, bit it in two, turned, and downed the rest in one gulp. Samaria arrived, very interested in the food. She found and ate a large piece of fish. More and more sharks were arriving, and excitement mounted. A huge piece of skin covered with a layer of meat hung in the water a metre or so in front of me at the surface, not moving in the still water. Why it was floating was a mystery.

By then, about two dozen sharks were whirling over the food. Trinket took a piece of meat and gobbled it, then another, and passed me, stuffed to bursting from the teeth down. Lillith passed closely and slowly, looking. She rarely came near. Fawn had disappeared, but finally he returned to the site.

The dark grey female who drifted through each session at the same time, from left to right, chose that moment to put in her appearance. But this time, she curved around the western edge of the site, and before she disappeared into the coral, I glimpsed her left side, just for a split second, in the distance. It looked exactly like the right side! The shy and secretive creature did not return. I named her Annaloo—Annaloo, the mystery shark. When I returned home and checked my records, I found that we had met one sunny morning very companionably, long before. I had mistaken her for Madonna and tried to hand feed her. The change in her behaviour was surprising.

The floating scrap still hung a metre in front of me. As I was meditating on Annaloo's theatrical passage, there was a shocking ripping sound. Where the meat had been, was a churning, opaque cloud of blood, bubbles, and fragments through which sharks rushed toward me at the surface, their mouths stuffed, as they grabbed the remnants out of the water.

As Filoh passed in front of my face I put out my hand in surprise, which I immediately regretted. My quiver of anxiety deepened when a shark rose toward my face as though to kiss me, the white underside of her nose materialising centimetres away, little mouth turned down. Again I raised my hand, drifting backward. It was Bratworst. She passed, mingling with the masses of sharks flying around the site, as I slowly relaxed, but moments later, she was back and did it again! A piece of fish meat was hanging in the water in front of my forehead above my line of vision, and this was what Bratworst wanted.

The sharks had reached an unprecedented level of excitement. Amaranth, Shimmy, and Droplet soared in and attacked the scraps aggressively, pushing them around. Amaranth picked up a mouthful and swam slowly away with several fish backbones hanging from each side of her mouth, pursued by a fleet of juveniles, while Droplet uncovered a long strip and flew away with it, followed by Christobel and Sparkle. They turned in the coral and reappeared some distance behind me, where Christobel shook off the last piece of the meat. Sparkle snapped it up and swallowed it whole. Kimberley, Valentine, Isis, Windy, Cochito, and Diamond were the latest arrivals.

Kimberley had been poking around in the coral on the periphery looking for scraps, at times vertical in the water as she reached down for something. Suddenly she picked up a fish tail, which has no meat on it at all, and shot away as fast as I've ever seen a shark go—and her an old lady! Far off in the coral she turned a-hundred-and-eighty-degrees, and came blasting back through the site as though suddenly crazy, still carrying the fish tail in her mouth. I could not have imagined she was capable of such speed. She disappeared into the blue beyond, followed by several of the others. Twilight was among the torpedoes shooting by. She was nearly as dark as Kimberley by then, and the snowflakes at the sides of her head were prominent. Again the sister sharks had come together.

Another unfamiliar and heavy dark female was among them. It took time to recognize Samurai. She had changed since the first time she had soared around me, slender and agile, not a year before. She had become enormous, and she was pregnant. Her growth had curved the lines of the inverted V in her colour line, so that it had become an inverted U. She showed no sign of the bold behaviour she had displayed when we had first met.

Most of the food was gone, but there were still bits to pick through, and the area was full of sharks, all of them excited. More came hurrying in as the sun set. Peri came and ate expertly. He had learned fast how to find and extract good food from the scraps. Storm floated by. Only once did she come to the food to fussily poke through it. I placed the last good piece out in the open, and she came and took it, then swam away toward the lagoon, trailing crumbs and fish as she accelerated. Sparkle Too, Flora, Valentine, Ruffles, and Glammer roamed through after the others dispersed. Flute and Piccolo, a pair of lookalike pups who had begun coming to the sessions together, were exulting over the food, snatching pieces and racing off with them with a lot of energetic shaking, now that the bigger sharks were farther away. Emerald passed by.

Nurse sharks carpeted the site, munching on fish heads and scraps, and covering much of the food that was left with their pale, plump bodies. The blackfins would do nothing to get them to move out of the way. I took the last fillet from them, momentarily resting my hand on one. The shark ignored me. As Flora approached, I threw it to her, but it frightened everyone, near and far. I tried twice. Ruffles finally took it and bit it up; turning methodically to get the pieces he had dropped, one after another. With difficulty, I found another good piece of food in the shadows near the massive coral formation bounding the site on the west. It had been dropped and had a bite missing. I replaced it in the site. Flora finally found it, picked it gently up, and manipulated it along its length as she swam away southward, finally pausing to bite and shake it. Diamond followed her to get the scrap.

Since the dry season began, the ocean swell had been rolling over the Pacific from the southwest, and my north coast had been protected from its onslaught. Though this made the sessions more comfortable and easy for me, an accumulation of organic matter in the still water spoiled the visibility.

During this period, I was holding the sessions two or three times weekly, since more and more scraps had become available. Though sometimes I was greeted by a crowd of sharks, most of the time, no one was waiting when I arrived, though always the excited fish filled the water. When the sharks did appear, they often just passed by as though not hungry or interested.

Sometimes, most of my two dozen regular sharks did not come at all. As it grew dark and I fed the fish, often no sharks returned though food still lay on the sand. I didn't mind; someone would find it later. As I joked with one of the women at the shop where I picked up the scraps, they came from the sea and were returned to the sea.

Annaloo, the mystery shark, floated through the site once or twice during nearly every session, always when my attention was diverted by something in the opposite direction. One time I was turned, watching something, and when I turned back, Annaloo was drifting by, right to left this time, showing the left side of her dorsal fin clearly for the first time. A few moments later she passed a metre in front of me from left to right, drifting so slowly that she seemed in a trance, her tail just fanning the water. I tried to memorize every detail as she practically hovered in place. Just before she disappeared, she was startled by my boat at the surface behind, and shot away at the speed of light. Though dreamy she seemed, she was clearly wide awake!

Flora still came near the end of the sessions. She seemed to have a schedule, which used to include meeting me at the beginning of the sessions, and now brought her to the area much later, at about the same time each night. I could predict when she would arrive. She was not timing her visits to cash in on the free food—there was never much left when she joined us, and I had the impression that she was just dropping in on her way by. Marco visited several times and he, too, seemed to be on a schedule. Time after time when he appeared, I looked at the sun and saw it landing like a balloon on the horizon. Since the sun took just two minutes to cross the horizon, such fineness of discernment was unexpected. He far surpassed me in his capacity to adhere to a long-term schedule. Just this one shark revealed a lot about them all. Some of the irregular visitors such as Eclipse still passed infrequently, and Splash and Droplet visited sometimes. A beautiful, pale young female I named Cinnamon appeared in the area for the first time.

I encouraged the increasing numbers of nurse sharks to move off the food that was swept under my coral formation by stroking them. They would lie there for a while beside me, then slowly move ahead, and I was able to put the food out in the open where the blackfins could access it. Other times they were so slow that I clapped my hands in front of them to get their attention. They would often pay me no heed though I was right beside them, scooping the food from under the coral. Then one evening one of them munched suddenly on a fish head right beside my ear, with a noise like thunder. Stunned by the power suggested by that noise, I decided to stop touching them, and that I should have been more respectful in the first place.

There was still the occasional special session. Once I brought only a small amount of food and nearly forty sharks attended. But when things were quiet, I was freed to spend more time feeding the multitudes of fish. For this, they waited impatiently, and sometimes for a long time when I was busy with the sharks. Whenever I roamed away from my lookout, the wrasses covered it; it disappeared in a cloud of fish. They had always tended to follow me around whenever I roamed through the lagoon, so I speculated that they followed large predators, and investigated places where predators had been eating.

There were so many fish waiting that I usually didn't have enough food to satisfy them. Often, I only had the scraps from one or two tunas, or only a few backbones and pieces of skin. For long periods, the fish shops gave the best of the scraps to people for their dogs or pigs, and often, I was stuck with nothing but skin and bones. So in my efforts to produce as much food for the thrilled little creatures gesticulating before me as possible, I eventually tried extracting the fish's eyes for them. These were surrounded by a layer of fat that attracted the sharks. Sometimes they came searching through the water at the surface when I did this —fat floats. The first time I tried, Splash appeared in the cloudy water at eye level, an unprecedented approach on her part, so I started to throw the eyes into the site for the sharks. Twilight, then Eclipse passed, swiftly searching them out.

I was waiting in hopes of seeing the spectacular pair again when a large, dark shark came up the scent flow to my face, and turned at the last minute. It was Madonna. I threw an eye for her, and after several attempts in which her fish eye was taken by a more efficient shark, she got one! She held it delicately in her mouth, then spat it out. She left and didn't come back. I was unable to examine the eye to see how much she had damaged it by this delicate manoeuvre, since Marco, who had been circling steadily, picked it up and carried it away. Avogadro, Ruffles, Marco, Pontiac, and Fawn, all males, were still roaming around at the surface amongst the floating fish eyes when I left.

After a session at the beginning of August, I was returning in high wind and darkness along the shore when I saw a movement on the surface, just a flicker. A bird was frantically struggling to stay afloat, already so deep that her beak reached up to the air with each breath. I picked her out and wrapped her in my towel for the rest of the journey.

At home, I put her inside while quickly rinsing and arranging my gear and showering. Then, at my treatment counter beside the kitchen sink, I rinsed the salt-water from her feathers and dried her with a hair dryer. She had fluffy white feathers, speckled with grey and gold, and huge, black eyes that filled her face, making her look like a little white owl. With her pointed black beak, it was obvious she was a baby fairy tern (*Sterna nereis*), though I had never seen a chick before. Her tail and wing feathers had not begun to elongate, and inside the fluff, she was nothing but a little aggregation of bones.

I tried to feed her, but her beak was soft, and I was afraid of injuring it while gently pushing a tiny fish down her throat. She must have starved for so long that her throat had nearly closed up. Fortunately, when I put some small pieces of food down in front of her, she ate them!

Fairy terns lay a single egg on the bare surface of a tree branch, and after it hatches, the chick waits there alone while its parents fish far out to sea. I had seen many of these unprotected chicks raised. They sat alone on the branch day after day, waiting for their parents to return from the ocean with fish. The parent placed the tiny fish on the branch in front of the baby, rather than inserting it into the chick's beak or throat. So it was natural for a very young chick of this species to eat on its own and not have to be fed by hand.

These lonely chicks could successfully face high winds that seemed sure to blow them off the high branch over the sea, but sometimes, they did fall. This one must have fallen long before, then been blown into the sea just before I found her. Not having developed her adult, waterproof feathers, she would have drowned had I not arrived in time. I named her Angel for her colour and the airy nature of her species.

She slept on the seabird perch in the corner of the house that had been furnished for my wild bird patients, and woke me in the morning with a baby version of the cries of the adults who were flying and calling to each other outside. But she never made that call again.

During the following week, Franck found an adult fairy tern who had been badly bitten by a dog. I treated her and she joined Angel on the perch. The baby soon began taking the adult's beak in hers for comfort. No longer alone, and regularly fed, she began to preen, which is a good sign in a distressed bird. But her breast bone was like a knife-edge.

As Angel slowly gained some strength, I took her for walks on my hand in the garden so that she could see the surroundings and the fish in the aquarium-like waters off the shore. Afterwards, I set her on one of the arching tree trunks over the beach, which the seabirds I cared for used. She liked that, and sat preening and looking out over the bay. I brought her in when the sun climbed higher, and the wind rose. Very gradually, she came back to life. She became used to riding on my hand, cooperatively stepping on by herself, and indicating where she wanted to go by inclining her body and looking that way. She began to vocalize with little chirruping cries to let me know when she wanted to go outside. Her tail and wing feathers grew rapidly, and I persuaded her to exercise them by suddenly lowering her when she was perched on my arm. After a few sessions, she began to beat and wave her wings on her own.

As the days passed, her wings and tail lengthened fast, and she changed from a ball of fluff into a sleek young bird. She spent longer periods in the perching trees where she looked avidly around, watching passing birds, preening, and exercising her wings. She preened much more outside. Sometimes, she called out in the evening when I was with her, as she sat on her perch — two long notes in a minor key. It was a call unlike her baby cries and very different from the harsh shrieks of the adults.

The adult fairy tern died as a result of the dog bites, and I put a mirror in front of Angel's perch. She sat looking into it. She would watch me coming toward her in the mirror, our eyes meeting, and then she turned to look at me approaching from behind her. She seemed to understand that she was looking at a reflection.

One morning three weeks after she had come, Angel flew and landed on the couch during our morning exercise session. I picked her up by placing my hand behind and under her feet, and she flew up onto her perch, where she made a neat landing. Later that day, while she perched outside, she began shaking her head in her bowl of drinking water, indicating a wish to bathe. I had been encouraging her to bathe in a shallow, glass bowl to keep her feathers clean, and help waterproof them. It was on the bench surrounding the deck, so I took her to it. But as I gently persuaded her to step off my hand onto the rim of the bowl, she spread her wings to keep her balance, and they lifted the astonished bird into the moving air.

Not knowing what to do next, she alighted on the slanting trunk of the coconut tree in front of her, which leaned away at an angle over the beach. Then she toddled upward, lightly fluttering to keep her balance. At the top, she found a perch, with difficulty, among the coconuts, and there she stayed, while I sat painting below where I could watch her shifting constantly as the tree waved in the wind.

She was usually tired after only three or four hours outside, and as the sun sank in the west, she became anxious to descend. She looked inquisitively downward, fluttering and adjusting her position for most of an hour before leaving her perch like a butterfly and hovering until she was able to pose her tiny feet upon the trunk below. Then she purposefully began side-stepping down. Sometimes she took fright and flew up momentarily, but landed and continued downward with a determination that suggested that she had planned this in advance. A third of the way down, she lost her concentration and took to the air. Uncertain and frightened, she managed to alight on a twig in her path, lost her balance, and flew back to the coconut tree, where she fluttered upward until she could find a perch. There, she huddled sideways, unable to continue.

I was lucky that Franck's student Teiuria was visiting. He climbed the coconut tree Polynesian style, and put the exhausted bird in his T-shirt for the ride down. But Angel kept trying to claw her way out, and Teiuria's precarious descent was interrupted repeatedly as he tried to adjust the frantic bird. Extricated with difficulty in the house, Angel sat stunned for two hours before I could coax her to eat a little bit, and days passed before she was willing, or able, to fly again.

Near the end of August, I upgraded my faulty camera to a better model. But the viewfinder was too small to see through, so for a long time, the only photographs that turned out all right, did so quite by accident.

Madonna was enormous with her pregnancy, and she manoeuvred clumsily. Her girth made her look far larger than she had the year before. Her skin was perfect, with a dark, velvety sheen, and no traces remained of the pale, unhealthy-looking mottling I had noted months before. However, her attendance had become sporadic, as if she were losing interest in me and my coarse fish scraps. I had finally to admit that the horrible cutting up I assumed she had suffered in a net had actually been due to mating. Since I knew exactly when that had happened, I would learn the sharks' gestation period when she birthed.

Sparkle, who had been so dark grey, turned brown. Carrellina reappeared. I had been concerned that she had died, since she had been coming to nearly every session, then one day was gone. Since she was a juvenile, I could not link her movements with the reproductive season. She had come in December, stayed for four months, then left for four months.

The spangled emperors, who had waited for me in the site so faithfully, no longer appeared. I feared that they had been fished.

CHAPTER EIGHTEEN

Madonna's Pups

On the last day of August, I went out to check the shallows in the curve of the reef close to the site for newborn blackfins, and found none.

Three days later, I went to see the sharks. As I anchored, a crested tern called out, hovering and looking down at me, the sun reflecting the turquoise of the lagoon up onto his snow-white breast. I threw him a morsel of fish, which he dove to snatch from the air, but missed, so we tried again. He called repeatedly, and in moments, there were two. I extracted more pieces from the scraps behind me in the wave-tossed boat and threw them to the birds, as more and more gathered. Soon there were six, then twelve. They tried to line up with me to catch the food in the air, watching as I prepared to throw.

The sharks started coming to the surface to snatch the falling pieces, and I was afraid that one would inhale a bird. The birds were plummeting into the water and snatching the pieces of food half a metre (20 in) underwater! But they were so quick that no shark could target one. They were deft at catching the pieces of food while they were still in the air, too, and distracted me for many minutes with their delighted cries and acrobatic flights. Finally, I felt I could keep my poor sharks waiting no longer and regretfully ignored the birds and slid underwater.

Sharks were flying everywhere, thrilled and expectant. They ate quickly, and when the first group circled away, I dove down to get the food out from under the coral, turned my head, and found Madonna at my side.

She had birthed! From a big, heavy shark, she had become a thin and sinuous creature overnight. The change was dramatic. When I had first seen her after she had mated, she looked at a distance as if there were a dark shadow over her, caused by the lacing of lacerations that had already turned black. Since then I had seen enough of these mating wounds to know that they turn dark after a few days, so she had mated at least five days earlier, probably not much more than that, since the wounds were still showing so clearly, and even the smallest had not healed. She had birthed at the very beginning of September, so the gestation period was roughly ten months.

Later, a pair of heavy, dark females cruised through. The parade of visiting pregnant sharks had begun again right on time.

Flora appeared as usual that day, at precisely the same time relative to the setting sun, and when the sun was poised upon the horizon, Marco came. Never had he come at any other time. He was with the magnificent Flannery, who had pale, shimmery patches on the white decorative stripe on his side, like Glammer.

I went looking for Madonna's pups in the nursery as soon as I had a chance. There were none there, suggesting that she had gone elsewhere to birth. It was incorrect to think that the female sharks cruising the lagoon around this shallow region formed in the curve of the reef were playing the role of protectors. And why had Madonna not birthed there in the nearby nursery where she spent so much of her time? It would seem logical for her to do so, but nature has no reason to satisfy human expectations.

So I paddled to the eastern end of the lagoon to the bigger, more sheltered nursery, where Franck and me had found baby sharks opposite the house where we had lived two years before.

A perfect, tiny shark soon circled close. Farther into the centre of the wide shallow area, I found a few more. Unfortunately, a thick layer of *turbinaria*, wrenched by waves from the reef, was floating in the region, apparently gathered up by conflicting currents that held it there in place. It effectively obscured the passing dorsal fins as I lay on my stomach in a few centimetres of water, pushed and pulled by the ocean's surge and craning my neck to look upward at them. The little sharks' dorsal fin tips were often above the surface, so impossible to see. Besides finding Madonna's offspring, I was intent on identifying as many as possible in hopes of seeing them later in the lagoon. Finding some who had unusual features could permit me to trace them from the time they were born. However, I was unable to find any unusually marked shark pups, and though I drew many dorsal fins that day, they were not accurate enough to use for a definite identification of an older shark years later.

I found a little group of three steel-grey pups who stayed together, and could have been Madonna's young, but there was no way to know for sure. She was not the only female who had birthed, only the first in my area. Though it seemed likely that she had used this nursery, I couldn't even be sure about that, and if Madonna's pups were at that end of the lagoon, it would be unlikely that they would ever attend one of my feeding sessions. None had easily recognizable markings like Trinket, Pixie, and Fawn.

It was disappointing since I had hoped to keep track of the lives of Madonna's offspring, to see how long they stayed in the nursery, whether they tended to stay together, how long it took them to grow up, and whether they eventually lived in the same area that Madonna did. Doing so could have revealed much about the sharks' society, and it would have been easy if she had put them in the nursery near where she lived.

Perhaps her need to scout out the distant area as her time of parturition neared explained why she had kept disappearing in recent weeks. Perhaps the many heavily pregnant sharks moving through the area were roaming to unknown nurseries farther west prior to birthing. (This later proved to be the case). What else could explain the many sharks passing through the western end of the lagoon when the commonly used shark nursery was on the eastern end? Perhaps the sharks came from different regions of the island. Unfortunately, since the kayak would not fit in our truck, I just was not able to extend my search farther than the immediate area.

I explored the corresponding region—the sheltered area within the curve of the reef—in the lagoon on the other side of the bay. Madonna could more easily have crossed the bay than gone to the other end of the lagoon; it was twice as far away. But this area was quite different from the well-used shark nursery. It had a much narrower shallow region, and there was a drop-off into deeper water on the lagoon side. I found no shark pups there, though it was a large area and seemed suitable as a nursery in some places. But the region appeared inhospitable compared to the rich ecological area formed by the wide coral lagoon where my sharks lived. The lagoon around it was narrow, it was surrounded by barren reef flats, and there was a wide, deep boat channel between the reef and shore. Nearly all the coral in the region was dead and the visibility was so poor that it was like roaming through a fog.

On turning back to my boat, a huge shape loomed out of the gloom. Astonished that a whale could be in the lagoon, I drifted closer until I could see two lemon sharks travelling nose to tail. They were dark grey, which confused me at first since lemon sharks are usually a paler, warmer colour. They circled me about four metres (13 ft) away then disappeared into the mist. It was the first time I had seen lemon sharks in such shallow water in the middle of the day.

In mid-September, I took some fish scraps deeper into that lagoon to see if a resident population of blackfins used it. The visibility again was foggy, and during an hour and a half, no sharks appeared. It did not seem to be frequented by the species, perhaps due to the desolate habitat. I wanted to investigate the sharks in a second lagoon someday, in order to compare them with my original group, and the lagoon on the opposite side of the bay was the obvious one to study. Without a motorboat, I didn't have the choice to go farther. So it was disappointing to find that it seemed empty of blackfins.

Angel couldn't have been more different than the sharks. Creature of air rather than water, she seemed more warmth and vibration than substance. After her fright during her first adventure on the wing, I taught her to fly confidently in the house so that her transition to the wild would involve neither privation nor trauma. At first, she was encouraged to fly from my wrist to her perch just half a metre (20 in) away. When she got the idea, I moved back a little each time until she was flying across the room to regain her perch. Soon, she was flying gracefully through the house. Landing seemed to be the hardest to learn, but after only a week, I could count on her to fly down from her favourite place high in the rafters when I called her, to neatly alight and take a fish from my hand.

So one calm morning, I carried her again to her perch overlooking the sea. After a few cautious flights from branch to branch, she settled down, and I fed her. She usually rested after eating, so I went inside for a few minutes, but when I returned, she had vanished. I searched each tree and eventually all along the edge of the bay, looking from my kayak. A white bird perched over the water, shines like a star in the westering light. But though I saw each neighbouring fairy tern come home from the ocean to circle and perch in its tree for the night, Angel was gone.

She didn't appear at dawn when the fairy terns began to fly, and I watched at intervals for her to appear as the day passed. It wasn't until 3:00 p.m. that I caught sight of her flashing by over the tree-tops. She seemed to be having difficulty descending beneath the canopy to her familiar perches. I stood on the lawn below, holding up a fish and calling to her as I had done in the house. She was perched high on the outside of the canopy of the enormous tree by the beach that shaded the house. But it was a long time before she flew down toward me, finally curvetting in beneath its shadows, to alight on a thick branch near the house. She peered down at me, but took many more uncontrolled and panicky flights, until a whirring of wings and several shrieks heralded the arrival of a female fairy tern, who was incubating her egg just above. She attacked Angel.

Crying in terror, the young bird flew toward me and alighted as close as she could get on a tiny twig that could hardly hold her weight. I put the ladder on the deck bench, and reached to the limit of my balance from the highest step to get a fish within her reach—I always offered food first. She ate, tottering and gasping in distress. So I slowly moved the end of her branch toward me, until I could slide my fingers under her feet and lift her. Then I smoothly descended the ladder and carried her into the house.

She relaxed on her perch inside, ate more fish, and drank copiously, alternately looking into my eyes with her intent, black gaze, and touching my face with her beak. She must have been frightened overnight alone in a situation that would never happen naturally to a young fairy tern. Her parents would have guided her on the wing and stayed with her, and she would not have had to find her way home alone. For many days, she rested in the rafters, flying down briefly to eat, and calling when she saw me coming or going from the room.

When she felt better, I took her outside, and she spent the day taking haphazard flights around the property. In the late afternoon, she was tired and hungry but again had difficulty descending from above the canopy to her perches on the arching trees beneath. When she finally managed to land on a perch within reach, I fed her and carried her inside. But though she was exhausted, she refused more food and hurled herself against the window, falling to the floor and trying repeatedly to get back outside! I tried to console her, but she ignored me.

As time passed, Angel established a pattern of exploration, rest, and feeding, as she practised her flying techniques. She still had trouble descending and landing and seemed to consider flying to be nothing but a problem. Much of her time was spent frozen with uncertainty on a high perch while she made and cancelled flight plans.

Her situation was especially tenuous at this time because her feathers weren't waterproof. A seabird's plumage must be waterproof, with all the microscopic barbules on each feather interlocking like Velcro to form an impermeable sheath. Until she was waterproof, she was in danger of dying of hypothermia if caught in a storm. So I was always anxiously watching for approaching rain.

One morning, I put Angel on her perch with pieces of fish beside her, while I took out a dolphin tour. When I came back at noon, she was in the big tree over the deck, resting. She was not interested in coming down to eat. All was calm, and the brooding female was resting on her branch.

Our new Polynesian gardener was raking up the leaves on the beach in front of the house under the tree where Angel rested. I kept an eye on her as she flew from branch to branch, rested, and preened. The gardener became interested in what I was doing and finally asked me where the bird was, so I showed him.

The next time that Angel seemed interested in coming down for a fish, and I held one up, calling to her, she didn't come, but the gardener did. Without saying anything, he looked at the little bird, and then climbed up the tree. The brooding female became alarmed and flew. Angel flew next, and the adult chased her around the tree, across the garden, and over the neighbour's house, far away from our property. Angel couldn't shake her off.

Finally, she alighted on a tree top high on the mountain and the adult fairy tern returned. I found her hovering over the beach, fluttering back and forth, while the gardener hefted rocks at her, watching her and calling to her. He was trying to kill her! I had explained the situation to him, but he destructively interfered.

I stood in the garden calling to Angel, scarcely visible far above on the mountainside. After about fifteen minutes, she spiralled down around the house and garden and alighted in her favourite big tree on the beach. The adult, doubtless deeply disturbed, had managed to resume her place on her egg. She flew out and attacked her. Angel managed to land in the tree anyway, and pecked at the older bird in self-protection. The adult landed about a metre away from her and the two stared at each other. Angel was exhausted and lay on her breast, her wings hanging down on either side of the branch. Finally, the older bird leaped at her and dislodged her. As they flew around the area I couldn't keep them in view, and by the time the brooding female returned to her egg, Angel was lost.

About an hour later, I caught sight of her in a large tree on the opposite side of the garden, looking down at me. When I held up a fish, she flew and landed low on a convenient, dead branch in a fruit tree between us. I went to get the ladder. The gardener watched, laughing so hard that I thought he would fall down and roll over. When I got back with the ladder and a fish, he was poking at my alarmed fairy tern with the end of his rake and tapping the tree with it. I told him he was frightening the bird, and he backed off. I set up the ladder, climbed up, gave the agitated bird the fish, and brought my hand up beneath her. She stepped on. I held her close beneath my chin, sheltering her with my hand, but she still tried to fly off when I walked through the back door, which was a dark and unfamiliar entrance. Inside, she fell asleep, her head turned backwards and tucked into her wing. Luckily, we found another gardener.

One morning darkened slowly, and by noon, rain threatened. Angel was in an exposed location in a leafless tree, and stayed there while the garden birds took shelter. It began to rain, gently at first, but steadily harder. Angel made bathing movements, and did not respond to my calls. As the hours passed and her feathers became soaked, her head and back grew dark with water, and she was soon trembling with cold. I waited for her wings to turn to spider-webs in the wet—if she couldn't fly, perhaps I could catch her.

I was afraid to alarm her with an effort to get her down while she could still fly, for fear that she would take fright and disappear from my view, and I watched for a sign that the storm was passing. But the sky was dark to the horizon, and as the day ended, the rain became a downpour. The air, which had been strangely still, began to stir. It was cold. I was wet and shaking with cold myself.

A friend had told me of a fairy tern she had saved who had died under these conditions because the bird had not been waterproof. My worst fears seemed about to come true. Angel was trying to shake off the water, compulsively preening and shaking herself as she shivered with cold. Sometimes, she spread her wings and shook them. It was hard to see in the grey light, but finally it seemed that her wings were full of holes, and desperation made me act on my best plan.

I got a very long bamboo pole with a tiny horizontal branch on the end of it, and raised it under her, trying to get her to step on. But she flew, circling the property and soaring past the tops of the coconut trees. She landed, to my intense relief, in about the same place. More time passed as I huddled out of the wind, wondering what to do. Hundreds of finches had come to eat at the bird feeder, but after a few minutes, they began flying into the big tree, to line up under its umbrella-like leaves by the hundreds. Glancing around to learn why, I was shocked to see an opaque grey wall approaching from the south. Angel had stopped bathing, and was sitting hunched, her face darkened by the water, her feathers clinging to her skin, her eyes huge, when it hit like a hurricane. She beat her wings to keep her balance as the blast struck her, but was soon blown away.

Spreading her wings, she caught herself, and flew strenuously across the wind sweeping her northward, slowly rising from just above the lawn to land in a tree beside the road. Had she gone farther, she would have been lost from view in the jungle beyond.

It was nearly dark. I climbed the tree as high as I could, and from a metre away, urged her to fly to me. She didn't react. Desperate, I climbed up and down several times in different efforts to coax her, but she was unresponsive. Finally I brought the bamboo pole again and, with little hope, manoeuvred it up behind her feet. She was lifted, and amazingly, I had her perched on the tiny branch on the pole five metres (16 ft) above! Trying desperately to keep it perfectly vertical, I moved it slowly free of the surrounding branches and walked toward the house with Angel balanced, fluttering at the top like some way-out circus performer. As we neared the house, I lowered the long pole very slowly and carefully in the direction of the back door, and managed to deposit her on the laundry basket.

She didn't move, even when I picked her up and carried her inside. Comforting her gently, I blow-dried her and placed her in a cocoon of warmth under a towel on a hot water bottle where she eventually recovered enough to climb out. Back on her perch, she grew more alert, began preening, and ate.

Angel slowly recovered, but storms continued to blast over us, and I dared not put her outside. I was torn between the need to protect her from death and to prepare her for life, and sad to see her depressed, like a cut flower, in the house. Daily, we worked on waterproofing her feathers. I encouraged her to float and bathe in a basin and often sprayed her with rainwater to keep her preening.

When the weather improved, I took her outside at dawn. During the morning I had a dolphin tour, and when I returned, she was high in a tree. She flew down crying anxiously when she saw me. I fed her and sat on the grass talking to her. Suddenly, she spread her wings and rose straight up until she disappeared into the blue sky above the mountain top. Twenty minutes later, she came swiftly down over the garden trees to alight beside me on the branch she had just left. At last she had discovered flight!

She flew from tree to tree for the rest of the day, and when evening approached, I fed her and slid my hand behind her feet to carry her inside. But she skittered away. After three tries, I had to put my other hand in front to impede her escape. The weather was unsettled, and I dared not leave her out overnight. Indeed, the next day was dark with rain, and Angel was tired. Stress, due to her unnatural life, had kept her from eating enough. She was thin and not as strong as she should have been. She just wasn't ready to live independently outside.

Time passed, and she flew more and more widely, enjoying flitting and soaring over the sea, usually within view. She practised hovering to drink from the surface, a skill that takes time and concentration for a young bird to perfect. She ate more and seemed well and happy. It became increasingly difficult to coax her inside. Eventually, she escaped my grasp altogether and began spending her nights on a favourite branch in the same great tree on the beach in front of the house. At first, some of the other terns chased her from her perch. Hearing her shrieking, I rushed out and threw fruits near them till they left her alone. Eventually, they accepted her, and each morning she joined them in flight, circling up and down the bay.

Still, she seemed to count on my emotional support. Once, after being closely chased by another bird, she landed beside me crying and agitated, constantly looking around, and too disturbed to eat. She stepped on my hand and let me carry her inside, which was fortuitous because rain was coming, and the downpour lasted for days, a period during which Angel rested much of the time in the rafters and spent long periods bathing on a pool of rainwater. By the time she resumed her life outside, she was waterproof.

She flew strenuously across the wind...

CHAPTER NINETEEN

Another Mating Season Begins

September passed into October, the water slowly warmed, and the atmosphere stirred with promises of storms as the wet season descended. Arcangela returned to my area, and another very beautiful young female came during this period. I called her Clara. The trickle of visitors gliding into the sessions grew steadily and I watched closely for those who had passed the year before. Instead, a continuous parade of strangers came through. How many could there be altogether, and where did they come from?

That month, I saw the shark with the S-shaped backbone again. This time, she cruised through the site once, her tail waving above her, her body undulating widely. As I left, a crested tern flew over the boat, lined up with it, and called out. I looked up and called back. The tern flew lower, and I explained that I had no scraps. Its mate had joined it. They watched for a while and left. Apparently, they remembered me and the boat, though I had only fed them once three weeks before. I decided to ignore them in future since I had enough trouble already with hoards of sharks and fish underwater to begin trying to cope with hoards of birds while I was trying to prepare. Besides, I was afraid that if I began attracting birds, it wouldn't be long before a shark would eat one. I speculated that it was not by chance that birds never alighted on the water.

It was almost imperceptible from one session to another, but the numbers of sharks who waited for me each evening grew. When I slid into the water, they shot away in all directions. But after the food was in place, they came swirling around it in such a state of excitement that the clouds of sand they raised made it impossible to see them, so I would drift to the side to watch.

Little Trinket returned after another long sojourn away. She was passing curiously as I pulled the boat into place to put in the food, but it was twenty minutes before she reappeared, swimming into the site from the direction of the reef rather than up the scent flow. Perhaps she had been listening, out of visual range, to what was happening at the site, and came to eat when things calmed down.

Throughout that month, Flora continued in her new pattern of arriving at the end of the session, about ten minutes before sunset, though there was no food left for her then. It was strange, when she had waited for me for so long at the beginning of the sessions and had her pick of the best of the scraps. The food could not be important to her. A small grey female joined us at this time. She seemed a quiet shark, no different from the others. Like other sharks settling in the area, she was a year or two from maturation. I called her Apricot. The pregnant females grew increasingly distended and clumsy in their manoeuvres. Samaria's entry into the site filled me with awe, so enormous did she now appear.

I checked the local nursery regularly but found no pups there. Each of the sharks seemed to become restless prior to birthing and often left the region entirely, weeks before the big event, so I could not pin-point the precise day of parturition in most of the sharks. In some of the more stay-at-home individuals, I was able to count the days of the gestation period, and it was always close to ten months. After a month and a half or two months, the shark mated again, some early in the season and some late. Madonna, for example, was one of the earliest in the season, both for mating and birthing, which she did in November and September, respectively, while Martha was one of the latest, birthing in February, and mating at the beginning of April. The others pursued their reproductive activities between those two extremes.

Increasing numbers of nurse sharks floated into the sessions, the biggest ones arriving as it grew dark. On one occasion, a nurse shark threatened a blackfin, turning sharply toward it as it passed, then circling, and watching it. I was surprised that these lazy-looking lumpish sharks, who spent their time lying around on the sand, turned out to be more aggressive than the large-eyed, swift-swimming reef sharks. I also saw a nurse shark snap aggressively at a reef shark and chase it. Maybe that was why the blackfins didn't bother them. They could be aggressive. After seeing this, I was careful to neither startle nor crowd the nurse sharks. Once, a nurse shark came to look at the boat at the surface, its oversized dorsal fin actually curving through the air as it turned to dive again. Another time I kept hearing a sound like a distant cannon going off, and looked around repeatedly above the surface to try to find the source of the noise, before realizing that the nurse shark munching a fish head on the sand to my right, was making it.

When my son, Peter, visited in October, I took him to see the sharks. It was the first time in nearly a year that someone had come with me, and there were big waves on the reef, strong current, and terrible visibility. Further, after the build-up I had given Peter, no sharks were waiting, and it took several minutes for them to come! For the first time I noticed their concern about me bringing someone else.

After about fifteen minutes, a white-tip glided in. It was a large one, and quite fearless, approaching us closely to look, then feeding, which had not happened before. Peter was able to take quite a few photographs. The white tip passed close enough to touch, and stayed around for a long time. Trinket came and started trying to eat things in her usual silly fashion, and when night fell, Flora came. She had several fresh cuts on the left side of her face, near her gills, that were not mating bites.

Peter dove often as he tried to photograph the carpet of nurse sharks that accumulated, until we were swept across the site by a wave, when the sharks vanished and did not return. His extra movements had alarmed them.

The fish were increasingly thrilled with my sessions, if that were possible, and would become impatient with me if I waited too long to start feeding them, swimming up to my mask to get my attention and fluttering around my hands. One butterfly fish spent much of his time in front of my mask during the beginning of the sessions when I was watching the sharks. Whenever I looked up, he was peering in at me as if to say, "What are you doing in there?" They knew the signals that indicated I was going to feed them and surrounded me when I swam down to pick up a fish head and carry it back to my lookout. With my view eclipsed with darting, flashing fish, it took all my concentration to remain aware of the sharks.

To my left was a deep grotto, from which emerged the squirrel fish—crimson with huge black eyes—waiting for a morsel. Beneath them were large purple groupers, decorated with stars, who would come up to take larger bits of food I floated down to them. Upon the ledges and in little nooks were rock cod looking at me expectantly. All these fish would creep ever closer to me as the session wore on and the moment in which I would feed them approached. They understood this routine, and the passage of time, and knew when I was ready.

When the sharks were satisfied and began to disperse, I would scatter a handful of meat to the squirrel fish and cod, a handful on the surface for the needlefish, several handfuls around me for the wrasses, butterfly fish, and goat fish, a handful to my right to the fish waiting there, and then I began all over again, trying to make sure that each patiently waiting creature got some. Often I was surprised by a shark suddenly gliding past my face. Usually it was Madonna, who had left after eating, but returned to see what I was doing. Otherwise, the sharks who still roamed the area ignored me.

I became aware of a bold Javanese moray eel in the labyrinths of my dead coral structure, who kept reaching up toward my arm as I fed the fish. People had told me that they would bite and not let go, and after seeing the prolonged, aggressive reaction of the one I had tried to chase away from the food—after he had already stolen several fish steaks—I didn't trust them close to me. Sessions passed, and the moray eel continued to lurk in the coral around me, surprising me at unwelcome moments. Once I found him in the hole in the coral to my left, where I gave the food to the waiting squirrel fish. He was curved up under the overhang, his menacing smile opening and closing centimetres from my elbow, yet out of my sight unless I leaned over and looked there. I had been putting my hand there repeatedly to drift the food to the fish. Once I was aware of him, I floated a handful of larger pieces to him, hoping to convince him of my benevolence, but avoided letting my hand get near him again.

On another occasion, I was concentrating on the sharks when I became aware of a sharp pain repeating at intervals on my leg. One of the black fish, of the same species who had bitten me and Franck at our diving jetty, was biting me, seeing that my back was turned. When I turned around in surprise, it ascended toward my face menacingly, hackles raised, and looked very upset about me being there, though after all that time it should have become used to me, I thought. Everyone else had.

One day, Angel was limping. When she came close, I saw that a claw was missing, and her foot was swollen and red. The next day she was limping and crying, but would not let me pick her up. Soon she was holding her delicate, webbed foot against her breast most of the time, and could scarcely bear to put it down. I was anxious to treat it as soon as possible, and when I was unable to persuade her to come in, night after night, I decided to catch her. That evening, I came late, so that she was especially glad to see me. I had a napkin hidden in my dress. Climbing up to her in the perching trees, I waited until she was swallowing her fish, then swiftly picked her up and wrapped her in the napkin to avoid damaging her feathers as I climbed down from the tree. She threw up the fish, but I got her safely into the house and cleaned and disinfected her injury while she was still wrapped.

It was an experience that she could never have foreseen, to be restrained, held tight, and treated against her will. When I let her go, after a bit of comforting, she flew into the rafters. I prepared her antibiotic, and when I tried to bring her down, a terrible struggle ensued before I could fold her wings and wrap her again, trying not to mess up her perfect feathers. She was so disturbed afterwards that she would neither eat nor preen and stayed on the perch she had used as a baby, where I tried to soothe her and offered her food. Finally, she consented to swallow a fish that had been rolled in sugar, one of my efforts to make amends for my disgraceful treatment of her.

In the morning, Angel called to me when I entered the room at dawn. I gave her a fish, she climbed on my hand, and I carried her outside, gently reassuring her. She looked at me, listening, as I walked to her perch, and she did not try to fly away, stepping onto it as usual. Soon, she flew off to soar with the other fairy terns, and her day passed normally, except that after a second fish, she would not come down to eat. She was interested—she seemed to be planning to come for the fish I held up, looking down at me, shifting her weight back and forth, stretching her wings, and aiming herself toward her feeding perch, but then instead of flying down, she would turn her back and preen. For two hours I urged her to come down for supper, but she would not. She had only eaten two fish, and missed two meals the day before. As dusk fell, dark clouds threatened rain, and lightning flashed over the mountain top. Finally, darkness obscured her sitting alone in the place she had chosen.

As the sky lightened again, Angel sat on her branch, looking around, and sometimes down at me. She flew out over the water, flipping through the air, vertical one way, and then with a lightning beat of her wings, vertical the other way, flicking back and forth and changing direction as she did so, flying for joy. Expertly now, she rose to a stop in mid-air, then dove and paused at the surface to drink. She soared with the other fairy terns across the bay and up the mountain slopes. I had to admit that she didn't seem to be in bad shape. However, she would not fly down to eat. Much later in the day, I found her waiting on her feeding perch, making continuous little cries. She ate and took a second fish, then rested there for a long time, apparently exhausted.

Late in the afternoon, she alighted with difficulty against the wind on her high sleeping branch, gasping with her beak open, and an hour later she was still there. She was holding her foot to her breast and looking unwell. She wouldn't come down to eat. Far below I called to her, offering her fish. She began to make the high-pitched cry she made when frightened. She cried continuously, more and more loudly, at times showing all the signs of imminent flight, then just not coming, and turning her back. She was weak and very hungry, having been eating less than a third of what she needed. I think she cried because she was hungry, but was afraid to be caught and brought in for a treatment, so she didn't know what to do. She cried louder as the sun sank, at times opening her mouth ninety degrees wide and shrieking with wings raised, as I begged her to come down. As night obscured our view of each other, she finally quietened. It began to rain very hard early in the evening, and when it faded away, high winds arose, roaring down the bay and shaking the house. I couldn't get her cries out of my mind.

The first light of dawn revealed Angel quietly resting and looking out over the sea. After half an hour, she flew, soaring high through the cloudy sky, dipping and turning at light speed, playing in the air. She flew with the other fairy terns, and when I came out, she alighted on her feeding perch and ate two fish. She came more often to eat that morning, and flew a lot, but rested in the afternoon. I began early to coax her down for supper. She cried again, and flew toward the feeding perch, but instead of landing, she hovered, circled over the water, and returned to her high branch. This she repeated at intervals until evening came, when at last she alighted on the feeding perch. Without climbing up to her, so she would feel safer, I handed her a fish from below. She snatched it from my hand and simultaneously rose up so high that she disappeared into the dark blue above.

She possessed unexpected power of spirit to choose starvation rather than risk violation. Later, I went to see her on her branch as night fell. She paid no attention to me. She didn't cry, and when I kept looking up and talking to her, she turned her back on me.

The next day, she was tired and rested. In the evening, she cried again as I tried for two hours to entice her down, and eventually she came, but on the following evening, she did not. Never again would she come easily for her evening meal, though in an effort to reassure her, I always handed her the fish from below instead of climbing the ladder to be with her as I had before, and, of course, I never grabbed her again. But the antibiotics I put in her fish killed the infection in her foot, and it quickly healed.

Angel became stronger and more adept at flying, and as this incident faded into the past, she ate more. She still counted on me for emotional support, flying down from her branch with little cries each morning and toddling anxiously over to me on my ladder at her feeding perch. I would lean toward her talking softly, and she would delicately touch my face, especially my eyes and mouth, with her beak. When she saw me outside, she still called to me as she had done when she lived in the house.

One morning she was trembling and seemed distressed. I softly stroked her neck and chin. She closed her eyes, tilting her head to press her face against my hand. Whenever I stopped, she cried. Touching a seabird is to be avoided since it harms the feathers' waterproofing, so this was one of the rare times I caressed her.

On another morning, I brought her the usual two fish in a bowl. She ate them, looked pointedly at the empty bowl, then at me, and made her little cry. When I held it closer to her, she looked in it again, then back at me with a louder cry, just as if to say, "Its empty!" I told her I could get her more, but she would have to wait. And she did. When I came back with two more fish, she looked in the bowl as I neared, to see what was in it. Then she ate the fish and flew away.

While searching for her late one day, I saw two birds circling like eagles, so high above that you could scan the sky and not see them. The next time I looked, they weren't there, but later, I saw one still turning high in the blue. As I watched, it seemed, strangely, to get larger and became Angel, holding in her wings and doing a guided free-fall straight toward me. She flew up to me and hovered briefly in front of my face before swooping under her canopy and alighting on her branch. At that moment, the female fairy tern flew to her newly hatched chick with a fish in her beak, and put it on the branch in front of him. He ate, and then they touched beaks gently, many times. Here was the instinctive basis for Angel's inclination to touch my face, and I remembered too, how she had held the dying fairy tern's beak in hers when she was still a baby.

One day not long after, I found her preening in the afternoon sun, and she seemed damp. Even her head looked wet. She wasn't hungry, and when I persuaded her to eat a fish, she could hardly swallow it. It stuck, and for several minutes, she used her head to try to push it farther down her throat. I realized she must have caught some fish herself that day. After all her troubles, finally Angel was achieving independence.

One morning at dawn, she fluttered from beneath the canopy, and flew straight out across the bay, feeling the wind and playing for a while, then she paused and hovered. Suddenly, she dove, then rose slowly from the water's surface. It seemed as if she had caught something. She flew on down the opposite shore of the bay, repeating the hovering and diving pattern several times, sometimes shooting back around over the water in a tight circle and diving again, as though making another attempt at a school of fish that had eluded her on the first try. With the languid pattern of lazy, climbing circles now so familiar, she soared up over the trees where she was joined by another bird then a second. Soon, there were four with her, and another group of four joined them until a whole flock was swirling there. I could still recognize her by her wing beat, a little faster than the others.

Later she came to her feeding perch, but wasn't hungry. She just wanted to be with me, and stayed for half an hour, touching my face and exchanging with me vocally in her way. Suddenly something startled her, and she fluttered momentarily above the branch. Our dogs were running toward the road, and when I looked, Angel looked too. The light fell on her from behind, encircling the perfect form of her rounded face in golden light. Her brilliant, black eyes gazed back into mine. How lucky I was to have this chance to know such a beautiful creature. That night, when I went out to offer her a last meal before nightfall, she called to me but would not fly down. I talked to her as I always did, until the gathering night obscured her.

The next morning, I flew to Tahiti for a doctor's appointment which had been long delayed while Angel depended on me. Now that she was fishing for herself, I felt I could go. However, that morning I missed greeting her at dawn and when I went out a bit later, she was not there. By the time I had to leave, she still had not returned, so I left three fish on her feeding perch, hoping she would find them after her morning flight. She was not there when I returned at 2:00 p.m., and the fish were untouched, except by ants.

By 3:00 p.m. I was worried. It was an unbearable anguish that the only morning I had not gone out at dawn to greet her was the very morning I should have. After nearly three months of compulsively checking on her to make absolutely sure that she was always looked after, at the crucial moment, I had not been there. I waited under her tree, looking out over the bay for her butterfly form flitting homeward, but she never came. Waiting as the sun set, I saw all of the fairy terns return to their trees, but the falling night brought its dark curtain down on my hopes, and the life of Angel.

In time, I concluded that a swamp harrier had taken her. This large hawk was introduced by some brainy soul to feed on the rats, failing to take into account the fact that rats are nocturnal, and these big hawks hunt during the day. The result is that a high percentage of young fairy terns are killed by them. Swamp harriers have eradicated fairy terns completely from some regions in the islands.

A few months later, I was brought an older juvenile fairy tern, who had become exhausted in a storm. When he had rested and eaten well for two weeks, I took him outside to the perching trees. On his first joyful, frolicking flight, he paused on a tree on the mountain, and two swamp harriers came over the hillside. One fell on him with such speed that I couldn't believe my eyes. There was a pause, then the great hawk lifted into the air and disappeared over the hill, the white wing of my fairy tern hanging limply beneath him. Hawks can tell at a distance if a bird is easy prey, and since then, I often saw them drawn to the playful flight of a young fairy tern.

The end of October brought turbulent conditions. Mountainous waves tumbled across the reef, turning the lagoon into a river. Through the cloudy torrents pouring through the coral, fish flickered from shelter to shelter, swept downstream whenever they passed through the open. The site had a fresh layer of sand on it carried by the waves, in spite of its distance from the reef and the depth of the water. In the intense surge, I couldn't let go of my perch to arrange my slate and camera properly, and clung there for dear life while peering through the gloom at the sharks, as they shimmied through the torrents. It took all my force to keep from being swept away, while the pressure of the waves on my lungs was close to intolerable. Bobbed around in the seething waters, I was soon light-headed and disoriented.

Feeding the fish was problematic, since the current instantaneously carried the food away, especially from the squirrel fish, who waited beside their coral home. With my back to the reef and the current, I couldn't keep myself balanced and feed them, too. Finally I let the current trail me, hanging onto my perch with my back to the site, while handing a crumb of food to each fish to make sure that each was able to get something. Unfortunately, the moray eel was back in place, and there was a second big one in the same labyrinth of coral. The effort to satisfy the darting, flashing fish, watch behind me for sharks, and keep track of the sneaky insinuations of the moray eels in the powerful current, seemed overwhelming.

True to her secretive nature, that was the moment when Annaloo came. She glided across the site, her tail beating gently against the current to keep her properly oriented on course. Instead of disappearing, this was the time she chose to circle high over the site then float down to the food. It was hard to see her in the cloudy water, and I drifted to the coral formation across the site, where I could face into the surge while watching, and stay balanced in the river more easily. Annaloo had several mating wounds. I parted the floating *turbinaria* as she passed behind me and peered through it to see her eating in the coral. A ray of late sunshine played over her as she readjusted her jaws repeatedly to get them back in place. Then she drifted slowly on and vanished.

I fed the last of the scraps to the fish before leaving, handing most of them to my favourite group of squirrel fish in the grotto to my left and not noticing the moray eel under the coral overhang. I was just releasing a bit of food when he reached into view to take it from between my fingers.

On leaving, I drifted with the current to get out of the turbulent waters, and find a safe place to climb into the boat. It was a surprise to find that downstream of the site, out of visual range, many sharks were passing, some of whom showed marked interest in me and the boat flying through their midst. Lightning came close twice as I travelled, and Flannery came to me from out of the gloom. A large whitetip shark glided beside me for a moment as I swept along. The sharks must have been socializing there. Then I was past them. Farther on, there was a deep and quiet place where I climbed thankfully into the kayak.

At a session soon after, night was gathering as I dug the meat out of a tuna head for the fish hovering around me. Throngs of needlefish hung pointing at me at the surface, and below them wrasses, goat fish, butterflies, angels, and countless others flitted in an effervescent cloud. Squirrel fish and butterfly fish hovered around my hands, cod lay around me on the coral, and groupers darted up from below. Suddenly a large shark approached through them, so I dropped the fish head and raised the camera in case it was somebody interesting. She swam on straight toward me, and when her face was half a metre (20 in) away, I took the photo. The flash didn't bother her at all, and still she came on! Her mouth passed a few centimetres from my mask as she curved by. Her body was so close to mine that I could have put my arms around her, and I waited with interest to see her dorsal fin pass my eyes. Indeed, she was a stranger. She hadn't understood what was going on when she encountered the sight and smell of my fish feeding.

She moved away through the coral, and I continued to feed the fish. Soon she did the same thing again, and I did the same thing, too, snapping her picture when she was very close. But this time, when she didn't startle at the flash, I raised the camera just slightly in a mild threatening gesture, but she didn't react to that either. Maybe she couldn't see very well, but at that distance, she should have been able to sense me just fine with her electro-sense. My heart was pounding and my stomach had tightened unpleasantly; she was not behaving normally.

But just at that moment, the sun pierced the clouds for the first time in the session, revealing the site arrayed in multi-coloured gleams and flashings in the play of flickering golden sunbeams. It occurred to me that this shark would be very easy to photograph—quite the opposite for example, of Annaloo, whom I had been unable to capture on film though she had been passing through for months.

So I swam up to each side of my new acquaintance and snapped a photo. She was such a memorable shark, so awe-inspiring to look at, her rhythmic oscillations so mesmerizing, and her behaviour so disturbing, that I decided then and there to call her Shenandoah. Her behaviour had not been menacing; she just had not recognized any personal space for me. But by then I had known so many sharks, of such a variety of temperaments, curious and shy, bold and sneaky, that I knew how peculiar the best of them could be.

Though continuing to feed the fish seemed to be out of the question, I was still surrounded by them in their thousands, fluttering around my hands and looking into my mask. Lacking the heart to disappoint them, I began to feed them anyway, while keeping a close watch for the reappearance of Shenandoah. She didn't come, and soon I became engrossed in trying to get my favourite butterfly fish to take a morsel of food from my fingers. Suddenly, Shenandoah's huge face appeared just in front of me, coming on fast. This time I handed her the fish head with a gesture I hoped she understood and slid backward out of her way. She curved sensuously over my perch and tried to grab the fish head. But she couldn't get her teeth into its smooth surface, so glided on. As I circled, my lookout became a swaying mass of rainbow wrasses all glittering in the sunshine, while around them flitted the other species, seeking crumbs.

I had just regained my lookout when Mephistopheles darted into the site. He was as macho as I remembered him when I had met him a year before, fleetly flying from one scrap to another, and changing direction at the speed of light. He didn't mind me at all, and I turned my attention to photographing him. He was greyer and darker than he had been, and I was able to check that the bite out of the middle of his face had not removed his nostrils, as I had often wondered. As he shot confidently by, I noted a long, light-coloured scar on his right side, and wondered whatever had happened to him earlier in his life. He was a creature of spirit to have overcome an experience that had left him with such terrible scars. He left the site often to return again several minutes later.

Since the nurse sharks demolished these big heads if I put them in with the rest of the food, I had begun keeping two or three on the boat for the fish feeding. I brought the last two I had, and put one in a handy depression under me in my coral formation behind my hand-holds, then ripped the meat from the other, and scattered it. The bright fish came flying, eerily beautiful in the shifting, golden light. For a while, we were tranquil there until Shenandoah glided through them again, hidden until we were nose to nose. She turned about a centimetre away. I dropped the fish head, drifted leftward and snapped photos of her as she tried to grab it. She knocked it down onto a shelf of coral half a metre above the sand, where it was noted by a nurse shark. He clambered up, using his pectoral fins like paws and, propped there, did his best to get a grip on it. He was so funny that I wanted to photograph him, but dared not get distracted from Shenandoah. She was making repeated passes over my lookout, trying energetically to get the other fish head, while wrasses fluttered around her, all lit from behind by the setting sun. I ran out of film, Shenandoah cruised away, and Avogadro and Ruffles came. I fed the fish a final time and left, wondering if Mephistopheles and Shenandoah could have come from the same place.

In November, four tiny grey sharks appeared in the local nursery, cruising the champagne waters of the breakers crossing the reef. One at a time, they flitted over to look at me. Perfect, satin sharks, they seemed to have awakened ecstatic in our world. They cavorted through the sunlit torrents pouring through the multi-coloured sea-flowers of their little world and vanished. I couldn't find them again.

Lillith finally birthed. The change in her appearance was startling when she reappeared, looking emaciated and sinuous as a snake. Apricot had become a resident, and at the end of the sessions, a very young female of delicate bronze, began investigating the site. She had a silver circle on her tail and soft gold flares in her white dorsal band. I named her Trillium, a pretty name for a pretty shark. Annaloo still passed in her ghostly way, always at the same time, always from left to right, and every time, she caught me looking the other way. She still did not stay and eat, though she had first begun these secretive passes in June, six months before. Glammer always put in an appearance about the same time. She and Annaloo were companions, yet Glammer's behaviour in the site and with the other sharks was normal. It was still inexplicable to me that my mystery shark was the same one who had come to swim with me and practically eat from my hand on that glorious morning when she first had visited so many months before. What reason could there be for such a change in her attitude?

In mid-November, I paddled out at the end of a windy day. As I prepared, the shadows of sharks passed steadily beneath the careening surface, and an occasional nurse shark came up vertically beneath me. Once, three came together. Flora was near when I slid into the water, and she swam slowly by and around me, looking. Her colour had paled from dark grey to light bronze, the same colour she had been before she mated. Madonna came next, and many, many more were farther off, cruising, waiting, and watching. One after another, they came up to me, and turned away.

As always, I took the anchor line and trailed the boat as I swam over the coral wall on the north side of the site to stand on the shelf of dead coral beside my lookout. From there, I had more leverage than I would if floating free to push the heavy scraps out of the well in the kayak. They fell in front of my perch, so that afterwards, I had a perfect view of the action while being protected from it on the coral wall. This procedure had evolved over time, and I managed every session the same way. I always threw the anchor in the same place so that when I finished with the boat and let go of it, it came to rest on its anchor line behind me, up-current, where I could instantly reach it in case of emergency.

The sharks swirled down to the food to eat. Storm seemed irritated when she arrived, pushing past a little shark in her way with a menacing gesture. Sybyl, who was only a metre long and very young, had a few mating wounds. I was waiting for a quiet moment to separate the scraps for them, but Nightwind rammed into the little pile and did it for me. It was interesting that the little one was the most forthright in this matter. Arcangela passed, dramatically pregnant. She had not visited since the previous mating season. Splash was cavorting around, and I looked for Droplet but did not see her. Annaloo floated into view. The boat had drifted in front of me, with the alternate turbulence of water and air, and intermittently banged me on the head, managing to entangle my camera in the anchor line. By the time I freed it, it was too late to photograph the mystery shark, and as usual, she did not return. Samaria slowly came up the scent flow. Her babies seemed about to swim out of her. Her lower abdomen bulged at each side, and looked squared off as though the babies were moving down to birth.

As darkness fell, the usual males began passing. I identified two new ones, Penrose, who had wild markings, and Cochise, who resembled Cochita and Cochito; the name would help me remember him. For the first time, I glimpsed the geometrically designed fin of another, who became Fermat. All were unknown males entering the lagoon where the females were gathered in the night, just as they had the year before.

Lillith came, moving more energetically than usual. She scouted about, poking at bare spines, though scattered bits of meat lay about. One nice one, was dark red in colour and away from the other scraps; it was easy to see on the white sand and presumably would smell bloodier, yet she passed over it, and a nurse shark came and inhaled it. Storm floated in, searched out, and devoured a large fish steak. She too, looked as if her babies were on their way out. She was enormous, and manoeuvred with difficulty.

The whitetip shark that had been visiting regularly, a large male with three dark spots in a line up his right side, came and stayed for about ten minutes, eating around my lookout, and when I fed the impatient fish, Mephistopheles arrived with another band of males, but it was too dark to see who they were. He only entered the site once on the far side, but his missing face glowed in the dark. Droplet came just as I was leaving, followed by Splash, who had joined her companion since her earlier visit to the site.

As I drifted home, a nearly full moon rose over the mountain, layering the water with silvery gleams and casting a violet glow across the night. As I passed into the bay, a flying fish flew by my face in the dark, its wing passing under my chin. I reflexively turned my head to see it splash lightly into the water three metres beyond. There had just been a flicker, a realization that something had flashed by incredibly close, yet not touched me, and delicately re-entered the sea before all again was still. I lived in a magic world.

It was no accident that we didn't collide. It's strange flight was well guided.

The incident inspired me to begin a painting to capture the exotic beauty of the bay in the moonlight as it appeared on those enchanted nights when I left the sharks. I sketched the mountains the next time I went there, while Martha circled the kayak, then cut them from the paper, laid them across the moonlit sky, sprayed with my airbrush, removed the paper, and like magic, the dramatic mountains appeared, surrounding the bay. It became my most beautiful painting, with the light arrayed across the mountains reflecting upon the water, and plumbing the depths of the bay. Later, I populated it with the spinner dolphins, in hopes of selling the prints on my dolphin tours. Those dolphins were often playing in the bay when I went out to the sharks. They rested during the day in those protected waters, and at night went cavorting into the ocean to hunt three hundred metres (1000 ft) down, deep in the shadows of the sea.

The following week I returned to the site in perfect conditions. Martha was passing when I slid in, and Cochita and Sparkle Too were drifting on my other side. The visibility was more spectacular than I had ever seen it after its recent cleansing, and the water had quieted so that the current was not a problem.

The sharks ate as usual. Samaria was there, hungry and slender! The ten-month gestation period was confirmed again, and this shark had not been gone long to birth. Carrellina and Bratworst had both mated, and little Nightwind had sprouted tiny claspers! She was a boy, I realized in amusement, watching him excitedly rushing around. Tamarack's claspers were finally of a size proportional to his body; it was ten months since I had first noticed them beginning to grow as little buds. Bratworst quickly left, stuffed, and Madeline soared into the site. She set about eating in her forthright fashion, looking young and willowy; she too had birthed. Jessica, missing since mid-October, followed her in, also displaying the now familiar sinuous look of a female after parturition. Storm was wonderfully slender too! I recalled how much she had eaten the week before, which contradicted what I had read, that sharks don't eat before parturition. And Storm rarely ever ate anything at the site.

Many sharks came through as the session advanced. Flannery and Avogadro appeared, now pale gold, suggesting that they had spent time out of the sun's rays in the blue ocean, with the mating season reaching its peak. The females were paler and browner, though from June to September they had been a rich, dark grey. Even Madeline was bronze, a sharp contrast from her usual charcoal colour. The sharks who had just birthed seemed especially pale. Yet Lillith and the occasional visitor, including Eclipse, remained very dark.

Windy and Filoh swept down for the same large chunk of meat at the same time, and Windy, the smaller shark, got it. I had read that to learn which dolphin was dominant, one had only to throw in a fish and the higher ranking dolphin would be the one who took it. But among the reef sharks, it was the fastest and smartest one or the one in the right place at the right time who got the food. There was no hierarchical structure to their society.

Fleur de Lis glided in, following an unpredictable and infrequent pattern of visits, and Trinket appeared. She usually appeared every few weeks, though sometimes not for months, and didn't look any bigger than she had when I had met her nearly two years before. At every visit, she showed her funny character. A few very small sharks came, one of which became Breezy (for his W-shaped dorsal fin tip line). The small sharks who first started attending the sessions were probably two year olds, but I never saw any of the pups identified in the local nursery appear with the lagoon population to verify this point. There was doubtless a high death rate among the shark pups, and it appeared that they didn't mingle with the other sharks until they had nearly doubled in size.

Then there was the factor that individuals varied greatly. Further, Lightning and Fawn had appeared to be very young when I first identified them, but were males, who scarcely seemed to grow as time passed. So they could have been older than they seemed at first.

Occasionally, a pup the size of a newborn flitted into view then disappeared again. These were so nervous they were impossible to approach, and they did not join in the feeding. They had likely come sneaking in to see what all the commotion was about but were uncomfortable away from their hidden refuges in the coral.

The food was soon eaten, and the sharks roamed in search mode, scrounging for leftovers while I fed the fish. Annaloo cruised by, covered with mating bites. I snapped a photograph of her for the first time in all the months that this sneaky shark had been passing through. The cuts were dark lines, so she must have mated about a week before. Many of her slash wounds gaped open. Flora passed through as she still did just before the sun set. She had filled out, but seemed to be maturing rather than pregnant, in spite of the terrible mating injuries she had received the year before. A beautifully proportioned, immaculate, platinum-gold male glided before me that evening. He was unusually large, his dorsal fin pattern matched on both sides, and he appeared to have been designed by an artist. It was love at first sight, and I named him Antigone. Whenever he came, which was only two or three times a year, he left me in rapture over his beauty. Dapples

appeared with two other males just before I left.

As in the previous season, unknown males passed at each session, often in groups of two to five, arriving after sunset. They scouted the site looking for odd scraps that had been missed by the others, and never paid any attention to the copious scent flow I was creating by feeding the fish.

I checked the nearby shark nursery before darkness fell completely. With so many sharks birthing, I wanted to see if more pups had appeared, while the sea was calm enough. But this shallow place on the reef was still turbulent, and the light was too dim to see much detail. As I drew near, a tiny grey shark approached, and there was another in the distance. Deeper in, Gwendolyn accelerated away and disappeared along the reef. She had not come to a session for six weeks, yet she was still in the area and still pregnant!

There again was the quantum mechanical conundrum of the observer affecting the observed. Even if I could freely and easily visit the nurseries at night to watch, it was unlikely that a shark would give birth within my view. As I paddled languidly home, the rising moon's light outlined the mountain from beneath like a cosmic spotlight. It was just beginning to wane. Gwendolyn appeared a few days later, and she had birthed. Much birthing and mating occurred, it appeared, in the light of the full moon.

When the waves unexpectedly calmed on the north shore, I went to check the nursery on the eastern end of the lagoon, swimming out from shore as I had done when I lived there. I moved along the reef, counting and drawing dorsal fins. The pups were scattered along a wide area of the reef, and it became confusing as the little groups grew to more than five, but by then, I had counted thirty-nine shark pups. They were curious, and some came a few times to look at me, so I could not be sure if the one I was seeing had already been counted or not. The water was so shallow that I had to bend my head down and twist it to see them circling around me underwater, and I couldn't look behind me at all. Finally, I just tried to get back to a region in which I could float.

After passing a long stretch of reef where no little sharks appeared, I came to the heart of the nursery. This was where the newborns had been when I had explored the region initially, and there I found many. After exploring, watching, counting, and sketching for some time, I raised myself to look around above the surface. A large, adult fin was proceeding slowly in my direction from farther along the reef. The tiny fins of little pups cleared the water here and there, too. I counted ten. I made my way carefully back to the edge of the nursery, looking above and underwater for the adult shark, but didn't see her again, and continued on. I tried moving into the shallows to find and count the newborns, but it was impossible, so I skirted that region on its lagoon side, watching from the edge to get an idea of how many pups there were in each area. I found more and more. They were everywhere. From above the surface, I had counted ten fins, but underwater I saw that most of the tiny sharks were deep enough that their fins were underwater. Could there be eight for every two whose fins showed on the surface? Or more? A conservative estimate would be forty to sixty pups in the region immediately in front of me, calculated by dividing the area into sections, bringing the count so far to nearly one hundred. However, there were many more farther still, and counting them accurately was impossible, since the area was so wide and shallow. There could be twice as many further on, or maybe not, so between one and two hundred shark pups was a very rough estimate of the pups in the whole nursery region. I wasn't concerned by this spread of uncertainty. I had found a place where the numbers of young corresponded roughly with the numbers being birthed by the pregnant females I had seen.

One especially tiny grey shark came to swim near my face several times while I lay motionless on the bottom looking at the little ones swimming around me. Its curious personality was already present.

I decided to go back earlier in the day, to try to see as many of the pups as possible, get a more accurate count, and note any with special markings that would allow me to recognize them later. Then I could learn the age at which they grew claspers, and the age at which they matured, if any ever attended my sessions, three and a half kilometres (more than two miles) along the lagoon. But within twenty-four hours, the ocean swell had grown, and the waves crashing over the area made it impossible to return for several weeks.

During one session, the lagoon was at its worst. Heavy clouds covered the sun, and in the underwater gloom, everything was surging, everything was trying to fly away, or slithering from shelter to shelter. The roiling torrents lifted spirals of sand like underwater whirlwinds that flickered continuously across the site. While a few sharks foraged in the scraps, a very young juvenile flitted through. Her dorsal fin tip was similar to Annaloo's, a graceful point at the leading edge on the left side. I sketched it and forgot about her. Many small juveniles passed and I never saw them again. Weeks passed, and she did not reappear.

Few sharks attended sessions in such turbulence. The extra effort necessary to navigate through the coral probably motivated the matronly females to retire to the calm depths of the ocean when the lagoon was agitated. Yet I was able to note that Madonna, Bratworst and Sparkle Too mated during this period in December, and Amaranth's pregnancy began to show. Flora appeared covered with deep mating wounds, laced with shallower cuts. She had not become pregnant the first time she had mated. Mephistopheles came again at nightfall while I was feeding the fish, almost precisely a year after my original sighting of him.

One evening, an extraordinary creature undulated near, just beneath the surface. She looked like a needlefish a metre (3 ft 4 in) long, as silver as a salmon. Whereas the beaks of the little needlefish were needle sized and shaped, hers was similarly pointed but fifteen centimetres (6 in) long, and lined with large, sharp, interlocking teeth. When she opened her mouth to snap up a crumb of the food I was scattering for the fish, her gullet was five centimetres (2 in) in diameter. She turned her head first one way, then the other, to gaze at me from each huge eye, and she did not approach.

I couldn't find her in my reference book so didn't know what to call her. Later, I was told that she was a crocodile needlefish (*Tylosurus crocodilus*), and that such fish were very dangerous, should you catch one and step on its mouth. I read that they attack and kill people in the water for no reason and kill more people in Polynesia than the sharks, a fact that was kept secret for the sake of tourism. This was a truly bad fish.

The crocodile needlefish came again near the end of the following session, too, always staying at least four metres (13 ft) away, which was fine with me. I found her appearance alarming, and had more than enough to cope with already, with the sharks, the demanding multitudes of fish, and the sneaky eels.

In the wind and current of the wet season, putting in the sharks' food was a balancing act. When the water level was high, just my shoulders were out of the water as I stood on my shelf of dead coral, bracing myself to scoop out the scraps. With winds piling up big waves, and yanking the boat around, it took all my strength to stay balanced. There were also the sharks to think about. Once I started to dump in the food, they zoomed excitedly around me, and with my head out of water, I couldn't see them, and they knew it. I liked to move quickly through this part and gain the safety of my perch.

One evening, I was just preparing to push the food into the water under such conditions when I saw one of the huge Javanese moray eels looking out of the coral beside my feet. I instinctively stepped back, and something soft yielded under my fin. I shrieked when I saw that I had stepped on a stonefish! The turbulence raised by my fin lifted it a bit from where it had been lying, and it settled gradually back again. It had not raised its hackles, and it was unhurt. So I knelt on my coral outcrop to push the scraps from the kayak. The stonefish stayed nearby for the rest of the session, and I kept a careful eye on its whereabouts. As I fed the fish, it came to rest a few centimetres from my left elbow. Though I knew that these fish only use their poison when they are stepped on or squeezed, I didn't feel comfortable with such a deadly fish so close. I remembered how they strike like lightning, and wished it would go a little farther away.

The yellow trumpet fish still stayed companionably near me and even drifted around the site with me, though he never appeared to be moving. Vertical in the water and about half a metre (20 in) long, he propelled himself subtly with his fan-tail and tiny rippling fins. Finally, I learned what was truly motivating him. Already edgy about the proximity of the stonefish, I was shocked when he suddenly shot vertically in front of my eyes and trapped a rainbow wrasse in his elongated mouth. I could see the wrasse's bright markings through his transparent yellow skin as he slowly swallowed it. Its companions fluttered near, watching.

CHAPTER TWENTY

Adventures in the Storms

The sharks always circled through my mind. I wondered what they were doing when I wasn't with them, and saw them in starlight descending into the oceanic darkness at night. I gazed out toward the lagoon where they were circling at the appointed moment of our rendezvous, posing to myself questions about them and wishing I was with them. In my dreams they could travel in air as easily as in water, and cruised through the landscape around me as I went about my dream life. One night, I was dreaming when a voice told me that one of the sharks I had seen was the local viewpoint of a divine being who had come to observe conditions there. The shark was conveniently shown swimming by, so I would know who was being referred to. Though I remembered the dream on awakening, I didn't take it seriously, not because I thought such things impossible—I did not know. But if a high level spirit had come to experience sharkness in the lagoon, I was not important enough to be informed.

Nevertheless, I had become involved with sharks originally due to a riveting dream, and I remembered the shark. It was the little one who had flitted through during the terrible storm, looking like a tiny version of Annaloo. The next time I came across the drawing of her fin in my notes, I smilingly wrote "Shilly," which was the only part of the divine being's long name that I could remember, beneath it.

After the Christmas and New Year's holidays, the fishermen went on vacation, so no fish were being caught and cut up for the hotels. As a result, the sharks and I had to wait to get some scraps, and thirteen days passed before I could return with food for them. They had not forgotten. As I approached the site, several sharks caught up with the boat and circled it. Underwater, the fish came streaming toward me from as far as I could see, and Bratworst was near. While pulling the boat to the site, Madonna swam across my path at arm's length, and Martha glided at my right side. She was slender and vivacious after having given birth—her cycle was the same as it had been the year before. Flora, Cochita, Isis, and Windy joined them, and they all swam beside and under me as I trailed the boat to the coral shelf and began to toss in their food.

The sharks torpedoed by. Amaranth had birthed as well. Christobel was not pregnant, but the once slender shark had developed a heavy build on maturation. Lillith had mated about one and a half weeks before. All the others seemed fine, many with mating wounds in different stages of healing. Ruffles, Meadowes, Avogadro, and Shimmy had turned from dark grey to pale gold, as they had the year before.

I fed the fish while Madonna glided lazily up the scent flow, time after time. As twilight fell darkly, I happened to look behind me, and found the little shark from my dream, Shilly, passing very close! It was weeks since my original brief sighting of her on the other side of the site. She was a beautiful little bronze shark, and it was amusing to see the way she had come sneaking around behind me to have a look, yet had not appeared in the site.

Near the end of January, I received a large load of fish scraps. It had been a long time since I had brought more than a couple of spines, heads, and some pieces of skin. After a few days of calm, on that morning a front began to blow in. The cloud cover darkened during the day, and as I prepared to leave, thunder storms were approaching over the northeastern horizon. On the ocean, it was raining. The food had come packed in six large sacks, which I just stacked on the kayak, thankful I didn't have to handle the prickly stuff. When I shoved the precariously loaded kayak onto the sea and settled myself, I found that it was so unbalanced that I had to lean slightly in one direction to stay upright, and my feet trailed in the water. The wind quickened as I paddled, and several times I came close to capsizing while crossing the bay.

The border was always treacherous in stormy weather because the water there was shallow, accelerating the current flowing out of the lagoon, the same way that a river turns to rapids. Meanwhile, incoming ocean waves rose up there, and broke over the outflowing torrents. Crossing this region was always suspenseful, sometimes more so than I would have liked. There was a deeper place I located by watching where the waves broke. A narrow channel crossed the border there and the waves rolled in without breaking. I glided slowly toward it, timing my crossing to coincide with a pause between waves, then as I shot into the lagoon, though not too fast if I was also fighting the wind, I was picked up and carried the rest of the way by the next wave. At the border, I was ten minutes from the waiting sharks.

Approaching the site, panting with the effort, and anxiously watching the conflict between wind and waves in order to gauge at which angle the kayak would line up when I anchored, I realized that the anchor line was under the sacks, and paused to free it. I had very little room to work, with my gear jammed between me and the sacks, and when I found my slate tangled with the anchor line and tried to extricate it, my camera nearly flipped into the water. The kayak was plunging like a frightened horse, and shifting broadside to the waves.

Engrossed with these precarious manoeuvres, I realized that the boat was surrounded by sharks. Looking back, I saw them following, two and three abreast, like children after the Pied Piper. More and more came circling, sometimes breaking the surface. Now drifting with the wind, I got my things sorted out, and paddled to the site, but by the time I freed the anchor and threw it overboard, I had drifted a fair distance away. A surprising number of sharks approached the boat in single file to sniff, and their remoras, wriggling with apparent glee, flitted around in the water under me. I watched each shark for a while, but through the tossing surface, I could only recognize the prominent rectangle on Carrellina's tip-line.

While putting on my gear in the cramped space, I thought about the best way to proceed, given how far I had anchored from the site, and the numbers of sharks already present. I finally decided that the fastest way would be to jump in with the biggest sack of scraps, dump it at the site, and bring the boat over to dump the rest after the sharks had calmed down. However, already agitated by the long battle against the wind and the trouble of extricating the anchor in the precariously balanced boat, I forgot about the part where I would be swimming with the sharks with a big sack in my arms from which blood would be billowing.

Being an absent-minded person who drifted through the world in an artistic trance, I often overlooked such things, which is why I made a point of trying to think everything out in advance on the shark expeditions.

I slid into the water, and turned around right away to face the sharks. While they surrounded me, I concentrated on handling the heavy sack into the water, and setting off for the site. Flora glided at my side, and Madonna, Bratworst, Martha, Spark, Carrellina, Cochita, Rumcake, Samaria, and Madeline, the latter two covered in mating wounds, swam in a crowd close around me. Halfway to the site, Flora circled in front of me, blocking me, so I had to slow down. I had already been feeling traces of anxiety, which had begun as I watched the approaching storms, and increased with my difficulties with the anchor line in the wind. Flora's close approach made it coalesce. Samaria kept touching me along my side, and as she slowly passed, others closed in.

I suppose that they were feeling what passes for happiness in sharks, but my quivers of anxiety grew. Many imperceptible signals passed between us as we approached the site, and I must have begun finning faster. So did they—my twenty-three sharks were anticipating a party. My increasing anxiety and their increasing excitement translated into a steady acceleration of our swimming speed. Arriving at the site, I turned the sack upside down very smoothly, and blood billowed into the water. The sharks were speeding around me, and moving faster all the time—I was starting to panic as they reached the speed of light, the speed at which they could no longer be seen. Frantically, I shook out the sack, unable to see in the billowing blood, and envisioning my hand being bitten off as they shot by snapping up the first of the falling pieces. I kept backing away as the scraps poured into the water, then leaped onto my perch, on which I knelt. I dared not let my legs protrude into the water solid with sharks.

It was a credit to them that they continued to respect the difference between the food and yours truly. Here was the proof that they could learn a routine—the food arrived in the boat, and they ate at the site after I placed it. Finding myself in clouds of blood, I managed to drift leftward to the other outcrop, where I tried to maintain myself against the current, and write down the names of those present.

In order of appearance there were: Flora, Madonna, Martha, Cochita, Carrellina, Bratworst, Spark, Rumcake, Samaria, Madeline, Trinket, Sparkle, Sparkle Too, Mara'amu, Meadowes, Ruffles, Chevron, Sybyl, Teardrop, Cochito, Flute, Pippet, Tamarack, Storm, Gwendolyn, Nightwind, Trillium, Arcangela, and Windy, plus some unknowns including a nearly black little thing excitedly eating something behind me. She had a few distinguishing marks which I could not note down in the mêlée.

I had taken two bags from the boat by accident, and one of them had dropped on the sand at the edge of the site. I waited for the sharks to calm down so that I could retrieve it, but there were always too many sharks around it. Once, Samaria grabbed it, and the woven thick plastic stuck in her teeth. But she shook her head a few times, and freed it, and after that, no shark bit into it.

The food was gone in moments, and when they went into search mode, I finned back to the boat, returned with another sack, and dumped it out. Amaranth passed me. How could everyone have come at once? Had they learned to count to seven?

Jem was there when I settled myself after dumping out the next bag of scraps, and the little ones had come to look for a bite as the large sharks dispersed. An unknown, large female who had mated within the last two days circled through the site in mid-water. I named her Anne Boleyn. She was with a companion whose tip line was a variation on the W theme, and who became Hurricane, the biggest of the wind sharks so far. The two companions stayed for the rest of the session. Shortly after they arrived, Eclipse soared through like a jet plane, chasing another shark who was shooting away with a scrap in her mouth. Eclipse always made a swift, hit-and-run pass from left to right.

The sharks got excited over a large piece of food in the third sack of scraps, and the whole group took off with the shark who had discovered it, shaking and dropping and snatching it up again. They disappeared from view, and I took advantage of the opportunity to move the anchor, pull the boat over, and push the scraps from the back of the boat. After that, the site was full of male sharks, including Marco, Kilmeny and Raphael, who had last appeared precisely a year before. I began drawing the dorsal fins of the others, most of whom I had seen once or twice before during the previous year. Two of them had swollen, uneven claspers, which could, I speculated, be evidence of recent mating. One was the little male with the geometrically designed fin pattern, called Fermat, and I was glad to perfect my drawing of it. But there were so many sharks present that I couldn't identify them all. There were many males visiting from the ocean and many female visitors, too.

A whitetip approached and I prepared my camera, but when I looked up he had vanished, and another six blackfins were coming. Emerald passed, and across the site, Annaloo and Glammer were scrounging in the coral. Annaloo still would not come near. Grace arrived after an absence of months, and we were all honoured by a visit from Flannery. A lovely golden female passed through once, whom I named Raschelle.

It was getting dark, and Lillith came. Many of the others were still roaming through the surroundings, and I finally began to feed the fish who were waiting, as always, in front of me. The clouds of needlefish extended from my mask across the site, all facing me. I had been there an hour by then, and they had waited a long time, so I got started. As the scent of crumbling fish meat filled the water, some of the sharks approached, but none troubled me.

There was a little moray eel in a hole under my face as I clung to my perch. So after I fed the patient squirrel fish in their chasm to my left, plunged a few pieces down for the groupers and cod, and sprayed handfuls of crumbs around me, I concentrated on trying to get a bit of food to him. He slid thirty centimetres (1 ft) of his body out of his hole, and took a large piece of food I wafted his way. Then he retreated.

Isis was passing when I looked up. A tiny nurse shark pup was lying on the last sack of scraps still resting on the sand, and nine nurse sharks were doing their exotic dance on the floor of the site. I got the rest of the fish heads from the boat and fed the meat in them to the thousands of fish, trying to make sure that each one got some, and keeping an eye out for sharks approaching too closely while the food was streaming from my hands. The last fish head was huge, and I couldn't get to the bottom of it. I was hypothermic, and wanted to go, but didn't want to miss anything either. I was energetically digging out the meat, trying to get it all, when I saw what looked like the three-metre nurse shark approaching. It looked huge in the gloom. It got bigger and bigger as it approached and I realized that it was dark grey and not a nurse shark. The only grey shark of that size that came to mind was the tiger, and my heart began to pound.

But when it entered the site I saw the shape of its face—it was the biggest sharptooth lemon shark I had ever seen. In less than two metres of water, the sight of it was astonishing. I was still compulsively pulling meat out of the fish head, but then I let it fall. The shark, looking as large as a baby humpback whale, curved down onto the sand and picked up a scrap at the edge of the site, which he carried away.

I had been taught that at the appearance of a potentially dangerous fish, especially when it was dark and one was alone a kilometre from shore, one should get out of the water, and of course that's what I decided to do. However, there was still that sack lying on the sand, still with the nurse shark pup lying on it, and I couldn't leave it there. So I swam over, shooed the baby off it, and picked it up. I was shaking it out, surprised to find that it contained bloody scraps, instead of fish heads, as I had thought, when the lemon shark came back. He turned away at the edge of the site, while I increased my efforts to get the prickly scraps out of the sack. I finished, and everything was quiet. But I couldn't leave when there was a chance of seeing that immense shark just one more time, so I waited tensely on my perch. After a couple of minutes, the great dark shadow approached again through the coral canyons. He reached the site, but when he became aware of me, he turned and disappeared the way he had come. I got back in my boat and drifted over the glassy lagoon, searching for another glimpse. But it was as if no shark had ever graced those dark waters, so empty and still did the coral garden now appear.

I could see no reason why so many sharks came, why they were so excited from the start, or why, for the first time in one hundred fifty-four feeding sessions, they had followed the kayak at the surface. It was the first time a lemon shark had appeared, and he was as big as they grow—the size of the one who had alarmed Franck and me while we were diving outside the reef. I was weak and shaky for the rest of the evening from my fright. At home after rinsing my gear and the kayak, showering, and starting dinner, I wrote my report on the session. I had written down forty-eight names, and described several more new visitors. Roughly estimating the number of strangers who had come, I concluded that at least sixty five sharks had attended. The usual number was half that.

But the next session was ordinary. It was a calm and beautiful day, with no wind, and the lagoon was glassy. No sharks followed the boat. Dapples came, and Cochito reappeared after an absence, with another vertical slash down his side. It was odd that he had been slashed twice in the same place in the same way, when I had never seen such an injury on any of the other males. Raschelle passed from left to right as she had once before, and Grace and Vixen came together. Vixen had first come at the same session as Grace almost exactly a year before, and they had been together a month later as well. Here was another set of travelling companions. Marco, with a deep cut at the right side of his mouth, came well before sunset for the first time in so many infrequent visits. Annaloo passed one time, left to right, in her meditative way.

I fed the fish, waiting in suspense for the lemon shark, but he did not come.

On a morning of rare calm, I went out to look for Madonna, who had been absent for two weeks. Anchoring just over the border, I swam toward the site, hoping to intercept any sharks who were there. I had a piece of fish with me, wrapped so well in plastic that no scent escaped.

The current was still so strong that I could scarcely swim against it. No sharks appeared, and I had nearly reached the reef when Martha came gliding beside me a metre to the left. Isis was with her. I circled slowly through the study area, the two sharks accompanying me, but Madonna did not appear.

Finally, I pulled out the chunk of food, broke it in two equal pieces, and handed one to Martha. It fell through the water between us, and she didn't see it. The same thing happened with the other piece, offered to Isis. The two sharks followed me expectantly. I was facing them, upright in the water, with my knees bent as I finned backward, holding the small plastic bag in front of me. Swimming side by side, each one's nose at the tips of my fins, how clean cut, how graceful, and how beautiful they looked in the sunlight. I thought of trying to get the large piece of fish off the bottom and offering it to them again, but it was out of reach, and I was reluctant to make any sudden and clumsy movements with them so close.

But before I had to do anything, Isis broke off, turned at a right angle, dove to the bottom, and swam unerringly to one of the pieces of fish, which she swallowed. Martha followed. She went straight to the other piece and inhaled it. Yet by then, we were quite far up-current from the two pieces. There had been no sign of a search; each shark had gone straight to one of the chunks of food. It was uncanny.

After, Martha swam close under me, and I followed her. Samaria was passing, and Tamarack was visible in the distance. I was only able to swim with Martha for five minutes, while she moved across the current—when she turned toward the reef, I was left behind. I never saw Isis again.

The next feeding session was in mid-February, 2001, and I had very little food. Many sharks swiftly appeared, their dorsal fins breaking and swirling the surface. They were excited, and I got ready fast. Once more, Carrellina was one of the closest, her unique dorsal fin easily recognizable as she swerved along the surface beside the boat with Sparkle Too and Martha. I splashed my fins on the water to encourage the sharks to move back so I could slide in. As I trailed the boat into place, Martha was with me, Cochito was on the other side, and many others were converging from all directions, to swim along with me.

The thirty sharks excitedly fed as their food fell through the water. I had saved the last of the scraps for the fish in a sack, which I held, folded, under my body. A nurse shark clambered up, and squeezed between me and my perch, trying to find it. When I snatched it away, he continued out the other side. I tossed the contents into the site, where Shimmy inhaled the one decent scrap; the shark food had already vanished. Fish grabbed the other bits, mostly skin, and began flying around with them, trailed by the rest of the hundreds present. The sharks all came shooting back from down-current and went after the fish, trying to get these little pieces of skin. There were no backbones and heads lying on the sand for the perch to nibble on, this time, and they were all impatiently looking for this missing food.

Christobel arrived, and the same scene was repeated as I threw them all the last few scraps left at the bottom of the well in the kayak, handful by handful. Each time I threw some, the sharks soared back from down-current and chased the fish, who raced with the scraps around the site, a situation that had never developed before. Fish waited in front of me, turning to look at me first with one eye, then the other, and the three-metre nurse shark lay on the sandy floor right in front of my perch, as if waiting for me to produce a fish head for him. I got the only backbone I had from the back of the boat, pulled off a handful of food for the groupers, cod, and squirrel fish waiting to my left, then dropped it into the site. It was instantly grabbed by a shark, and in a moment several more were trying to take a bite from it. I was so disturbed by the unsatisfied, milling fish and the frantic sharks that I left early. Madonna had not come, and I noted Isis's absence as well.

However, the following session was different. The day was calm and sunny, the current was slow, and the lagoon glassy; it was easy to see that there were no sharks following the boat. When I arrived at the site, they glided languidly into view one by one. To my relief, Madonna came. She and the others cruised with stately calm, drifting over the food to inspect it, carefully choosing a bit, and eating. Watching Flora tasting a piece of skin while nosing into a belly strip and ignoring it, I puzzled again over the wild behaviour they displayed sometimes. They were obviously the most reasonable of creatures.

Flora was again among those waiting for me; when I slid underwater she was nearly always at my side.

As darkness was falling, two dramatic new visitors arrived with Arcangela. One of them brought an exceptionally large and pale remora with her, who left her and came over to eat with me. I was feeding the fish at the time, and this large creature ate from the fish head as I shredded the meat into the water. He put his head deep down inside with my fingers, and I was worried about accidental bites. However, he behaved as if he had always eaten with a person with skin the colour of the meat, and never made a mistake. He stayed as long as I fed the fish, and became stuffed. Once he was attracted by Sybyl, who was circling nearby close to the surface, but he quickly left her, returned to me, and later rejoined the same large shark with whom he had arrived, the next time she passed. I called this shark Celeste, after Galileo's other daughter, since she had come with Arcangela.

Was it possible that remoras distinguished between individual sharks and had preferences? Perhaps a remora who usually remained with the same shark would learn the shark's habits, and adjust his or her own life-style accordingly, rather than being subject to continual surprises and unwelcome displacements with strange sharks.

A nurse shark got his head stuck under an outcropping of coral, and wiggled back and forth for some time before managing to use his fins to push himself backward, out from under the overhang that pressed his head against the sand. He had gone in after a fish head, which had been swept underneath earlier in the session.

A whitetip passed, ghostlike in the dusk. When I was about to leave, I found that I had another little bag of fish chunks, forgotten in the back of the boat, so I dropped the anchor again, went back, and began scattering crumbs for the fish so that the carpet of nurse sharks, including the three-metre behemoth, wouldn't inhale them all. While I was doing this, Kimberley came up the scent flow with Mordred. I threw them the scraps, delighted to see them. Twilight had appeared earlier, and I had wondered if Kimberley would show up too. But I wouldn't have seen her if I hadn't gone back with that little, forgotten bag.

Toward the end of February, I went out under a darkly overcast sky, no blue having shown all day. A three-day storm was settling down. Upon the lagoon, a light west wind arose as I slowly approached the site, watching. No sharks were following the boat, but while I put on my gear, they appeared one after another. When I slid into the water on the eastern end of the site, due to the unusual west wind which had pushed my boat there, I was astonished to see the water full of sharks, all facing me as far as I could see into the coral. In the middle of the site, the three-metre nurse shark hung in mid-water like an improbable centrepiece. I pulled the kayak to my coral shelf, swept the food overboard, and the sharks began their riotous feeding. One of them grabbed a piece, and they all converged on it in a radial formation, rising to the surface as they came together. They disappeared into the coral chasing each other over this one piece, which was when the nurse sharks and fish took the rest. When the sharks returned, they went into search mode and scooped up everything that was left, remaining unsatisfied.

The three-metre nurse shark was sweeping around, monopolizing various backbones and fish heads and stirring up the sand. He was swimming with his mouth wide open, but veered away every time I approached him to photograph his strange look. I found this frilly fish the size of a horse endlessly fascinating, but nearly impossible to photograph well, due his size, the dim light, and the extreme fore-shortening that occurred with frontal views with the basic camera that I had. From the side, he was too big to fit in the picture.

Rumcake, Cochita, and Sybyl arrived with a young male I called Darcy. There was still a spine with a good coating of meat in the boat, so I got it for them. Cochita approached it as it fell through the surface, and soon, all the sharks were taking bites and passing it from one to the other. I drifted after, as they moved slowly westward to where the three-metre nurse shark landed on it, his huge flag of a tail waving slowly above as clouds of sand rose around him in the dark waters. In and out of this pale green fog countless sharks were flying. This scene of utterly alien beauty was breathtaking, and I tried to find a way to photograph it, though no photograph could ever capture the eerie reality, all in motion. Samaria got a bite, and as she swam away, her mouth still open, she rammed straight into a coral formation, and was brought to a stop. Ouch! She left after that. Finally, two sharks got what was left of the fish spine and carried it, one after the other, back to the site.

When Arcangela came, the gigantic nurse shark turned menacingly toward her as she glided above him; she turned quickly away. It was the fourth time I had seen a nurse shark menace a blackfin, but never the contrary. Nor had I seen blackfins menace each other.

I fed the fish until it became too dark to see, feeling increasingly disturbed by the waves, current, and opaque clouds of fish impeding my view of the movements of the sharks, especially the busy nurse sharks. The three-metre one kept circling around me. Finally, I picked up the anchor and drifted, hoping to see more blackfins down-current. I did see shadows passing in the stormy light, but they were far away, and the current carried me on. Moments later, the three-metre nurse shark appeared, following a smaller three-metre nurse shark, suggesting that there was more than one of these mega-creatures around. The two were winding among the big coral structures in the area, the male following the female. I couldn't follow them, since the boat and I were being swept along by the current. Yet their path crossed mine three times. Their graceful flight, fins twirling, flags waving, was just beautiful. Finally they appeared crossing toward me, and after the female passed, the male saw me flying toward him on the surface, and crouched down, pressing his fins against the sand. Then, with a flickering of his pennant tail, he accelerated and disappeared.

I drifted to the edge of the lagoon, where I suspected the sharks went to hunt in the evening, but found no one. A cloud of needlefish had stayed with me all the way. I was sorry I had nothing for them and found, on close examination of the back of the boat, a tiny piece of meat, which I threw into the water. One of them ate it.

February and March were stormy months, but each time it seemed that the benevolent spirits of the air parted the curtains of rain, and quietened the winds, so that I could see my sharks on schedule, in comfort and safety. Often, dark clouds covered the sun, bringing light rain in drifts as I went out, and darkening the sessions. The streaming lagoon created a dim, stormy scene in which the sharks were seen only fleetingly, shimmying through the torrents. Nurse shark pups sought refuge in the lee of coral formations.

The nurse sharks were aggressive. I had just a few pieces of skin and a spine or two most of the time, so there was little for them on the sand for so many hoping to eat. They crept vertically up the other side of my coral lookout, and suddenly their heads would appear, eyeing me hopefully, just in front of my face. When two of them shoved their heads at me at the same time, one on each side, I became nervous. With the mask limiting my view, and the many fish obscuring the scene, coping with everything in the strong current took all my attention.

I was followed everywhere I went by throngs of needlefish who stayed as close to me as they could get, looking solemnly into my eyes whenever I glanced their way. Sometimes I had nothing to feed the fish, but since I fed them very well when I could, which was quite often, and they were able to get a little of the shark food each time anyway, I told myself I shouldn't be too awfully worried about it. However, the needlefish seemed terribly concerned about me ignoring them, so I always cleaned every little crumb from the back of the kayak for them.

The day had been sunny and mostly calm prior to one of my excursions, though at times, violent squalls had passed without warning. The sea was restless, and as I neared the reef, I saw that the waves, which had diminished a little in size the day before, had grown alarmingly. The current would be a nightmare at the site. So I decided to find a suitable deeper area farther from the reef, where the current would be weaker. A big advantage would be the elimination of the problematic nurse sharks and the milling, unsatisfied fish.

This worked. The water was deep and clear of fish, even needlefish, and the ones who were there didn't hound me. There were two dead coral structures, appropriately placed for me to hold onto, and even a shelf I could stand on to dump the scraps. My sharks had no trouble finding me. The scraps were gone in moments, and the choice pieces from the last tiny bag were swallowed by the next three sharks coming up the scent flow; those behind got nothing. I had one fish head and fed the fish who accumulated slowly, eating scattered across the much wider area. The scent drew arriving sharks close enough for me to see them.

Several nurse sharks were looking for food, and one was swirling around me, eyeing the fish head. At least some of them came to the sessions up the scent flow from unknown distances, maybe even from the bay's depths. Yet the smaller ones slept at times under the large coral formations around the site in the day time, so it was unclear which of the nurse sharks lived in the lagoon, and whether some, especially the very large ones, came from deeper waters.

Sparkle Too stayed for the entire session, as she often did, and Apricot remained in the area long after the other sharks had left. Cruising through the region as darkness fell, I found Shilly drifting toward me for a close look; weeks had passed since the tiny shark who had appeared in my dream had come.

I cancelled the following two sessions due to storms. The week after, as I was starting to get ready, another storm began blowing out of the east. In half an hour it still hadn't materialized, so I left, thinking that if I could just get out there, I could always get myself home, even if I had to swim. On the way, the wind gusted strongly from four directions, and on the lagoon, there was a steady east wind. There were tumultuous waves on the reef, and the lagoon was like a witch's cauldron.

Searching for the deep water site I had used the time before, I lined up the landmarks which I had noted, anchored the plunging kayak, and slid underwater. But I was in a different place. There was no dead coral shelf I could stand on to brace against the wind and current, and nothing to hold onto for stability. The lagoon flowed like a raging river, and in the gloomy torrent, a throng of excited sharks were approaching me fast. I had to push the food from the boat, trailing in the current while clutching it with one hand.

Madonna was close by and Valentine was with her! Hard as this session was, just seeing Valentine again made it worth the struggle and stress. She had been gone for three months. There was a long slash on her right side, and mating wounds about a week old, but it was hard to see her well as she zoomed through the fog with all the others in chaotic motion. I found a dead coral structure down-current to hold onto.

Twilight was there, and the dramatic shark, Samurai. She swept in, grabbed something, and left. The beautiful, golden Antigone kept drifting by like a vision. Watching him while hanging on, desperately bracing myself against the sluicing water, and half drowning when swamped by a wave, I was transported with joy. As the light faded, the current became so bad that I dared not let go, and I worried about how I would ever get the anchor and myself back into the boat. The visibility got worse, as the increasingly powerful waves raised spinning sand storms. I waited and watched as the sharks dispersed, till only Sparkle Too occasionally shimmied past me.

Ruffles and Meadowes arrived at nightfall. Once dark grey, they were pale gold! Since the water at this location was deeper than at my usual site, and the coral was lower and more spread out, I could watch them roaming through a wide area. They scouted nearby at first, then just swam slowly together, sometimes side by side, sometimes in single file. They followed a slow and meandering path of circles, loops, and figure of eights, in a similar pattern to that of the shark pups I had seen on the fringe reef. I watched them until there was a lull in the current, whereon I grabbed the anchor, rushed to the boat, and flew into it.

In the world above, there was black sky to the east, and a vivid, almost circular rainbow shining before it. That long wait in the gloomy flood shot with silvery lights, wrenched by the sea's mystic power, as the golden shark companions roamed below me, remained in my memory, touched by the glory of the supernatural rainbow that overhung us. It was the last time that I saw Meadowes in good health.

The following session, I went to the deeper site again, and was relieved to locate the original place where I could stand on the worn coral shelf. Reassured, I started to pull out the food, then remembered that I had forgotten to check that there were no moray eels or stonefish where I was standing, so looked down. I saw a huge head near my foot through the bloody water, and screamed, then realized that it was just a fish head!

This session proceeded normally, and instead of Meadowes and Ruffles gliding together as darkness gathered, Annaloo and Glammer were roaming. They came at the same time, and due to my greater altitude and view, I could watch them cruising together, side by side, or one after the other, much of the time. They stayed for about twenty minutes, passing close to me several times, scouting for food, leaving and returning in the pattern typical of the sharks. It was the first time in eight months that Annaloo seemed relaxed. The first time she had drifted through, Glammer had been on the periphery. Yet due to Annaloo's shyness, this was the first time I was able to see them together roaming naturally, as if no one was watching.

Three belly strips lay near my lookout. The belly strip was the slice of poorer quality meat from the large fish's ventral surface. The strips were long, thick and wide, and had enough meat in them to feed a small family. They were often sold for human food, but the island peoples were spoiled when it came to fish, so rarely bought them. I was pleased to get some for the sharks since they were among the few scraps that did offer nourishment. Yet, the blackfins showed a surprising lack of ability to locate these choice morsels. They seemed to be the most inept of animals at finding proper food, in spite of their exaggerated reputation.

Instead, the belly strips were inhaled one after the other by the nurse sharks, who had no trouble finding me at the deeper site. Their thoroughly stupid appearance seemed to be all show. In fact, they appeared to be more capable than the blackfins in many ways. It was mind-boggling to consider how much of my sharks' precious food they consumed. There seemed to be no limit to the volume they could suck in. They were a terrible nuisance.

As it got dark, Chevron came up to me at the surface, seeming a little agitated, followed by Sparkle Too, who turned closely around me as I fed the fish. I was paranoid because of the approaches of all the nurse sharks and was looking back and forth between the sharks, the nurse sharks all around me, and my hands, to make sure that nothing was getting close enough to bite me, while having great difficulty maintaining myself against the current. At the same time, the boat, which I had already moved a number of times, began banging me on the back of the head with each wave, so I left the two fish heads in a depression of my coral lookout and went to move it. When I came back, the three-metre nurse shark was clambering up my coral perch to get one. When he peeked over the top, his eyes bulged in surprise as we stared at each other.

I began to feel high tension after that. There were nurse sharks undulating in slow-motion throughout the huge volume of this site, their tails waving between the floor and the surface; some had risen vertically under a fish head, which like a mask, replaced the nurse shark's. With the fins of the fish head curving outward above the wide-finned, pale body swaying slowly back and forth, the scene appeared as a macabre dance. Unfortunately it was impossible to photograph due to the bad visibility, through which no flash would work, and the submarine darkness, which enhanced the ambiance of the bizarre. I finished and threw the fish eyes to Sparkle Too, who got at least one of them.

Then I stayed watching for just a few moments before leaving. Suddenly, the large silver remora I had met a month before appeared, flitting back and forth at the speed of light in front of my mask, then around my neck, and torso, as if he were just thrilled to see me again. I was really not thrilled to see him, however. He badly startled me. Exhausted and unnerved, I just wanted to get out of the water. Him darting at me so fast with his mouth open, disturbed me along with everything else, and I ignored him at first, since I had nothing at hand to give him. I was much more interested in looking for the shark who had brought him.

But after a few minutes, when he went to try to get a few crumbs from a nurse shark, my heart melted, and I found a fish head that still had meat in it. I had to swim down to the nurse shark to attract the remora's attention, but when he saw me, he eagerly joined me, rose with me to the surface, and ate with his head beside my fingers, as if we were the very best of friends. When the fish head was empty, I threw it to Sparkle Too, who was still closely circling us. It was very dark. I put the anchor in the boat and drifted away, passing Chevron and Bratworst down-current. But the remora was still wriggling happily around my body, so I got back into the boat; I didn't want to displace him. As I took off my gear, he appeared under first one side of the kayak, then the other, his long graceful tail waving slowly out from under it as he turned. We were quickly drifting away, and when he did not leave, I paddled slowly back to the site, watching to make sure that he was coming. There, sharks were passing in the black waters, and though I didn't see when my remora left, I knew he would find his way.

One evening toward the end of March, I was watching the sharks swirling over the food that I had just placed for them, when one of the larger females approached me and kept on coming.

She came so close that I put a gentle hand on her head to make sure she turned. This was strange behaviour for a shark I didn't know, especially when there was plenty of food available for her. When she curved away, I recognized Celeste, as pale grey as Arcangela, who soon appeared as well. They had come together again. Many big females were still passing through, though the reproductive season was almost over. As twilight fell, a group of very large, older sharks swept in with Lillith, Annaloo, Glammer, Kimberley, and

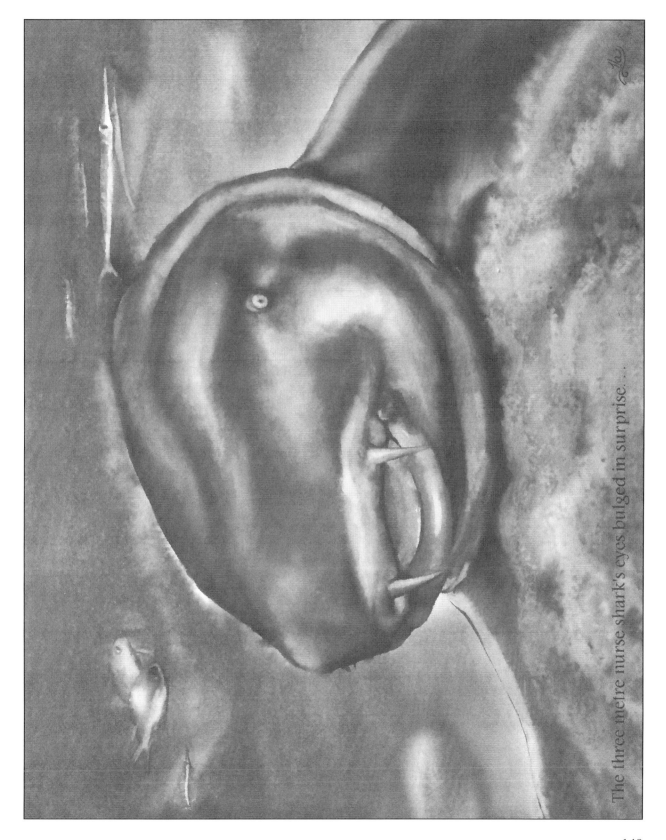

The three metre nurse shark's eyes bulged in surprise....

Vixen. Since all the scraps were around my perch, and the visitors were swooping aggressively around them, I held out my camera, aimed at their dorsal fins, and snapped photos. So in spite of the darkness and their numbers, I was able eventually to sort out who they were.

All were rare visitors from far away, including a pair of comrades who were to visit together again just once a year at this time in the future: Clementine and Odyssey. They were very old, blotched, and scarred. Odyssey's gill on the right side was damaged; the separation between the first gill opening and the second had been cut out, leaving a hole five centimetres (2 in) across.

Early in April, I paddled out to the site over water the colour of a black pearl. Storms were passing to the north, and the day was glittery, yet hazy, all edges blurred. I had been carrying the scraps on the boat in the sugar sacks in which they had come. This was so much easier on my hands than extricating them, and packing them in the boat, and saved me half an hour when I prepared the kayak to leave. But having the food in sacks made the boat less stable. The centre of gravity was higher than it was when they lay in the well in the back of the boat. On the previous occasion, a motorboat had passed me so closely when I crossed the bay, that the sharks' food had nearly fallen off, so this time, I carefully strapped the five sacks on, feeling very conscientious about it. But having loaded the boat on the beach, I wasn't aware until it was afloat that it wasn't balanced. As I paddled out to the sharks, I had to lean far over on one side to keep it level. Crossing the wide expanse between the end of the bay and the edge of the lagoon, another motor boat zoomed close by me, and I nearly capsized.

I drifted when I got to the lagoon, looking down through the glassy waters at the coral and fish. After a while, a dark shape under me caught my eye—a big shark was swimming directly under the boat! Startled, I watched, and when she meandered out on one side, I recognized Madonna's dorsal fin. She drifted under the boat again, swimming with it like a remora. When she meandered out on the other side, the dark circle above her left pectoral fin became visible. She went back underneath, and as I became aware of other sharks, and hastened toward the still distant site, I began to believe that she was no longer there. It was not easy to believe. Usually, sharks avoided being right under the boat. After a while, I felt the boat touch something. On verifying that I couldn't have bumped a coral formation, I decided that some of the fish scraps must have shifted. The bump came again, and a moment later, Madonna's side appeared right under me. She had nudged the boat. If I had reached down, I could have stroked her. She stayed under the boat, as far as I could tell, until I reached the site.

When I slid into the water, as carefully as ever, the boat overturned, and I tried desperately to right it. But the shark food, packed in sacks that had filled with water, were hanging underneath it, strapped to it tightly. It was impossibly heavy. I located my camera and slate, which were already flowing away, and hooked them over my arm. The water was clouded with blood seeping from the sacks, and the sharks were darting around looking for the food, which they could smell but couldn't find. Their normal routine was not proceeding as anticipated.

I concentrated on freeing the sacks, reaching underneath and feeling along each strap to the catch and releasing it, forcing myself to focus on what my fingers were finding and ignoring everything else. It was impossible to see anything, with white-caps rushing over me and the tossing water opaque with blood and sharks.

When the last of the sacks fell to the sand, I flipped the kayak upright, swam sideways out of the blood, and held onto a coral formation to survey the situation. Greenish clouds hung in the coral, and through them flew the sharks, eerily emerging and disappearing. Blood was billowing from the sacks, and they couldn't find the food. They seemed excited and confused. I counted thirty-five present, including the magnificent Clementine and Odyssey.

Avoiding the bloody water, I took a sack that was relatively in the clear, dragged it to the site, and dumped it out. The sharks began to feed. One by one, I brought the other sacks and finished placing their food in the site, which became a cyclonic vortex of sharks.

The little black thing that had appeared at the excited session in January was there, and it was possible to note her distinguishing marks and draw her dorsal fin. I named her Brandy. Storm and Samaria came with an anxious Droplet.

Cochito appeared with another gaping slash ten centimetres (4 in) long, behind and parallel to the first one that he had received two months before. I had noted almost precisely a year before, on March 31, that he had a slash behind his gills, and wondered if he was aggressive in the mating season and had been slashed by another shark. But it seemed counter-intuitive that another shark would slice him vertically, all the way from his back to his belly. Such a slash from another shark would likely be angled and more horizontal than vertical. I remembered the long cut in Marianna's bloated side, which had looked so much like a knife wound. Could he have been slashed by a knife? The possibility of attacks on the sharks was worrying. Still, why would the same shark be cut with a knife the same way three times, and no other? It defied reason.

As it began to get dark, Marilyn cruised through. At home, I confirmed that I had seen her just once, a year before, on March 31. This sighting was on April 2. Her return, at exactly the same time of year, was another striking example of how closely sharks could follow a schedule over the annual scale.

A pile of spines, untouched by sharks and still covered with meat, lay in the place where the boat had over-turned. The incident showed to what extent the sharks anticipated my routine: I took their food to the site, and then they ate. They had waited until the table was properly set for them, before beginning to eat. Then, though behaving as if they were starving to death, they had ignored these perfectly good pieces that had been in the wrong place! I spread them out in the site, and fed the fish intermittently. A strange female passed through whom I named Algebra, due to her mathematically perfect dorsal fin tip line. 'Geometry' didn't sound poetic enough, and I already had a Carrellina. Juveniles were feeding, and the big sharks were placidly roaming the area.

Down-current, drifting idly, I found Gabriel drifting idly too. He had visited only four times in two years. As I looked him over, he moved to a pillow-shaped formation of worn coral, positioned himself, and used it to scrape his ventral surface. It was an interesting addition to the sharks' repertoire of behaviours, to see that they surveyed the landscape to choose a formation in the environment that suited a need. It wasn't tool use, but only because the shark hadn't modified the coral pillow—it was a choice made among a variety of possibilities. As well as seeing a shark turn on its back to wriggle in the sand, I had sometimes seen them slam their sides against a sand bank. Here was another example of choosing an appropriate surface in the environment, from among many, for a specific purpose.

Two remoras came close as I fed the fish. One ate from my fingers, while the other stayed in the centre of the site, snapping up crumbs, and wouldn't approach. This one dashed off when it saw a shark in the distance, then turned vertically in the water, and returned to the site.

Clementine and Odyssey returned as darkness fell. Clementine swam straight up the scent flow I was creating, and nearly collided with my face. Jessica, who rarely came, glided in then too. I suspected that she had her home range quite far to the east and sometimes came to the border to hunt in the evenings. Marco was circling the site counter-clockwise at a distance, as he often did for long periods, before approaching.

Kimberley and Twilight arrived together! It was so dark then that I could hardly make them out, but the two sets of twin white snowflakes and unique dorsal fins were unmistakeable. Their resemblance to each other was as impressive as was this new confirmation of their companionship. Lillith glided in behind them, but by then the sharks were only movements in the dimness, and it was time to leave.

At least there was still a bit of food lying on the sand for them.

CHAPTER TWENTY-ONE

Further Investigations

My job taking tourists to see wild dolphins ended when a contract was lost, and suddenly I had some free time. I decided to spend more time with the sharks individually. Having been observing them to find out what they were like as individuals, it would be interesting to learn more about where they were at different times. Did they go out in the ocean? At night? When they had left the study area, where had they gone?

So on a flawless morning, I paddled out to the mirror of water held in the embrace of the reef, anchored near the border, and swam slowly into the lagoon toward the site. The magic garden, so packed with sharks when I came at night, was deserted, except for tranquil fish hovering in the fairy glades.

Finally, Bratworst came swimming very slowly. She passed three metres away and disappeared ahead, then every few minutes as I went on, she passed again. I held a bit of food out to her, planning to give it to her and then swim on with her, but she didn't seem interested. Finally, she approached, but when I tried to float her the food, it dropped onto the sand. By the time she went into search mode and looked for it, the current had drifted it, and the fish had eaten it. I put a second piece down for her, and she ate that. Then she searched all over the area, looking for more pieces, before circling twice and heading east. I followed, but she swam into the current and I was gradually left behind. I slowed and drifted, looking around.

Bratworst came back. She glided past me and we set off again, but this time when she left me, she did not return. Shimmy was roaming near, and a juvenile passed, then Madonna soared up to me and circled away. When she came again, I handed her a piece of food, but she failed to see it, and it fell to the sand. She returned, and I floated the morsel to her again. This time, she turned, plucked it from the water, and swam away, slowly swallowing. Her mouth was distorted open as she languidly left with me in pursuit. She made a large circle before heading toward the reef flats at the border. I was able to keep up for about five minutes, but she increasingly accelerated and finally disappeared from view. I went on in the general direction she had taken, hoping to find her again, and arrived at the sharks' little nursery. There were no pups there, so I drifted back toward the kayak, looking around, but Madonna had vanished, and no other shark appeared.

As I neared the boat, a remora came swimming quickly along in mid-water as if it were going somewhere. It briefly considered me and swam on. Thinking there must be a shark in the area, I followed it. But swimming fast is no way to find a shark, and I didn't want the remora getting attached to me. Yet when I turned back, it came. Holding onto the kayak, I glanced around, and Martha appeared around the back of it. She raised her nose delicately toward it as she turned to me. No doubt water containing traces of fish fluids, was draining from the holes in the well. I drifted beside her as the remora fluttered over her body. She was badly bitten all over. The bites were not fresh, but they were not outlined in black either, so she must have mated about three days before.

I stayed with her as she swam languidly through the coral south of the boat. The remora flitted against her. Every few minutes it swam halfway to me and then returned to her. Sometimes, it left her to investigate something in the area then quickly rejoined her, wriggling over her back as if happy to be with her again, before taking up its preferred swimming position between her pectoral fins. Martha sometimes rose in the water to examine something nearer the surface, and once her dorsal fin broke through. She went south, then east, in a wide counter-clockwise circle as she wound through her coral range, and when she turned toward the north, and I was swimming against the current, I could scarcely make any headway so she gradually left me behind. When I lost her, I was not far from the boat, and I hoped she might have paused near it, but when I got there, she was not in sight, and no other sharks appeared. So I let the strong wind that had arisen sail me home.

Later, I thought of how placid Martha had been on this first excursion, even for her. Was it because she had been in pain? I had often recorded, while observing them, that females with mating wounds moved very slowly, and seemed less alert.

I went to the lagoon every few days, and roamed with each shark who came. Sometimes I was able to stay with her for fifteen or twenty minutes, and sometimes she quickly left me behind, often coming back, before leaving me behind again. Food was not necessary, since the sharks knew me and came to me anyway, through curiosity or camaraderie. The big females' natural cruising speed was slower than that of the males or juveniles, so they were the easiest to swim with.

During the following sessions, I was honoured by repeated visits of the magnificent Clementine and Odyssey, and Twilight and Kimberley, who arrived together just at the fall of night. Marco was still in the area too. His visits in the past had frequently coincided with Kimberley's. Could that be chance, or were they from the same region and roaming together very loosely? Twilight and Kimberley did not always come at the same time, but I would see them in the area, sometimes together and sometimes separately, during a period of about four weeks on most visits. Then they were absent again for several months. Glammer and Annaloo came regularly and had begun feeding with the other sharks more naturally. When Glammer came, I could nearly always count on seeing Annaloo and vice versa. The society of the sharks appeared as a gigantic puzzle of which I could only place the odd piece. So when I did discern some relationship, it was gratifying.

Toward the end of April, I was watching Samaria swimming under Chevron as she swallowed a scrap, and trying to wrest part of it away from the smaller shark, while thinking of leaving. I'd brought very little for the sharks to eat, and they were circling. They often approached the surface at the back of the kayak to sniff it as if to see if it might contain more food. They seemed to have no trouble understanding that I used it to come to them from beyond the silvery ceiling of their world, and that I kept their food in it.

I put the anchor in the boat, and began to drift with it, but before I had gone far, Kimberley emerged from the gloom down-current. I waited, still drifting down the scent flow. A minute passed. Then Twilight appeared, following Kimberley's path. The two sharks were not within visual range. Trailing the boat behind me, I finned toward Twilight and was able to approach and swim with her. She went toward the site, but turned to pass down-current from it. Had I been on my usual perch, she would not have been visible. Then she turned back. Kimberley appeared, equidistant on my other side. She had apparently already crossed the site and circled back. The two immense sharks were curving toward each other, as though following an arc of the same circle. They met in front of me, passing close by each other. I took in every motion they made, but saw no sign, and wondered whether something more was exchanged between the old friends than a human eye could discern. Had there been communication? Perhaps the close meeting itself was the communication, affirming for each shark, "We are here."

Twilight languidly cruised back and forth down-current from the site in figure of eight patterns, not minding me staying with her. Time after time, she overtook another female shark who was already present, apparently by pursuing her trail of scent. Had she been targeting the other shark's vibrations, it was unlikely that we would always join her by coming up behind her. Each time, the two big females passed close beside each other, and continued on their separate ways. As Twilight swam placidly on a sinuous path toward the east, I sensed rather than saw a change in her and instinctively stopped. As soon as she felt my pressure off her, she turned around and swam back toward the site, brushing past me as if I weren't there. I followed.

Many sharks were present, and as we passed the site again, Kimberley came into view. Two big dark females materialized in the gloom in front of Twilight. I managed to get close enough to one of them to recognize Lillith. So she was present again at nightfall, and if I had stayed at the site, I would not have seen her. Perhaps she passed by for social reasons since she was so rarely there in time to eat.

Twilight's actions illustrated that sharks have a way of locating each other when out of visual range. I remembered the shark pups I had seen, playing in front of our house in the early morning, responding to each other and travelling together. Perhaps as baby sharks in the nursery, they became companions. Coming from the same mother, they would already be familiar with each others' scents. As time passed, they moved out through the lagoon, staying in loose contact through their swimming patterns of huge circles and figure of eights. As well as Twilight and Kimberley, companions who closely resembled each other were Ruffles and Meadowes, Flute and Piccolo, and Cochito and Cochita. Annaloo and Glammer did not.

One evening at dusk I went to the border to look for sharks. Those from farther up the lagoon often passed by at sunset, and I wanted to investigate. Twilight was roaming there, but not Kimberley.

Near the site was a large female, and I followed her for a while, but she was deliberately keeping ahead of me, and I knew she would accelerate if I pushed her. So I returned to the site, and found that she had returned too; it was Glammer. I swam after her, but she lost me, swimming straight west toward the border. I cruised around a little longer, and saw Droplet passing and the shadows of sharks in the distance, so drifted back down to the border. After many minutes of watching, a female came into view. I moved as quietly as I could toward her. She wasn't happy about it, but by remaining very quiet, I eased myself close enough to recognize Glammer. She had come to the border when she left me.

As I went on, one of the three-metre nurse sharks passed me from behind, undulating with surprising rapidity toward the bay. I hastened after him, in time to see him go straight out of the lagoon, and plunge down the slope into the depths. After all this speculation, I now knew that at least the big nurse sharks went at times into the bay!

By then, the accelerated current over the shallow reef flats was carrying me away. The bottom dropped off below me into blackness. But at the surface, hovering as if in meditation, was the crocodile needlefish who had attended a few sessions some months before. I recognized the two parallel scars on her side. She gazed solemnly at me as before. I didn't go close enough to frighten her, and climbed back in the kayak, wondering if that was her usual abode and whether she, too, had discovered the scent flow at the border and followed it to the sessions. That explained why she had arrived about half an hour after I started, though not why she had never come since.

Arcangela was waiting for me at the start of the next feeding, and came to look into my face before swooping down over the food. There was a new female visitor, too, whom I named Charm. She approached once inquisitively. They and the residents swirled around the food. Suddenly, a large shark left it and swam up toward my face. I wondered who it was and why she was coming over, unable to see her fin design from in front. When she changed her trajectory to the side after nearly touching my nose, Martha's dorsal fin passed my eyes. She still cared to leave the food to come to me! Whatever went on in those shark minds?

A short time later, Celeste came up to me at the surface, so I accompanied her as she went on. But she didn't like it, and began accelerating to avoid me, so I left her alone. By the time I returned to the site, she was back too. An oddly thin Meadowes was scouting about looking for scraps, while Ruffles followed Shimmy nose-to-tail. Later, he followed Fawn in the same way. This pretty little shark was growing tiny claspers! Feminine Fawn was a boy!

Afterwards, I drifted to the reef flats at the border and saw no one except the three-metre nurse shark, who hadn't been coming to recent sessions. The nurse sharks were much more plentiful at the sessions during the wet season, and given the male I had seen following the female, it was conceivable that their reproductive cycle also took place in the wet season. Perhaps they put their young in protected places in the lagoon, maybe under large formations of coral where hollow cavities inside afforded protection for the little ones, who searched for food in the vicinity under cover of darkness. However, though it made sense, nature is too complex to speculate about. Though I had seen nurse shark pups in the lagoon, I had never seen more than one at a time.

CHAPTER TWENTY-TWO

The Mirror Experiment

Among the many questions I had about the blackfin sharks was the relative importance of their eye-sight as opposed to their sense of smell. Everything I had heard about sharks suggested that their sense of smell was all important, and even that their brains were devoted to the analysis of odours and little else. But given the range of behaviours I had seen, I felt that this was impossible and that eye-sight was important to them.

So, on one of those still and golden mornings as the dry season began, I set off for the lagoon with a large oval mirror for my sharks to look into. Six chunks of fish meat lay in a bowl so that no fish liquids could escape into the water. I anchored west of the site.

Underwater, Bratworst was cruising slowly. She didn't approach as I unstrapped the mirror, pulled it into the water, and set it up against some dead coral, setting it so that it was nearly vertical, but leaning safely. An approaching shark would have to pass close to it due to its placement among the coral formations. I put two pieces of food in front of it and retreated.

Martha glided slowly up the scent flow, but shied away when she saw the mirror. It took her attention off the food she had targeted. She made a tight turn in front of it, moving with sudden, strong thrusts of her tail. As she came around to it again, she positioned herself on one side of it with a fin pressed against the coral to support her. She pushed her nose against the mirror's edge, apparently examining the strange phenomenon from the side; one of her eyes was in front of the mirror, one behind. She glided past, took another turn around the immediate area, and continued her investigation with similar movements, approaching it, shying away, turning in front of it, and then making a second approach from the other side to touch the other edge. This was far more intelligent behaviour than I expected. It is considered a sign of intelligence when a dog looks behind a mirror when shown one. Here, a fish was doing that, not just once, but once on each side!

Blackfin reef sharks cannot stop in one place the way the nurse sharks can. They must keep water flowing over their gills for an adequate oxygen supply, and they do that by continuously moving forward. Thus, circling is the only way they can stay in one place. To stay in front of the mirror and look into it, they had to circle there. I had never seen such tight circling before.

Finally, Martha took the food with a powerful drive of her tail, accelerating and veering past the mirror as another shark appeared in it and seemed to nearly collide with her. She returned to get the second piece.

Bratworst started swimming around behind Martha as I returned to the boat for more food. For a while she was occupied trying to get a scrap out of the coral. I was trying to stay in position to see the mirror, without making any noise in the water, and she kept doing little turns right under me in a search pattern on the bottom. She did not approach the mirror.

Madonna arrived, and she too, became skittish at the sight of her reflection. She made some turns in front of the mirror, as Martha had, but not for as long, and she did not approach the edge. When she circled away, Martha ate the food. Bratworst seemed to be nervous of the mirror, and on the only two occasions that she approached it, she turned and accelerated away in an apparent fear reaction. She didn't get any food.

Madonna got the last piece, finally satisfied from her investigations that there was no danger. The two big sharks left, and only Bratworst continued to cruise the periphery without approaching. Jem passed back and forth down-current but did not come close.

The mirror looked like a magic window in the lagoon, and many fish were attracted to it. Some of the goat fish swam in front of it, gazing into it. I could see their eyes in their reflections looking at me. Once in a while one would touch the mirror then shoot away. A mullet began floating in front of it after swimming around and across it several times. After a long time the other fish lost interest; only the mullet continued to gaze at its reflection. I put the mirror back on the kayak, and swam around looking for a shark, spending long periods just drifting with the current and looking, so that I didn't make any noise in the water. I went to the reef, drifted back, and found no shark in the area.

In mid-May, I took the mirror back to the lagoon, placing it as before, with two pieces of fish in front. No one appeared for many minutes, so I increased the scent flow. Madonna materialized at my elbow, with Martha and Madeline behind her. Bratworst glided in from another direction. Each shied away from the mirror at first sight, but were much less interested in it this time. In the presence of Madeline, Martha and Madonna were more motivated by competition for the food, than interest in the mirror.

Madonna passed the mirror first, and startled away. Martha followed, and ate one piece of the food, then the other, delicately touching the tips of her pectoral fins to the sand as she reached her head toward the second one. She ignored the mirror, apparently remembering her conclusions about it from the former session.

Madeline swooped boldly down and shied from the mirror. She returned, and circled briefly in front of it, then began looking for the food that was supposed to be there.

I got another two pieces of fish from the boat. The sharks accelerated toward me, showing no patience with my desire to do an experiment with them. Windy arrived, already reflecting the excitement of the group in her swift approach. I put the food in front of the mirror, and Martha ate both pieces again, while Madeline and Madonna cruised around the area and circled fast in front of their reflections. When I went for more food, Martha zoomed up to me so fast that I let go of it before getting it to the mirror, and Bratworst tried to extricate it from the coral formation into which it had fallen.

Bratworst swam with the others past the mirror, only once accelerating away from it, as they all looked for food. The others still turned in front of it intermittently. Strangely, in spite of the presence of the larger and more aggressive Madeline as well as several other accumulating sharks, Martha got five out of the seven pieces of fish I had brought without accelerating. Bratworst got the one piece of fish she retrieved from among the corals.

Jem passed, but didn't come to the mirror, and didn't come back. The group was excited, soaring around, and zooming past the mirror. When they finally dispersed, I got out the last pieces of food, and was swimming to the mirror to place them, when they all soared toward me under the surface. There was little current, so the scent could not have reached them, beyond the curtain of visibility, so fast. They had come because they had heard me go to the boat.

I was startled at such acute perception; it was the first time I had seen it so clearly. Once again it was Martha who ate the food, locating and inhaling each of the three pieces, practically stationary in the water. The sharks cruised the area much longer this time to see if I would produce something more.

Finally, I accompanied Martha as she meandered away through the coral. She had a small remora attached to her back, but in a moment it flitted over to me. Then it wriggled after Martha, first settling between her ventral fins, then attaching to her pectoral fin near her body, where it stayed.

She travelled in a path forming large figure of eights. Each time she returned to the centre, she passed directly in front of the mirror without adjusting her trajectory, or showing a reaction to her reflection suddenly appearing at her side. This reinforced my impression that she had investigated the mirror and decided that the image it showed was meaningless. And this, in turn, suggested that she was thinking. She also seemed to have a very finely grained mental map of her range to return to the same place so precisely.

Her figure of eights were skewed, so that the path she traced when viewed from above was the shape of a rough cloverleaf, the last oval taking us onto the reef itself. But we arced back, following the pattern, and after a last pass in front of the mirror, we set off toward the east. This time she followed a fairly straight line eastward and increasingly accelerated. When she began to fade into the veiling light, she suddenly shot away and vanished.

I had just stopped finning to rest, gasping for breath, when Samaria appeared from the east and circled me then returned from whence she had come. I followed, but she drew ahead and I lost her. When I started slowly back, a large remora joined me. It swam nearby, leaving occasionally to investigate something in the vicinity and returning. A large female shark passed. I was trying to get close enough to recognize her when a little pup flitted by. It was one of those born in the local nursery the previous season, my first sighting of an identified newborn in the lagoon. But I never saw it again.

Just as the kayak's anchor line came into view, my remora sped on toward it and swam excitedly around beneath the boat. A cloud of fish emerged from the surroundings as if they had been waiting for me to return. I rinsed the plastic bags that had held the food, hoping they could get a crumb, and was able to find a little chunk of fish for the remora.

I climbed into the kayak and drifted west with the rising wind. The remora stayed under the boat until the water got so deep that I couldn't see the bottom. By then, it was staying well down under the surface most of the time, coming close to the boat every few minutes. Finally I couldn't see it any more as the bay deepened beneath me. But I wasn't worried about it. The day was young, and it had lots of time to find another shark.

Besides, increasingly I believed that animals have more powers and capacities than we assume.

CHAPTER TWENTY-THREE

The Range Experiments

Anxious to find out how far my sharks travelled, and to gain a better idea of where the visitors were coming from, I divided the lagoon into five sections, each about the size of my study area, with the intention of holding a series of feeding sessions in each one.

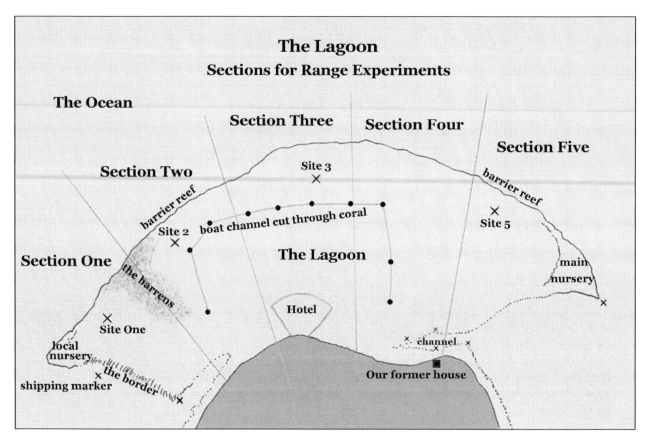

To start with, I took a bit of food about half a kilometre (a third of a mile) along the reef to the east, to see if I could pinpoint the home ranges of some of the regular but less frequent visitors such as Samaria, Madeline, Venus, Droplet, and Splash, whom I suspected lived in that direction. While getting ready in the boat, a large shark passed, so I slid in. It was Martha! I tossed in the few scraps, and she cruised very slowly around, not at all hungry. She approached me to see if I had anything else, and when I didn't respond, she ate a bit of skin. After a few minutes, Ruffles drifted through, also uninterested. They glided around, and after another few minutes, Madonna arrived and swam to me.

The area was quite deep, with little coral. As far as I could see in all directions, a dark, flat, seafloor of shattered coral formations, spread off into the distance, with only the occasional tiny coral structure to relieve its monotony. It was a unique area in the lagoon. Heavy cloud cover laid a pall of darkness upon the scene, which appeared in shades and shadows of blue-grey.

Shimmy, Jem, and then Madeline arrived, and they all cruised around together, but didn't eat, and soon vanished into the deepening gloom. I waited.

A bad storm hit suddenly. It was so opaque that I couldn't see the island, but underwater a strange white light glowed as zillions of heavy raindrops bubbled the surface. I waited. Once in a while a shark passed, but it was always one whom I had already noted, and finally, there was no one. After an hour, I was becoming lonely. The food still all lay on the floor. I tried to feed the fish, but no one came. Finally, three sharks appeared, swimming slowly toward me. Ruffles glided under me, Avogadro curved around, and a big female circled away while still too far to see. She passed again a couple of times in the distance. But they soon left. Finally, when the torrential rain and wind subsided a bit, I gave up and left. On the way home, I battled through one of the worst storms yet.

Since I had skirted around the study area on the way out, it seemed odd that my sharks could have arrived so quickly, unless they could hear the sound of my hollow plastic boat from quite a distance. The sharks who had come were my familiar sharks—I hadn't discovered a new group, and the experiment had led to only more questions. But, though I didn't know it then, I had stopped in the wide, deep, barren region east of my study area where the sharks liked to go to rest during the middle of the day. Therefore, it was not surprising that some of the sharks who knew me had come.

The following week, I took some fish scraps to the next section along the lagoon as shown in my diagram: Section Two. In an appropriate place, about the same distance from the reef as my original site, I laid the fish scraps on the sand. The visibility was good, the coral was radiantly healthy, and many fish hung in the shadows.

I waited a long time before a grey female I had never seen before began quietly passing. After about fifteen minutes of appearing and disappearing around me, she took a small scrap and left. After another long wait, a small male passed by, and came again every few minutes. A long time later, a juvenile appeared. There were long periods between the passes, and I became seriously cold. None of the sharks were familiar. Sometimes one came slowly straight toward me where I was holding onto a dead coral structure beside the scent flow; sometimes two or three would approach me at once, but apart from the first shark, they were all juveniles. The food still rested on the sand, and each time I thought that they had all gone and I could go, two or three reappeared.

Then Avogadro came gliding slowly through the cold turquoise, followed by Mordred. They didn't eat, just glided over the food, circled around a few times, and left. The two males had attended a feeding session the night before at Site One, a kilometre away! Here was evidence that these two roamed together, at least sometimes. Otherwise, the session was disappointing.

In the past year, commercial shark feeding tours had been held in the middle of the lagoon, opposite the new hotel that had recently been built there. Given my experience with feeding sharks, I figured that those who lived in Section Three could easily be met and identified very quickly, just by holding a few sessions at the commercial feeding site.

On a flawless morning in early June, I followed the shore around the island to the hotel—so I could be positive that none of my sharks heard my kayak passing—then crossed the lagoon to the feeding site near the reef. The east wind began to rise strongly, and it took a long time to get there. Three tour boats were just leaving as I arrived, so it seemed a good time to put my own scraps in and see the sharks that the feeding had attracted.

The site was on the inner edge of the barrier reef, and was floored with broken coral. Apart from a few low coral structures, it was a barren plain stretching off into the gloom. The water was deeper than average—three to four metres (10 to 13 ft). Sharks were hard to see coming in the lines of light running over the grey floor, and in the poor visibility, they tended to disappear into the background. There was a large throng

159

of the oval black fish with streamers on their tails, who tended to aggregate at the ocean dive sites, and expansive schools of Hardwicke's wrasses, but none of the lagoon fish who were always present at my usual site.

As I dropped the scraps onto the floor, the fish seized pieces and excitedly ate. The water churned and became opaque with fish in a sphere around the scraps. Soon, a large female shark came languidly circling. I drew her dorsal fin and gave her the name Cheyenne, since her fin tip was a similar style to Cochita's. As time passed, one shark after another casually swam up the scent flow, circled lazily around the area, and around the food, while I drew and photographed. They were all large, dark females.

Finally, a male came who resembled Cochito. It was Cochise; he had attended one of my sessions. I observed him with satisfaction. At least I had discovered something after such an effort to go so far. A magnificent female came then with a W-type fin tip. Such a dark and striking wind shark merited a dramatic name, and I called her Tornado. Mordred came tentatively approaching me. This time, Avogadro was not with him. It was five days since he had been at the Section Two site—Site Two—and six since the feeding session at Site One that he had attended. Did he do this circuit daily, or was he moving slowly along the lagoon? There was so much to learn.

Back in the boat I drifted with the wind, looking down into the coral. I was going fast, and when I warmed up, I decided to drift underwater to see more of the area and identify any sharks visible as I silently flew. I got in close to where I had held the Section Two feeding. The coral, fairly thick and healthy in deep water not far from the reef, continued, and I saw Simmaron, a new juvenile who had been visiting the sessions, and a little farther, Dapples glided.

Soon I reached the deep plain where I had held the recent session. After fifteen minutes of sailing over the barren region, the water became shallower, and the broken coral floor turned to white sand, studded with large coral structures widely spaced. This gave way to colourful coral growing thickly in the bright light of water just half a metre deep. Not making any noise as I flew over it with the wind, I saw many tiny shark pups turning together in an open glade. When I investigated, some of the little ones came over to have a look at me before they disappeared. They were the size of newborns.

Here was another piece of the puzzle. After leaving the nursery, it appeared that these very young pups sheltered in the lagoon, in shallow regions intricately overgrown with thick coral, where they were safe from predators, and the water was less turbulent than the nurseries.

On past the familiar region of the site, where large coral formations alternated with strands of sand, I saw a few of my sharks. Amaranth cruised slowly past me from behind. I admired her pale mauve colouring, her complex dorsal fin pattern, and the unmistakeable signs of age in her scarred hide, well lit for once, by the floodlight above. Then the water became shallower with a floor of worn, broken coral as I crossed over the border. As the seafloor fell away into the blue, I climbed back in the kayak.

Three days later, I returned to Site Two, but the session proceeded just as the first one had. I waited endlessly, becoming bored and cold, except for one interesting event. Mordred was there within two minutes, and Avogadro was with him. Neither bothered to eat anything, though I had managed to procure a good load of bloody scraps for the occasion in hopes of encouraging the residents to come into view. After they left, half an hour went by in which only a little grey female passed in the distance. When she grew up I came to know her as Cotlet, but then, she was just another unknown, sleepy juvenile.

Finally I loaded up the scraps again and went to the site opposite the hotel, Site Three. On anchoring, I tossed a few crumbs overboard for the fish, and the water boiled. Happy to see some action, I quickly got ready. Underwater, a long-horned cowfish (*Lactoria cornuta*), was hovering near amongst the excited fish. When my eye fell on it, it drifted slowly down. I grabbed a handful of scraps and threw them in.

Underwater, an enormous barracuda glided toward me from the clouds of fish. It was nearly two metres (6.5 ft) in length, and had a great heavy head and canines like those of a wolf. Never had I seen a barracuda close up. On ocean dives, much smaller ones than this had appeared in a school at a distance. Later, I speculated that this individual was in the lagoon alone since he was old, but I didn't know. I had heard that some shark attacks were initiated by barracudas, so watched with interest to see if he was going to bite me. But he turned and drifted away, then returned, and approached me again. Every time he approached, I had to watch, since each time, he came closer. Sometimes when the fish were eating, he suddenly accelerated into their midst and swallowed one of them. I was watching this performance, when a movement caught my eye, and I was face to face with a medium-sized, bronze female shark, who had sped in behind me. She circled tightly, looking for the piece of food the fish were eating, often brushing me. It was Vixen, the first female I had seen there that I knew from the Site One sessions, two kilometres away.

Among the confusing motion of the excited fish, the barracuda and the sharks, three big females suddenly appeared. One of them was Tornado. Another sailed by at eye level looking at me, and with the floodlight above, my camera for once caught a beautiful, clear photograph. I drew her dorsal fin and named her Willow. A few more magnificent females came one at a time about every ten minutes and circled around the food long enough for me to draw their dorsal fins and photograph them. But none were sharks I knew.

Willow's photograph turned out to be the best one of the species I had, thanks to the good lighting, and it was later used for publication in the newspaper and any journal or presentation requiring a photograph of a blackfin shark. But I never saw Willow on subsequent visits to Site Three, so in spite of the wonderful souvenir I had of her, she became another mystery shark.

I drifted back underwater parallel with the reef, and found that this deep region soon gave way to large, healthy coral structures, on strands of white sand, which extended past Site Two. There, the coral became thicker and very beautiful. Eventually, it thinned, and the water deepened as the wide, barren plain stretched before me. I began to refer to it as 'the barrens' for the sake of convenience.

Twilight passed in the distance, and Samaria approached for a closer look as I flew, clutching the wind-blown kayak. Twilight's remora left her and joined me, often lightly touching my legs. After a while, it got lazy and attached itself to the front of my right thigh. When I reached the shallow garden where I had seen the shark pups turning in a coral glade, I found a place to anchor up-current. I had brought a bag of left-over seabirds' fish; they would be just right for the little ones.

There were many other juvenile sharks roaming through the brightly lit fantasy garden: Teardrop, with a bite on her gills, Shilly, Lightning, Muffin, Filoh and Breezy. Several other little sharks passed nearby as I looked for an appropriate location. These sharks of two to three years of age probably preferred the regions of thick, protective coral, while the big sharks found deeper areas with large, widely separated coral formations easier to swim through. Perhaps that was why, I reflected, I had seen no juveniles at the deep and barren Site Three.

Choosing a likely place up-current from the tangled thickets of coral, I emptied the bag of fish onto the sand and waited, feeding my remora. Storm unexpectedly passed over my shoulder from behind, paying no attention to the food. I broke open the little fish to release the scent, and moments later, some tiny shark pups appeared, circled, and ate, while I delightedly looked them over. It was so satisfying to have found them in their haven, having wondered so long where they went after leaving the nursery. Presumably, there were other shallow regions like this throughout the lagoon, away from the nurseries' troubled waters, where they were living protected lives.

I didn't stay long because I was sure that the tiny sharks would feel more comfortable coming to eat if I wasn't there. I was reluctant to give them the impression that people supplied them with food, given my experience with the cruelty of some of the residents toward sharks, especially baby sharks. It was a pleasure to drift homeward on the wind, cutting short my third feeding session of the day. I had been in the water for many hours.

Three weeks later I returned to Site Three. The scraps were bloody, and I was looking forward to an exciting time. Rays of sunshine rushed over the barren floor as I waited, fed the fish, and watched for the barracuda. But no one came! The scent flow from this central region moved eastward—I was past the place where the lagoon's flow divided as it moved from the reef to exit by the western or eastern end. After I had been there for an hour and a half, four tour boats arrived and began the official shark feeding. I moved to the side and watched, sad for them that just one female and two males came.

This seemed very odd, especially when I had such bloody, smelly scraps. So I used them to trail a scent flow, crumbling a piece of fish in my hand as the wind swept me westward toward the familiar border area. But no sharks appeared. At the western side of the barrens, a female and two males were swimming, and I stopped to put some scraps in the water.

The sharks I had seen didn't come, but the juveniles, Filoh, Teardrop, Shilly, Lightning, and Breezy, and two strange males came cautiously forward. The scent flow was streaming through the shallow region where the pups sheltered, which explained the appearance of the juveniles. Farther on, Sparkle Too glided, but she passed without giving me a moment of her attention. At the site I put down the rest of the food in anticipation of seeing Martha, Madonna, and Bratworst. But they didn't come either! No one did! So I left it for them to find later. After six hours underwater, I was immobilized by spaghettification and wanted to go home.

The relative absence of sharks throughout the lagoon that day seemed inexplicably bizarre, so ten days later, I returned to the barren Site Three. The scraps were bloody enough to create a good scent flow, though there wasn't much to eat. Again the barracuda wasn't there, but the fish all came as I put the food in the water. I used the fish head I had to feed them and enhance the scent flow while waiting for the sharks.

About every fifteen minutes, a large, grey female appeared and circled calmly around the food while I drew her dorsal fin and photographed her. One was Clara, who had come once to Site One. Otherwise, one shark after another was a stranger. Much later, two sharks came together and one looked familiar. Looking her over, I noted a dark freckle near her dorsal fin on her left side. The photograph of the shark I had named Merrilee, two years earlier, came vaguely to mind.

Then two dramatically marked males, who looked like brothers, came circling around the food. One had a large, dark patch between his dorsal fins, and other unusual markings. I called him Penrose. The other appeared to be his brother, so I named him Hawking, after the other famous British cosmologist. But no other sharks came, though I waited until I was trembling with cold.

Back at home, I looked again at Merrilee's photograph and my records. It was actually possible to see the black freckle she had on her back in the photograph. If only I could see her again and double-check that the mark was in the right place, I could be sure that I had finally found Merrilee. Penrose had been in the study area on November 17, 2000, but not Hawking. Another male I had seen had also visited my study area. I had drawn his fin on both sides at session eighty-five. This suggested again, that the males roamed more widely than the females.

A month later, I managed to get away to look for Merrilee, and went this time in the evening, hoping that the sharks would be more plentiful, bold and active then, and that I could learn more about them. But in spite of the creation of a bloody scent flow, no one came for twenty-five minutes! Finally, a male passed far off down-current. Bored and anxious to attract any passing sharks, I continually shredded meat from the fish heads into the water, starting with the eyes, which never failed to bring the sharks at Site One. There were very few fish feeding. After forty minutes, the first big female appeared. She was a stranger. Other sharks in the distance approached and turned while still only shadows. An hour had passed, and it was getting dark when an unknown male came. I was to see him again years later, near the border, just one time.

When the sun touched the horizon, a large, dark female came and circled around me. She was the first to come close enough for me to see her well enough to draw her tip line on both sides. She didn't eat, and left. More sharks appeared then, but did not come close enough for me to see them well. I had a few impressions and waited for them to return and approach closer, but they did not. I was in an agony of cold.

When the sun slipped beneath the horizon it got suddenly darker, as if a curtain had fallen over the scene. Two large sharks glided lazily into view. One swam near and her dorsal fin looked exactly like the one I was looking for! Then she turned toward me and I looked for the black mark on her back. It was there, identical to the one in the photograph of Merrilee! It was her! She actually swam right under me as if she, too, shared my thrill at our meeting again. After all the waiting through such a disappointing session, to find her against such tremendous odds seemed miraculous.

Merrilee's companion was decorated with irregular black circles, about fifteen centimetres (6 in) across. Never had I seen this remarkably coloured shark before, and I never saw her again. But Dapples' mother had to be somewhere!

A male swam to the food and ate, the only shark who did. It was Cochise. Another circled curiously, and at home I identified him as Bree, who had been at the session with Merrilee upon our original meeting!

Here was another clue that not only female companions travelled together, but that males from the same region could accompany them. It was an unexpected and thrilling find, which more than made up for the boring session, tortured by cold.

I returned sporadically to this site when I could. There was always a long wait before the sharks came one or two at a time, but I identified a steady parade of strange females there. The casual way they circled the food made it easy for me to draw and photograph each one on both sides. Given how many travellers I had identified at Site One during the season, the mystery was enhanced, and suggested that my visitors might well be coming from other parts of the island, rather than farther along this lagoon. Perhaps these female sharks did not come to my study area at the lagoon's western end because they used the large nursery at the eastern end.

None of the sharks attended the sessions there regularly as Martha, Madonna, Bratworst, Gwendolyn, and others did in Section One. Since commercial shark feedings had been held there daily for more than a year, this pattern was even more surprising. It seemed that the sharks who appeared there just happened to be passing by, and that juveniles, even older ones, avoided places where the water was deep, and large predators such as a two-metre barracuda, could roam.

One shark I got to know there was a memorable, bronze female, who treated me like an old buddy from the start. I called her Innisfree. Charm, who had visited the study area once, appeared one time. I saw Cheyenne again, and one day, Hurricane appeared as soon as I put the food in the water. Since I had been seeing Anne Boleyn regularly at Site One, the sighting confirmed that the companions were not travelling together at that time. Months passed before I saw Merrilee again, which emphasized the amazing luck I had had, in finding her the one time I had gone specifically to look for her. The barracuda swam in, looked around, and left once in a while, but he never again approached as he had on our first memorable meeting.

I went for the last time, a year after the date of my first session there, and again, the trip out was a marathon struggle against the wind. I arrived trembling. But I wanted to see which sharks were using the area by then. It was the hour before sunset, the hour of the sharks.

Cochise was passing when I slid underwater. He investigated the food, ate, and cruised casually near me for a long time, so I was able to study him. He was a perfect shark with not a mark or scar of any kind.

Much later, as it grew dark, three more males approached, gliding in together. One was the largest male I had seen so far; he had visited Site One and was called Danny. The beautiful Antigone swam at his side, and the third shark was Dante. He had come to the sixth session there with Flannery, who soon appeared too, followed by Cochise. All five males soared together around me. No one was interested in the food; this was purely social. *Will wonders never cease*, I thought to myself, slowly turning in the shadowy waters to watch them as the night gathered around us—it was the only time a group of male sharks had treated me that way, though females I knew often did.

Not one female had come.

Except for that one sighting of Mordred at my very first session, none of the sharks who lived in Section One had ever appeared, nor had any of the infrequent visitors such as Kimberley, Twilight, Splash, Droplet, or Dapples. With the exception of Mordred, the few sharks I had recognized had visited Section One only once or twice. I had not shed any light on the origin of the countless visitors to the Site One shark sessions.

Chapter Twenty-Four

The Dry Season Returns

I carefully avoided Polynesian fishermen when they were in the area, due to increasing ill feeling toward white people. With the rise of the Independentists, the political party that had burned down the international airport to protest nuclear testing just a few years before, racism had increased. The idea that a white woman was visiting with Polynesian sharks could have been met with an intensely negative reaction, and I was alone, a kilometre from shore with the sharks.

Even my practice of seabird rehabilitation had been criticised. Seabirds were bad because they ate fish. The Polynesians held seabirds responsible for the holes made in oceanic fish by cookie cutter sharks, and they killed the rare and endangered Tahitian petrel to make fishing lures from its feathers.

So I developed the most secretive approach to all my activities related to sharks. When collecting scraps, I hid them so no one could see them once I was outside the store, and I substituted the word 'fish' for 'sharks' when talking about what I did with the scraps. I was studying, photographing, and feeding the fish. Sharks are fish.

This factor added a high level of stress to my shark pursuits over the years. I was threatened with death by one man who didn't like me going in the lagoon in which he fished. There was an incident when a neighbour had come and thrown gasoline around our garden while shouting racist slogans. When our dog had come out barking at him, he had thrown gasoline on the dog. By the time Franck arrived, the man was striking the match to light it. Franck's interference resulted in the match going out, and distracted him from lighting another. But he continued to yell racial slurs and throw gas until Franck persuaded him to leave. The *gendarmes* didn't even come.

So we felt a covert hostility around us that could emerge without warning. In general, I ignored it, considering it part of living in a racist country where we were in a ten-percent minority.

Though Franck had long had access to the Internet and e-mail through his work, it wasn't until this period that I had a computer to use and e-mail for the first time. I was a slow learner, but managed to type the name of the species of my sharks into a browser and performed an Internet search. I only had minutes to stay connected while Franck collected the messages from the server in Papeete, so when the only titles that appeared on the screen in connection with blackfin reef sharks followed the words 'shark attack,' I concluded I had made a mistake.

I had to wait two weeks before I was able to look again, and this time I typed in the Latin name of the species, *Carcharhinus melanopterus*. The same list appeared. 'Shark attacks!' I read, all down the page beside the Latin words. I had time to click on one of the entries and waited many minutes in suspense for the page to load. A famous shark attack file opened and informed me that *Carcharhinus melanopterus* was responsible for fourteen attacks in five hundred years, with zero deaths. I just had time before I was disconnected to write to the author, a professor at the University hosting the file, whose e-mail address was displayed.

"Is this not a scientifically negligible number?" I asked him. "And why is there no other information about this species, as if this information is the only thing about them worth reporting?" The man, I learned later, was a famous one, often quoted in the media on the subject of shark attacks. He didn't answer me.

I had briefly been in contact with Richard Johnson, author of *The Sharks of Polynesia,* and he had given me the names of two scientists I could contact about my sharks. I had the names written on a scrap of paper, Arthur Myrberg and Sonny Gruber, and during the following weeks, during my brief forays onto the Internet, I searched for them. Professor Gruber did not appear anywhere in my searches, but after many enquiries to the wrong people, and when I had nearly given up on getting in contact with the rest of the world about the sharks, I received a message from a woman letting me know that she had forwarded my letter to Professor Myrberg.

Soon, I received an e-mail from him asking why I was contacting him. I explained that I had tried to get to know the local blackfin reef sharks as animals and individuals, and over the years, I had seen a lot. Then I asked him if he could kindly clarify a few things for me. He wrote back with many more questions, which I answered as best I could. He sent more questions than ever, which disheartened me considerably. He wanted to know how I had arrived at all of my conclusions about the sharks' reproductive season, and when I explained, he questioned my answer. It was quite impossible, he told me, for most people to count sharks in a nursery, and how, exactly, had I done it? I described dividing the length of the nursery on the reef into sections, counting the fins above the surface and estimating the numbers hidden beneath, then doing a certain amount of multiplying, adding and estimating, finishing with a very rough estimate. Reading it over, I laughed at myself, and added defensively that since people elsewhere doubtless knew all about the reproductive cycle of the sharks, I was just doing this out of curiosity.

Then I asked him again what I really wanted to know: why did the sharks approach my face when I entered the water? It was so much like a greeting. What did it mean? Were they using the electro-sense, since they could not be using the sense of smell?

When he wrote back, he suggested that I try to organize my data into tables and graphs and publish it.

I found this disappointing. I had no idea how to publish, no wish to, and had been sure for so long that if only I could get in touch with the rest of the world, I would learn all the intriguing answers I was seeking. I had never doubted that the simple facts of sharks' lives were known. They were so common, and aquariums would have been able to learn all about them too, in the comfort of their facilities.

That dry season of 2001 brought calm conditions. The waves on the reef were small, and there was no current. Often when I entered the water no sharks were visible.

The fish were thrilled when I returned to the original site after the storms ended. More fish than ever came. Astonishing numbers of dozens of species met me when I appeared underwater. Countless squirrel fish congregated in the coral on my left and rock cod assembled around the hand-holds on my lookout. Fishes filled the water in front of me, making it at times impossible to see the sharks. I alternately fed them and cruised around, whereon they covered my perch, looking for bits I had dropped, and then followed me, in a giant, glittering cloud, making it problematic to sneak up on shy sharks cruising down-current. When I found a spine or scrap with some meat on it, I would feed them some crumbs, then return to my lookout to give some to the red squirrel fish, who remained in their grotto by my lookout. They were my favourites. They watched me approach, looking up at me and what I was carrying with their huge, black eyes. Then they darted out to take crumbs from my hand.

The resident females who regularly attended the sessions at this time were Madonna, Bratworst, Martha, Gwendolyn, Flora, Valentine, Venus, Christobel, Samaria, Madeline, Storm, Lillith, Amaranth, Sybyl, and Sparkle Too. The males who frequented them were Avogadro, Ruffles, and Meadowes. Shimmy and Peri were just maturing. The juveniles of three to four years old were Diamond, Nightwind, Fawn, Piccolo, Flute, Brandy, Trillium, Cochito, Cochita, Windy, Teardrop, Sparkle, Simmaron, Apricot, Lightning, and Chevron.

Valentine returned from her seasonal wanderings at the end of May, having been absent during most of the mating season. Annaloo snagged a fish-hook on her ventral fin, which trailed a short length of fishing line, but it was an isolated incident. When Lillith arrived on her infrequent visits, always as night was falling, she immediately became the centre of attention. The other females followed her around as she cruised the area, such that sometimes, half a dozen of the biggest female sharks were following her. Once when she did a hundred-and-eighty-degree turn in a narrow coral canyon, they did too! No other visitor made the impression that she did. She was developing tiny, white freckles in different places which I speculated were signs of ageing.

Twilight came two nights in a row with her remora, and it stuffed itself until it was swollen to bursting on both nights, then departed collapsed on Twilight's back. She let me accompany her, and I swam with her whenever I could, though all she did at those evening sessions was socialize—she didn't take me anywhere.

An enormous school of vertically striped, yellow fish, called convict tang (*Acanthurus triostegus*), dwelt in the vicinity and often passed through. They presented a moving wall of shifting, flashing shine, sometimes alighting delicately to feed, whereon they draped the coral formations with waving, shifting light.

The huge, old leopard ray continued to come sporadically, but no longer did the spangled emperors wait for me. None of the large fish I had once seen appeared any more. New juveniles drifted in, and some lingered in the area. One was Eden, who eventually became a resident. Marco and Kilmeny occasionally visited. They were the only two big males who came regularly, though infrequently, independently of the reproductive season.

Meadowes returned again from the ocean with Ruffles in late May. I was shaken by his appearance. Where he had been thin before, he was now emaciated. He resembled a tadpole. His head was too large and his body concave, tapering to his tail. There was something wrong with him, some wasting disease. Or perhaps he had swallowed a fish-hook while out in the ocean. He came at dusk, and there was no food left for him, nor did I have anything in the boat to offer. He was obviously famished and rummaged through all the places where scraps were often swept under the coral into crevices. But there was nothing.

I returned with food for him two days later, and found him waiting underwater. He was the only shark present. As I pushed the scraps from the boat, he began to eat from the fall of food before it touched the sand. He and Ruffles had never bothered to come right away to the sessions. They had always arrived later and picked through the scraps as though they didn't really care about the food. They just came because it was there, and to socialize.

One of the largest stingrays I had ever seen was scrounging around in the sand. I dove and swam along the bottom to get closer. Meadowes swam low above him, and the two were backlit by the glow from the setting sun. The spell-binding vision of beauty swept me with nostalgia.

So I made special trips to get food to Meadowes apart from the regular feeding sessions.

Trinket returned, after another one of her mysterious disappearances. Watching her lovingly as she flitted about, I was astonished to see that she, too, was growing tiny claspers! A boy! And still she seemed hardly to have grown at all since I had first seen her! Pardon me, him.

That night I kept saying to Franck, "Would you believe it? Trinket is a boy!" Whenever I mentioned the word shark, he limited me to a sentence or two, so was unable to understand why this information could be so thoroughly amazing. It took some time to start thinking of my funny little shark as a male. Trinket had been one to two years old when I had first met him, and was growing claspers two years later, at the age of three to four. They would not be fully developed during the coming reproductive season; it had taken ten months for Tamarack's claspers to grow to full size. Assuming that he mated the next year, he would mature at the age of five, or six. I would need to follow the development of sharks identified in the nursery, until their maturity, to be sure.

The females I had been able to keep track of had not become pregnant the first year that they mated. But though Sybyl was still very small, she appeared to be an exception; she was pregnant. There seemed to be a year of ineffective mating prior to the beginning of the reproductive cycle in most sharks. During that year, females grew swiftly, and transformed from sinuous, slender adolescents, into much bigger, maternal-looking adults.

Observing the rate of development of both males and females suggested that five years or possibly six, were necessary for an individual to become fully mature.

Little Trinket still had a lot of growing to do. I should have guessed by his size that he must be a male.

The females grew faster, which made sense since they became bigger in the same length of time. Diamond was the size Flora had been when I met her, and Flora now resembled the other big females. She was well filled out, rounded, and had become very dark. The flares of colour into her white band were prominent and outlined with a hard, dark line, whereas formerly they had been a soft flare of bronze, like a flame against the white. There had been a remarkable change in her in the two years in which she had matured.

I had found rare eye-witness accounts describing blackfin reef sharks mating in shallow water. They told how the two sharks undulated as one animal, while the male grasped the female with his teeth. With this information, that the male shark holds the female, often by her pectoral fin, to stabilize the pair during copulation, it became clear that the males were just the right size. If they were larger, and especially if they were larger than the females, their claspers would not be properly positioned for mating this way.

The crocodile needlefish began to come again. About half an hour into the sessions—I guessed that this corresponded to the arrival of the scent flow at the border—she would drift into view, a silvery shadow in the surface shimmer. She hung in the light, her tail sweeping gently, turning first one eye toward me, then the other, as she slowly approached. Though it had been January when she last had come, her large size and general appearance were unmistakeable; two parallel scars on her side confirmed her identity.

She hovered down-current watching, occasionally accelerating briefly to snap up a crumb with her pointed mouth. I couldn't help but watch. An animal of her size with such teeth was impressive. She seemed to remember me very well. The previous December she had stayed at least three metres away from me and only ventured closer when taking a morsel of food. She had never come closer than a metre away. But on her return in May, as the dry season started, she began coming within a third of a metre (a foot) of me, always looking intently, with those large, questing, trusting eyes.

Shilly came sometimes, and occasionally, she was present at the beginning of a session. She was never a regular visitor, but she became less shy and more comfortable with me. There was another small, pale juvenile called Ali, with whom she sometimes travelled. Sometimes a third little female, Ondine, was with them. Diamond suddenly disappeared, and though I waited for her to come again, she never did. It was another mysterious loss of a healthy animal.

One evening as darkness fell, a nurse shark was undulating toward the site, when it spooked and accelerated away, passing Shilly closely. The tiny shark also accelerated and rose, arching her back vertically in a series of rapid jerks as she swam. It was the first time I saw a blackfin perform this display repeatedly; it is highly developed in the grey reef shark, and called an *agonistic* display. However, this word implies aggression, while I only saw this posturing display as a result of fear. I had seen a juvenile blackfin perform this display once before, also after being frightened by a nurse shark—it had jerked vertically just one time. The rapid motion seemed adapted to make the pup harder to grab by a predator. That seemed the most obvious reason for the evolution of such a reflexive reaction.

One day, I was resting in the kayak and drifting with the east wind homeward across the lagoon from Section Two. I had been in the water so long, and was so cold, that the wind felt icy, even though it was the middle of the day and I was under a tropical sun. Looking absently into the lagoon, I saw a small, bronze shark begin circling the boat. She seemed to be following me, circling several times over a period of many minutes. Then she came up to the surface to look at the boat, moving very slowly and turning parallel to the kayak before going down again. It was Shilly. No leakage of scent had occurred, either going or coming. Since I was not paddling, she could not have come to the sound typically made by my kayak, unless she recognized the sound of the waves lapping against it. I found her attentions remarkable.

As a result of swimming with Martha between the sessions, she became increasingly intimate. Once as I fed the fish, the clouds of them that surrounded me, the poor visibility, and the fluids in the water, prevented me from seeing anything else. Martha was circling me closely, sometimes approaching directly to nearly touch my nose with hers. Then Apricot brushed past me from behind. I stopped feeding the fish for a while, but when I resumed, Apricot as well as Martha closely approached. I moved off my lookout to avoid them, still holding onto the fish head. When Apricot followed me, I handed it to her. She shied away, then grabbed it, but dropped it right away.

I circled around the site as Martha circled me. When Apricot approached again, and kept on coming as I backed away, I fluffed water into her nose with my fins. She turned. Martha stayed with me as I looked around down-current to see if anyone special was lurking out of sight there, to avoid Apricot for a while.

Apricot distinguished herself as a nuisance, though she was just one of many little sharks who had become familiar with me. Nearly every time a shark nosed toward me through the clouds of fish, it was her. When I went down-current the juveniles would often circle curiously and sometimes come straight up to me; as it grew darker, even the shiest of them grew bold, as though they felt they had an advantage in the darkness. But Apricot sneaked up from behind, whipping away when I looked at her, then returning. Sometimes when I turned to face her, she dodged and accelerated forward, a new reaction. Yet she was so small, scarcely a metre long!

I stayed until twenty minutes after sunset to see all the visitors who might pass by. By then, it was very dark underwater. Hurricane and Eclipse visited one night and cruised around the area with several of the residents, which was worth waiting to see. At that time, with their deep grey colouring, they were nearly invisible. I was less comfortable with Apricot's behaviour in the darkness, but by then, I was stuck with her.

There was the occasional unusual session. In late June, there was a tropical depression over Bora Bora, and high winds with torrential rain prevented me from going to the lagoon for a few days. A week after the storm, the weather was calm with light clouds and gentle, eastern breezes. Martha followed the boat across the lagoon to the site. When I slid underwater, multitudes of sharks were visible, ranks of them, on and on into the blue. Flora and Martha, as always, were nearest. I paused to look over the incomparable scene, and one at a time, they swam up in greeting, turning away just before our noses touched. Then one after another, Samaria, Madeline, and Madonna performed this greeting gesture too.

As I pulled the kayak into place, Chevron, Christobel, Teardrop, Ali, and Rumcake joined them to swim close beside me. It was touching how they seemed to wait while I swam to take the anchor line to pull the boat to the site. Then we would all move there together in one big mass. Flora swam beside me and under me, and sometimes with her nose against my fins as she would with another shark. I swept in the scraps and watched.

Madeline chased the fish away from the food, following each of them with her nose as she shot forward. Then she found a backbone with scarcely a crumb of meat on it, picked it up, and manipulated it in her mouth until she had a proper grip. She began to flicker, then violently shook her scrap, rising in the water and breaking through the surface. Swarms of fish and several accelerating sharks followed her. She dropped the scrap then whipped around to pick it up again. I had drifted toward her, and she startled when I pushed the water with my hand to right myself. But she snatched her scrap anyway, moving fast.

Ruffles and Meadowes came. There were still a few scraps left in the passageway in the coral to my right, and Meadowes found them. Gwendolyn joined them, and as she moved to leave with one, she slammed straight into an outcropping of coral, and stopped cold. That nose full of sense organs must have hurt! It seemed that the sharks couldn't see well directly in front of them.

Storm, Lillith, and Amaranth came in from the east a little later, and as night fell, a few more gathered, including Jessica, who came earlier than usual. When all seemed quiet, I started to feed the fish. Sparkle Too spiralled closer, then began swimming through them, trying to bite them. So I cruised down-current. There, a young, pale shark crossed my path and circled. I held my breath, drawing the first side of her dorsal fin. She turned right on cue and I drew the other side. Looking curiously, she still circled closer, then brushed me softly as she passed, before disappearing. I never saw her at the site, and never saw her again.

Back at my perch I fed the fish, and Sparkle Too resumed her antics. Meadowes was still poking around looking for food, and I searched for something good for him. All that was left were the two small chunks that could be extricated from each fish head. So I methodically collected the heads, picked the little chunks of meat out, and began throwing them into the site. This excited the sharks, and caused them all to zoom toward me. Meadowes could not compete for the bits of food, and I was unable to get one to him.

Chapter Twenty-Five

Meadowes

As the dry season advanced, Meadowes grew more emaciated. He appeared more often at the sessions, and was with Ruffles less. He was noticeably weaker and no longer participated in the initial rampaging around the food with the others. His skin was becoming duller and darker and had lost its velvety sheen. On two occasions, he banged straight into a small outcropping of coral as he searched underneath the coral formations for food.

While exploring with Martha, I had found that a large number of sharks went to the barrens during the daytime. On one occasion, just drifting with the current, I encountered Trillium, Gwendolyn, and Cochita cruising together, along with a large fraction of the residents, who were gliding in pairs or small groups, over the empty floor. Hurricane was slowly cruising there. Cinnamon, a pale and graceful shark, approached me to circle, looking. Even Kimberley passed once in the distance, and later I saw Twilight.

It seemed that in this deeper area, where there were almost no coral formations, and the broken coral floor had been swept flat by the surging sea, it was easier for sharks to swim unobstructed. So they went there to rest. They seemed semi-conscious, and were slower to react when I neared. If sharks from the coral on both sides of the barrens went there to rest during the day, that would explain why I saw sharks there who rarely came to the site.

In the barrens, it was hard to see both sides of a shark, so identifying new ones was difficult, though I could recognize ones I already knew. I had begun taking some food with me in case Meadowes appeared, and was delighted one morning to find him approaching alone when I slid into the water. He swam beneath me and fed before other sharks became aware of what we were doing—no sharks were expecting me to bring food to the barrens. Fortunately, Meadowes found me, and had the intelligence to seize the opportunity to eat. The sharks tended to be wary if I produced food outside of the usual routine, and it was the first time in months that I had seen him in the morning.

At the next evening session I had very few scraps, but brought pieces of decent food from the freezer for Meadowes. Flora was passing when I slid underwater; how she could be beside me each time at that moment was incomprehensible. For once, she was the only shark present. She circled me steadily as I placed the food, and the others began coming as I moved the kayak back and swam to my perch. After a brief swirling of my thirty sharks over the food, it was gone. Martha took a fish head and paused in the water column to shake it against a coral to her left. She could extract no meat and dropped it there.

All these sharks who had found almost nothing to eat, glided through the sunbeams, and sometimes came to me. I watched, waiting for Meadowes. Sharks kept arriving and joining the other sharks until there were close to fifty roaming restlessly around me.

Finally, Meadowes arrived. I grabbed one of my bits of food from the back of the kayak, and held it above water level while waiting for the moment to throw it. As he entered the site for the second time, I tossed it. He seemed to understand that something had happened and began searching underneath my perch. The piece of meat was mobbed by fish as it fell, but they helped him to locate it, and to my surprise he got it! I tried another, but it was pounced on by a nurse shark. The third was swallowed in mid-water by Martha. The sharks were rising to the surface, sniffing, and it was obvious that they suspected me of something. They surged toward me to see what I was up to. I waited, and when they dispersed, checked for scraps that had been swept under the coral.

Amazingly, a belly strip, a thick strip of meat, was lying on the sand in the corridor between my lookout and the next coral formation. I tossed it into the site, hoping that Meadowes would get it, but Samaria picked it up, and swallowed the whole thing.

Meadowes found something else to eat, and after that he roamed around the area, passing through the site every few minutes. I extracted more food from the fish heads and succeeded in getting a few to him by dropping the pieces of meat on the sand as he approached. He responded well. He did not shy away at my efforts, and accurately searched out the pieces. Many were taken by other sharks, but considering how many sharks were following me, trying to locate the source of every sniff of scent, he found them surprisingly often.

I fed the fish as he circled back more rarely, and the light faded. Apricot kept coming straight up to my hands as I scattered crumbs of food, but she seemed to have accepted my fish feeding at last. As the glittering clouds of fish approached to flutter around my hands, I scooped out every bit of meat that I could from the fish heads, and finally extracted the eyes. Bratworst ate one, regurgitated it, and spat it out, then later ate another. When the sharks got too excited, I drifted around and saw Shilly, Lightning, and Breezy waiting down-current. Apricot came with me, often circling my head half a metre away. As I swam along the bottom, following Meadowes from the site after he ate another little piece, Madonna glided in front of me. I tried to feed her the last bit of food I had, but she ignored me, and didn't return.

Meadowes glided through one more time before disappearing. A large and very beautiful, silver-white remora was flitting around, ethereal as a fairy. Occupied with studying the sharks, it paid no attention to me.

Some days later, I tried again to give extra food to Meadowes. The weather had been stormy, but improved enough for me to get out to the lagoon. I had very little food again, so was not pleased to see a line of sharks following my kayak as I neared the site. When I slid underwater, Meadowes was gliding toward me, so I quickly put the food in place and watched. The sharks indulged in their usual joyful zooming as they ate, but Meadowes did not. He got something to eat but swam weakly, looking unbalanced. Where the other sharks were smooth arcs no matter what angle you looked at them, in him those arcs were flat or concave.

Each time he swam in the site, I followed, trying to assess his condition and figure out if there was anything I could do for him. His skin was blotchy, and his once white belly was dark; he looked as if someone had smeared him with mud. There was no doubt that he was gravely ill, and I realized that he was going to die.

My slate drifted off on its own one time, while I was with him, and he spiralled up underneath it to touch it with his nose. He seemed to have learned that I would feed him, which would permit me to medicate him, so I began wondering about trying to save his life. The trouble was that I had no idea what was wrong with him, so deciding on a medication seemed senseless. Many possible causes of his plight came to mind.

The next day, I took food to the barrens to see if I could feed him alone there but could not find him. So I put the food back into the boat and went to the usual site.

The order of the sharks' arrival was completely disrupted by this change in approach, suggesting again that they took note of me crossing the lagoon from the border. There were no sharks at the site, and when I put the food in place, the little ones, Tamarack, Fawn, and Rumcake came excitedly to eat alone. Sparkle Too soon soared in, having followed me from the barrens; she had been circling as I loaded the food back into the boat. The rest of the sharks I had seen in the barrens trickled in one or two at a time, along with Martha, Madonna, Bratworst, Gwendolyn, and a few other residents.

Meadowes came, looked, and left. I swam to the boat for a handful of food for him, and held it above the water while waiting for him to return. After a while my muscles began to ache and I changed hands. A rock cod bit my finger as soon as I put it underwater, having scented the food on it.

Many more sharks appeared, including Annaloo and Glammer, whom I had not seen for some time. Old Lillith was searching for scraps in the coral across the site, and I was tempted to feed her but did not, because I needed the little food I had for Meadowes. But he never did come. As night fell, I threw his food into the site and sat on the kayak, wondering what I could do for him.

Finally, I wrote an e-mail to Arthur about his condition. He replied with his usual interested questions, trying to gain some perspective on the animal's predicament and what I might be able to do about it. Finally, he suggested that I give him an antibiotic. Tetracycline, he told me, had been used successfully in sharks.

The problem was that I would have to give Meadowes a dose daily for five days. I didn't see how that would be possible. Arthur suggested that I put the medication in a piece of fish and tie it up with a fine thread. He told me that the shark would be able to regurgitate the thread by turning its stomach inside out, and said it would be the first time that a person had medically treated a wild shark.

So I bought some tetracycline at the pharmacy, calculated how much Meadowes should get as a daily dose, prepared two doses tied up in chunks of food, and took them to the barrens. But Meadowes was not there that day, and I was obliged to wait for him at the usual site at sunset. I had managed to get a small load of scraps to use to distract the other sharks while trying to medicate him.

Many of the usual sharks were not there that evening, but Meadowes was coming across the site as I began pushing the food off the boat. When I looked underwater, he had a scrap in his mouth and was slowly shaking it at a fraction of the usual speed. He gradually rose vertically and broke weakly through the surface. I got one of his pieces of medicated fish and followed him around trying to get his attention, but I couldn't. Even when I held the food in front of his eyes and floated it toward him, he seemed not to see or sense it. He looked much worse. His eyes appeared dull and dark like the skin around them. His once-white underside was a dull brown.

Martha arrived late. I had saved two chunks of fish for her, got one, and swam toward her, calling to her steadily. She came. When I tossed her food near her, she turned and took it. I got the second piece and threw that one to her too. It fell between us, and I drifted out of her way as she targeted it, turned one-hundred-eighty degrees, and took it. Then she stayed close beside me, but when I returned to my lookout, she seemed to understand that that was all, and she left. But my glee at the success of my ploy faded, as Madonna and Bratworst arrived with a crowd of other sharks in a long, ragged line from the west. Since they came from there rather than up the scent flow, I speculated that they were returning from the ocean.

I fed the fish, disturbed by the appearance of one of the two-metre moray eels at my left hand, and when Meadowes suddenly appeared, I got his medication. But he left and didn't come back.

The next day I started again in the barrens, drifting slowly along its length, trailing scent and watching. But Meadowes was not among the sharks who gathered, so I went on to the site, followed by the others, who remained beyond visual range. As I drifted in, fish poured from the coral in all directions. I swept the food out, and anchored the kayak in the proper place. The sharks who had followed came soaring in, in a party mood, while I watched and waited for Meadowes.

Martha glided in with him. But uncharacteristically, she didn't stay! It would have been impossible to feed him in her presence. Once she had gone, I got his medicated food from the boat, and handed it to him as he swam by. But he didn't catch on. The food was carried away by the fish, and I went to get the second dose. Again I let go of it just in front, and to the side of, his face as he passed, so he could easily see and sense it, but he didn't seem to notice. Time after time, I had to recover it from the masses of fish carrying it away. I waited, holding the wrapped medication above the surface, for him to return, but he was gone.

At home, I rinsed my gear and kayak and sat on the deck to recuperate, muse, and gaze across the sea. The last of the light was fading in the western sky, and the night was silent. There was not a ripple to the horizon, not a sound. It was one of those supernaturally still evenings that falls at times upon the islands of Polynesia when the sun has set and the trade-winds have stilled. I sank contentedly into a daydream, running through one scenario after another for getting Meadowes' medication into him. After a long time had passed in which the only sound was the trilling of waves every few moments, lapping upon the beach, I had the impression of a sound, just a whisper, so faint that it was almost beyond the reach of hearing. But a car was coming along the bay, and the sound of its passage rapidly became the only one audible, so I sank again into my reverie. As the sound of the car faded, again there seemed to be the faintest sound far away, still almost too inaudible to detect. I went on looking out across the flawless sea, still reflecting on the splendour of the silence of the night, when the distant murmur once again intruded. I stood and listened toward the south. It sounded like wind somewhere. Another car passed, masking it again, but afterwards, the noise was no longer a murmur. It had gained substance, and now I could hear it approaching. Still the sea lay still as a dark mirror, and I alternately gazed upon it, and listened for whatever was coming from the south as the whisper steadily turned to a roar.

When it hit, it was instantaneous. The atmosphere howled, the trees bent and screamed, everything that wasn't attached took wing, and the chairs on the deck went whizzing by as I stared, appalled. What if I had still been on the sea?

Again, I was shocked by the fickleness of it, the danger always under the surface. It was a week before I was able to return to my sick shark due to the wild, black storm that had hit the island.

I had to fight against the last gusts of the storm on the way back out to the lagoon, but the wind died away for a brief spell, and an oval of glass opened just as I arrived. This had happened before, as if the reef itself sometimes sheltered that area from the wind, under very special conditions. The site was full of sharks.

Cochita was undulating around the back of the boat at the surface. Recently she had been the first shark to follow the boat when I arrived in the lagoon. She appeared to be in her first year of reproduction, and more and more, she and Cochito appeared separately. She seemed to be making the region her home range, while he was likely roaming more outside the reef.

The usual scene of delighted anticipation met me underwater. Martha and Flora were already at my side, and others moved forward from all directions, when I swam to the anchor rope and trailed the boat into place. While I pushed the food into the sea, the water around me was solid with sharks.

When I looked underwater, Meadowes was one of the closest sharks, and I was shocked all over again by how bad he looked. He grabbed at the only fish head that I had brought, along with a few bits of skin. The scraps were already gone. I had kept aside some small pieces of food and began throwing them, hoping to get one to Meadowes. But, he didn't have a chance. Martha and the alert ones who flew after her snatched them up.

While Meadowes looked for something to eat, I got his medicated piece of food from the boat. As he approached me, looking, I wafted it before him. He shied, and at that moment, Martha swooped down my arm from over my shoulder, and inhaled it. She paid attention if I went to the boat, because I often returned with food.

I felt in the boat for another little piece of meat, but it was all gone. It was sad to see the horribly emaciated Meadowes looking for something to eat. As the other sharks began to disperse, he remained, apparently starving and still looking for a morsel, any crumb, to eat.

There was a small piece of intestine in the well of the kayak, and I tried to give it to Meadowes. He was circling, and as I moved around, he often approached me. When I passed him the scrap on one of his approaches, he didn't sense it, and it floated, instantly attacked by fish. I tried again, and Meadowes seemed to understand that there was something, but he missed it, and continued to circle. I tried again, and became aware that Bratworst was at the surface behind me, followed by several more sharks; I got out of the way as Meadowes came close too. Bratworst saw the piece of food and lunged onto it at the moment that a large grouper emerged from the coral, snatched it, and vanished with it into his hole. After so very much effort, hope, and worry, I had failed to feed him something! I promised myself I would make up for it the following night.

The next day it poured with rain. I must have done the session the evening before in the eye of the storm. After fretting about it all day, I set off through the downpour with two medicated pieces of food, neatly tied up with thread.

I went underwater just south and west of the site and looked around. No scent had leaked—no one was there. Cochita soon appeared, and circled me, but she didn't linger. I drifted watchfully to a coral glade south of the site.

There, I poured in some of the water I had used to defrost the chunks of frozen food, and waited. Shilly, Eden, and Brandy were appearing and disappearing in the distance again, suggesting that the little juveniles might already be present at the beginning of the sessions, waiting out of visual range until the big sharks left. Tiny Sybyl was there, too, fatly pregnant. I was already surrounded by a cloud of fish of many species who had come from the site. Cochita reappeared and circled me closely, soon joined by Bratworst, who did the same. Cochito, Jem, Windy, Samaria, Valentine, Ruffles, Madonna, Christobel, Teardrop, Sparkle, Sparkle Too, Shimmy, Filoh, Chevron, Gwendolyn, Amaranth, and Apricot gradually accumulated. I drifted slowly at the centre of a swirl of sharks, and waited for Meadowes.

Finally, I saw him coming, and grabbed one of the pieces of medicated fish. He was arcing away, and I followed, holding the food above the surface. I could scarcely see him in the distance and was hurrying after him, trying to keep his shadowy silhouette in sight, when he began a long, slow turn. I finned to intercept him, and finally our paths began to converge. As he neared, I threw the bit of fish toward him. I had his attention. It fell in the coral and was instantly mobbed by fish. He made a wide circle as I retrieved it and threw it for him again, and he circled closer as if he were beginning to understand, though he couldn't have smelled the food so quickly.

I fetched the medicated food, and threw it once more, deeply regretting that I had not treated him a lot sooner, when he still had the energy to make an effort. This time, he turned toward the falling food, and I saw him target it. Sparkle was zooming toward it, and it was surrounded by fish, but I let Meadowes have his chance to get it. Sparkle dove triumphantly onto the food with the rest of the gang strung out behind her. She missed. Meadowes glided beneath her and gobbled it.

At that moment flying torpedoes closed in from all directions on the place where the food had been, encircling me in a whirling vortex of irritated sharks. We were far from the boat and down-current—it was daunting to see how aware they were of events underwater. I swam back to the boat. When I got there, I was alone, but they were all coming behind me and I was again in a mass of orbiting sharks, all trying to be close to me when there wasn't room for them all. Then Martha zoomed over my shoulder and turned into my face. I decided I didn't need any further reason to go home. I leaped into the boat and pulled my feet in after me as fast as I could. The surrounding water was solid with sharks. I sat there zinging with joy to have succeeded, against such odds, to get some food and medicine into my poor shark.

I prepared and went again the following evening; Meadowes needed a dose of antibiotics daily for five days at the very least.

This time, there was a scary wind from the south, the unpredictable and dangerous *mara'amu*. So I blew out to the lagoon, surfing on the waves, and swept back and forth as the kayak tried to sink into the furrows, always on the verge of being swamped by the wind.

Strangely, no sharks were present underwater at the site. Only Peri passed as I pulled the boat over. But when I threw in the few scraps I had scrounged to distract them while I gave Meadowes his medication, they appeared from nowhere, swirled over it, and it was gone.

Storm, Lillith, and Samurai all soared in together, Samurai on Lillith's tail and Storm to one side. Kilmeny was with them. Many more began cruising in and around the area while I waited, in suspense, for Meadowes. Annaloo and Glammer came together, and Pippet, Jessica, Madonna, Fleur de Lis, Simmaron, and Apricot came as I began to feed the fish. I had saved two fish heads for that. The advantage of feeding the fish, of course, was the strong scent flow it generated, which I hoped would encourage Meadowes to come. Madonna was obviously disappointed to have missed out and repeatedly swam up to my face, so I got one last scrap from the boat, and tried to slide it to her. But Simmaron swallowed it seconds before she got to it. Madonna always was a bit slow.

When I went back to feeding the fish, two moray eels were at my lookout, munching contentedly on the fish head I had left there. So I got the other one and tried to give some crumbs to my favourite squirrel fish without disturbing the eels.

It took Meadowes half an hour to come, and by then I was cold, worried about him, and worried about the wind. It was beginning to get dark. Meadowes swam slowly across the site and bumped into the little outcropping of coral under my perch, the same one he had bumped into before. I got his medicated food, and held it above the surface while I followed him around and tried to float it to him. He ignored me. When he was hunting around in the coral under my lookout, I went over and dropped the food there, but it drifted into the coral, and I had to get it, and try again. It was covered by fish, and sharks were approaching me from all sides, including Meadowes very closely on my right. I let it go and drifted, not moving a muscle. Meadowes swept down and took it. Terribly relieved, I watched him eat it with difficulty, weak as he was, swallowing repeatedly as he disappeared into the coral. I followed him as he swam away southward, while Mordred appeared and glided with us, not far off. Gradually, Meadowes disappeared into the gloom, and I returned to the kayak. I didn't even have a scrap for the crocodile needlefish that night, who arrived as I was leaving.

Because of the wind, I had to swim all the way to sheltered waters near the boat channel before I could get back into the boat, and even then, I was blown wildly back toward the lagoon so that I had to paddle frantically, still with my fins, mask and snorkel on, to keep the boat steady and make slow headway toward the shore. There I sat on the beach to rest, as night fell. When I continued home along the shore, the wind died down enough to make the last part of my trip easier.

In hopes of fooling my suspicious sharks so I could give Meadowes his third dose of antibiotics the following evening, I went to the barrens in the late afternoon on the chance that in the deeper water, where the sharks would be widely spread out and not expecting me, he might come to me as he had before.

I poured a bit of scent in the water as I paddled up the barrens, and more as I anchored near the reef. When I slid into the water, a shark was approaching me at the surface, and I laughed to see that it was Apricot—the brat.

Sharks were cruising everywhere at the limits of visibility. Drifting down-current for forty minutes, holding onto the boat containing Meadowes' medication, I encountered many, and many came with me. But Meadowes was not there.

I got back in the kayak, and the wind blew me away toward the reef as I pulled off my gear, and dried and wrapped myself in my towel. I was already shaking with cold. I paddled an erratic course toward the site, while most of these sharks, being by now wise to my tricks, kept me in sensory range and followed. When I slid back into the water, I saw no one, but the moment I poured some scent into the lagoon, they all materialized from behind their blue curtain and began to circle me.

They circled, and I watched, waiting for Meadowes. If they got too close, as Bratworst did repeatedly, I glared at them. As darkness fell, they dispersed. Only the faithful Sparkle Too continued to circle steadily around me about two metres away. Trinket had arrived and was gliding back and forth. Everyone else was long gone. I was happy to see the little Trinket again, and Sparkle Too was so patient, that I decided to give them the chunk of fish that I had been saving to distract the others while medicating Meadowes. So I reached into the boat, took it out, tore it in two, and threw it. Sparkle Too wasn't paying attention, and Trinket didn't seem to see it. But out of the blue soared Apricot, Sparkle, Bratworst, and several others. Apricot went straight to one of the pieces and swallowed it; Sparkle inhaled the other.

They circled me suspiciously for a while longer, then left. There was no one. I waited, as the twilight darkened, reluctant to go when there was still a chance of seeing Meadowes.

Night had enveloped the scene with shadows when Madonna and Flora appeared and began to circle me, Flora clockwise and Madonna counter-clockwise. It was at least fifteen minutes since they had left, and far longer since there had been any scent in the water. I sang to them. For a long time, they turned, placidly, close around me while my heart melted. For that moment it was worth nearly freezing to death.

I wrote to Arthur about the difficulties of the situation and he responded immediately:

Sparkle missed...

Dear Ila,

The most important task for you is to never place yourself in harm's way. Being in the water, alone, with sharks and reasonably far from help is something to seriously ponder. Do not place yourself in greater harm by attempting to treat a sick shark if doing that causes you to take chances. Do not attempt to feed any shark by hand especially if there are other sharks around. You must be able to see around you (i.e. in all directions) rapidly, and that is impossible to accomplish when you're concentrating on getting the food to the sick shark. Another shark will suddenly try to get the food from a direction that you're not expecting. You're correct in that sharks can rapidly learn to associate any actions with a food reward, and your presence is likely one of those actions as well as where the food appears on your body. Also, you are trying to feed as the sun is going down or is down.

That is a bad time to be in the water with sharks, as is the early morning. These are the periods when predators are most often active, and there are other species of sharks that will come into your lagoon at such times. The sick shark is not worth your hand or any other part of you. Stop feeding at dusk; it's simply too dangerous, especially since there are many sharks around in your lagoon. Feeding from a boat is dangerous enough when you're alone. Your great interest in the biology of blacktip reef sharks is commendable, but do not let yourself be conned by the sharks into thinking that they are not dangerous, especially when they are seeking food. They're one of the world's best groups of predators and any predator must be carefully dealt with. Don't ever push your luck with them. Please don't try to accomplish something on your own that likely would take a team of individuals to accomplish. I would stop feeding and even going into the water for a while. Let time pass to allow the sharks' learning patterns about you feeding them to be forgotten. Many may take a rather long time to forget, but time is on your side, and when next you meet them without food, they'll most likely habituate to your presence and not be concerned with handouts.

Trying to help your sick shark is something that few would ever attempt with any wild animal. . . let alone a shark. You tried diligently, and it may recover based solely on your efforts. You did your best. . . Now let things return to normal and stop feeding in the water alone, stop swimming with sharks at dusk or early in the morning, and always be cognizant where the sharks are around you (which is almost impossible to know when there are a few since they can move so rapidly).

Be careful. . .

It's probably time that you sit in your home or in the yard and compile and gather together your many notes and begin to make concise tables and graphs of times and individuals and happenings. The black-tips will always be where you last saw them. I've been in the water with many sharks over the last thirty years, and I've never done many of the things that you have done with the sharks in your lagoon. You're to be congratulated for what you've accomplished. But please remember. . . it's better to be safe than sorry.

Sincerely,
Art

This letter deeply concerned me, but I was too involved to stop. Martha had not come again and I couldn't abandon Meadowes. So the next evening, though there was a strong wind from the northwest, I paddled out to the lagoon and found it calm when I arrived. I had enough fish scraps with me to occupy the sharks while I gave Meadowes his medication. I went to the barrens, poured scent all the way up its length, then anchored and slid underwater. Bratworst was circling.

When Madeline swam up to me, I handed her a scrap and she inhaled it. Madonna began coming up to my face, and I threw her a good scrap too. But she didn't see it fall through the water, and searched for a long time before she found it, whereon she picked it up and dropped it. I went over to look at the depth of her teeth marks while she circled the region in a bored fashion. There was no trace upon it that a shark had touched it, so delicately had she taken it. Later, she came back to her scrap, plucked it from the bottom, and again she dropped it. More sharks were arriving, and after waiting a while longer, I went on to the usual feeding site, trailing scent.

There was no one waiting in the water there, and again I noted a big difference in the order of appearance of the sharks during the hour in which I waited worriedly for Meadowes.

The sharks I had seen in the barrens trickled in, with some from the immediate area. They weren't excited and didn't seem very interested. Only Muffin ate anything, though I had picked up scraps that day so there was lots of food this time. In ones and twos and sometimes little groups, more arrived to circle, pass, leave, and return, and sometimes poke around in the scraps.

Night was gathering when Eclipse and Hurricane, with a large dark companion, soared in from the east, and whipped toward me just for a moment, as though to confirm that it was me. They swept through the site, then circled the area with several of the local females who had come back, it seemed, for this meeting. The three friends were enormous with pregnancy. The new visitor appeared to be twice the volume of Madonna. I called her Keeta, because her dorsal fin pattern was like Cochito and Cochita's. Annaloo and Glammer joined the group, and Antigone, golden and beautiful as ever, glided in. They all shot around in the gloom, and Flora came when it was so dark that I could scarcely recognize her; she was the forty-second shark of the night.

But Meadowes did not come, and neither had Martha.

Conditions prevented me from going the next day. By then I was concerned about not having seen either of the sharks who had taken the medication since twenty-four hours after each had done so. In case my change in routine was to blame, on the following evening I went straight to the feeding site, spilling scent through the lagoon as I went.

No sharks followed the boat, and in the water, no sharks were waiting. It seemed strange to be alone while putting the food in place, but gradually, the residents accumulated. The food was quickly eaten, and they began dispersing. For several minutes, no sharks were there. I had saved aside some extra scraps and threw some to three stingrays who came and circled in mid-water the way the sharks did. Then I fed the fish, and a group of anxious little sharks arrived with Glammer. Annaloo was behind them. The stingrays were still flying near, and the big one who had been coming sporadically for the past months approached my lookout, and circled me, then munched a bit on the few scraps left lying upon the sand.

But neither Martha nor Meadowes came. I looked for the floating bodies of sharks in the glassy waters of the lagoon as I went home, and again in the morning before the wind arose, but I did not find them.

That night I went back to the site, and again, no one followed the boat, and no one awaited me underwater except Droplet, who must have been there by chance. As they had the night before, sharks trickled in over the next hour, but neither Martha nor Meadowes came.

I continued to look for Meadowes and Martha or their bodies, twice a day until Franck and I left for a vacation in Australia a week later, but I could not find them.

We returned three and a half weeks later. The next day I procured a tiny load of fish scraps, and my anxiety about the sharks moved me to buy them a box of sardines too. I couldn't get out there soon enough.

The bay and lagoon lay flawlessly still for the first time in memory, the lagoon a perfect window into a supernatural garden. As I crossed toward the site, a shark came along the bottom at right angles and shot away when it saw the boat above. Soon, some were visible swimming through the coral, and a small one began following the boat.

Then a big, excited shark was circling the boat fast at the surface, stirring it into purling waves. It was Martha! I was intensely relieved and as excited and happy as she was when I slid into the water with her and the rest of my beloved sharks.

All of the residents were there. Their bit of food vanished instantly, and I started throwing the sardines to them. They rose to the surface to catch them as they fell—how beautiful they were! They were thrilled, zooming in from all directions toward the treats raining down from above. Meadowes did not appear, but it was no surprise. I had known that there was little chance to save him.

Martha's strange disappearance had simply been ill-timed. It had been one of the exceptional times that she had left her home range for more than a week, apart from mating and birthing. She behaved as always, swimming to me often, circling close around me, swimming under me, and remaining on the alert to come to me the moment I went to the boat. She got lots to eat, and so did Valentine, who was also behaving as if she were pleased I was back.

The sharks stayed around long after the food was gone, as they had done before, when I had been absent for a long time, but this had been the longest period I had ever left them. I stayed watching until they vanished into darkness. Drifting homeward, the multitudes of stars of the southern sky appeared and glowed into the night of the earth with a cold and alien brilliance.

A few days later, I went out in the morning to see Martha. In the wake of my scare over her disappearance after eating Meadowes' tetracycline, I wanted to indulge my pleasure in finding her alive. Anchoring near the border, I went underwater and called to her, finning in the direction of the site. She appeared within five minutes, and I spent a beautiful morning swimming with her, and drifting through the barrens, where she eventually left me. Samaria materialized in the same place I had found her before. Cinnamon and Shilly, with their little group of juveniles, were in the place they had been before, too, at the edge of the barrens nearest the shore. They all swam over to me, and Shilly in particular circled me closely. Floating silently in the current watching, I was amazed to see that sharks were everywhere, hidden in the lines of light rippling over the barren floor. Many that I identified that morning rarely came to the study area.

Haunted by Arthur's admonition that I should be arranging my data into tables and graphs, I began examining my notes for patterns, and interesting evidence on the lives of the sharks. But I didn't think I had enough to present a complete picture. It was a relief to grant myself another year to watch the sharks' visiting and roaming patterns, and try to figure out what it was all about. I was ashamed that, against Arthur's professional advice, I was continuing to visit the sharks.

Though what he said sounded reasonable enough, the bond that I had formed with the sharks went two ways. It showed in the way they behaved around me and the way they responded to my behaviour. And by then I loved them, as I had loved Merlin, as I loved my birds and dogs.

CHAPTER TWENTY-SIX

Poachers

Gwendolyn had her babies in the second week of September, so I went to the shark nursery in the nearby curve of the reef, to see if she had put them there. Having seen her prowling through it before giving birth, it was possible. I found just one tiny shark, grey like Gwendolyn, who came over to look at me. Sketching both sides of his dorsal fin, I named him Spring.

There was a soft southern wind as I headed for the lagoon a few days later. But as often happened, the *mara'amu* rose as I travelled, whipping the sea into greater and more powerful waves behind, pushing me broadside, and threatening to swamp the boat. The oceanic waves rose high over the shallow border region, and crashed into the lagoon in snowy breakers. I flew over them on luck alone. Surfing onward, I lost control of the kayak and, unable to see anything through the choppy surface, was blown far past the barrens while trying to find the site. I finally threw the anchor in to avoid being swept onto the reef. Waves washed over the boat as I put on my gear and slid underwater.

Though I had been trailing scent, only Tamarack meandered by in the thick coral in which I found myself. From its appearance, I concluded that I was on the other side of the barrens in Section Two. Swimming west, pulling the heavy boat into the driving wind and waves, was difficult. Jessica cruised by for a look. I had speculated that her home range was beyond the barrens. As that desolate plain opened around me, Cochita circled curiously, and other sharks passed in the distance. I could always tell which were my sharks—they came to see me.

It took over half an hour to get to Site One, struggling against the wild yanking of the boat, and blasted by waves. As I drew near, there were sharks everywhere, speeding up as they became aware of me approaching from the wrong direction. Cochito flew up to me and away, as I put the anchor in the usual place, and the rest came zooming in from down-current in long lines. Tamarack swam in from behind. He must have come with me all the way from Section Two without ever showing himself. Cochita had come, too, but Jessica did not appear.

It was clear on this occasion, that the sharks had been waiting there already. What was it like for an animal to wait, night after night for our rendezvous? Were they disappointed when I did not come? Did they associate me with good weather?

The wind was blowing the kayak too strongly for me to hold it in place while I put in the sharks' food, so I went back to move the anchor. Sparkle Too and Chevron circled swiftly around my head as I tried to estimate the exact place for it that would result in the boat balancing in position. When it settled over the site in front of my perch, I clutched it and pushed and threw handfuls of the prickly scraps into the water. It took time and all my force to get them out of the jerking boat and into the site while the endless waves washed over everything. At last, I finished and drifted to my lookout to rest, watching the sharks.

They were picking over the food as if it weren't good. I had brought them a heavy load, and watched, amused, bounced about by the waves as I wrote down all of their names. They fussed and endlessly circled.

When Martha arrived half an hour later, there were still scraps in the site, waving back and forth in the surge. She poked about among them, then passed again and again without eating. When I called to her, she turned toward me, and my slate floated from my arm, and hung, drifting slowly, in the water. She rose up to touch it, and descended past me, brushing me with her fin.

As usual, the swirling of the sharks had swept the remains of the food under the coral around the site, and I waited for a chance to dive down and pull the pieces out from underneath. In a moment when the sharks were down-current, I was about to go, but Martha soared back. Every time I decided to move, she came straight toward me at eye level, having apparently sensed the faint signals of my intentions. When she circled beyond the site, I hastily scooped the scraps out from under the coral and was rising to the surface when she sailed past me from behind. She picked up a meaty scrap and spat it out!

The sun was setting when Lillith came. For once there was lots of food left. Glammer, Shilly, and Simmaron cruised in from the east with Danny, the very large male. Vixen glided in behind them, so pregnant she was swollen up like a balloon, an impression heightened by her still being so small. She was very grey now, in sharp contrast with the foxy red she had worn when I had met her. Anne Boleyn was with her, also fatly pregnant. Since I had just seen Hurricane two kilometres away at Site Three, she appeared to be travelling this time with Vixen, whom I had seen there as well. The pregnant females were beginning their restless roaming through the lagoon.

Madeline arrived and Keeta zoomed in behind her; she had the young and slender look of a shark right after birthing. I was poised to photograph her when a female whitetip, looking like she had undergone several degrees of magnification, glided straight to the scraps that had swept underneath the coral formation where I had paused at the eastern end of the site. She was pregnant and about two and a half metres (8 ft) long. A large male, much smaller than she was, had come with her, but he didn't approach. Awed, I tried to take in every detail of her enormous square head, big eyes, and face, as she moved around me searching for something decent to eat. She picked up various scraps, busily nibbled, circled and returned several times to feed with Madeline and Keeta upon the scraps there. I could lean out over her and watch from less than a metre above. The other species of large sharks in the lagoon, such as nurse sharks and lemon sharks, had very heavy builds, but whitetips are slender. Even though she was heavy with pregnancy, this enormous female retained all of the sinuous beauty of her species. I couldn't take my eyes off her. I was sporadically feeding the fish, and she came up to me often, though whitetip reef sharks were usually so shy! Then she swam slowly away toward the west, haloed in the last orange sunbeams of the day. Never did I see her again or any whitetip approaching her size.

I had used up my film photographing her, but the camera had begun to stop advancing after only nine photos had been taken, on some of the rolls of film. Thus, I lost not only all the photographs of this remarkable meeting but also all of my photographs of the mirror experiment, among others that were irreplaceable. Since I bought films of thirty-six exposures, this meant that three quarters of each affected roll was lost. I was sure, and Franck too, that I had failed to put the film in properly. This conclusion caused me to lose hundreds of photos before I was convinced that it was the camera, and not me, at fault. I had very bad luck with all my underwater cameras.

With the arrival of the crocodile needlefish, the sun settled upon the horizon, looking like a bowling ball rolling along that indigo curve. I finished feeding the fish, especially the giant needlefish, whose attentions were increasingly fascinating. As darkness fell, several gold males soared in, another sign of the advancing mating season. Droplet followed them, swimming fast. She was still pregnant.

Luckily for me, the wind had fallen enough that I could paddle against it, but it was a hard fight to get home in the falling night.

The wind continued for several days. When the atmosphere calmed, I found the floaters of an enormous net stretching along the outer third of the border. It reached from the tiny channel I used to enter the lagoon, to the shark nursery. I anchored and investigated. It fell several metres to the seafloor where it was entangled all along in the coral. Countless fish were waiting to cross, like wild horses stopped by a fence. It was impossible to do much to help them. I tried to weigh down the top part so they could swim over it, but it was so long and heavy that that was impossible, and I nearly drowned myself trying to make openings on the bottom for the fish to use, and free some of the countless ones of different species caught in it who still lived.

The net traversed the region used by sharks and big fish as a submarine highway to the ocean. Many of them could be caught! Already, I was mourning my crocodile needlefish, whom I believed spent much of her time hunting just there in the current that flowed from the lagoon.

At home, I started making phone calls. After being passed from one department to another for an hour or two, I managed to contact the person in charge of enforcing fishing law on the island. She drove out to see the net, and we met there on the shoreline to survey it. She estimated that it was four times longer than the legal limit and told me that all nets are supposed to be removed each evening by five o'clock. Together, we visited some of the local fishermen and learned that this one had been there for many days. She said that she would have the net removed, but days passed, and it stayed in place.

There was a small jetty on the shore opposite, and I found that a group of young men brought several garbage cans full of fish from the net twice each day by motorboat. There, they were loaded into two trucks and a van. A second boat was anchored at the net, where others worked, stocking the fish who had been caught. I took photographs of these activities back to the fisheries representative, and she identified them as well-known poachers. Unfortunately, she told me, she didn't have a boat, so was obliged to wait for the authorities to come from Papeete to have the net removed. Unfortunately for the fish, they had other priorities. It was soon clear that there was no mechanism in place to enforce any fishing laws. Anyone who wished to make a lot of money easily was free to plunder as they wished.

The poachers were a mean-looking gang, and I was afraid of them. Further, it was illegal to tamper with an illegal net. The poachers went to the net in the early morning and late afternoon, so I went at dawn. It was exhausting to dive down, manipulate the heavy net, and try delicately to extricate the dying fish without damaging them. I proceeded in a panic, looking frantically around each time I surfaced to make sure that no one was near, that no motorboat was approaching. On both sides of the net, large groups of fish waited like traffic at a red light, as if at any time the net would be lifted for them. My concern for them drove me to make every effort to create gaps for them, but it was always in vain due to the impossible weight of the net. I often stayed down so long that I was afraid I didn't have time to get back to the surface, and when I finally reached air, I was at the end of my resources.

Slowly I moved along the net's two hundred metres (660 ft), doing my best for the living fish as I went, sick at the sight of it hung with dead ones. However, no sharks were among them. While it was in place, I dared not encourage circulation in the region by taking food to the sharks and causing a scent flow, and all my extra time and energy was spent in saving fish from the net.

Finally, after two weeks, during which the poachers must have earned a fortune selling their illegal catch to fish markets, stores, and hotels between there and Tahiti, the net disappeared. A letter had finally been sent to the poachers asking them to remove it, and warning them to stop their illegal activities.

I went out to survey the damage, visit the sharks, and check the nursery, since the net had been attached to one end of it. Baby Spring was still there to my relief, and came to look at me again, still alone.

As I swam to the site, Martha met me, and I threw her the treat I had brought for her. It drifted down through the water and almost landed on her. She didn't notice and seemed sleepy. She swam away and did not return.

Filoh, a young chocolate-coloured female, came speeding up behind me and found the bit of food. I swam slowly on to avoid disturbing her, but when she saw me leaving, she came to swim just in front of me as I went eastward. In the deep region of large coral formations before the barrens, I found a huge, pregnant female drifting, virtually still, just above the sand. This was the closest that these sharks came to sleeping. She came gradually to life as I drifted toward her, turned broadside and circled me slowly. I drew her fin.

Then I tried to manoeuvre to see her other side, but she countered each of my subtle motions with one of her own, as if reluctant to take her eye off me for a moment. After circling, she turned her back and swam away. I got just a glimpse of the other side of her fin during one swift undulation. It was a long time before I unravelled which rare visitor she had been. This was the problem on shark encounters without food to distract the animal's attention. The shark kept the eye with which she had spotted me fixed upon me until she decided she had looked enough. Then she sped off, so I was unable to properly identify her.

I drifted on down the barrens, and saw no one. It was such a puzzle, that sometimes there were sharks moving through it like people in a park, and other times, it was abandoned. Swimming slowly back through the sun-filled coral, I saw Martha investigating something at the surface and watched her for a while. But she paid no attention to me and swam away eastward.

September was passing, and I didn't have much of a record for that month, or for August either, when we had been away. So I was thrilled to be given a huge load of bloody scraps, maybe more than I had ever had, a few days later. As I often carelessly did, if there was no reason to do otherwise, I let scent trail from the well of the kayak as I paddled out to the site. Yet no sharks followed, and none appeared while I prepared.

Underwater, Flora approached. She was pale gold. I had recent photographs of her showing how dark grey she was, but she had birthed. Of all the sharks she seemed to undergo the most dramatic colour changes at the times of mating and birthing.

It took time to dump the scraps overboard, and when I ducked underwater again, the resident sharks were nosing over them and snatching ones they liked. Then each shot off, wildly shaking and streaming fish, particles, and other sharks. Groups of sharks were cavorting through the coral, shaking out bites, sweeping up the remains, and soaring out of view only to reappear again, showing every sign of shark delight. Chevron and Sparkle Too, both females just maturing, appeared to be companions. Fawn was with Rumcake, both three year olds, and Madeline was with her friend Samaria, both older ladies. Sparkle, Amaranth, Valentine, and a group of the young males, including Tamarack, arrived within the next few minutes.

For once, all got a good feed, even Amaranth, who normally came too late to eat. Little Simmaron was excited and chased the fish away from the food. With a deft movement, she snapped one up, hardly changing her rhythm. It looked so easy, given the abilities of sharks to accelerate instantaneously to light speed, and I wondered why they didn't catch the fish in the site more often.

I had often thought of the remarkable crocodile needlefish, and had given her up for dead in the net. But the graceful, silver fish drifted slowly toward me, half hidden in the ruffled waves. She wriggled a little bit to move forward, with her head turned enough to see me with one eye, then glided and turned her head to look with the other as she wriggled again. In this way she came on as I tried to affirm that she was my familiar visitor. Finally, the two scars on her side came into view. For the first time, she came to me, looking, and I looked back at her as she began to circle around my head. She stayed there right beside me for the duration of the session. It had been two months since I had seen her. It was not the first time a fish showed me extra attention when a long time had passed since our last meeting. It seemed that this meeting marked the moment that she acknowledged and accepted my benevolence. Ever after, we were friends.

Sharks continued to arrive as the first group dispersed. But when I drifted down-current, I found them all still there, socializing out of visual range after eating. I had never seen any fighting among them, nor signs of one shark disliking another. They had friends, but no enemies. I started to think of this behaviour as "partying behaviour," unable to find a better word to describe it.

The crocodile needlefish stayed close as I fed the fish, sometimes coming right up to me at the surface to snap up a bite. When she opened her pointed mouth to take a morsel, she revealed a throat nearly the same diameter as herself. She was well equipped to catch and swallow large fish. When she accelerated, she frightened everyone around her. The little lagoon fish shot away, creating a dazzling, submarine firework exploding in silence.

She accompanied me to the boat to get each of the two fish heads that I had. While we were alone there, I offered her a chunk of meat in my hand. She watched until I drifted it toward her, whereon she came and swallowed it. When I cruised around down-current, she swam beside me with my usual entourage of fish. While I fed the fish from my perch, I tossed the larger chunks to her. She accelerated and took them, staying close and solemnly watching me as I watched her and the ever present sharks.

The red disc of the sun slid below the horizon, and rays of fire, green, and violet shone high into the

The Crocodile Needlefish

zenith for just a few moments before fading, as a palpable darkness fell. The sharks were becoming invisible in the twilight waters as they scouted around looking in and underneath the coral for dropped scraps. Five minutes later, Shilly came in from the east, followed by a large, dark, and emaciated shark, who swept aggressively onto the few stripped spines remaining. I watched closely, pencil poised, for a glimpse of her dorsal fin, but it was unclear. When she turned sharply in front of me, I saw that a long, triangular section had been cleanly cut from the leading edge. It appeared to be a knife-wound. I drew what was left of her fin tip, but there was not much there. How anyone had managed to nearly cut off her fin was a mystery—another machete slash from a boat that essentially missed?

I was trying to get a photograph when a little pup appeared very close, so I snapped a photograph of it and scared it to death. It did not return, and neither did the injured female. It was too dark to see anything more. I picked up the anchor, and put it in the boat, reflecting that neither Chevron nor Sparkle Too had come; the companions must be away roaming somewhere. A last glance beneath the surface found the serpentine, dark shark floating by, close enough to touch. She knew me. Who could she be?

Paddling slowly away across the glowing surface, one more time I watched the universe expand in wonder as the sky grew transparent, and the first light of other worlds appeared in the deep abyss above. And me, there I was, so small, upon this insubstantial disc of sea, in the centre of it all.

(Cosmologists tell us that the big bang is equidistant in all directions in space and time, so we really are in the centre).

CHAPTER TWENTY SEVEN

Shilly

I checked the nearby nursery in case Flora had birthed there, and found three tiny, grey shark pups with Spring in the champagne water, swirling through the miniature corals of the nursery. They came forward to within a metre of me and remained, swimming near. This was unusual for little ones in that nursery and made it easy to draw each newborn's fin on both sides. They were vivid, shiny sharks about forty-five centimetres (18 in) long. I couldn't be positive that they were Flora's or Gwendolyn's, but it was likely. Regardless, it was good to have positively identified a group of newborns that I could hope to watch mature.

It grew windy as I went home that day, heralding the beginning of a week of violent storms. I was prevented from going to the sharks on my scheduled evening by torrential rains and near hurricane-force winds, and was a bit nervous about it when I managed to sneak out there during a moment of silence the following night.

I had three garbage bags full of backbones from the *mahi mahi* fish (*Coryphaena hippurus*) and nothing else. The sharks did not favour those. There were no ordinary scraps from which they could pick and choose and at least come away with a snack. I had half a *saumon des dieux* (*Lampris luna*) too, which was an enormous, silvery oceanic fish, literally translated as *salmon of the gods*. It had orange polka dots, orange fins, and tail. The meat of the *saumon des dieux* was relatively hard and dense, and difficult to pick off the bones for the fish. Neither did the sharks like it very much. And due to having to wait for the weather, all this had become somewhat rotten. I cut as much of the meat off the bones as I could to make it easier for the sharks to eat.

The atmosphere was still, beneath a low, dark sky as I glided over a hazy sea toward an indistinct horizon. The lagoon reflected the hovering gloom, shot with gleams, where, as if by magic, the dim light had concentrated. As I neared the site, Trinket came cavorting by with other sharks, vaguely seen through the rushing surface.

I dropped one of the heavy bags of backbones on the seafloor, to give myself enough space in the precarious kayak to put on my gear and secure the paddle. The fish went to the sack, and some of the sharks swam over it, then ignored it, and waited for me. When I slid underwater, they came and accompanied me to the site. Sparkle Too and Chevron glided, this time, at my right side. Braced perilously against the current, standing on the ledge under my lookout, I chucked the huge backbones overboard, and an opaque cloud of sand, fluids, and fly eggs grew underwater. The usual sharks swirled through, and I drifted downcurrent to be sure of seeing everyone present. The crocodile needlefish was with me. I found a few small scraps of meat the right size and threw them to her one at a time. She swept in to snap up each one. When she came to look into my mask when I returned to the lookout, I smiled and wiggled my finger at her. She stared at me solemnly from those huge eyes.

Several of the sharks had birthed, including Glammer and Bratworst. Annaloo had disappeared when Glammer left to birth, and I hoped she would reappear soon. But even though they were hungry, the sharks would not touch the *mahi mahi* backbones, and after looking over what I had brought, they spent their time circling through the area and did not eat. The only thing that I had to use to feed the fish was the half *saumon des dieux*, which still had some meat on it. It was lying on the sand on the other side of the site and had been ignored by the sharks, but since it was outside the retreat of a large moray eel, I had to wait until he was occupied with a couple of backbones he had also taken. Then I swiftly retrieved it.

When I returned to my perch, there was another of the huge Javanese moray eels in the grotto to my left where the squirrel fish and groupers waited, looking at me expectantly. It was not easy to scrape the hard meat off the *saumon des dieux,* but I persevered with all the hundreds of characters impatiently waiting and watching me, and distributed the food from crocodile needlefish, to little needlefish, to squirrel fish, to groupers, and then to the multi-species cloud flashing about in front of my face. Then I wafted a large piece past the moray eel, hoping to earn his good will. Several sharks came over, since they still hadn't eaten. Bratworst came leisurely up to my face with Lightning and Tamarack right behind. So I left the bony frame of the big fish there and went to the other outcropping to my left for a while. Jessica and Anne Boleyn arrived as the sun sank beneath the horizon.

Suddenly, a very thin, dark shark glided past my face, and when she turned, I recognized Madonna. I was happy to see her, but shocked by the change in her. The last time I had seen her she had been so fat! She had birthed, but her back was nearly flat, the way Meadowes' had been—she had changed dramatically in this short time. But many *mahi mahi* spines were still lying around, and I saw no reason why she and the others couldn't get a bite to eat from them if they were so hungry. A few sharks had taken bites from the edges, and in any restaurant, *mahi mahi* was an expensive dish. So, I considered that it should be good enough for a hungry shark.

After a while, I resumed feeding the fish. It was quite dark, dark enough that it had taken some time to recognize Madonna. I was concentrating on whether the squirrel fish were able to eat the pieces that I was freeing for them, if the needlefish were getting enough, making sure that the crocodile needlefish was getting her share, and watching the two-metre moray eel in the hole, whose jaws were a few centimetres from my left hand. This was one of the big ones, as long as the big sharks, with a heavy head and jaws. I angled my body away from him. Once in a while, he retreated, and the space filled with groupers and squirrel fish, all waiting for crumbs to come their way. So I would whip into action and do my best to free up a few more bits of food for them, pushing larger chunks deeper for the groupers, whose purple bodies were covered with stars that shone in that dim light.

The giant moray eel reappeared, then. I watched him, thinking that I wanted to leave. He watched me too. Our faces were close together, nearly at the same level. There was a palpable eye contact, and I began to wonder what was in his mind. Then he moved forward, his eyes fixed on mine, and gently took his side of the frame of the *saumon des dieux*. He began to pull it ever so slowly, without breaking eye contact with me, so I had the impression that he was asserting his wishes, but respectfully. It was as if he were asking if he could have it, the whole thing, and waiting to see my response. It was about thirty-five centimetres (14 in) square. I let go of it. He recoiled and carried it gracefully into the grotto.

He had seemed polite and intelligent and far from the vicious and aggressive animal described in my reference book and by others when referring to moray eels. At that moment, Madonna passed centimetres under my body, having come sneaking up while I was communing with the eel. Then she turned toward me. Bratworst, Lightning, Jessica, Breezy, Tamarack, and others swept with her, but I was transfixed by Madonna, now zooming straight toward my face too fast. I splashed the surface in front of her and she didn't respond. She was still coming, and I began flying back away from her. She was at the surface with her nose practically touching mine, my body beneath hers, and the other sharks converging around her. I clapped my hands together in front of her nose, and when she didn't respond to that either, and was still coming at me, I panicked, splashed the water with my fins, and changed direction. She veered off while I swam to the kayak and flew in like a dolphin. The anchor was still on the sand, and I managed to pull it up from above.

I took off my gear and glided over the site where no sharks were passing now. I wasn't sure how to think about what had happened. Had it been a misunderstanding between Madonna and me? For years, I had fed the fish and no one had bothered me. Madonna's behaviour was surely generated by her obvious hunger, and the darkness. She was thin as a snake. Having just had her babies, she was under the influence of a hormonal change of state. Sitting there in the falling night, musing, I heard a call. A crested tern had paused above me, questioning whether I didn't have something for him to eat.

I felt bad for poor Madonna. So when the storm subsided three days later, I went to look for her with a few good scraps, saved in the freezer for just this sort of emergency. At the site I waited for her while picking up the few vertebraes scattered around. Little Shilly had been following, and she came for a look while I was busy there.

I swam on, and a few minutes later, she reappeared, flitting along in front of a large, grey shark. It was Madonna! It seemed amazing to have found her so easily, especially without a gang of her friends. I threw the food where the two could see it as they swam past me, and was surprised that Shilly got it. Madonna showed no sign of aggression. She placidly followed the shark pup, who accelerated, shaking the food. Then Madonna returned, circled, and ate the morsels that I threw her, one by one, until she was satisfied. She looked a bit better than she had right after birthing. She was slender but didn't look emaciated. Her back was slightly arced rather than flat.

After she had eaten, I stayed with her as she followed the same route that Martha often took, winding slowly southward, before turning east. Like Martha often did, she drew ahead when we reached the barrens, and I lost her. Shilly swam nearby as I returned to the kayak, then went ahead. In spite of her being so small, and having a tendency to swim swiftly, I was able to stay with her for many minutes while she swam in a large circle. Samaria appeared in the same place I had seen her on many other occasions. But Shilly accelerated and left me behind when she entered her refuge of thick coral, near where I had seen the babies.

I drifted, watching, and upon one of the deeper sandy strands that cut through the region, a large, pregnant shark was resting. It was Emerald; she had visited the region just once that I knew of, well over a year before. She circled me slowly then disappeared in the direction of the reef, showing the other side of her dorsal fin just briefly as she turned tail, so that I could confirm her identity. She displayed the same pattern as other unfamiliar sharks I had surprised—the reluctance to let me out of view of the eye with which she had spotted me. When she did so, it was to accelerate away.

I went to the lagoon whenever I could in the mornings to swim and explore. Most often, it was Martha with whom I roamed, but sometimes other sharks let me swim with them. Sometimes, the sharks I knew accompanied me, singly or in a small group, as I wandered from the border up toward Section Three. There were many puzzles to which answers came hard. Why were there sometimes no sharks anywhere, when other times large groups came soaring up to me, as if they were just thrilled to have found me? Flora was consistently present in the evenings, but I never saw her in the lagoon during the day.

One calm, sunny morning I slid into the water close to the border and swam to the reef near the nursery. Following it eastward, trailing the kayak, the coral was so thick at the surface that it was hard to manoeuvre. There was the occasional stream, like a miniature barrens, floored with broken coral. Gwendolyn passed, and a juvenile swam out of the champagne water of the breaking waves and disappeared. When I swam deeper into the lagoon, traversing the wide thicket of coral where the pups lived, more sharks appeared, and Gwendolyn passed again. When Bratworst swam over, I anchored and threw them some pieces of food, but the sharks didn't bother to come to eat, so I put it back into the boat. Bratworst seemed tired and sleepy. Gwendolyn had never approached me since that first time I had met her, when she had startled me by shooting up my body from behind with Bratworst.

I swam on across the barrens, where many sharks were moving slowly in twos and threes in the distance.

On the other side of this wide, grey expanse, Section Two began, an exquisite forest of living coral. In a sandy valley Amaranth and Samaria were slowly circling above the sand, resting. Amaranth turned toward me and swam lethargically past. The supernatural garden beyond was hung with pale and colourful fish. The coral was healthy, and the water so crystal clear that I could see sharks cruising far into the distance.

I was trailing scent and regularly checking behind me. Jessica turned widely around me in the same region I had seen her before. Gwendolyn suddenly swam straight toward me, turned a couple of metres away, and disappeared again before I could throw her the bit of food in my hand. I had already tried several times to give her a piece but she didn't cooperate! She had been following me out of view for nearly an hour.

Suddenly, a little shark came wriggling excitedly toward me in the gleams of the sunlit surface. It could only be Shilly. She must have followed the scent trail a long way and was so thrilled to have finally caught up with me that she was cavorting along almost too fast for the eye to follow.

If only you could have seen her.

Arriving at my face, she orbited me like a little electron in the quantum mechanical fashion of now-you-see-it-now-you-don't. I went to the boat to get her a piece of food, and when I tossed it to her, she snatched it. The second piece fell a little farther away, and though she swam straight down after it, it disappeared into the coral, and she tried without success to get it. I had to dive down and retrieve it. When I threw it again, Bratworst zoomed in and snapped it up, while Gwendolyn, Droplet, and a stranger also closed in, all now looking wide awake!

I wanted to spend this time with Shilly, who was more intriguing each time we met. The adult sharks were circling widely, waiting for me to put food on the sand, while little Shilly was again flitting straight to me at the surface. I got the best piece of food from the boat and threw it to land beside her so that she would be able to see it easily. Far from being startled by the splash so close, she snatched the food with a lightning gesture and went away gulping. Seconds later she came shooting back. I was surprised that the little thing wasn't stuffed already.

Though the other sharks had had longer than Shilly to learn to target a bit of food I had tossed to them, they remained in search mode on the bottom. I got the last piece and threw it to Shilly. She zoomed to it as it splashed down a metre away from her and inhaled it. Not even Martha could have equalled her performance in the presence of other sharks, since Martha never moved so fast. I was astonished by her accomplishment. Happy to have seen her, and rewarded with such an unprecedented display of attentiveness, I got back in the boat, and drifted briefly, watching Shilly circling me at the surface. By that time she was about three quarters of a metre (2.5 ft) in length, including her tail.

Though Keeta and several other sharks had birthed and been in the area shortly before and after parturition, I kept finding the same four, tiny sharks in the local nursery. By mid-October, they were about six weeks old, and they began to explore the region around it. When I went to check on them, they were in the lagoon nearby, often in the thick coral along the reef some distance from the nursery. Each time, I verified their identities. They still came to look at me as I drifted through the area, joining and parting from each other, yet remaining in a little group, and treating each other as companions. Here was another bit of evidence that companionships could begin in just this way, and that some of the close companions I had known, particularly those in which the sharks resembled each other, could be siblings. A relationship that began with close proximity in the nursery, after close proximity in the uterus (the sense of smell being the first cue of familiarity), would be reinforced when the sharks began travelling and adventuring together in the unfamiliar and frightening wider environment.

As November came, I acquired an excellent load of scraps for the sharks and let the blood from them trail behind me, gleefully anticipating treating them for a change. The sea lay calm with a light breeze from the west just touching the water and granting me an easy journey. I watched for the sharks to gather behind the boat, but no little fins appeared. No sharks were at the site, and as I got ready, no one passed near the boat for the first time since I could remember. Underwater, no one was visible, and as I placed their food on the sand, no one appeared.

When I returned to my perch, Bratworst entered the site, and gradually the residents trailed in. Valentine and Martha arrived from the east, so perhaps this evening they had not been between the site and the border where they could hear me coming. Just as I was considering this possibility, the crocodile needlefish swam in from the east as well. Martha always came to me when she arrived, and this time, Valentine came with her.

I watched from the eastern end of my coral wall for better visibility, where two Javanese moray eels waved from crevices close-by. The scraps were being swept under the coral, there, and many sharks swam

up past me after snatching one. They quickly ate and left; no one stayed for very long. Jessica came much earlier than usual, as if she had come with the group from the east. She cruised around with Madonna and picked at the scraps. Brandy zoomed in, chasing the fish. Storm arrived; her range was eastward too. But she didn't go to the food, and cruised the periphery for several minutes, then left. I had seen her eat only one time, just before birthing.

Sparkle Too, enormous in pregnancy, was stuffing herself. Her belly was so distended that her ventral fins were pushed out three centimetres (just over an inch) from her body. When she turned her posterior end toward me, I could actually see up inside her; she was open! She seemed to be in the beginning phase of the birthing process.

Brandy swept in, dove on the food, and greedily fed until she was stuffed full from the teeth down. Little Pip circled around her. Droplet arrived, snatched up a backbone, and dropped it. She left, then returned ten minutes later, moving twice as fast as usual. She grabbed up another scrap and dropped that too before speeding away. It was a long time since I had seen her graceful friend Splash, and was becoming worried about her. Droplet had become huge in the three years I had known her, especially now at the end of her pregnancy. And she was still a fine bronze in colour. Keeta glided in from behind me with a beautiful, graceful motion.

As it grew dark, Samaria swam slowly into the site. Eden, Christobel, and Apricot were still gliding through occasionally, looking thoroughly bored. They were not in search mode. A lot of food remained on the sand, including actual chunks of meat, yet there were periods when there was no one there at all, and no one came even when I fed the fish.

The crocodile needlefish roamed far more freely with no sharks in the site. She usually remained behind me, suggesting that she was afraid of them. This time, she swam beside me each time I toured around down-current. The region was empty of sharks, though the scent from the pile of bloody scraps flowed away like smoke. Finally, Sparkle passed me, heading for the site. But she only wandered the area, and paid no attention to the food.

Later, I was feeding the fish when the shark with the sliced-off dorsal fin appeared and slowly ate. She was Annaloo's size and colour and had reappeared the same time as Glammer. Further, Annaloo had not returned after birthing, and this shark had just birthed. I noted some details, which I could check at home against photographs of Annaloo, and confirmed that the injured creature was, indeed, my mystery shark.

It was almost dark when Lillith came, loosely accompanied by Ali and Fleur de Lis. Annaloo and Lillith glided around opposite sides of a large coral formation at the same time, and, centimetres from collision, both startled violently, and accelerated away in opposite directions.

Later, Annaloo came to the food alone. From my perch above, I peered closely at her injury, wondering if it was healing and how she had been so badly cut. The wound was red; it had been white the first time I had seen her hurt. As I gazed down in concern, it was as though she perceived my attention. She looked at me, then left the food, and rose up to slowly approach my face, gazing at me as I gazed back. Then, with that infinitely slow way she could move sometimes, she drifted over my shoulder. She had never made such a gesture before, and it was eerie the way she had seemed to respond to my empathy for her.

Down-current, Flora passed, coming from the direction of the border and the ocean. She did not stop to eat. Then Madeline soared through in the opposite direction as if she were headed somewhere, possibly to the border. She did not stop to eat either. In between tours through the area, I fed the fish. They were very excited, freely filling the site, with no sharks there to disturb them. The crocodile needlefish circled close beside me. I had already given her many pieces as she accompanied me throughout the session and she had taken several bites from the chunks I had thrown for the sharks too.

A little moray eel was looking out of a hole in the coral near my right hand, and as the fish got excited and started to agitate in front of him, he oscillated wildly back and forth, trying to bite them. He didn't seem to pay any attention to me, but the next day, I had a painful bite on my finger and thought of little him.

Kimberley swept in, suffering an alarming fit of pique. She shot around the site, and when a little shark poking in the food extricated a backbone from among the scraps, she went after her. When she couldn't catch the end of the scrap and snatch it, she snapped at the juvenile. Then she coiled downward, twisting her head to bite at the fish spine, which was now drifting away. There was scarcely any food on it. Droplet arrived, still agitated, and when Gabriel glided in and picked up a piece of food, she took chase. The two disappeared into the coral. There were a few large visiting female sharks with them, but it was too dark to see who they were.

Keilah had taken five sessions to join in a feeding session, so it must have often happened that these older, shy females stayed down-current, so that I never knew that they were there. Watching Eclipse shoot in, grab a scrap, and speed away again, I realized that she could have been hanging around down-current for ages with friends I never saw, and then just flew in for that one hit-and-run appearance. By drifting quietly down-current, I saw a lot of sharks that never came to the site—this time, some visiting males came startlingly close to look at me. One of them was Cochise from Site Three—it was the first time I had seen him there.

The fish seemed satisfied, and the site emptied. I cleaned the last of the crumbs from the kayak, giving a treat to the crocodile needlefish. She took it and swept away toward the west, holding it in her mouth. I waited for her to return, drifting, then when she came back, gave her one last piece, which she also carried away west. Gazing after her into the gloom, a movement to the left caught my eye. A large shark was undulating toward me. I thought it was the three-metre nurse shark coming. We were approaching each other quite fast, since I was drifting in the current, and by the time I saw that this was a sharptooth lemon shark and not a nurse shark, I was nearly above him. He was smaller than the previous lemon shark visitor, but still several times my volume, and I was painfully aware that I must smell of food. Several long remoras undulated around him.

He saw me flying toward him on the surface seconds after I saw him, and turned abruptly away. I decided to go, swam to the boat, and climbed in, pulling my legs in quickly. As I took off my gear, I could see him clearly through the still waters, and he drifted by the boat right beside me at the surface. He had come to look. How beautiful he was. He moved peacefully around the site and I was glad that someone, at least, would appreciate the good food I had left on the sand.

The wind strengthened as I departed and as I rinsed the kayak at home, it began to roar down the bay, raising white-caps I was glad to have missed. The following day, it was blowing close to hurricane force, reminding me one more time, never to bet on the wind.

I saw Sparkle Too just a few days later, and she was slender again. She must have had her babies that night after I had seen her. But no newborns appeared in our local nursery. Lillith had birthed in those few days too, and Madeline's streamlining was becoming distorted as her time approached. The parade of visiting, pregnant sharks intensified in November, bringing some of the magnificent, older sharks who passed just once a year.

When the wind died enough for me to go out during the day, I went in search of Shilly again, but the lagoon was empty of sharks. Only Lillith idled by me as I crossed the barrens. I went on into Section Two where an enormous old female I had met only once, two years before, appeared. We circled each other, and she let me swim along with her as she performed two long counter-clockwise circles in the form of a bent figure of eight. Then I paddled home with the treat I had brought for Shilly still in the kayak.

Annaloo's severed dorsal fin was healing well by mid-November.

Whenever I could, I went out in the daytime to see what I could see, and continued to be mystified by the unpredictability of the sharks. Sometimes, only males and juveniles were roaming sleepily through the coral, and no female was to be seen anywhere. Did they all go off together sometimes? Where? I had never seen any of them even a kilometre farther up the lagoon, and had confirmed, too, that there were no sharks sleeping in the depths of the boat channel at these times.

The question of what they did at night was also a puzzle, and one calm, cloudy evening I set off to look for Martha, hoping that if I could swim with her, her evening pattern would reveal something further.

I anchored within the border and found no shark visible when I slid underwater. But as I did my three-hundred-and-sixty degree survey a small gold shark appeared in the distance and swam toward me, so I drifted until I could identify her—it was Shilly. How remarkable. What was such a young shark doing on the border in the evening? I returned to the kayak, which was already far away due to the river-like current, got a piece of food, and held it above the surface, waiting for my little prodigy to reappear. When she came, I threw it to her, but failed to take the current into consideration, and she missed it. I retrieved it from the coral, and, as she sped around excitedly, threw it again. This time, like lightning she turned and inhaled it, shooting away and leaving a trail of pieces behind her. She darted back and circled around me. Entranced by this very close view of her, I called to her softly. She orbited nearly too fast to see. Finally, she swam under me, and as she moved away, I went with her, and she performed a large circle, which completed back at the kayak.

There, Flora was angling through the coral toward me, and Bratworst was farther off. Valentine drifted by. I had the food in my hand when Shilly flew toward me again at the surface and I threw it to her. She swept it up and should have been full. By then, sharks were everywhere. Ruffles was underneath me, and Windy was circling closely. They were obviously expecting a feeding session, though we were far from the site, and they had enjoyed their usual party just three days before. I held the boat, watching, as Madonna sailed by, and Shilly returned to flit around me.

I threw the last piece of food, and it drifted into the coral. Both Shilly and Madonna soared up to my face. When they passed, I dove down and put their food on a patch of sand. Madonna circled, swooped down, and took it. When I raised my eyes, Martha was sailing into the picture and out again.

As she exited right stage, I flew after her, glad to be away from the claustrophobic circling of the sharks. Martha headed west to the border, and just as I began to thrill at the prospect of us going to the ocean together, she began a clockwise circle, which brought us back to the boat. On she went, and I went too. This time, her clockwise circle was larger, and as she turned against the current I was hard pressed to keep her in sight. By the time she returned to the boat, I caught up with her. There, sharks were still prowling around, and several approached me. Martha and I swam away, deep into the lagoon, but once again, she slowly executed the largest circle yet and again we were in strong current. This time, when she swam against it, I just couldn't keep up. When I lost sight of her I grasped a dead coral, to rest.

Breathing as hard as I could, I could not get enough oxygen, and it was a long time before I began to recover. When I finally opened my eyes, Shilly was at my elbow, moving infinitely slowly, which was possible in such a swift flow of water.

At the end of my strength, I swam to the boat, where I put in the anchor. The kayak was swept immediately away, pulling me with it. I took a last look around underwater before heaving myself in. Martha was passing, gazing at me. So I swam away with her again, trailing the kayak. She was heading southeast, and we again did a huge circle in the same area. When she lost me, she was heading in the direction of the nursery, but since she was continuously turning, that didn't mean anything. I got in the boat, and sat looking through the surface of darkened glass. In a moment, Shilly appeared, circling the boat as I drifted away. It was a remarkable coincidence that this relationship had developed with Shilly, of all of the sharks I had met—the shark that I had dreamed was special.

A few days later I was able to get some scraps for the sharks, and returned to feed those who had missed out when I had gone off with Martha. It was sunny and calm with a light wind. Few sharks were waiting, and none had followed the boat. As I prepared, I rinsed my weight belt, which had some fish crumbs on it. Flocks of little blue fish rose to the surface and a shark charged through them.

Underwater, Martha appeared at my side, and Cochita, Christobel, and Flora were gliding to join us. I called to them as always, and placed their food, as the tornado of sharks began swirling. Accidentally, my slate floated from my arm and away over the scraps. As I reached out my hand to take it, Lightning shot out of the cloud of scent to snap at it, but seemed to realize at the last moment that it was not food, and curved back down. Shilly glided in, gave a fish spine a shake, and left. She paid no attention to me. She never did at the feeding sessions.

Samaria flew into the site, slender and hungry. I had never seen the old shark zooming so fast. She had birthed.

Pippet joined in. Every few seconds, she twisted and thus, suddenly and unpredictably, changed direction. I saw no wound or reason for her restless flicking, and she was eating. This odd behaviour continued until she left.

The crocodile needlefish arrived, half hidden in the silvery surface, and repeatedly swam past my eyes, looking solemnly in her gentle way. She seemed to be trying to attract my attention away from the sharks. She had another cut on her side. Until I went to the boat to get her pieces of food, she waited behind me. She never circulated above the sharks in the site.

Vixen had birthed. Yet I had seen her pregnant very recently, so what possible reason could she have for being at the western end of the lagoon, if she had put her young in the nursery on the far eastern end? Her home range was in the vicinity of Section Three, so it would just have been a short trip for her to go east to the big nursery and then home again, without swimming two kilometres west, before and after the birth. Her case convinced me that the passing sharks were going to some unknown nursery farther to the west. Perhaps, like salmon, they returned to the place of their birth to bring their young into the world. (This turned out to be true on both counts.)

Once a foxy red, then darkest grey, Vixen was now pale ochre. She roamed around aggressively, chasing the fish and looking for food, and she stayed a long time. When Madeline appeared, I threw in the last of the scraps. She turned, took one, shook it, and spat it out. The rest of the sharks snatched up everything else. As they dispersed, I fed the fish. The site was carpeted with nurse sharks, tossing around, and rising vertically in their efforts to scrape bits of meat from the spines and fish heads.

Javanese moray eels moved like anti-gravity snakes from one coral formation to another as they manoeuvred to steal a scrap. The tiny eel who had appeared in front of me at my usual perch was in a hole beneath my hands again, though this time I was at a different coral outcropping. He must have placed himself there on purpose. Again, he lithely snaked back and forth before my eyes, at least half of his slender body out of the hole. I handed him bits of food in turn as I tried to get a little food to everyone.

While I drifted down-current, Anne Boleyn came to circle me. Then Cinnamon, very pale, slowly passed. In the past year, she had grown from a little shark to nearly the size of Madonna. I had often seen her while passing through the shallow far end of the barrens, though she rarely came to the feeding sessions. On the other hand, Cochita, who seemed to be the same age, didn't look much bigger than she had when she first came to the area. Yet Cochita was bigger than her brother, Cochito, though they had been identical when they had moved into the region. That made sense, since the females needed to grow a lot more than the males to reach maturity.

As I fed the fish for the last time before I left, the little moray eel began again to behave like an insane snake in his efforts to bite them. He actually got himself entwined in my fingers as I fed the fish. Nothing I had ever felt could compare with his softness. What a strange little creature he was!

It was almost too dark to see anything in detail, and I watched quietly for a while as I usually did before leaving. The sharks had left, and the fish had gone to their holes in the coral to sleep. A shark proceeded through the coral on a trajectory that would take him five metres (16 ft) to my right. But when he saw me, he came over for a look. It was Fermat, the young male who had visited the year before. After a brief look at me, he continued on his way.

As I returned to the boat, Shilly swam by very slowly. She was up-current from the site, and apparently had come just to see me. I never saw her eat.

CHAPTER TWENTY-EIGHT

December

A continuous parade of visiting females attended the sessions, as the reproductive season advanced. Some were shy, and some were comfortable with me cruising around them for the obligatory sketch, and hopefully a photograph. Bands of thrilled males soared in as sunset came, and they were harder to positively identify. They zoomed around looking like identical, grey torpedoes, and few had distinguishing features to use to pin down their identities. Each shark had to be named once it was identified, so it could be entered into the Table of Sessions on the computer. Franck was sure that he could order the meanderings of the sharks by using their names as strings, or identification labels, and running them through a computer program using neural networks to mathematically analyse their activities. I was thrilled and worked to correct all mistakes to make sure that the thousands of sightings could run flawlessly through the program.

By then I had identified more than three hundred and fifty individuals and could recognize most of them on sight. I believed that the long hours of observation of them, their habits, their visits, and their body language put me in a better position to tell what seemed random and what seemed to be a trend, than a mathematical algorithm. It was difficult for me to believe that the sharks were functioning on any predictable schedule. When a group of males appeared at sunset, their coming together in the ocean and passage through the lagoon was probably more spontaneous and dependent on their circumstances than on any outside imperative, just as the passage of a group of young men through town on a Saturday night is not predictable by mathematical cunning. We know that the men are roaming on Saturday night because it is the weekend, and the sharks are there at sunset because it is the time they are more active biologically. Both groups are on the move due to their level of testosterone. But no mathematically based computer program could reveal such reasons.

As December began, I acquired just a few scraps, along with far too many fish heads. So I took the meat from the heads before leaving, and threw in a few extra pieces I had saved from better loads in the freezer. These, I cut up, so that there would be pieces for everyone, then headed down the bay toward the magical place where the sharks awaited. It was a calm and cloudy evening. Madonna undulated languidly behind, with her nose nearly touching the kayak. As I prepared at the site, many sharks steadily circled counter-clockwise. I put a hand down as Martha passed, and she turned briefly toward it. Flora did too, and passed under the boat at the surface. For once I stroked her.

As I slid in, I banged my knee so hard on an unseen coral that my consciousness diminished. Martha passed in front of me and I tried to call to her but just made a choking sound. Sharks surrounded me, but I was concentrating so hard on just getting to the site as waves of greyness and nausea swept over me that I didn't pay much attention. I wanted to get back in the boat, but my knee was not working. I gained my ledge and pushed the food into the water, which churned beneath my hands as the sharks fed. They caught the falling scraps, and when I looked underwater the food was gone, and the sharks were in search mode.

Pippet was there, behaving normally again. Tamarack flew by with swollen claspers; he had mated. His claspers had grown to full size about a year before, but I didn't know if he had successfully mated the year before or if this was his first year of sexual activity. The females didn't usually become pregnant the first year they mated, so the precise year of their maturity was blurred. I didn't know whether this was the case with the males.

Trillium and Shilly were there too, poking around in the coral looking for fallen scraps with a few other juveniles. I began feeding the fish, accompanied by the crocodile needlefish. Each time I went to the boat, I threw her a piece of food. She waited, watching, for me to toss it.

The little eel emerged from his hole, grabbed some meat in the fish head I held, and yanked with all his force. I wafted a piece his way, which he snatched, then he searched around my fingers for more. There were two big moray eels in the hole by my left elbow, too, and I drifted some larger pieces to them.

The scene was gloomy with heavy cloud cover and poor visibility, as the sun left the sky and the usual visiting males came. Lillith soared in with Amaranth. Glammer was searching the coral in the periphery for fallen scraps as was her custom, since she rarely came for the main part of the feeding. When Madeline arrived, I threw her a treat. She caught it as it fell, and I began tossing the last pieces into the site, delighted to see that Lillith perceived the falling food, and deftly snatched one. But she nearly collided with another large shark, and changed direction, flashing past my eyes, and accelerating as she went over my shoulder. The little sharks were quick to perceive the falling pieces and ate most of them. Twilight suddenly soared into the centre and excitedly joined in.

The crocodile needlefish was confused about the pieces falling in the site, and I could see her indecision about going deeper to try to get one. She would dip down only so far after a falling morsel then retreat behind me. I threw a few pieces to her there, understanding her fears; a shark could so easily bite her in two.

It was the evening of the full moon. Kimberley had come during the previous full moon, and this time Twilight was present. Perhaps they both had been in the area, and I had only seen one and maybe not. Dapples had also come, and he had attended at the same time as Kimberley and Twilight on earlier visits.

It was nearly time to leave, when a shark approached on my right, and I waited to see who it was. Exactly as he had before, Fermat turned toward me as he drew level with me, then returned to his original trajectory, as if to say, "Oh, that's you there."

The site was alive with excited fish as I scattered the rest of the pieces. Some visiting males were flying around the carpet of nurse sharks who were munching and crunching the fish heads I kept tossing in. They were adding considerably to the scent flow by chomping on the eyes, and letting all that dark and fatty fluid into the water. The scent flow was a murky plume that disappeared, fading, in the direction of the bay.

I sensed the movement of something colossal idling up the scent flow, and peered through the fish and murk to see what was coming. A sharptooth lemon shark swept into the site. All of my fish and sharks, startled by the arrival of such a huge and unexpected guest at our little party, shot directly away from him, so the sight of this stunning animal was highlighted by thousands of shining streaks radiating away from the place he had entered. This flight of fish away from the centre, when startled by a predator, is called a *flash expansion*. Letting go of the fish head, I moved left away from the scent flow. The shark circled the site, looking all frilled with his fins extended for the tight turn. The site was five metres across, and he was nearly four metres long. He paused to pick up a fish head from the sand and disappeared into the dimness from which he had come. I went to my lookout, dropped the fish head on the sand, and waited uncertainly.

It was so dark that all I could see were shadows and movements. After a minute or two, the dim silhouette of the lemon shark passed on the other side of the site, going deeper into the lagoon. He had left the scent flow and was roaming freely. I didn't want him surprising me from behind, so got into the kayak, threw the last two fish heads out in the site, and watched. But the lemon shark did not appear again.

I limped around for a few days as the pain went out of my bruised knee-cap. The incident was a reminder of the unexpected things that can go wrong, sometimes at the worst moment.

As the period of the full moon passed, the sessions were less eventful. But for weeks, each time I was at the site at nightfall, Fermat appeared, swimming toward the reef on a trajectory that took him past me about five metres (16 ft) to my right. Each time, at the same distance from me, he turned to come over to within two metres of me for a look, before continuing on his way. He too, was following an extremely fine-grained schedule, and he was a visitor in the area.

My records showed that nearly every visit by a lemon shark had occurred during the period of the full moon; they must have needed the light to manoeuvre in the labyrinthine coral. While in other parts of the world, correlations with the moon are usually associated with the tides, in Polynesia, the tide is solar. The sun pulls the ocean toward it, so the water level rises during the day, is highest at noon, and subsides during the afternoon. But it is scarcely noticeable since it rises and falls by only a few centimetres, not enough to have any effect on the movements of sharks. Water levels are dependant on oceanic conditions and the volume of water pouring over the reef.

December brought storms and an ocean of mountainous indigo, with snow-capped peaks. I slipped out to the sharks in moments of calm, but the lagoon was a river that breached the seal on my mask if I turned my head. Nevertheless, there were some magical sessions when everything briefly paused, and the lagoon, washed and refreshed with new sand and water, was at the height of its supernatural splendour.

It was during one of these spells that I was feeding the fish, and Shenandoah, as big and fat as any shark I had seen, drifted, ghostlike, behind me. I waited, not wanting any more nose to nose encounters with this unusual shark, and threw a few scraps into the site. A little pup, scarcely more than half a metre (20 in) long, flitted in at the surface. I drew only one side of its fin before it saw me and turned away. It returned and came close enough that I could almost see its dorsal fin pattern, and was fairly sure it was one of the four newborns I had watched in the local nursery in September, one of Spring's companions.

Chevron, Flora and Valentine appeared now that fresh food was out. Chevron had been missing for weeks, and I could scarcely see the faded marks of her mating wounds. This was her first reproductive year. I had first identified her when she was roughly two years old, during the first, daytime sessions I had held. Then she had disappeared until she was almost mature, and had settled in as a resident. She had become a heavy, reddish-brown shark, and was easily identified by the perfect chevron shape of her white band.

Shenandoah appeared across the site. She turned, passed through it one time, and then made a very long, slow circle around it before disappearing westward. When she didn't return, and my usual impatience set in, I resumed feeding the fish. A big shark approached, and I held my breath, but it was just Madonna. When she passed, Mephistopheles appeared. He had been at the same session as Shenandoah when she had come before too. Dapples was often in the area when I sighted Kimberley or Twilight. Could Mephistopheles and Shenandoah have a similar connection and both be visiting from the same area? I watched him as I absently scattered odd handfuls of food for the fish. He circled tightly over the food, always moving faster than any-one else, his poor nose a perfectly symmetrical concave shape against the paleness of the sand.

Suddenly, a shark whipped straight up to my face. I moved sideways to avoid her. She was an unknown, serpentine female and she shot up over my perch the instant I left it and grabbed at the fish head in a de-termined way, her body arching. I could see the muscles in her head and neck flexing. Her dorsal fin tip was unusual—I had never met her before. Her momentum carried her onward, but instead of circling, she whipped her tail at a right angle and zoomed up to me again! I raised my hand, and she turned enough to pass me.

She was covered with fresh mating wounds, and her right pectoral fin was curved under her body with a big bite out of it. She was thin, the colour of dark chocolate, and she began to follow me around. I surmised that she had become adept at bullying fishermen out of their catch. It was impossible for me to pay attention to anything but her and her repeated charges. After I had to push her away, finding her alarmingly vital and strong beneath my hand, I decided to go. But even then, she stayed with me and swam swiftly back and forth on the other side of the boat while I was putting in the anchor. I had to charge her myself to make her back off, so that I could feel safe taking my eyes off her long enough to hoist myself into the kayak.

There were a few fish heads left, and I pulled chunks of meat from them, and threw them into the site for her. It was interesting that she had appeared at the same session as Shenandoah and Mephistopheles, given that Shenandoah had also been unusually bold and fearless. Maybe they came from a place where they ha-bitually threatened fishermen. That both sharks showed this unusual behaviour suggested that it might have been learned. Experience with many sharks, even those used to the commercial shark feedings at Site Three, had not revealed any who tried to push me into feeding them in such a determined manner, particularly when they had not met me before. But it had revealed that visiting sharks who appeared at the same time were often travelling companions and could come from the same place. I named this shark Janna. Her dorsal fin pattern was so unusual that she would be easy to recognize if she visited again, even if her behaviour improved!

Martha disappeared for a while, and one day, I decided to look for her. I was no sooner underwater when Madonna glided into view from behind me. Due to the limits on the underwater view imposed by the mask, she was passing me before I became aware of her. I tossed her a treat, and as she turned toward me, I pointed at the place where it was falling onto the sand a metre away. She went over and took it. I didn't see how she could have understood my gesture of pointing with my hand, but she behaved as if she did. She was notoriously poor at finding the pieces of food I threw to her, which was another reason I was surprised that she understood this time.

Sparkle joined us. She had some two-week-old mating wounds; it was her first season of reproduction too. I gave the sharks two pieces of food each, and they meandered away while I went on to the site and picked up the few scattered vertebrae left in view from the sessions. Gazing across the coral garden running with golden lines of morning light, I crumbled a piece of food for the fish. A stranger passed. While drifting toward her, drawing her fin, I saw that she was not alone. A little shark was peeking through the veiling light, flitting in and out of it. It was Shilly. The two sharks were very calm. Shilly roamed around me, and I gave them a few pieces of food. When I threw the last piece, a strange male suddenly sailed in, snatched it up, and left! It was Penrose from Site Three. He had passed through the study area just once, early in the mating season the previous year.

For the next hour, I swam in long curves through the region with Shilly, sometimes almost losing her against the current and wind. But since she always came back to me, we would eventually go on together. Finally, she disappeared into the thick coral beside the barrens where she seemed to prefer to be, and I returned to the boat. The placid, new visitor was still there and, surprisingly, let me accompany her on a long, circular, clockwise path through the lagoon. Back at the boat, she accelerated away. It was odd that a stranger would have made an association between me and the kayak or the site. But she did leave me back in the place where she had found me.

The sun and wind had risen high and I was about to go when suddenly, my beloved crocodile needlefish was approaching! She was a little shy of me in the daylight, keeping more of a distance, but came closer when I found some little chunks of food for her in the boat. But Martha had not appeared. I assumed that she had not returned and went to check the nursery before going home.

The waves had suddenly fallen dramatically, and I was able to check the whole region without being repulsed by the usual pounding breakers. The four little sharks who had appeared there in September, including Spring, were gone, but in the shallowest waters were several newborns! Since they were curious and kept coming to look at me, I was able to get detailed drawings of five out of seven of them, and returned, delighted, to the boat.

The ocean swell grew so high overnight that its distant roar as each wave exploded upon the reef awakened me. But I was unexpectedly given some fish scraps the following morning, so returned to the sharks, though I dreaded coping with the waves and current. The ocean surged over the barrier reef in rolling cascades, sending generous whitecaps furling past the site. Underwater, the pressure of them was almost too much to bear. But Martha was gliding in front of me, so that made it all worthwhile. And amid the whirl of sharks who met me was Carrellina! I hadn't seen her since April 8, and here she was again in December, following the same pattern as she had the year before! She had grown a lot, she wasn't pregnant, and she hadn't mated. Her rich, golden brown colour had faded to pale ochre, and she had a bite out of her left pectoral fin.

Martha left quickly, and after the mêlée calmed and the sharks dispersed, I tried to feed the fish, but the current swept the particles of food away so fast that I soon gave up and left. I had found Martha, and Carrellina had miraculously returned after eight months.

Torrential rain fell that week, but the sky began to lighten as I prepared for the next weekly session, and the rain diminished to drips. There had been a leisurely south wind all day, enough to pleasantly blow me out there, but as I went, it seemed to be increasing behind me. Each time I looked behind, more and larger white-caps seemed to be rushing toward me, though in front, everything looked gently wavy with ever lightening skies. I even started to see blue behind the cloudy veil in places. I was drifting as I reached the lagoon, and suddenly the white-caps caught up and became huge waves breaking around me. I had to paddle fast to keep from capsizing as they washed over the boat. The kayak shot into the lagoon, and I was in agitated waters with howling winds and sudden sunshine. Towering waves crashed upon the reef and sent waves rioting across the lagoon, colliding with those surging in behind me. Accompanied by the wild wind from the south, the place was a witch's cauldron.

Underwater, Martha was mercifully at my side. I braced myself to push out the food, consisting of old, unsold chunks of fish, that I had cut into pieces. Holding the boat while throwing the chunks was problematic, and many of the sharks didn't seem to understand, and searched on the bottom. Or, one shark would get a piece, and all the others would take off after her, while the rest fell to the bottom to be stolen by nurse sharks and fish.

The visibility was terrible. The sunshine vanished as if it had been a mirage from the start, and in the cloudy light, with the water full of particles swept from the reef and twirled into tiny tornadoes by the current, I could only clearly see what was happening right around me. I counted only thirty sharks, and scattered some food for the fish before heaving myself to safety, glad that at least I had been able to see them, in spite of the conditions.

We were leaving for Canada to spend Christmas with my parents, and my son, Peter, so it was my goodbye visit to the sharks. However, my sense of relief to have survived once again was short-lived. No sooner had the thought crossed my mind, than I found myself in danger of being blown onto the reef. I plunged back underwater and had to swim a kilometre into the lee of the island before I could get into the boat. The rest of the trip home was an all-out fight against the cold south wind.

Things were worse when we returned. The storm that had descended as we left was named Hurricane Vicki, and it departed or died the day we returned, having raged the entire time we had been gone. I went to the lagoon to see what the conditions were like and found that the current had become impossible to cope with. The visibility was barely two metres. Bratworst appeared and swam with me. I got some food to her with difficulty and continued battling onward through the tumultuous environment. She disappeared, and in a few minutes, Madonna came rapidly up from behind. I threw her a bit of food. Time after time, she missed, and I picked up the piece and threw it for her again. Cochita appeared, but the chunks of food I put on the sand were carried away by the current, so I had to relocate them in more sheltered places. Cochita finally took the piece I placed for her, and Madonna got a second one as it fell through the water. I went on to the barrens for a look around, and checked some of the other regions favoured by the sharks, but found no one else.

It wasn't surprising considering the energy involved in slithering continuously through the torrential coral canyons. It would be extremely unpleasant compared with cruising the deep and silent blue of the ocean, so it seemed reasonable to speculate that the big female sharks, especially, took refuge in the ocean during storms. I decided to give the lagoon time to settle down before holding a feeding session.

By the time it was calm enough, three weeks had passed, and the period of the full moon had come. I was looking forward to an exciting time with the sharks, and plugged the drain holes in the well of the kayak before leaving. I wanted to be able to place the food without being in the centre of a thrilled pack of hungrily orbiting sharks. Due to the strong current, I went to the deeper region which I had used the year before under such conditions. The difficulty there was to find a clear sandy area with a high, dead, coral formation from which to watch, by looking in through the tossing surface. I planned to take the food in bags to the bottom to empty them onto the sand, where the scraps wouldn't be so easily carried away by the current, and searched back and forth in the agitated water for just the right spot. Sharks were accumulating, so finally I just anchored upstream from a sandy area. Martha, Ondine, and Rumcake were the only sharks visible when I slid in. Martha rose up in her greeting gesture and I called to her as she nearly touched my nose with hers. When she descended, I took the bag of scraps off the boat, dove to the bottom, and deposited them on the sand. Martha began looking them over, and sharks soared in from all sides. Holding onto the kayak, I started to throw the better pieces of food, and they soon realized that treats were falling from above, and snatched them as they fell.

Much of the food was swept down-current, and there was a lot of activity there that I could only dimly see. No tall, dead coral structure was visible that I could hold onto, so I still clung to the boat, throwing scraps a bit at a time and watching the sharks. Little Trillium was trying to get a bit of food that was, strangely, floating. After many tries, she popped her nose through the surface and grabbed it. Shilly got one or two of the pieces and approached me tentatively at the surface before she left.

A pair of young, mature females I hadn't seen before appeared. Both were pale ochre. All the sharks seemed very pale, as they had each year in December. I stopped throwing food and started to draw their fins. One had mated about a week before, and her fin pattern was a dramatic shape, which I endeavoured to get right while being bounced about on the agitated surface. She appeared and disappeared as she circled. Her companion's fin was equally interesting, and I became engrossed in the two of them, delighted with their willingness to stay near me, though they did not know me. Watching them circling closely around me and passing underneath, I relaxed, feeling that everything was going smoothly. It was time to increase the scent flow, and I joyfully pulled out the plastic plugs in the back of the boat.

As the blood poured out, I realized belatedly that it was emanating from the place I was clutching. There was still no structure visible that I could to hold onto, and I hadn't thought much about it while throwing the food and watching the sharks. While I considered the situation and surveyed the region again through the billowing blood, a sharptooth lemon shark glided in. He was bigger around than a draft horse, and that smile seemed exaggerated as I gazed at his approaching face. He picked up a backbone and carried it away, shaking it. The sun was still at least seven degrees above the horizon; I couldn't believe he had come in broad daylight.

I crept up the boat away from the blood flow, trying to stay calm, and at the same time remembering that lemon sharks like fish heads. So I paused in my retreat to throw the four I had brought into the middle of the clear area. But in the current, they drifted up against some coral almost directly underneath me. The same thing happened to more of the scraps that I decided to throw in, since those I had placed had disappeared. Mephistopheles swam under me, swiftly joined by Madeline, with a group of youngsters and males. I could scarcely maintain myself in the current, and my arms were already aching with the effort of clutching the kayak.

The lemon shark circulated. I felt uncomfortable with a predatory fish of such length and volume cruising nonchalantly about. I well remembered the dentition of lemon sharks that I had seen up close while diving with Franck, when one had approached us with his mouth open. Still, during the periods in which the lemon shark left, I took more of the food to the bottom. Storm had come, and I was hoping that she would eat something. But each time I looked up, the lemon shark was coming back, and I returned to my place as far from him as I could get. During one of his absences, I moved the anchor farther up-current, but with me hanging onto the boat, it drifted back, and I ended up in the same place, right over the shark food.

The lemon shark emerged from behind a coral nearby, his mouth open as he leisurely chewed on something. In front of him was a little nurse shark pup, who was lying on the bottom with his back to the behemoth, munching on a backbone. The tiny creature was as pink and small as a newborn human baby. The huge shark very gently nosed along the side of the baby, who ignored him! This display of tolerant non-aggression eased my mind considerably. The lemon shark remained visible for most of the rest of the session, and my need to keep track of his whereabouts was a serious distraction from the blackfins. I wanted to cruise around down-current, where most of the scraps had flown, and where most of the blackfin sharks were concentrated.

Once when the current increased, the lemon shark became excited and began to speed around. Such power! I watched him in awe and crept even farther up the anchor line, though with the torrents streaming over me, my muscles were sorely tried. As he disappeared from view again, I rinsed out the boat, to reduce the scent coming from it. When I put my head back underwater, I was nose to nose with Apricot.

Finally, the lemon shark appeared to have gone—it was fifteen minutes since his last appearance. So curiosity overcame self-protection, and I drifted away, writing down the names of each shark as I turned. I had only done one three-hundred-and-sixty degree rotation, when the lemon shark crossed the site to my kayak, doing so for the first time, as soon as I had left it. While I had been paying attention to him, he had been paying attention to me! I was astonished to see him accelerate away from the little craft, which was smaller than he was. How anything so big and armed with such teeth could be afraid of me was a puzzle.

Down-current, I drifted among the blackfins, intermittently catching hold of a coral structure for support, but none were high enough to allow me to rest, so I soon regained the kayak and resumed clinging to it.

The crocodile needlefish came late, and seemed glad when I threw her a scrap, but shied away from the next treat I threw. I waited impatiently, holding her last morsel above the water, struggling frantically in the current, and watching for the lemon shark. But she disappeared, and a curtain of gloom dropped over the scene as the sun sank beneath the limb of the earth. The current had strengthened and the visibility became worse. It was no longer possible to see the sharks in enough detail to identify them. And still the lemon shark passed. I picked up the anchor, leaped into the kayak, and pulled off my mask. By the time my view was clear, the lemon shark was sliding slowly past my boat, his dorsal fin above the surface, so close that I could have stroked him.

I left feeling strongly that I had completely missed the point in being nervous of him in spite of his size and dentition. He had behaved like a peaceful, curious, and intelligent animal. But still, still. . . If ever I had an accident alone a kilometre from shore as night was gathering. . . How could I afford not to be cautious?

I looked up the lemon shark of the Indo-Pacific in my books. In one it was said to be harmless and sluggish, while the other said that it could fly into a rage if provoked, and sometimes even if it wasn't. This book said that the sharptooth lemon shark was very shy, taking flight at the slightest movement of a diver and waiting until it was alone to eat if food was brought. At other times, it would chase a person from the water, and circle for hours around the coral structure upon which he or she had taken refuge! Fish could certainly get mad at you and stay mad, as I knew from the time I had tried to persuade the Javanese moray eel to stop stealing food. An angry sharptooth lemon shark would be a nightmare.

Perhaps this individual had come to my sessions more often than I knew, but only at nightfall after I had left. Perhaps he had finally decided that it was safe enough to attend openly. No scent had escaped from the boat prior to the feeding itself, which supported the idea that the big shark was already familiar with my activities. There had not been time for the scent flow to reach the ocean when he had come, so he could not have found and followed the scent flow to get there. I was worried by this development, and the effect that a predator ten times my size would have on my ability to collect data at future sessions, but I had forgotten that it had been the night of the full moon! Indeed, the magnificent shark did not come again until another full moon signalled to him that a trip to the lagoon was in order.

Each night, I sat beside the sea and thought about the sharks, their behaviour, and what it meant. I was becoming increasingly obsessed by the question of how to put all of my information into graphs and tables, as Arthur kept reminding me I must. But the quantity of data I had, with hundreds of sharks coming and going at different times, relative to their special gatherings, the mating season, storms, and the light of the moon, was all very complicated to arrange in my mind. And every time I went to see the sharks, just the wildness and vitality of their lives defied any sort of description, particularly one that reduced them to numbers in a table or on a graph. Daily, I browsed my notes for hours, my mind circling the problem like a shark in a search pattern, until finally, I managed to find one small solution, then another, and my paper began to take shape.

Chapter Twenty-Nine

Accident at the Spawning Event

I had been unable to pinpoint the spawning of the black surgeon fish the previous year, if indeed they had spawned. But one noon time, as I crossed over the border after the passing of the hurricane, I was surprised to see tiny ones, whom I had assumed were immature juveniles, appearing to spawn in the place used by the adults, with the sun shining straight down on them. The adults in the coral nearby ignored them. The following evening while I was with the sharks, as far as I could see through the coral, files of the solemn black fish moved steadily toward their spawning site. After sunset, they returned, following unseen and fairly straight paths back through the mazes of coral.

So on the first evening of the year 2002, I went to see the strange event again, and which of my sharks might be in attendance. The weather was calm, and I left early. I wanted to see if the fish were spawning anywhere else, and explored by kayak all along the border. But beyond the fishes' chosen place, there was strong current over shallow flats, breaking waves, and a drop-off into the deep near the end of the barrier reef where the nursery was. Not one black surgeon fish was visible anywhere.

The conditions at the spawning site were unique. The border with the coral lagoon was deeper, so the force of the outflowing current was minimal, and from there, the sandy floor fell gently into an approximately circular region about thirty metres (100 ft) in diameter, and five metres (16 ft) deep, where there was a scattering of large coral structures. This was the stage upon which the dance took place.

I slid underwater and found the fish beginning to gather at the expected time, just before six o'clock. A disturbed ocean continued to pour over the reef and through the lagoon, picking up particles that spoiled the water's clarity, and even in the deep pocket of the spawning site, I had to fin continuously to keep from drifting away.

While watching, I moved through the area, trying to gain an overview of the extent of the event. But I was always above the thickest gathering of fish, who seemed to consider me a provider of protection.

In one place, a long string of coral formations jutted much farther toward the bay than the others, and the surgeon fish had accumulated thickly around it. There was a young shark hunting there. I explored along it to its end, near the slope into the bay where the gathering of fish also ended. They seemed to prefer to do their spawning ritual near coral formations. But dense clouds of the solemn little creatures also streamed rhythmically out over the sand. As the time advanced, they began flowing in one direction or another together, fluid as running water. And wherever I went, fish poured toward me.

They grew amazingly numerous. The evening before, they had passed me in their many lines for more than half an hour—it had seemed as if every black surgeon fish in that end of the lagoon was going to the spawning site—and the concentration that gathered now was staggering. Still they came swimming from among the corals of the lagoon. They streamed to me until they were so thick that I couldn't see through them, and I stayed still, watching as they milled and flowed together. Sometimes they streamed toward me from out of the shadows of the bay, but it wasn't possible to see what had spooked them. Other times, a shark passing down-current caused them to pour inward.

The current relentlessly pushed me toward the clean demarcation line between the fish and the clear water toward the bay, where most of the sharks were cruising. Beyond, the gloom of the deep water was unrelieved as the sandy floor fell into the depths and vanished. Jem passed, closely followed by a whitetip shark.

Other blackfins appeared and cruised slowly by as the minutes drew out. One of them was Shilly—I was surprised to see her so far from home. She recognized me, paused, and came to circle me as if it were an obligatory greeting, spiralling in to arm's length as I turned with her, looking into her eye fixed on mine. Then she continued on her way. Antigone, the striking, golden male who passed through so rarely, glided into view and approached me, too, as if surprised to see me there, though usually he paid me no attention. Then Lillith passed, a great, dark shadow. Quite a few of my sharks seemed to have come to the border for the mating dance of the black fish.

Circumnavigation of the area revealed that the fish remained within the bowl-shaped region bordered by the reef-flats, the shallows of the lagoon, and the limits of the coral formations, where the water deepened into the bay. In spite of their streaming about, they were very specific about the place they wanted to be. I slowly finned back among them until they filled my view in continuous, sinuous movement in all three dimensions around. Their colours were slowly turning rose-grey as their excitement increased. Sometimes, they all rose in the water in a wave very gradually, then descended again. As the sun approached the horizon, the tension among the fish became a palpable force.

Once, as they collected densely below to begin their wave expansion upward, Madonna appeared. The flowing purple fish descended, as she sailed over them, a dramatic sight in the mysterious light, all in motion. She turned briefly toward me as she passed, and behind her the fish rose again in an enormous wave. It was the beginning of their dance.

As if at an invisible signal, in every direction, they shot upward in small groups to flash, and eject their eggs and sperm at the surface. It was just 6:30 p.m., and the sun was down. The fish were shining that pale, pearly rose colour, accented by their black fins, eyes, tail, and the crescent of their gills.

I was finning gently against the current to keep myself stationary, but at some point, I just stopped finning, hypnotized by the fervent dance of the pearl-coloured fish that surrounded me, without a gap to see through.

A long time seemed to pass before I saw that we were all moving very fast over reef flats in shallow water. But the fish continued their dance, and I continued to watch them, despite concern that this was not in the plan. Then my visibility was suddenly zero. I couldn't see my hand in front of my face and looked above the surface to learn that my boat was far away. Underwater, I could see nothing, and started swimming across the current and up it, trying to find clear water. But downstream from the spawning, the water was opaque, and I had to continue blind, checking my direction by glancing above the surface.

When I reached clear patches in which I could see, I found that the fish were swimming in a panicked column at my side, shooting ahead, pausing to normal speed, then shooting ahead again in unison as regular as a heartbeat. The column was about a metre wide, perfectly straight, and vanished into the murk ahead and behind. A pregnant whitetip shark was shooting through the column of fish in front of me, and a large blue jack fish (*Caranx melampygus*) came darting through them from behind, passed, crossed, and disappeared. I swam with the column of fish, now very concerned for them and the distance we had drifted from the spawning site. When we got there, the event was over—only scattered groups were still flowing together and flashing to the surface. I followed the last of them toward the lagoon. They were black again, with purple tinges, as they hurried away in a loose line. It was still only 6:40 p.m.

Two male sharks came through. One approached me and circled to look, and the other continued on his way.

By 6:45 p.m. there were no black surgeon fish visible at all.

The following night, I arrived at 5:50 p.m. There was an outrigger canoe with three Polynesian men in it, anchored just a bit seaward of the spawning ground. It occurred to me that every black surgeon fish that lived in the area could be netted right there while spawning, so vulnerable were they at that time. I anchored at a distance, and located the gathering fish underwater.

Again, there was no question that the densest group of fish was always around me. They began to change into their pearly colours and rise together in rhythmic waves, starting at 6:10 p.m.; the actual spawning began, again, precisely at 6:30. There was less current, and they stayed in the vicinity of their coral landmarks. The row of coral extending far out into the sand was an important one. Apparently, just as many fish attended as had the night before. The water throughout the spawning area that I had scouted out the previous night, was again solid with them.

The men in the boat were not far off, and were standing up, peering toward me, surrounded by the darkened water and leaping fish, but they didn't approach. The last fish were still gathering together to fly to the surface and spawn at 6:48 p.m., while the rest streamed into the lagoon in their little files and vanished in the coral. Finally, bang on 6:50 p.m., the event was over.

I saw no blackfin sharks at all, only a pair of whitetips. Maybe the strange event of the spawning fish attracted their attention at first, but once they knew what it was, they lost interest. Maybe they stuffed themselves so much the first night that they weren't hungry any more, but I doubted that. My sharks couldn't get stuffed enough. Maybe they were suspicious of the outrigger canoe, and the noises emanating from it, though that also seemed unlikely.

I went to see the event each night that I was not involved with the sharks, and each time, it began at exactly 6:30 p.m. The grand, initial spawning gradually waned in excitement, the fish became imperceptibly fewer in number, and it ended at 6:50 p.m. So it appeared that our mishap in the current had badly disrupted the spawning on the night I had let myself drift away.

The incident troubled me, since I was implicated in what had happened. From watching the event over and over again, I concluded that the fish had been using me, instead of their coral formations, as a reference point. When I stopped finning, and began drifting, so did all the fish spawning around me, and those around them, as well. To us, all sailing along together, and seeing only each other, we seemed to be in the same place. The current in the neighbouring shallow region was very fast, so once we started, we were all swiftly displaced. This theory was reinforced by the way their geometrically perfect column, shooting in panicked pulses back to the safety of the lagoon, had formed at my side.

On the fifth night, I roamed through the lagoon in the vicinity, to see the extent of the area covered by the filing fish coming to spawn. However, many of them were turning pearly rose and going through the motions of their spawning dance above coral structures in the lagoon and not continuing to the border. Others gathered into lines and moved toward the spawning site, then clustered, turned colour, spawned briefly over a coral, and returned the way they had come. At 6:40 p.m., I returned to the spawning ground and found some black surgeons, but it seemed that the event had essentially ended, and only a few were still spawning.

I investigated the thick cloud of spawn and circumnavigated it. It was huge, and stayed in one place, rather than dispersing with the current, in spite of the continued turbulence in the lagoon. It appeared that the fish had chosen the one location on the border where not only was the water fairly deep, slowing the current, but the surge from the bay counteracted the outflow from the lagoon, holding the spawn right there in position, caught by the opposing forces of the sea!

On the sixth night, there was no sign of fish filing through the lagoon, and on the border a very few of the black surgeon fish in the spawning area were turning rose, clustering over the coral, twirling together, and occasionally darting to the surface to emit a spurt of eggs and sperm. No cloud formed this time, so the event seemed to have ended with a few individuals still spawning, while the majority had lost interest.

But the following week I went to see the sharks three times, and saw again large numbers of the solemn black fish filing across the lagoon, then returning after sunset. I was unable to get away to investigate further, but it was apparent that the event continued after a break of a few days. By January 20, it was over.

Chapter Thirty

The Change

There was always a long, quiet time after the Christmas season, when the fishermen went on holiday, boats were over-hauled, and the fish shops were closed. So I visited the lagoon when conditions permitted, without any food for the sharks. When some scraps became available, it was well into January.

There was that familiar thrill as I waited for the sun to begin its descent into the west, and began getting ready. Remembering the way the lemon shark had come so early to the previous feeding session, I blocked the scent flow from the back of the kayak. Finally, I loaded the scraps into the kayak's well, and aimed the boat down the long bay seaward. The volcanic plug of the island passed slowly, as I pulled the heavy craft through the wind sweeping around the mountainside, across the ultramarine bay, over the border, and onto the emerald waters of the lagoon.

The wind died as I neared the reef, but the lagoon was swirling past the coral formations like a lazy river. I anchored carefully at the site, gauging the place the boat would stabilize—close to my lookout, rather than sweeping onto the big coral structure at the western end. When I slid in, it was Sparkle who swam up to me, just as Martha usually did. She had mated about three weeks before—her scars had almost healed.

The current was strong but manageable and I pushed the food into the water. For the first time that I could remember, Storm appeared right away. The enormous creature was pale golden, and signs of ageing were evident that did not show up in the darkness in which she usually arrived. She inhaled a belly strip, shook it, and dropped it, still whole, some distance away. I retrieved it. The resident sharks were feeding, and I wrote down their names. Little Brandy was chasing the fish again, and Amaranth glided into view, having recently birthed. She virtually hovered in the current as she looked for something good among the scraps. Madonna joined her.

When Martha arrived, I tossed her a few chunks of food I had saved for her, and she deftly took them one by one. When she dropped one, it was snatched by a nurse shark. The blackfins did not seem to realize what Martha and I were doing, but the sluggish-looking nurse shark was upon the fallen scrap the moment it hit the sand. The irritating animals were rapidly accumulating as they always did in the wet season. One had taken over what remained of the scraps when Christobel soared in and darted up to it, while many more sharks glided in behind her. But the nurse shark did not move.

Several rare female visitors, who came only once a year, arrived with two companions I had not identified before. Some had recently mated, and were covered from gills to tail with bites. The usual golden males from the ocean were roaming through, rarely coming close enough to positively identify on both sides. Trying to trap them into showing their other sides was difficult, and rarely worked. But sometimes it did, and I was rewarded with a new identification.

The crocodile needlefish stayed with me as I roamed around trying to see everything, and down-current, as the daylight faded, the orderly files of black surgeons hurried by to spawn at the border. I fed the fish, watching the scent flow for the lemon shark, but the only blimp that ambled out of the gloom was the three-metre nurse shark.

It was nearly dark when Bratworst arrived, and scouted around looking for the last bits fallen in the coral. Ruffles, ruffled as usual, came too, followed a minute later by Mordred. It was still just light enough to see them and my familiar little juveniles poking around in the coral. There were a few scraps left in the back of the boat, and I threw them to Ruffles before I left, pleased to have seen them all, and that all had seemed satisfied.

As I drifted back over the lagoon in the light of the afterglow from a sun far below the horizon, I could see the little fish underwater filing back from the border. It was touching to watch, how solemn and orderly they were. One could well wonder how such an event had evolved.

Arthur had investigated the spawning of these fish in Israel, and told me that each young fish learned the route, and the spawning site from the older ones—it was an important example of social learning. He sent me a copy of the article he had published about the event, which described almost exactly what I had seen. However, there were differences due to the geography of the areas, the depth of the water at the spawning sites, and the extent of the regions from which the fish came.

The following day I was given some more fish scraps, but had no time to use them, so to poor Franck's irritation, I crammed them into the bottom of the fridge. That weekend, two of his friends from Paris arrived for a visit, and I asked them if they would like to come out to the lagoon with me to see a lot of sharks. I wouldn't have suggested it had I not noticed that they had their own gear, so assumed they were experienced divers. They seemed pleased to accept, and I tried to make the routine clear to them before we left, pointing to the sinking sun as we dawdled over a few beers on the beach, and trying to have them understand how far we had to go before that sun set.

It was an evening of supernatural calm as we paddled across the silken surface to the site. It took much longer than usual to get out there with two inexperienced kayakers in the other boat, but I didn't plan to stay long. I anchored in my usual place at the usual site, and tied the second boat to mine, while encouraging my guests to get ready. The lady said it would be better to put on her equipment in the water, and I pointed out that underwater, she would need to be ready for the sharks, so she must prepare on the boat, difficult though that may be. Seeing her distress, I slid in and went to help her. It turned out that she had never gone snorkelling before, even though she had seemed so pleased at the prospect of swimming with sharks!

When she was ready, I supported her as she slid into the water. She clutched me as if she were drowning, while I checked her mask and made sure that her snorkel was comfortably adjusted. I looked into her terrified face underwater, and pointed to Sparkle Too, who was coming over to look at us, trying to get her to look around. Still, she seemed frozen with fright just at being in the water. I was becoming very concerned about her. I had to actually unfurl her fingers from my arm and encircle them around one of the handles of my lookout, encourage her to float, and do the same thing with her other hand. Luckily, her boyfriend was able to manage by himself. Once she was positioned, though, she was fine, and I pointed out some of the interesting sharks as we watched them feed and roam through the area.

Two interesting things resulted from this brief visit. The first was that the sharks present were ones who had not come two days before when I had done my main feeding. The sharks who had attended then were now absent. It seemed that roaming was not associated with looking for food. You went out after eating at home.

The other thing was that though nurse sharks had become so numerous that they carpeted the site at the sessions, and often slept nearby, not one nurse shark appeared that night. The same thing had happened each time I had brought someone else, with the exception of my son Peter, but they had left while he was there. On the two occasions that Franck had accompanied me, he had not been treated to the sight of these strange creatures that I had described doing their alien dance all over the sand. Dumb though they appeared, the giant sharks were paying attention.

Given what happened later, it seemed unbelievable that I had been with these inexperienced people, with these sharks, in the heart of the stormy season.

The conditions deteriorated again after that, and the lagoon began to riot. The heavy surge carried tumbling white caps far past the site, and I had to use the deeper region farther from the reef for the sessions. But even there, the current was so strong that I had to carry the scraps, handful by handful, to the bottom, just to keep them from flowing far away the moment they hit the water.

Madonna, Bratworst, and Martha still formed the core of my group, usually joined by Flora, Cochita, and Valentine. Valentine was a magnificent dark grey shark. She spent long periods roaming, especially during the reproductive season, so it was always a pleasure to see her return. I had never been able to learn where she went.

And now that she was back in the area, Carrellina joined them. She behaved again as if she had never left. At her first session with us, it was she who rose to greet me when I slid underwater, only her. There was no doubt that she remembered the sessions, the routine, and me from the previous period she had spent with us, from December to April the year before. By the second time I saw her, she had mated—she had attained her first reproductive year.

Martha had left early in January to birth, and I saved a special piece of fish to give her for a treat when she returned. She was usually gone about two weeks for parturition, so I could pinpoint the time she would be back. Each time on her return, she circled the boat excitedly at my approach and swam up to my face as soon as I slid underwater. Flora had left to birth too, and then there was a session in which she rose to greet me, thin and so pale that the flares in her white band were scarcely visible, so at first I didn't recognize her.

At that session, the sharks ate languidly, and left with food still lying on the sand, though I had brought little. For half an hour no one was there, while I waited, hoping that darkness would bring a new group up the scent flow from the many fish heads being devoured by the nurse sharks. But only Gabriel came at the end, with a companion I didn't know.

When I returned the next night with a few scraps and Martha's treat, I found the current worse than ever, but the visibility was perfect. Martha was rising to meet me as I slid underwater, and I called to her. She swam by very slowly—I watched her eye looking. She was slender and swift, giving the impression of being much younger, than the stout and inept creature she had been before she left.

The enormous school of yellow perch streamed toward me as I pushed in the scraps while holding onto the kayak and twisting in the current. Martha circled slowly above the food, scenting what had already fallen, and not accelerating to take the falling pieces from higher up. Then the others came. Carrellina was there, and so were Chevron, Madonna, and most of the residents. My thirty odd sharks cruised languidly around the huge space revealed by the deep, clear waters, finding the scraps that were drifting down-current.

I waited to give Martha her treat until she was alone, then called to get her attention, and threw it. It landed between us, and the current took me flying toward her as it sank and she approached. By the time she opened her mouth and engulfed it, it was almost against my side. She passed beneath me, and drifted away. The other sharks placidly ate. Annaloo drifted slowly in the shadows down-current, as the others began to disperse.

Finally the lagoon stretched empty into the gloom. I fed the fish, and spent time with the giant needlefish. Down-current a little shark came poking about—it was Shilly. She paid no attention to me. Ruffles was also looking around for missed scraps. I had saved a small treat for Shilly and got it, but she disappeared. So I fed the fish around me while waiting for her. Darcy passed on the way to the feeding site. When Shilly and Ruffles returned, I threw Shilly's piece of food, but she didn't see it. It fell in the coral, and she continued to come forward while Ruffles went into search mode in the vicinity. I tried to find her bit of food as she closely circled, but could not, and finally she left. After that, there were no sharks visible anywhere, and I gazed out over the extraordinary panorama revealed by the transparent waters—all the fish eating, or going about their affairs, flowing together and away. And through them coursed files and files of the black fish returning from the border, where they had started spawning again, apparently the night before. Looking out from my high perch in the deep water, the view was tremendous and very beautiful.

It was the last time the behaviour of my sharks was calm, in the way they always had been. Suddenly, after three years, there was a bizarre change.

My next visit took place several days later, in the shadows of passing squalls. One was deluging on the other side of the reef, disturbing the air around me. Another was off toward the west, darkening the light. The island was piled with clouds. It had been heavily overcast and calm all day, but as I paddled across the lagoon, a nasty east wind arose, whipping the grey seas up into white-caps. Rain threatened from the approaching storms, and the lagoon was like a river. The ocean waves appeared to have risen again, so I went again to the deeper site.

As I got ready, I paid little attention to the sharks circling under the boat. I was very tired and worried about the conditions, wondering how bad the storm was going to be. When I slid underwater, Sparkle Too was at my side. Normally when I got in, Martha, Madonna, and a few others would swim calmly up to my face, but this time, all the residents did so, all together and very fast. I was instantly in the midst of a mass of excited sharks whipping around me. Martha coiled around me like a remora.

I swiftly ducked under the boat, switching from the down-current side—I always jumped in on the right side—to the other, so at least the boat would be between me and them. Then I hurled a few scraps into the water and looked to see the result. Some of the sharks were approaching the food, but at the same moment, Carrellina darted up to my face and away at the speed of light, and the bodies of sharks right around me suddenly blocked my view of anything else. I pushed more food in and looked again under the surface. It took a moment to understand what I was seeing. Martha was a few centimetres below my face, gliding forward, and just to her right, Apricot was passing. Somewhere in between the two sharks, was my body. Sharks filled my view, too close to focus my eyes on. But over them I could see the scraps sweeping away from the boat in the current.

I began using all my force to push the heavy masses of scraps out of the opposite side of the kayak, finning as hard as I could to raise myself out of the water, and brace myself against current, waves and wind. I was so concentrated on doing this as fast as possible that it took an extra second to realize that my foot had touched something, and at that moment, the instep of my right foot slammed into something firm and solid. All my force had been behind the blow. Horrified, and expecting the shark to turn and slash, I looked underwater to scrutinize the situation. Martha was just below me.

I watched closely for her reaction to having had her side kicked in, but there was none. She curved down sedately to look over the food. The rest of the sharks were descending too.

I felt a chill, and retreated up the boat to watch. The sharks were eating, catching the falling food, and Martha had not changed her easy rhythm, showing not the slightest sign of being mad at me. After a while, I hurriedly scooped out what remained of the scraps, trying to hold my feet up under the boat and clinging to it, but that didn't work very well.

I finished and wrote down their names. Present were Martha, Madonna, Carrellina, Bratworst, Apricot, Sparkle Too, Flora, Gwendolyn, Christobel, Windy, Teardrop, Mara'amu, Ondine, Ali, Trillium, Cochita, Lightning, Muffin, Fawn, Tamarack, Brandy, Pip, Pippet, Jem, Breezy, Spark, Fleur de Lis, Rumcake, Jessica, Antigone, and Shimmy. Since I only saw Sparkle Too when I first went underwater, there could have been sharks who ate and left before I could identify them. Brandy snapped up a fish, and kept chasing them.

As the excitement calmed, and the sharks went into search mode, two strange males came roaming through. One turned out to be Thrasher, and I was able to draw the other's fin accurately, and named him Becquerel. As I drifted, following them around and drawing them, another very pale male passed me, and I started to draw him too. It was soon clear that it was him who was doing the circling of me, not me of him, and I recognized Ruffles, the lower line of his white band so pale that I couldn't discern the features by which I usually recognized him. I went to the boat, got a bag of extra scraps, and placed the food on the sand so that the current wouldn't scatter it. Ruffles came immediately and ate. Lillith soared into view. She ate and roamed through the area. Piccolo and Brindy, the littlest ones, also came and ate.

They left me gazing through an empty lagoon. I had been tossing the odd scrap to the crocodile needlefish, and began to feed the fish to keep the scent flow going. Eventually Storm and Mordred arrived, and a strange male, who had been passing at the visual limit, came closer with them, so that I was able to draw his fin tip on both sides. By then, half an hour had passed. Above the surface, it was raining.

I cruised around, alternately feeding the fish and looking down-current for sharks. Annaloo passed but didn't come to eat. Even the nurse sharks had retired, though normally they were all over the remains of the scraps. A large female approached slowly through the murky surroundings, and I recognized Samaria. She took a scrap, dropped it and left. After another long period in which the darkness gathered, I found Shilly down-current. But though I called to her, she kept on going and did not come to the food.

Even my faithful crocodile needlefish left. I waited and waited, but no one came. It was disappointing, with all the food left lying on the sand. Finally, a big shark approached up the scent flow, and Madeline circled the food and took a scrap. Then she left. I waited until it was too dark to see, then put the anchor in the boat. As I drifted over the site two young males were cruising through, too far to identify.

Above the sea, the weather had mercifully fallen still. Sunset colours were glowing in the clouds. The storms had finished their mischief while I was in the silence below. As I paddled home, I reflected that part of the problem was the deeper site. My usual site, where over a long period I had established a routine familiar to everyone, was valuable for keeping control when the sharks were excited. There, I was protected by the coral formation along the up-current side when I dumped out the food, the sharks used swim-ways between the coral structures, and their feeding area was a defined circle of sand. It was just not physically possible for them to surround me as they had done while I was clinging to the boat in deep water.

That night, Franck took me to a party. Mostly shark food—raw fish—was served in a variety of forms. Just smelling it made me lose my appetite. I found a glass of wine, and sat trembling and thinking about Martha. The men were discussing the heights of the waves they had ridden on their surfboards, but my vision was within.

What had happened? Martha shouldn't have been especially hungry, yet she had been between my legs, or I would never have touched her with one foot and kicked her with the other. What if I had struck her in the gills where I could have really hurt her? Would she have slashed me? I had heard that sharks can give a warning slash, which can be very deep. In my wildest dreams of worst case scenarios, never did I imagine that I would accidentally hurt one of them. What was breathtaking was that I had kicked her so very hard and she hadn't reacted.

Choose any human, I thought to myself, looking around at those present, and *kick him or her like that, and see what would happen!* No matter what the circumstances, Martha had never bitten! No shark ever had. It seemed that the instinctive fear I felt when I was with sharks in unexpected circumstances, was misplaced. The only conclusion was that these sharks did not bite nor react with aggression in any way that we understood.

At the word *shark*, I flashed my attention back to the surroundings. The surfers had mentioned *shark hour*, the hour before sunset, when everyone got out of the water. For surfers, my sunset rendezvous was *the hour of the sharks*.

I checked my records at home and found that this session corresponded to the one held in the dark of the moon the year before, in January 2001. Sixty-five sharks had come, and the water had been solid with them around me when I had arrived underwater. It was an interesting correlation with the manic beginning of this session. But the week before, everybody had been so reasonable and placid, that such a sudden change seemed fantastic.

In my records of the dark of the moon in January 2000, I had written how surprised I had been to find fourteen sharks swimming coolly by me when I arrived underwater, not showing the slightest reaction to my splashy entry. It was the first time they had not waited behind their curtain of blue for me to place the food and withdraw.

So there had been a new development in their behaviour then too. Three years in a row in the dark of the moon of January, they had been especially bold and excited. But why they would suddenly change at that time was a mystery. There had been no group of visitors present this year, which had been the case in 2001, and which I had assumed was the reason for the excitement.

The event was not related to food either, since the black surgeons had been spawning. Any shark who wished, could stuff him or her self on any night of the week. Brandy showed, nearly every time I saw her, how easy it was for one of the sharks to snap up a fish. And they had left the pile of food I brought lying on the sand.

I was called to pick up more fish scraps a few days later, and while preparing to take them out to the sharks, I plugged the holes in the well of the kayak so that no scent would escape. The wind on the way out pushed me relentlessly back so that not for a moment could I rest. Pulling against wind and water with all my strength was scarcely enough to advance at all, and gusts threatened to blow me broadside. It took an hour to get there, and I threw in the anchor without even trying to manoeuvre into an appropriate position. But I remembered to check one more time, that the fleeing rope didn't flip any of my gear into the sea.

The sharks were waiting, apparently thrilled. They turned tightly around the boat at the surface, as I quickly put on my gear. Martha repeatedly came over to sniff. When I slid underwater, they all swam to me again. I concentrated on getting their food to them as fast as possible, while trying to keep out of the scent. Again, they surrounded my body as I lifted myself to scoop the food out of the boat. As they fed from the fall of food, it vanished. Martha tried unsuccessfully to get a scrap out of the coral, then tried to get a fish head which had landed in an inaccessible place. Finally, she approached me with several others, while I tried to keep out of their way, often finding myself circling Martha in an effort to avoid her circling me.

Besides Martha, there were Madonna, Carrellina, Flora, Apricot, Cochita, Gwendolyn, Sparkle, Chevron, and the usual juveniles and males. I found Annaloo meandering down-current but couldn't get away from the lines of sharks approaching me at the surface. Apricot, Martha, and Carrellina, with the juveniles, Lightning and Ondine, did this most persistently. Apricot kept swimming above and against my body when I was down-current, which bothered me since I couldn't see her there above my back.

When the sharks went into search mode, I found a few scraps swept beneath the coral, and put them on the sand. Martha zoomed in behind me, followed by a line of sharks. Then she pestered me when I drifted down-current to look around, and kept going to the boat to sniff it. One after the other, the sharks went to sniff the boat, and I knew that if I went to it myself, they would accelerate over with me, so I didn't. From the way they kept circling watchfully past me, rather than leaving, I sensed that they were waiting. Finally, Martha took a swipe at the boat, her nose going through the surface, as she made a sharp movement, and produced a string of bubbles. It appeared that she had snatched at one of the straps hanging from the back of the kayak, but when I examined them later, they were perfect.

I drifted down-current, and Shilly came. I called to her. Her behaviour was unchanged—she approached at the surface, circling rapidly then arcing away. Amaranth was gliding through the coral there, and I swam with her briefly. Back at the site, Madeline, Glammer, and some juveniles were scouting around for something to eat.

Martha repeatedly approached me, and when it wasn't her, it was Apricot or Lightning. This was unprecedented behaviour happening literally from one session to another. I couldn't get to the boat without being harassed by them and the others present, who followed their example. Finally, when they accidentally left me a clear passage, I reached the boat and leaped in, pulling my feet in so fast that I nearly overturned. The crocodile needlefish was circling me, but I just couldn't feed her. I didn't have anything extra set aside for her, and unexpectedly, it was impossible for me to dive down to get a scrap and break off pieces for her. I was too unnerved to care.

I sat there for about ten minutes in the wind still raging over the sea, and began to feel silly. No sharks had appeared around the boat for many minutes, so I slid back in to see what was going on, and if any interesting sharks had come. But all the same sharks were still there. They hadn't left! As soon as I appeared, they streamed toward me, and this time it was Tamarack who led the crowd at the surface. I grabbed the anchor, threw it into the kayak, and scrambled back in as fast as I could.

The next time I went, I was uncomfortable from the beginning. Powerful winds from the east exhausted me again as I struggled out; I could scarcely control the boat or advance. As if mocking me, the sky was clear and cloudless, the sunshine bright upon the sparkles of deep blue waves, shot with gold. The ocean swell rolling over the reef was weaker, promising less current, so with relief, I went to my usual site.

Anchored in the proper place, I watched the sharks. They came surfing toward me on the waves and circled fast. Two flowed together over my lookout, right under the surface, as if fondly thinking of me, then raced to the boat. I got ready, and it was hard to jump in. I was filled with dread, and had brought a bandage for the first time to staunch the blood in case I was bitten. In addition, the terrible winds made it impossible for me to throw in the food over the site before I got in myself, since I had been blown away from it, while freeing the anchor. It was very hard to control things when so many unexpected problems could arise.

I sat frozen in uncertainty, watching the sharks for several minutes, then just jumped in. Martha, Flora, and Chevron were coming toward me, their serpentine motion slow. I began to breathe easier. They swam beside me as usual, as I pulled the boat to the site. Due to my concern in the wake of the previous sessions, for the first time, I knelt on my lookout formation to push the food into the water, then grabbed my camera and slate, and settled to watch. Martha, Flora, Chevron, Carrellina, Cochita, and thirty other sharks were in their usual shark tornado, eating in the site. I was pleased that the older sharks fed well. In minutes, just a few scraps remained, buried under the pile of nurse sharks underneath my lookout. Martha kept coming over to see me. Time after time, she approached me, slowly.

Once she arose in the water, as she passed in front of my lookout. When her face was opposite me, her eye appeared above the coral edge, looking straight into mine. Scarcely advancing, with each gentle wave of her tail, she moved her nose closer to me. First, she examined the slate I held, and I moved it so that she could see it better. As she passed around my arm, looking, her nose nearly grazed it with each undulation. She did seem truly curious. When she reached my side, she paused in indecision, and I lifted myself as she slowly sank and passed under me. She remained in my immediate space like this for some time. I went to the boat, threw the crocodile needlefish a couple of pieces of food, and pulled it over so I could throw the few scraps left, into the site. The sharks rushed to get them, and Martha left with one.

They all drifted away, and some visitors came, languidly checking for scraps. Cochise and a companion I had seen at Site Three were among them. No one interfered as I fed the bright cloud of waiting fish. They were very excited. I showered my crocodile needlefish with treats and attention to make up for my neglect the week before.

As darkness gradually obscured the site, Keeta came, and I swam around the area with her. Shilly appeared and approached, but left without eating. The moon was close to full for this session, so it had not been the lunar influence that had facilitated the pushy behaviour of the sharks during the previous one. On the other hand, many of the sharks who had been present at that session had not come to this one; they tended to roam when the moon was full.

CHAPTER THIRTY-ONE

Carrellina, Chevron, and Sparkle Too

February came. At the next feeding session, I had only a small bag of scraps, complimented by a few bits of food I had saved in the freezer. The quantity I was able to take with me had always varied widely. But many times, sharks missed sessions, and other times, they came too late and didn't eat. There was rarely enough actual food for them to do more than pick over the scraps, and get little bits. Many times when I had brought lots of food, they had left it lying on the sand. So the amount of food had never caused problems. I took what I had. The extreme cases were those in which I had taken medicated food for Meadowes, and only scent for everyone else. Never had a pattern been established in which everyone had a meal regularly.

This was in contrast to the commercial shark feedings outside the reef which were held as often as once or twice daily. But even there, little food was brought for many sharks—none got much to eat.

The evening was fair, with light clouds and no wind, though storms were passing on the ocean. I went to my usual site. In the water, I turned swiftly to see all around me. Valentine, Madonna, Flora, and Chevron were shooting toward me, with countless sharks coming behind them. I held my place, looking calmly back at them, as they zoomed up to my face and away. Behaving as if their display had not the slightest effect on me, I concentrated on finding the anchor cord, and making sure it didn't get caught in the coral as I pulled the boat into position, while the sharks massed thickly around and under me. Bracing myself against my lookout, I reached up and pushed the food into the site. There was a furore beneath my hands as it fell through the surface, and when I finished and gained my perch, the site was filled with zooming sharks who often approached me. I waited for them to finish with the regular scraps, and calm down a bit, then, one at a time, I threw the little chunks into the middle of the site, hoping that the more aggressive ones would be calmed. Valentine had that emaciated look that comes after birthing. She snatched a good piece of food as it fell, so I was happy about that.

The sharks whirled and I watched them. Carrellina was there. Madonna, Bratworst, Sparkle Too, Samaria, Christobel and Twilight had all mated during the period of the full moon, all within a week, judging from the freshness of the wounds. Many of the other residents and some visitors were there too, for a total of about thirty sharks. Everything was gone when Martha soared into the site. I put the scraps I had saved for this moment on the sand, hoping that they would take her attention. Instead, time after time, she led swarms of sharks toward me at the surface. Then they all went to sniff the boat.

Annaloo passed slowly by on the western periphery, but she did not come to eat. Things appeared to be settling down after the initial excitement, when Madeline, Ali, Brandy, and Amaranth came soaring up to me at the surface, turning away just at my face. This sort of approach by these individuals was new. The greeting gesture was always performed slowly. Meanwhile, Lillith, with little Brindy, were cruising placidly in and around the site, not finding any scraps left over, but unconcerned and not bothering me.

Keeta appeared and joined Martha in zooming up to me every few minutes along with the others. Watching them closely, I estimated that she was twice Martha's volume, though Martha was a good-sized shark. Within minutes, Keeta was charging me by herself. I was astonished. It was inexplicable that suddenly, and for no apparent reason, there could be such a change in their collective and individual behaviour. Wild animal behaviour tended to evolve gradually over time, in the absence of sudden changes in their circumstances.

The sharks milled around when not streaming toward my face, but Martha did it non stop. Were they copying her? Then why had this never happened before in three years? Martha had always made close passes. Apricot, Tamarack, Lightning and Breezy, all juveniles, had occasionally harassed me briefly when I was feeding the fish, but this time all the sharks were doing it, and not stopping. There was always someone soaring up to my face at the surface.

Then they all swam away toward the border, as if they were in a hurry, but moments later, they were all back, so they must just have turned one of their circles together. They sniffed the boat, zoomed up to me, and once more milled around me.

My fish were fluttering near, filling the site and hoping to be fed. The crocodile needlefish waited too. I had saved a few bits of food for her, and the next time the sharks circled away, I got one, threw it to her, and brought a fish head to scatter a few crumbs for the fish. That was all they got. The sharks, with Martha in the lead, all reappeared. I pushed the fish head toward Martha, and she attacked it, while the rest of the group flew around again, and once more began to soar menacingly up to me as if they were sure I was withholding food that they should have been given.

The fish were still begging me to feed them, surrounding me and looking into my mask, while I, in consternation, observed the sharks. Eventually, they dispersed and I made another attempt to feed the fish. But the sharks all came back at top speed, Martha straight to me. I moved out of the way and waited for an opportunity to get into the kayak. When the way was briefly clear, I picked up the anchor and rose up to the boat as Martha and the others came sailing back. I leaped in and threw the fish's food into the place where they waited. The water boiled with sharks, and as I slowly drifted away, many sharks stayed with the boat, circling and sometimes coming to the surface beside me. Brindy, a tiny pup, came so close I could have touched him. Many sharks were underneath him, steadily circling, and sometimes sniffing the boat. They didn't leave me for at least ten minutes. Intensely disturbed about their dramatic and worrying change in behaviour, I was mulling it over, looking down at the sharks, when a storm descended on me. As I battled my way home through the deluge and gusting wind, I vowed to give my sharks the surprise of their lives by bringing the greatest predator on earth with me when I returned. Frighten me? I would show them!

Accordingly, I asked a family friend, Bou Bou, who owned a dive club and a video camera, if he would be so kind as to accompany me in order to film the behaviour of the sharks while I put the food in the water. I felt that if I had it on video tape, I could analyse what they were actually doing in the wonderful comfort of my living room and come to a better understanding of what was going on. But, in fact, I was afraid to go back alone.

Bou Bou arrived as planned, and we set off in the kayaks to videotape the session. I had a case of sardines, since I had been unable to get any fish scraps on the day he could come. The weather had been calm and sunny all day, and I expected good visibility too. I wanted to go to the regular site so that the routine would be as normal as possible for the filming, but as we moved across the lagoon, it became clear that the current had become much worse, and so I looked for the deeper site farther from the reef. Not finding the precise place, I chose a deep, sandy area at random, and threw in the anchor.

Bou Bou planned to film from a position on the sand, which provided a better view than the surface, and began to put on his scuba gear. I slid underwater. Martha shot up to my nose and away. Just within the limits of visibility, I momentarily saw the rest of my sharks. Then they vanished, and one could not imagine that there was a shark anywhere in the area.

As I looked around, a cloud covered the sun, and gloom descended. The visibility was worse than it had been for weeks, and the current was too bad for me to swim around to choose a good place for the filming. I analysed how to use the open, sandy area we were in, visualizing how the sharks would approach through the coral and what their swim-ways would be, depending on where I put the food. That would determine the best positions for Bou Bou and I to take.

When Bou Bou slid underwater I indicated my plan, and once he was settled, balanced, and ready, I swam down-current out of camera range trailing the kayak, and took out the plastic plugs blocking the scent. Then I swam slowly back toward him, while the yummy odour of the sardines poured from the back of the boat, and was rapidly swept into the blue where my sharks were lurking. I swam slowly, waiting for them to come. This was the moment I wanted to catch on film, approaching surrounded by sharks. But no sharks appeared.

They were pretending to be shy!

After waiting for a few minutes, I threw in a handful of sardines. Ten minutes later, the dead fish were still rolling in the current on the sand, and there was no sign of any shark in the area! We waited unmoving, and I wondered what Bou Bou was thinking about my descriptions of my bold sharks, when it was quite obvious that there were no sharks in any direction.

Then they suddenly appeared. In two long lines led by Martha and Carrellina, they came threading through the coral, and both shot up to Bou Bou's face so fast that he fell over backward. He disappeared in a crowd of excited sharks milling around us.

Later, he told me that what impressed him the most about my sharks was their size and their numbers, since he was used to the smaller males outside the reef. But I had failed to get the critical scene on film. The sharks seemed confused. They never shot up to the surface to grab the sardines as they would have done had I been alone, as they had done before on the one occasion I had brought some. Instead, they roamed around in the coral, picking up the little fish after they had been swept away. Bou Bou filmed while trying to balance in the current.

The residents were all there, and Arcangela, one of my favourite yearly visitors, honoured us with her presence, cruising calmly around, and often passing in front of the camera. She had just mated, and had a hole through her dorsal fin. Bou Bou observed her with awe. He had never seen a recently mated female, proof that the females did not attend the shark feeding dives on the outer slope of the reef.

One interesting sequence involving Martha's attentions was captured on film. When the sharks were dispersing after all the sardines were gone, I dove down to pick up some that had drifted under the coral. Martha was leaving, but the video showed her turn around as I dove down, and come to swim at my side. We stayed together as I picked up a few fish, but as we passed Bou Bou, she suddenly found herself between him and me and startled, accelerating so fast that she instantaneously vanished from the screen.

Another sequence showed the sharks giving the chain attached to Bou Bou's kayak a vigorous bite and a shake when they passed it, where it arced down through the water near him. This was due to a mild electrical current created by the metal in the salt water that their electro-senses picked up. It seemed that the stimulus was too strong for them to ignore, though they could tell when they grabbed the chain that it was not food.

More sharks cruised in, including Shilly, but she was very reserved, would not approach for a last fish, and left immediately. I fed the fish, and Madonna came, rising perpendicularly to my hands to take the sardine I was crumbling. Bratworst zoomed up too close too fast, too, but shied away when I finned rapidly backward. Unfortunately, Bou Bou was too close to capture both me and the sharks at once. The crocodile needlefish was wary, and only made a brief appearance into visual range. I swam toward her and threw her a couple of pieces of food, but she was too far away to show up in the film.

The three-metre nurse shark passed off in the coral, and Bou Bou went after him, filming. He told me gleefully that it might be the first sequence of a nurse shark ever taken, because no one ever saw them out of their holes! I was happy for him and kept my own opinion to myself—that it was a very good thing that these incredible nuisances had not come.

Bou Bou was enthusiastic, and assured me that compared to the sharks outside the reef, my sharks were very nice with me. They had certainly been on their best behaviour for him! But I was already worrying about the next time, when I would have to return alone. His film was a wonderful souvenir of my sharks, and showed my favourites in colour in action. I imagined having it to watch, remembering them, for the rest of my life. But it vanished from the place I had put it beside the television set, where it was waiting to be analysed when I had some time alone. The mystery of how it disappeared from the house was never solved. It was the only time anyone took pictures of me with my beloved sharks.

Storms began to rage again upon the sea, but the following week when I was given a load of scraps, I tried to sneak out between them. A squall blasted me as I crossed the exposed region approaching the reef, but afterwards, silence fell, and the sun emerged for the session. I went to the main site and followed my usual routine. Chevron, Sparkle Too, Cochita, Carrellina, and Madonna appeared beside me underwater, but no one charged or came too close. I was focused on getting to my perch and climbing on. Kneeling there, I pushed and threw the food into the site, then looked underwater. Lillith was shooting past my face. It was rare for her to come at the beginning of a session. Thirty-three sharks were present, including Martha, who, for the first time, never approached me, and Arcangela, who stayed until I left.

As night came, Cinnamon arrived, and Annaloo cruised slowly on the periphery. Down-current, I found some rare visitors who did not come to eat. Everyone was calm. Back at the site, Darcy, Keeta, Ruffles, Breezy, and Ali were in search mode. A whitetip shark passed through, a ghost, and disappeared. I was able to feed the fish and give my crocodile needlefish, who had been circling around my head, lots of attention and bits of fish. The little ones, Breezy, Apricot, and Lightning came up the scent flow and appeared among the fish in front of my face from time to time as they often did. I felt calm and happy.

There were many nurse sharks present, a carpet of them on the sand, and as I fed the fish, hastily scattering the meat from the fish heads in the water and dropping them on the sand, a flower of seven of them formed, all their noses in one fish head, tails waving at the surface. Wishing I could photograph that, or even paint it, I fell into an artist's trance, marvelling at form and motion, the pale sharks, the violet shadows. I continued to scatter handfuls of meat crumbs into the water for the clouds of fish, automatically, spellbound by beauty.

One of the three-metre nurse sharks approached me, then, for no reason, he accelerated away. Other large forms were moving in the clouds of sand. Though my view was obscured by the hundreds of flashing fishes feeding before my eyes, I sensed that there was too much agitation in the site. Dropping the fish head, I drifted leftward, just as a large lemon shark shot in. His body rose vertically, as much as a shark the size of a baby whale can in two metres of water, as, with volcanic thrashing around, he tried to get its jaws around the fish head I had just dropped, now jammed against the base of the coral wall.

My regular camera was not working but I had bought an underwater disposable camera for special occasions like this, and I had promised myself a photograph of a lemon shark the next time I saw one. So I raised it. But instead of rampaging around the site while I joyfully snapped photos, the shark came around behind my lookout in his repeated efforts to get the fish head. He was moving much faster than normal, too fast for a huge shark in a coral garden—worse than the proverbial bull in your china shop—much worse! I knew an agitated shark when I saw one!

He proceeded around the back of my protective coral formation toward me, and I faced him, camera raised, wondering when he was going to realize that I was there. His head was covered with long, deep scars, I saw in growing alarm, as he swept closer. I squeezed my legs up under me as he began to pass beneath, and before his face could disappear, I took the photo. With just the movement of my finger, the huge shark reacted, so that the photograph shows him already turned and rising in the water to sail over my coral wall in flight. How such a big animal got himself turned so fast was a wonder, because I was sure I would have a picture of his face from above, close up and straight on.

Faced with this violent reaction, and aware that he was already in some sort of a huff, I went into a panic and nearly drowned myself trying to free the anchor from under a coral shelf, where it had become stuck. No matter how I manoeuvred it, I couldn't get it out. When I finally succeeded, I was so desperate to breathe that I left it on the sand and flew up through the surface and into the kayak, whipping my legs in behind me in my terror at the vision of a big, spiky mouth pulling me back down into darkness at the last minute.

It was counter-intuitive that this enormous animal was frightened of me at all. When I saw myself in

Bou Bou's video, a small, spidery creature on the surface, I was reassured that no shark would bother to bite me. Still, those instinctive terrors kept coming back when I was there with the reality—especially in the dark.

I had to creep over the bow, trembling, catch hold of the line, and pull the anchor in. Then I freed my paddle and drifted over the site. The fish head left on the lookout had fallen in the squirrel fish's grotto, and the nurse sharks were writhing over my perch, their tails breaking the surface as they went down after it. But no great shadow moved anywhere I could see in the darkened lagoon.

This was not the full moon, but two days before the dark of the moon, so the lunar connection with lemon shark visitations had to go under review.

The following week, it was calm and sunny, but the oceanic assault on the reef poured dazzling cascades across the lagoon. Even at the deeper site, where I threw in the food from the boat to simplify matters, it was hard to maintain myself against the surge of the invading seas. Only the big scraps remained where they had fallen. The sharks hunted the rest down-current. Martha, Carrellina, Chevron, Sparkle, and Sparkle Too, the sharks who tended to harass me most, were not there. I drifted to a tall coral formation where I could hold on facing into the current, overlooking the sharks. Marco and the beautiful Antigone were present with other visitors.

When the sharks thinned out, I fed the crocodile needlefish, who had become impatient and was circling my head. An unknown female blackfin came alone to eat, placid and unconcerned about me as I sketched her dorsal fin on both sides. Then I fed the fish and watched as Twilight came and ate, and other visitors, male and female, cruised in and around as night slowly fell. When it was time to go, the current increased to the point where I could not move. I was holding on with all my strength, just to keep my snorkel above the surface. In the dim and cloudy light, rivers of convict tang, the pale yellow, vertically striped fish that move in immense schools through the lagoon, were flashing silver in remarkable shining patterns as they moved nearby, separating and merging in a dramatic display of synchronous movement. In the twilight waters they looked ghostly, silvery instead of gold. Their ebb and flow went on and on through the underwater torrents. Finally as my strength was giving way, the current eased enough for me to swim against it to the kayak, and lift myself in.

That week was calm and sunny, with small clouds marching in formation across the South Pacific ocean. They sometimes dumped a shower below, but not on me, as I paddled out in sunshine the next time I got scraps. The waves on the reef were low, and the current was negligible, which eased my mind greatly as I neared the sharks. Madonna followed me across the lagoon to the site, soon joined in gentle undulations by Sparkle Too, and others beneath.

I anchored at the usual site, and Cochita, Chevron, Sparkle Too, Carrellina, and many others circled the boat as I prepared. Underwater, I found Madonna, Cochita, and Chevron beside me, but not Martha or Flora, who were usually the closest to my side, and more trusted. Beyond, the coral was thick with sharks, but they were cruising calmly. I took the time to move the anchor, and pulled the boat into position so I could push and throw the food into the water while standing on the dead coral shelf, bracing myself. Masses of sharks swam with me. The waters were high, and it was hard to stay balanced in chest-deep water, with the upper edge of the kayak above my eye level.

With the food in, and the sharks eating, I watched and wrote down the names of those present. Arcangela and Jessica were with the residents. Martha was spiralling through the water in front of me, taking one bit of fish skin after another as they drifted down. More sharks swept in from the coral surroundings. After the food was gone, I waited for them to leave, so that I could feed the fish, but they continued to patrol the area. When I drifted down-current on my usual tour, Tamarack, Lightning, Apricot, and Ondine came, leading long lines of sharks up to my face. I lay back in the water, facing them, finning gently to stay balanced, as well as to keep a barrier between us.

Back at my lookout, I watched, perfectly still, while Carrellina intermittently orbited my head, and Martha made one close approach after another. Time after time, Chevron passed to my right, watching. I began to feel a tension in the sharks' persistent watchfulness and their relentless circling around me and the site, focused on me as never before. But I felt calm this time and faced each shark motionless, hardly breathing. I dared not move a muscle as I waited for them to go.

Time passed, and nothing changed. The fish filled the site and were particularly thick in front of my face, where many of them began looking into my mask or swimming swiftly back and forth before my eyes. The crocodile needlefish passed me, gazing solemnly. I gazed back. But mostly she waited behind me. Shilly appeared and circled around me. I had a treat for her, but dared not produce it, since I sensed that if I did, or tried to feed the fish, the tension would break and all the sharks would charge.

I waited unmoving for the rest of the sunset hour, while slowly the night gathered, the sharks circled, and the tension built. Finally, Martha left, but no one else did, not even Arcangela. A couple of times when I was left relatively alone, I went to the boat, extracted a chunk of meat from a fish head, and tossed it to the crocodile needlefish, but I didn't dare do more than that; each time I did, there was a rush of sharks back into the site and straight to me.

The tiny eel, presumably the one who touched my hands before, kept coming out of his hole to touch my fingers with amazing softness, no matter where I put them, but never when I was looking. I felt sorry for him and the fish, waiting like excited children for the treat I always brought them. But I was helpless to feed them. Though Martha had left, and Apricot for once had not come, the other sharks who could trigger a mass charge, Chevron, Sparkle Too, and Carrellina, passed and passed and passed me, hypnotically.

Finally, taking advantage of a brief respite in which these three were on the other side of the site, I leaped into the kayak and threw the meat from the fish heads in over my lookout, where the fish were waiting. The water writhed with sharks. The splashing and power they displayed, boiling the waters, was daunting. I tossed in the fish heads afterwards and the nurse sharks joined in too, in waters too dark to make out the wild mêlée beneath.

CHAPTER THIRTY-TWO

Spooky and Kim

I was looking after two seabirds at the time, Spooky and Kim, and twice a day, I took them out upon the bay for exercise. Sometimes in the evenings, we drifted so far that we were opposite the place where my sharks were circling, waiting for their sunset rendezvous. As the birds drifted and played, I gazed to where the waters of the site glittered beneath the blue sea air, musing about them and their changed behaviour.

I felt intuitively that the sharks had learned when I was treating Meadowes that I could be tricky. Possibly, a series of factors had come together to cause them to behave as they were now doing, including the excitement of the mating season, the warm water of the southern summer, and the presence of Carrellina, whose unusual character could be facilitating their behaviour. Several others, all in their first reproductive season, including Chevron, Sparkle, and Sparkle Too, seemed particularly ready to be influenced.

Each night and morning, drifting while the birds played upon the sea, my mind examined the situation as an octopus examines a jar with a crab in it. I had no one I could discuss it with. Indeed, since I was the only person who had watched the development of the current situation, how could anyone else advise me? I did not know what to do, and the subject was never far from my mind.

Kim was a Tahitian petrel (*Pseudobulweria rostrata*), a large species, dark brown with white underparts, known to nest on only a few islands, and about which little is known. He had been brought to us when he crashed into the mast of a sailing boat, and fell to the deck. His left wing was injured, and an X-ray had revealed a subluxed wrist joint. For two weeks, he had lain on a bed of stretched netting, a seabird bed, with his wing immobilized. Since he would not eat on his own, I hand-fed him whole fish, treating him with exaggerated gentleness to reassure him.

When his injury had been given enough time to heal, I put him on Merlin's inflatable pool twice a day, to encourage him to begin using his muscles again. Its strangeness didn't bother him after an initial investigation, and he began to enjoy bathing and resting on its surface, and looking around at the trees full of birds, and the shining sea. He made no efforts to escape, and seemed a surprisingly calm and reasonable bird.

Spooky had been brought in soon after Kim. She was a young brown noddy (*Anous stolidus*), the largest species of the tern family of seabirds, dark grey in colour, with a little white crown encircling her forehead. She was so starved that her beak was soft, and she didn't seem to have a throat. It appeared to be blocked, but investigation revealed that the tissues were just tight from starvation. She was sleepy and drowsed. Though near death, in the following days she was able to digest increasing numbers of tiny fish. Resting on a perch at the window, she looked out across the bay through the arching trees that Angel had used. She could see Kim when he was in his pool, but the two birds were never together.

When Spooky felt stronger, I verified that she could not fly and put her outside on the bench surrounding our deck, near where I was painting. She looked around for a long time, then began to play in her water bath, lost her balance, and fell to the deck floor. Once there, she saw that she had more options. She pattered about, peering over the edge on each side, obviously considering an escape attempt. Finally, after a short, experimental flight, she took to her wings, soared out over the water, and curvetted into the neighbour's garden. I rushed along the road to see if I could see her, looking on their beach, the grass, and in each tree, before spotting her on the roof of the neighbour's car. It was embarrassing to peer through the hedge in view of passing traffic, but I was frantic to catch her before she flew out of reach. Time passed, and she began to preen, more and more intently. Finally, drowsy in the sun, she closed her eyes and extended a wing; moments later, she changed position and unfolded the other wing, slumping as her eyes closed. I walked swiftly up the driveway. She saw me at the last moment, but was too surprised to try to escape, as I plucked the straying bird off the car.

With her safely back on her perch in the house, I decided that she must learn to take fish from me, eat them by herself, and practice her flying in the house before she could go outside again, just as Angel had done. It took a week of encouragement for her to eat on her own. The first time she managed to take, manipulate, and swallow a fish by herself, she was so pleased that she ate one after another. She would explore the house when she was alone. Once, I looked up from my painting outside to see her sitting on a lampshade, watching me, on the other side of the glass. She made many clumsy flights around the house, and obviously had no experience or skill in landing, which is the hardest thing for a seabird to learn. She was no longer thin, but remained small and delicate. Her long neck and wings gave her an especially elegant look.

When she was ready to go outside, I intended to put her in the trees arching over the beach where I had fed Angel, so that she could look around, and take short flights before sailing any farther. But she wriggled from my grasp as I crossed the deck, and flew far out over the water, gaining altitude, to land, after many minutes and much difficulty, high up in a coconut palm. For two days, I called out to her to come down for the fish I held out, but she was either unwilling or unable to descend. On the third day, when she swooped down in the morning to sip from the surface of the sea, I intercepted her flight back to shore, holding her fish toward her. She soared past, and since the first convenient place to land was in the trees arching over the beach in front of the house, the problem was solved. There, I could easily feed her, and from then on, she used no other perch. Through storms and deluging rains over the next days, she ate well and remained dry and warm, in spite of my concerns about her waterproofing, under such extreme and unforeseen circumstances.

Kim's case was worrying. If he was to fly again, exercising his wings had to become central to his life. So I took him out on the bay in the kayak. As we floated away from shore for the first time, he stepped up on the edge, looking around and out across the water. The other petrels I had cared for had single-mindedly headed toward the ocean, but he didn't even look in that direction. Soon, he hopped overboard and floated quietly, gazing toward the mountains and arranging his feathers. Then he raised his wings and ran across the water, flapping hard, trying to take off.

However, the left wing could not extend, and curved downward in a floppy arc. After a few tries, he headed in to the beach. I returned him to his pool, where he bathed and calmly floated until nightfall. He jumped out just as darkness was falling, and I hurried to intercept him before he could rush toward the sea. That night, he practised his efforts to take to the air, running, wings flapping wildly, all the length of the house till I fed him and put him on his couch of netting. The Tahitian petrel, like the shearwater and other seabirds, becomes active as night falls.

The next morning, I took him out to swim on the water at the shore, but he shrieked and tried to climb up my arm as I set him down, so I didn't force him. In the evening, he was comfortable about going out in the kayak again, and swam steadily away from shore, looking around. He preened, and bathed, and made only half-hearted efforts to fly. When it was time to go home, he neatly avoided me. It took some time to manoeuvre the kayak so that I could scoop him up. Then I wrapped him in a light cloth to keep him still, so I could use both hands to paddle home. Spooky sat watching in the perching trees above as we landed, and I carried him to his pool.

The next day, he looked ahead eagerly as we left, and Spooky flew down to watch. She swooped repeatedly over us, and then over Kim as he swam. I was sure she was lonely. Since she had come to us already grown, she hadn't formed much of a bond with me, yet she had lost everything she had known in her short life. She was still eating very tiny amounts, and flew only occasionally out over the water, returning quickly to her perching trees. She looked at me intently when I talked to her while feeding her, but I doubted whether I was much of a comfort to her.

Kim was crossing the boundary between hospital patient and pet, strange as it seemed. He had learned the family routine and followed it, coming out of his bed in the morning to be fed when Franck had his shower, waiting near the kitchen for me, and going to his pool after. When he felt it was time to go out on the bay, he jumped from his pool and walked to the edge of the deck to look across the water. At other times, he asserted his wishes for privacy, arching his back, flaring his tail, and giving an alien, wild call of defiance when one of us approached and he didn't want to be disturbed.

Each evening he leaped from his pool at exactly the same moment, and waited at the door for me to open it, if I was not already present. Then he trundled down the hallway to his bed. He hated me to wrap him in a cloth, which was necessary to protect his feathers from the oil on my skin when I carried him. So I found a compromise in placing one hand in front of his feet to help him balance, and one behind to lift him, always the same way. He stepped backward onto my hand, and never tried to jump off, thus retaining his dignity, and keeping his feathers immaculate. When I occasionally stroked his chin, he closed his eyes, leaning against me and angling his head toward me.

Spooky began sweeping down from the trees over the beach each time we left in the kayak. She looked first for Kim in the boat, and if he had already hopped overboard, she went wheeling above him, dipping down so closely that he reached up with a shriek in self defence. One morning, I even found her waiting on the kayak when I went to get it ready. She flew with us for a few minutes, returned to shore, and often came back later.

Though Kim enjoyed our outings, he rarely tried to fly. Then, one windy day, he dashed up a wave, took off, and caught the air in his wings to glide, just for a moment, before splashing back into the sea. I was barely able to keep up as he climbed repeatedly into the air and flew unsteadily down the wind, always pivoting back down into the sea, when the wing he could not extend failed him. He needed the wind to lift him, and that was why, I realized, he no longer made any effort when it was calm. So I tried to get him out for exercise when it was windy, but in high winds, I lost control of the kayak. There was a fine line between appropriately exercising Kim, and losing him.

When it was time to go home, he kept a close eye on me, noticing the earliest signals that I was about to start trying to catch him. He was ready, and that was really the only time he engaged in serious exercise by running across the surface, flapping wildly until he was far away. Sometimes it took me a long time to successfully intercept him, catch hold of him, and lift him gently into the kayak. But, once with me again, he folded his wings and settled down, making no effort to climb out as I paddled home, so I no longer needed to wrap him to keep him still. When I ran the boat up the beach, he jumped into the fringes of the waves, and toddled up the sand as I pulled the boat up. When I placed my hands under him, he climbed on, balancing with his wings, as he rode to his pool.

Sometimes when Spooky was with us, she seemed to be thinking of landing on the boat. But she always returned to her perching trees, then flew around us again later. Within a few days, though, she did alight, with her usual delicacy, upon the bow. There she sat, looking around at me, Kim, tiny things floating by on the surface, and the printing on the boat. Once in a while, she flew around us. Once, she alighted upon the water momentarily beside Kim. But mostly, she stood on the bow, observing him. This confirmed my feeling that she was lonely, and wanted to be included.

Her company changed our outings from an invalid's exercise session, to a happy group lost in play, as we floated far out on the sea.

CHAPTER THIRTY-THREE

The Bad Sharks

It was only three days after the evening of tension with the sharks when I returned again to see them. Kim had to miss his outing that evening, but seemed relieved to settle down early to rest in his bed. The weather was calm and sunny with light clouds. Spooky was resting as I pushed off from shore, but soon, she alighted on the kayak. With a relaxed and contented air, she looked down at the water streaming by, gazed at the shore, and studied each tiny object on the surface. I was increasingly worried about this development, which had never happened before with any other seabird, and I suggested a number of times that she should go home. I even jogged the boat slightly, hoping she would take to her wings and go. But she easily adjusted her balance. Twenty minutes later, we entered the lagoon. After staring down into this coral wonderland for some time, she flew far away across the waves breaking over the barrier reef, to disappear above the sea.

She was gone for a long time, while sharks began to gather behind the boat, a line of little fins above the surface. There was Cochita, Flora and Madonna, with several juveniles lower down, and more joining them as I progressed. Sparkle Too passed the boat at the surface, moving fast. I sincerely hoped that my bird had gone home, but suddenly, there was a whir of wings, and she alighted again, this time behind me on the shark food. A little later she flew up, and I looked back to see that Sparkle Too was swimming beside the boat, centimetres under the place Spooky had just vacated. No wonder she had been alarmed! She flew so far that I lost sight of her, and I found the site and got anchored as quickly as I could, so that if she did return, she would find the kayak resting tranquilly. She habitually perched on the anchor line, and throwing in the anchor with her on the front of the kayak would have frightened her. I didn't want anything to scare her, in case she really couldn't go home by herself. She was not very strong.

Sharks were milling around the boat at the surface. Carrellina, Sparkle Too, Martha, and Chevron repeatedly passed with their dorsal fins showing above the surface. Madonna glided back and forth right under the boat, and Flora steadily circled it, clockwise, two metres away, as she always did while waiting.

I was nearly ready when Spooky returned, this time alighting innocently behind me on a fish-tail in the scraps, where she sat gasping with her mouth open, obviously stressed and anxious. However, she closed it as she looked down at the water, now dense with sharks just centimetres below her. I felt infinitely precarious with the water full of sharks, the kayak so loaded I could move with only the utmost care, and this most fragile of creatures perched trustingly just above the water, staring down.

I was ready, and wanted her off the shark food without scaring her, so I delicately undid the straps holding it. She flew low across the site, just centimetres above the surface, and the water actually shivered as the thousands of fish startled. I watched her fly, blissfully taking in the bird's elegant form gliding above the surface shot with emeralds and gleams of gold, and the surreal quivering of the sea with her passing. So I was not sure how it happened, or whether it could have been a coincidence, that just at that moment, the sharks attacked the boat.

The heavy weight of the loaded boat with me on it was bashed with shocking force first one way, and then the other, as the sharks slammed it from multiple directions. The surface had disappeared—all I could see was sharks emerging at high speed, twisting, bashing into the boat, and flowing together as more replaced those shooting away. Then they came out of the water to snap at the food in the kayak's well behind me, like great whites you see in films. I could hear the sound of their jaws snapping shut, and one got a good bite of the remains of a *saumon des dieux* that overhung the water a little bit. Obviously, the sharks could see through the surface, and were aware that there were things above! They knew, though they had never seen it from above the surface, that the food was in the boat, and they knew where in the boat it was.

I had discussed this matter over and over with Arthur Myrberg. Though he acknowledged that these marine animals had some idea of the space above the surface, he would not believe that sharks could possibly understand that I was in the boat, even when I had just disappeared from the water into it, even when they were circling and waiting for me to come underwater. But I was seeing before my eyes that they understood the situation between me, them and my boat very well! They knew that before I put the food into the water, it was in the boat!

Carrellina, I was sure, was the instigator of this attack, and I could pick her out from among the sharks by the shape of her unusual fin tip—the square projecting from the tip line. But Chevron was just as guilty. Sparkle Too was taking part. Flora and Cochita were there, though it was unclear whether they were involved in slamming the boat. Eden passed at the surface. I had been on the verge of jumping in, but started throwing in the food instead, hoping that the sharks would leave the boat and start eating. But Chevron continued to shoot to the surface to slam the boat, even when all the food had been thrown in! Martha spiralled up and descended again. Surely she was looking for me. Martha could not have been involved in the attack.

The sharks were moving so fast that it just wasn't possible to identify each one. The heavy blows came mostly from beneath, and I had to turn around in the precariously balanced craft, twisting at the waist to see the back of the kayak where they were shooting up to try to grab the food. I dumped in the sack of extra scraps, and rinsed the scent from the well where the food had been, stupidly scooping sea water up with my hand to do it.

At last, the sharks sank to the bottom, and began searching on the sand. I used my fins to move the boat away from the area above the food, then slid underwater. Spooky had disappeared. I hoped that she had concluded that this was quite enough, and had flown home.

Underwater, the first shark I saw was Merrilee, from Section Three, passing through the clouds of sand raised above the site. She had visited the study area only once before that I knew of. Many sharks were feeding beyond, and paid no attention to me as I drifted, writing down their names. Along with Merrilee, many visitors were present. Jessica and Amaranth flew by. An ancient, hugely pregnant shark named Raintree, whom I hadn't seen for two years, circled over the food. She could compete, in terms of sheer volume, with Keeta. Emerald, Vixen, and many more rare visitors were swirling through the site. I noted who had mated and who had birthed, and wrote down fifty-six names, twenty more than usual. I looked above the surface, and no slender, dark bird was perched on the boat. If she had left by herself, I would not have to worry about getting her home in time for her to fly to her sleeping perch, which would mean leaving half an hour early.

It was one of those special sessions, strangely heralded by the attack. I could not think of such violence as anything else. Had they slammed me like that underwater, I could not have escaped. Never could I have imagined that their fit of pique could extend to knocking me senseless; I had been afraid of their teeth! Watching my beloved sharks gliding peacefully about—Droplet passing with her mating wounds healing well and Christobel acknowledging me with a lazy approach—the sense of unreality that had descended when Spooky flew over the site, deepened with a wave of incomprehension.

The sharks who had been there at the beginning were drifting away. I looked above the surface to make sure that no bird was expecting me back in the boat, and there was Spooky, looking fragile and anxious, perched in the centre of my own seat. Amazingly, she didn't startle when I grabbed a fish head from beside her to feed the fish. Nor was she concerned about my appearance, wearing a mask, and emerging suddenly from the water. Birds were normally highly sensitive to a change in my appearance or dress. One handicapped bird, for example, who was frightened of the dangers outside, flew down from the trees and into the house if he saw me dressed to go out, rather than in my usual *pareu*. Others sounded alarm calls if I appeared wearing red. So for a half-wild seabird to accept my appearance wearing mask and snorkel, with my normally frilly hair plastered down, was a surprise. It was convenient though, that she was so cooperative. Perhaps she had been watching me as I swam around, so understood enough of my strange transformation to accept it under such extreme circumstances. She had never seen me in the water before. Never, since I lost Merlin, had I snorkelled or swam from our beach.

The height of the sun indicated that I had little time left, if I was to get Spooky safely home, so I checked that I had recorded the identities of all the sharks, took a tour around down-current, and briefly fed the fish. Christobel charged through them. I spent some time with the crocodile needlefish and fed her, too, but the sharks all returned when I began crumbling meat for the little squirrel fish, and circled me in their new, demanding way, now joined by a macho group of golden males. The sun was low, so I picked up the anchor.

Spooky flew when it clinked into the front of the kayak, and I climbed in. She alighted on the bow as I began to paddle, then took off for a last flight out over the waves on the reef. But soon she came winging back, low out of the glittering west, to circle and alight upon the very tip of the bow. There, she settled for the home trip and looked more and more tired as we went, her neck slowly shortening, as she sat watching the shore go by. When we approached our beach, she didn't seem to understand that we had arrived, and didn't jump off until I was obliged to get up to hold the boat steady against a wave that knocked it sideways. Then she startled and flew up, but missed her perches, and disappeared into the layer of branches above. I had barely got her back in time.

The first thing I always did on my return was to rinse the kayak. This time, I examined it closely. Concerned as I was about the sharks' attack on the boat, I was disappointed not to find it decorated with sharks' teeth. But I was far more disappointed to have missed the last half of the shark session. The idea that my data would remain incomplete to that extent was an inconsolable anguish. Fifty-six sharks had been present, just in that first half, far more than usual. More would undoubtedly have come at nightfall. Worse, was the realization that my curious bird would expect to come each time now! My unbreakable rule had always been that no bird could interfere with the shark sessions. Their schedules had been arranged to leave me free, always, at the hour before sunset, to spend with my sharks. And now this!

When I entered my data into the computer that evening while the potatoes cooked, I was in deep thought. Moving slowly down the long list of sharks, I noticed that I had written beside Sparkle's name, *distinguished herself as a smart shark* after her performance when I was trying to get medicated food to Meadowes. Now I wrote underneath, *distinguished herself as a bad shark*. Then picking up a loose envelope, I wrote down her name and headed the list, *Bad Sharks*. This was not very scientific, but it was simple, and I was feeling sorely tried. Above Sparkle's name I wrote, Carrellina, with a number one beside it, then added Sparkle Too and Chevron below. All of them were in their first year of reproduction.

The next evening, Kim was avidly awaiting our outing, and stepped firmly onto my hand to be carried to the kayak. He sat in front of me in the boat, looking around as I paddled free of the shore waves. Spooky had been waiting too, and swooped repeatedly over us until Kim jumped from the boat into the water. Then she dove down to harass him while he raised his beak and wings and shrieked at her. When she settled on the bow to watch him, he began playing with a leaf, then paddled steadily seaward into the wind. Sometimes he drifted, bathing. His feathers all came loose, floating free in the water, as he energetically rinsed them, moving in all directions, to give each part of him, and every feather, the care it deserved. But when he shook the water out and righted himself, his plumage was perfectly aligned and dry. Sometimes his lame wing sank farther and farther underwater, unbalancing him, and eventually he found himself upside down for several seconds, his little feet kicking in the air. Then without the slightest sign of agitation, he did a back flip, and was once again perfectly dry, and upright. How comfortable he was in the water. Finally, he paddled glumly on toward the ocean which beckoned and shone in the brightness beyond the bay.

I was thinking about the sharks. Paddling sporadically to stay near Kim, and keep myself steady in the wind, I could see the distant shimmer near the reef where they were circling, thinking of me, as I was thinking of them. It was the hour of our sunset rendezvous. The attack on the boat had confirmed that my feelings had been right, especially considering the session before in which I had spent an hour holding my breath, scarcely daring to move a muscle while they circled me, all of us poised. I was convinced that they believed I could produce food if I wanted to, and was deliberately withholding it. The mêlée that had ensued when I had thrown in a few crumbs for the fish from the kayak on leaving, had shown what a knife edge they had been on. The attack on my return two days later suggested that their attitude had lasted throughout the intervening time, whatever the ultimate trigger had been. It had taken place two days before the moon was full, once more invoking the lunar connection.

No matter how I thought about it, the violent slamming of the boat always appeared as a negative emotion, something akin to rage, expressed toward me. I kept coming back to the conviction that the seed of change had been planted when, day after day, I had tried different ways to get medicated food to Meadowes, sometimes only bringing scented water, but no food, for them. They had known that I had food, but that I was not giving it to them, and now they behaved as if they thought I could be tricky. Perhaps the attitude they had held for some time was bearing fruit in action, now, as a result of the warmer water, the excitement of the reproductive season, the many excited visitors among them, and the return of Carrellina. It seemed that fish could hold a grudge—the attack on the boat had flowed from the tension that had built up during the session two days before it. Of that I had no doubt.

Their desire for food could not be their motivation for slamming the kayak. The way Chevron, the same shark who had circled me with special intensity the session before, had come again and again to slam the boat, even after their food was in the water, provided evidence of this. She had not gone to the food in spite of the scent of it filling the water, but had remained focused on slamming the boat, as if she were finally venting intense feelings of shark-anger or frustration.

In the first seconds of the attack, big sharks had accelerated toward the kayak, and slammed it from different directions at the same time. Somehow, they had acted in synchrony. To move the heavily loaded boat with the impact I had felt, required daunting power—speed and big sharks. Further, there had been no pause. After the first sharks had slammed it, more had slammed right behind, and more behind them. Still, I remained sure that Martha, Madonna, Bratworst, Flora, and the other sharks who seemed most attached to me and swam with me, would not have joined in. Bratworst had been the bold one when I had first met her, and she had been in her first reproductive year then, too. All the sharks who had troubled me with their bold behaviour in recent weeks, with the exception of Martha, who was simply over-intimate with me, were in their first reproductive year—Carrellina, Chevron, Sparkle Too, and to a lesser extent Sparkle and Christobel. I was sure that these were the instigators, and that others, such as the excited juveniles, Apricot, Lightning and Breezy, had joined in after they began. I kept wondering what would have happened had I chosen that moment to go underwater, which I would have done, had Spooky not been with me, and distracted my attention for a few extra moments.

But I did not know. Their behaviour toward me underwater had not been different. Nevertheless, I could not escape the conclusion that it was evolving in the direction of less respect and greater aggression. If the evolution continued, would not a similar action, directed toward me instead of the kayak, be one of the natural next steps?

It was inexplicable. There had been a session in which their behaviour had been normal, just as it had been for years. Then at the next session, it had changed. It had been that sudden. The urge to ask an expert for an opinion was overwhelming. I was sure that this was an instinctive human reaction due to our social evolution, to consult with others when troubled. It probably had survival value!

But who could I ask? By then, I had realized fully that no one understood the behaviour of sharks. No one had tried to understand what they were really like as animals. No matter whom I tried to confide in, they would skip over the details of the history of the situation, the contributions of individual sharks, and how it had all evolved. They would cut me off and just tell me to stay out of the water, adding something like, "They're sharks, Ila!" I winced when I thought of how Arthur had told me, when Meadowes was still alive, not to go back. And he did know something about shark behaviour, when the vast majority of people who wrote about sharks did not.

In researching references for my manuscript about the sharks' lives, I had been sent a variety of published scientific articles on the species. Few had been written about these sharks, so I had nearly all that were available. Only one concerned their behaviour, and it was devoted to convincing the reader that the species was dangerous—Randall and Helfman 1973, "Attacks on Humans by the Blacktip Reef Shark (*Carcharhinus melanopterus*)" Pacific Science, Vol. 27 (3) 226—238.

The article was based on anecdotal evidence, and slanted every encounter with a shark toward the possibility that the incident was a shark attack. For example, it called the typical greeting gesture individuals performed—swimming up to one's face then turning away—a shark attack. The authors baldly stated that the shark would have bitten had it not been hit, slashed up with a machete, or blown up with a power head, before it was able to do so.

In an additional anecdote, it described how the senior author had been frightened more than once by blackfin sharks appearing, and termed their behaviour aggressive though the sharks did not approach. While he was killing the fish in the coral outside the reef from my study area with rotenone, he described a shark of about three quarters of a metre (2.5 ft) long coming into view. This startled him, because he said that its swimming was "erratic" and "rapid." He decided to exit the water. The juvenile in question could have been under the influence of the poison he was using to kill the coral fish. The only times I had witnessed erratic swimming was the direct result of fear in the shark, and so far, I had only seen it in very young ones. Could the little animal have left the nursery due to the rotenone being swept over the reef and into its refuge? Such questions were not considered.

The author thus admitted that his personal fear of sharks influenced his judgement. He then stated that the shark had charged and bitten the boat's anchor while it was being pulled up, implying that this act was further evidence of aggression toward humans! It is well known that certain metals in sea water generate an electric current that attracts sharks' attention due to the strong stimulation of their electro-sense—Bou Bou had filmed such a reaction. But these so-called scientists left out that fact.

The only actual cases of bites by blackfin reef sharks, were on the feet and lower legs of people walking through shallow water. The sharks seemed to be attracted to the splashing, and mistook the walking feet for fish. These incidents were rare and confined to certain regions such as Caroline Atoll; some of the incidents described in the article occurred there. In contrast, in his book, *Together Alone,* Ron Falconer described his life with his family on Caroline Atoll, including the behaviour of the blackfins chasing after people's feet in the shallows. But he understood that they were attracted by the splashing, and never thought of their chasing from one foot to the next as an attack. His dog and children played with the sharks, and no one was hurt. His descriptions as an objective observer starkly contrasted with those of Randall and Helfman.

Apart from the lack of any actual attack on a person underwater, considering how common these sharks are, I was concerned by the interpretation of the sharks' behaviour. In many cases cited, the shark was attacked before it came near the person. The authors did not distinguish between the natural behaviour of the sharks, and their reaction to being slashed up by a machete, or picked up by the tail. In each case the shark's behaviour was cited as a shark attack, and presented as proof that the species was dangerous.

It was worrying that such openly skewed reasoning would be published in a scientific journal. The strong and irrational bias against sharks apparently extended to scientists, who have a duty to humanity, and the search for objective truth, to remain open-minded.

I had managed to pick up a second hand copy of the book *The Shark*, by Jacques-Yves Cousteau and Philippe Cousteau, on our last hurried trip to Canada. Given the reputation of these pioneers of marine wildlife appreciation, I had no doubt that I had acquired a source of real knowledge about what sharks are like. But the authors depicted sharks consistently in negative terms, such as "mad hordes" and "monsters" as if they were not just ordinary animals.

From a cage, the Cousteaus had filmed sharks attacking a human-like dummy that they had stuffed with pieces of fish. For several pages a long, shocking description of the violent attacks on the dummy was given. It was referred to by its cute name, when detailing its many injuries, and the horrified emotional reaction of the witnesses was described as if they were really seeing a person being attacked! Later, the brief mention that the experiment had not worked when the dummy had been stuffed with anything but dead fish, was easily skipped over. The reader had already been convinced by the authors that the sharks automatically tore apart the diver-like construction because it looked like a human. The implication was that the sharks would have treated any human who had gone outside the cage in the same way.

Yet the entire team were aware that the sharks were only biting the man-like object because there were pieces of fish concealed inside, which they could smell. The episode was a deliberate creation to give readers the impression that sharks were dangerous and stupid, and attack people without reason or warning.

To learn whether reef sharks were sedentary or not, the Cousteaus stabbed tags into one hundred and ten sharks in the Red Sea, then left for three weeks. On their return, they threw a huge hook baited with a kilo of meat into the sea, and described the fisherman as a hero as he caught one of the tagged sharks. The authors thrilled with the death of the animal with a graphic description of how the shark inhaled the food and convulsed in shock as the hook ripped into his insides. The nauseating description went on for two pages as the authors gloated over the agonizing death of "their old enemy." Enemy? Then they described how they caught all the sharks that they had tagged in that region and killed them all the same way! They learned that the sharks on the reef were sedentary. Then they concluded that this result was not conclusive. But they had killed close to the entire population of sharks for this inconclusive conclusion.

A few pages later, they described another "experiment" in which they killed all the sharks on a different reef, with the exception of a few small juveniles, just to see if other sharks would move into the territory. Yet, they described diving in the region for only a few weeks to check to see if sharks were moving in, which was not nearly long enough to observe the evolution of a population of sharks. Everything was described as if their anecdotal experiences established the truth, when in fact, their activities were unscientific, and their conclusions were convenient assumptions.

Every description reflected the same attitude to sharks in bombastic descriptions. Sharks were given human qualities such as "toleration of rivals" and "knowledge of being the master." The authors reported trying to start a "feeding frenzy" by firing a spear through a shark they were feeding. But instead of devouring the victim and providing the hoped-for photography of blood, teeth, and violence, the sharks withdrew and circled much farther away, as if they had become cautious. Cousteau admitted surprise that the sharks behaved as if they had understood that he and his team were actually dangerous. The account stated that four species of sharks were present.

I remembered how my sharks had hidden when Bou Bou came, as if they were afraid of him. Never would I have thought that if I shot one, the others would have gone into a "feeding frenzy," and eaten my victim. Any animal would have understood that they were under attack when one of their numbers was shot; they are hyper-sensitive to danger. I have yet to find a vertebrate species that lacks warning signals of alarm which are broadcast among them and intercepted by other species in the region at the first hint of danger.

This episode gave further proof that the Cousteaus did not consider sharks as ordinary animals with the capacity to perceive reality. That they operated in such a fashion, yet kept a good reputation, revealed the bias against sharks shared by their audience. I had put the book aside in disgust at the appalling waste, and injustice to sharks, in the name of science and profit. The prejudice expressed was similar to the hatred toward snakes, but in snakes, the bias was acknowledged; people knew they had an irrational fear and hatred of snakes. In the case of sharks, recognition of the prejudice was lacking. It seemed that writers believed that the animals really deserved to be despised, hated, and killed.

My searches on the Internet reflected the same attitude, a basic assumption that sharks were more like sea monsters or robotic killing machines than animals—essentially the image portrayed in the horror film, *JAWS*—yet I could not find any description of efforts to find out what they really were like. So there was no one who could help me to understand the behaviour of my sharks. I had created the situation, and I was the only one who could judge it and decide what to do.

I sat and mused, watching the birds. Kim was paddling, drifting, and bathing while Spooky played. She picked up tiny things from the surface to lift them high, drop them, and pick them up again, then bring them to the bow to toss and manipulate, always swooping low over Kim before she landed.

The wind began to gust more strongly, and Kim arched his wings, trying to leap into it from the tips of the waves. I followed closely as he took off and glided. After several attempts, he succeeded in flying over the waves for a few minutes, steadily drawing away as I paddled frantically to keep up. Luckily the wind died down just then, and I was able to catch up with him. I guided the kayak between him and the wind, into which he was forced to face, poor soul. Seconds later, he floated within reach, and I scooped him up.

He settled on his breast, and by the time we approached our beach, he seemed to be sleeping. But as we swept into shore, he climbed onto the edge of the kayak, jumped off into the waves, and surfed up the beach, while Spooky shrieked at him. He waited while I pulled the kayak up and handed a fish to Spooky, who was impatiently waiting to get one, before she flew to her perch. Then he stepped onto my hand to be carried to his pool.

Each day at dawn, Spooky flew from her sleeping perch to the branch where I fed her, and delicately took one tiny fish. Then she waited, often perched impatiently on the kayak, for Kim and me. She flew to him with piercing cries as soon as he appeared on my hand, and soared out above us while we launched the slender boat. As we floated away from shore, she alighted on the bow where there was a fish for her. Often she dropped it in the water, and tried get it back, but she still lacked the necessary skill, and finally waited for me to get it and toss it to her. While Kim swam lazily, preening and playing on the surface, I threw her little bits of seaweed, which she tried to catch. Sometimes, she would toss them back, so that I could catch and return them, and sometimes, we actually played catch! But mostly, she stood poised on the bow, watching Kim glumly paddling along. Often, she flew to circle above him and sometimes harassed him unmercifully, hovering in front of him, facing him, just out of reach, and shouting at the top of her lungs at the crippled bird. He raised his wings with a shriek and snapped at her, while she nimbly evaded him.

Occasionally, she startled him so badly that he dove, and before long he tried to fly out of the water with enough force to gain the altitude to catch hold of the elusive bird, which he persevered in trying to do.

Spooky now began alighting on the surface to try to swim beside the irritated petrel, but she soon lost her balance on her tiny round breast, and rose up to try again before soaring away to play in the breezes, pick leaves and flowers from the surface, and toss them in the air. She always swooped low over Kim before alighting once more on the boat. There, her attention turned to me, and she came toddling over for another little fish. One by one, she carried them back to the bow and kept me in suspense by half swallowing them, tossing them up, catching them deftly, and doing it all over again. When she lost one, however, and I produced another for her, she swallowed it immediately as if contrite. These small fish were so precious that we could not afford to waste even one. When it was time to leave, Spooky paid no attention to the violent swerving of the kayak as I manoeuvred desperately to catch Kim. She balanced neatly and acted bored. Once he was on board, she kept an eye on him and shrieked at him if he approached her as I paddled home.

Petrels, like all seabirds, spend their lives at sea, in the air, or on the surface, and their bodies are not made to support them on land. Their legs are too small, and their skin is not tough enough. As a result of all the walking Kim did, his jumps from his pool onto the wooden deck, and from the kayak onto the sharp sand, he had developed sores on his feet and legs. I began to treat and wrap his tiny legs each day, and at first, this upset him.

When I came to reassure him afterwards in his pool, he threatened and shrieked at me with raised wings. But soon, his posturing turned to play as he bit at my hand when I reached out. It was soon clear he was biting to make contact, not to hurt me. After that, nearly always when I came to check on him in his pool during the day, he drifted slowly over to touch my finger, then playfully try to catch hold of it. I tried to get past his defences to poke him, while he energetically avoided me, and rushed in to grab my finger. He became very expressive with his eerie, wild cries, as he lifted those long wings and rushed at me. It seemed remarkable that this bird, who would normally be flying the trackless oceans using unknown instincts, could share this sort of amusement with a person. Other times, he drifted over to me, floating as close to me as he could get, usually just below my face, while I softly talked to him. It was touching to see him entering into his new life more each day, yet I was becoming convinced that he was not going to recover his ability to fly. And a seabird who cannot fly has no future.

Meanwhile, Spooky slept and rested in her perching trees. She would not go anywhere without us.

One day that week I went to the lagoon just after sunrise, to look for Martha and Shilly, and spend some time with them. Spooky swept out after me, and it was fun to have her with me. But she seemed nervous out on the lagoon, and didn't explore. She rested upon the bow, unmoving, with her mouth wide open.

I fed the fish at the site, then got back in the kayak, and reassured Spooky while we went to the barrens. In the shallows where the little juveniles lived, I drifted, glimpsing them here and there, but none of the large females I usually swam with. It seemed strange that initially I had always been able to find Martha, Madonna, and Bratworst, just by going to the site in the mornings, and now it was rare to find them. I went to the region favoured by Shilly, and quickly she appeared and circled me. I was able to swim with her for a while, but since she was so small, her cruising speed was just a bit fast for me.

By then Spooky was shading herself with one wing, and was obviously stressed, so I took her home.

She slept during the middle of the day, but was ready in the evening to accompany Kim and me for our outing. The birds played as usual, but suddenly, as Spooky hovered, shrieking, just above Kim, he caught hold of her wing, yanked her out of the air, and stood on her. Only the tips of her dark, little angel wings were visible, as he stared glumly into the distance, unmoved in the murderous moment. I hastened over, but before I could get the kayak close, Spooky popped up like a bubble and rose slowly into the air on those continuously vibrating wings. Undaunted, she flew tightly around Kim, and landed beside him, as if her fondest wish was to swim upon the surface the way he did. She soon lost her balance and flew on, but continued diligently practising, losing her balance and using her wings to regain it, until she actually bounced on the water. Once only, she managed to fold her wings for about a second and floated elegantly. The rest of the time she held her wings extended upward, fluttering them rapidly to keep herself upright.

One afternoon while I was painting on the deck, Kim came over and touched the things I was using with his beak, a sign of the strengthening bond between us. I had begun to love him, and to wish that I could just be with my birds each evening.

I had become frightened of the sharks. I had always thought to myself that if one is frightened of them, one should not be there, and over the years in different circumstances with them, I had needed to learn to control my mind and my reactions sternly. Even in the beginning, those I had confided in had warned me that I was going to get bitten, and when that happened, without help, I would probably die. But I had followed my own instincts and had turned out to be right. I had never been bitten, no matter what had gone wrong. My mind circled over every detail of their new behaviour, seeking understanding. Yet in spite of this continual musing, I had come to no conclusion on the day that I acquired more fish scraps.

So I tried to tackle the situation in a practical way, pinpointing each thing that could go wrong, and planning a way out in the event that it happened. The period of time in which I was most likely to be bitten was the one between my entry into the water, and the placing of the food. I could avoid that situation by throwing in the food before going underwater myself. I could also minimize the number of sharks there at the very beginning by ensuring that no scent leaked into the water beforehand. The lack of scent might also help to minimize their level of excitement before the feeding, at least among the visitors. I doubted that this would have any effect on the residents, though, since the underlying reason for their behaviour was not food-related. Having thus provided for my own protection, I prepared with more serenity and confidence than I had been feeling all week.

It was a calm, sunny day, and Spooky was off playing with a crested tern when I left. She had begun flying on her own a little more in the past days, so several minutes passed during which I believed I would be able to get away without her. Then, with a whir of wings, she flew past my ear and landed on the bow of the kayak. I threw her some little bits of fish, but she didn't eat them. Though she flew briefly sometimes, to pick a tiny object from the surface nearby to play with, this time she stayed close.

The waves on the reef had dwindled, so the lagoon lay relatively still for once. Madonna followed me across the lagoon, joined by the usual gang as I drew near the site. As they circled, they spiralled closer and closer to the boat, until they were swimming altogether against and under it at the surface.

I was able to get the anchor dropped while Spooky briefly took to the air, but she stayed on the boat while I prepared, and the sharks circled. I was especially worried that this one bird who accompanied me to the shark sessions, was also the only seabird I had looked after who loved to play and bath on the surface. Kim only did so because he could not fly. I hoped that Spooky would have the sense not to alight in the sharks' site. How often did Martha and the others rise to the surface to nose a floating object? They would love a hot little bird! But she just looked and stayed where she was.

As I prepared, there was a slight bump against the boat, then silence. But as I was about to start throwing the food in, one after the other, and then all at the same time, the sharks began again to slam the boat with alarming force. I was fairly certain this time that Carrellina, Sparkle Too, and Chevron had initiated the assault, then the others had joined them.

Spooky rose slowly, hovering, staring down into the waters roiling with sharks.

They began to leap from the water again in an effort to grab the food. It seemed to take forever to pick handfuls of scraps from my precarious perch. My seat faced forward, and the food was directly behind me. In the narrow boat, there was no room to move. I had to twist my upper body to reach behind me, and throw the food I managed to grasp into the site, while the sharks bashed the boat and leaped at the food from the water. Watching and hearing the sharks' jaws snapping shut with loud clapping sounds, it was obvious that in a kayak, it was not easy to do anything while keeping all parts of one's body out of the edible zones.

I had thrown quite a bit of food in when Martha suddenly broke through the surface beside my right elbow and snapped her jaws closed on a trailing bit. The power of her movement shocked me. She had come so close to catching my arm in her teeth. Martha! I was terribly disappointed and concerned that she was involved; it was starting to sink in that all of my beloved sharks were involved.

For the very first time, I didn't want to get into the water, and sat waiting. Whenever I tried to make the boat move away from where the food had landed, by waving my fins back and forth with my hands, the sharks were attracted, and they continued to come to the surface even after the food was all in, and I had rinsed the scent from the back of the boat.

Finally, I slid in. No one paid any attention to me—they were all the same sharks I had known and loved for so long. Shilly came and circled around me. Gazing at her suspiciously, I wondered if she too was involved in the attack, since every shark I had identified attacking were ones who habitually made close approaches to me. However, I had begun to wonder whether their behaviour had less to do with aggression, than impatience, to get the food by themselves, rather than waiting for me to give it to them. Slowly, I began to feel better, and relax. Indeed, they were like excited children helping themselves to birthday cake without waiting to be served.

I kept an eye on Spooky, who waited nervously, perched on the back of my seat, while I was underwater.

The excitement faded, and the session passed as usual. Later, Kimberley came. I had predicted that she might come during the period of this full moon, since I had seen Marco in the area. A theory that leads to accurate predictions is a good one!

Strangely, while the sun was still far above the horizon, everyone left, and I was able to feed the fish without harassment. The crocodile needlefish accompanied me as she did each time, and I was able to give the squirrel fish who hovered near, and all the others waiting, handfuls of crumbs of the fish meat that I had prepared for them at home. The nurse sharks pursued their usual activities. There were more of them than ever. Down-current, I found Kimberley with Raschelle, a young female I had identified the year before. The shy creature swam slowly by me, looking. In the distance Flora passed, heading back from the border.

Spooky waited, untroubled, when I reached into the boat to take handfuls of scraps for the fish, and when it was time to go, she did not fly up when the anchor clattered in. As I climbed in myself, she circled

Spooky harassed Kim unmercifully...

to alight on the bow, and we began to drift away. Soon she took to her wings, and I lost sight of her for a while, but as I neared the border, she materialized from the direction of the reef, flying low over the waves, to alight in her usual place upon the tip of the bow. She rode looking around, sometimes flying to bounce on the surface and pick up a tiny toy. But she would not eat. Finally, she rested on her breast, watching the shore passing as we drew closer to home. I thought I had left in plenty of time, but she had trouble seeing well enough to fly to her perch, though I could still see everything easily. Birds see more fine detail than we do, but not as well in dim light, which is why I was so concerned about getting her home well before dark.

One glittering morning, Kim paddled all the way to the ocean, nearly two kilometres (1.25 miles) from home. Spooky was alternately teasing him, and trying to float beside him, when five noddy terns swept over the boat and flew around us, clearly curious about Spooky. One seemed ready to land on the boat, and dived repeatedly over her until she flew up. But she only circled and landed again. They left as we went home.

I always took food for her with us, and when she was hungry, she pattered toward me, looking at my hands. She ate only about two thirds of the quantity that my other noddy terns had eaten, and she still could swallow only the tiniest fish. If it was too big for her, she waited for me to break off the head, then she ate it, apparently understanding that I had fixed it for her. She was still flying very little by herself, compared to other birds of her species I had raised, and seemed to be more attached to me than any of them had been, perhaps due to the companionable time we spent together with Kim on our outings. Once, she nearly landed on my head when she saw me on the beach. She spent all day on the property, either in her trees, or on the rocks at the shore, looking down into the water. Though I watched closely, I never saw her leave for a long flight in which she could have met others of her own species, even at dawn, when my other noddy terns had routinely taken long flights.

One Sunday, strands of oil streaked in rainbow colours over the bay's surface. By the time I noticed, Kim had already filmed his feathers, and Spooky was flitting through an oily patch too. Her immediate reaction was to bathe, splashing with her wings, and ducking under the surface in the rhythmic manner of bathing birds. Luckily, she chose a clean stretch of water. I was awed at the consequences for wild birds flying over oiled waters if their first reaction was to bathe. Had she been in a strand of oil, each filament on each feather would have been glued together so that they would have failed to enmesh, as feathers do, like Velcro, to create a waterproof sheath. The frantically preening bird would have been soaked, cold, poisoned by oil, and soon dying.

Spooky was able to work the tiny amount of oil out of her feathers, but Kim required a waterproofing bath. I did it as gently as possible, but to him the treatment was an unforgivable affront. He never played with me again.

During our outings, the birds were increasingly hilarious. Spooky finally achieved stability while floating, presenting an elegant silhouette, with her long neck and her pointed tail and wings. Kim dove as Spooky flew over him, to explode from the water, shrieking, and try to snatch her from the air. She flew around him, hovering above him, and shouting at him, then landing in the water to paddle. He hastened toward her to catch her when he saw her on the surface, and she flew when he caught up with her. But there came a time that she just accelerated, and they swam side by side, finally acknowledging their friendship.

When I gave Spooky a fish, she took it, half swallowed it, tossed it out and up, caught it, half swallowed it tail end first, repeated each step a few times, then took to the air with it, tossing it, catching it, flying super fast, darting and changing direction like an aerial acrobat, dropping it, and plunging into the water to pluck it back, only to rise again. Finally, after three drops into the water and plunges in after it, she ate the fish. She flitted across the surface, bathing and playing, finding things and tossing, dropping and catching them, and sometimes just floated still, balanced perfectly on that tiny rounded breast. She was so delightful the whole time that I sat back, watching. Floating there upon the sea, watching those birds together, was all the reason I needed to be happy.

One early morning, Spooky flew away and didn't return until late afternoon. She retired to her sleeping perch, but flew to join us on the kayak when I took Kim out later. She was hungry and came for several fish, which she ate immediately without playing with them, then returned to the bow, where she rested on her breast. She was a very tired bird, and flew to her perch as we approached the beach, not waiting for me to paddle in. This became her new pattern. It seemed she had even begun fishing for herself. She was too tired to play in the evenings during our outings on the sea, though she still accompanied us on the bow of the boat, resting. It wasn't long before she began staying away overnight.

Kim's routine remained unchanged. Each morning and evening we went out upon the bay, and when we glided into the beach, he jumped off the boat as I pulled it up onto the grass. He walked up the beach, wings held out for balance. Sometimes, he received an appreciative sniff from one of Franck's huge dogs, who came to the shore to meet us. He lay waiting on the sand, for he could not climb onto the deck on his little legs, and as soon as I could, I carried him to his pool. He rinsed, and gave his feathers a final coiffing, then floated silently in front of his mirror. At exactly the same time each night, he paddled, wings raised, to the edge of his pool and hopped out. I tried to be there, having put the boat away and showered, and softly talked to him and stroked his chin as he pushed against my finger, guiding it to where he wanted me to stroke. Then he stepped onto my hands to ride into the house. I set him down inside, and he walked partway up the hallway while I prepared his bed and his supper. Often he had to wait for Franck to have a shower before he could retire on his seabird bed in the shower stall. We kept it there so he could drink from droplets falling from the faucet; this pleasure of his convinced me that the species drinks rain as well as sea water.

One night, he was waiting while I was rushing around getting things done to prepare for our evening, and in the darkness of the hallway, I accidentally stepped on his foot. He was quite invisible on the dark wood floor. He screamed, and I carried him to the couch, trying to examine each little bone in anguish as he tried to bite, but I could find no real damage. Finally, I just held him, and he eventually calmed down. I fed him and put him in bed.

Courtesy of Franck Porcher

CHAPTER THIRTY-FOUR

Thinking it Out Again

Two weeks had passed since I had been able to get fish scraps, and it was mid-March. I had not stopped considering the shark situation and advising myself that it must be re-examined. Eventually in accompaniment, a song arose from the past to play over and over in the back of my mind: "I'm reviewing the situation" and ending, "I think I'll have to think it out again!" from Lionel Bart's "Reviewing the Situation" in the musical, *Oliver*. I returned to the sharks filled with misgivings, no farther along in comprehending their change of behaviour. All I could think of was that I should put the food in before I slid underwater myself.

It was sunny and calm with small passing clouds as I headed out to see the sharks. The ocean dribbled listlessly over the reef, and the lagoon lay still as a plate of glass. Spooky didn't come; she was probably soaring over the ocean with her new friends. I had to go around some Polynesian fishermen in my path, so approached the site from a slightly different direction. No sharks were visible there, and only two came as I prepared. I thought of just sliding into the water as I always had, but more arrived, so I played it safe and threw in the food.

Only Lightning and Muffin came to the surface as I did so, and they just looked and did not slam the boat or leap from the water. I saw Fawn rise below, and Tamarack deeper down. All of these were immature males. I moved the boat up-current from the site and slid underwater, where I found a group of males and juveniles, plus Martha and Cochita. None of the juvenile delinquents, Carrellina, Sparkle Too, Sparkle, Chevron, and Apricot, were present! This explained the tranquillity in the site. Many of the juveniles there had been involved in the attack the previous times, I was sure, but their failure to slam the boat in the absence of the bad sharks reinforced my conviction that certain individuals initiated the new behaviour, then the others joined in.

Madonna came late. I was with Martha, floating to her a piece of food that I had saved for her. She turned and took it. That was the moment that Madonna zoomed around a coral, and they collided, hard, head-on. They were within arm's length in front of me, and Madonna turned to me, but passed on by. It was a good example of how absolutely unexpected and unpredictable things could happen.

Odyssey appeared with Darcy and several juveniles from the east, and later Kimberley arrived. The two elderly sharks swam together, cruising the region, and looking for scraps. Lillith passed over my shoulder from behind and circled. The usual juveniles and males were roaming the region, intermittently bursting into excitement as they found some little scrap. I occupied myself trying to get accurate drawings of both sides of a new visitor while Madonna repeatedly approached. For the rest of the session I tried to get a piece of fillet to her, taking one from the boat, waiting for her to pass, and throwing it in her path. She never did catch on, and each time, a juvenile swooped in from the periphery to grab it. Poor Madonna never was one of our brightest lights.

Jessica came cruising through the coral, and as twilight fell, several strange males appeared, including Danny, the biggest male I had seen, first identified at Site Three. It was a pleasure to see him again, and I drew the dorsal fins of those with him while trying to get a bit of food to Madonna. It was a relief to feed the fish without the sharks bothering me. I did so a bit at a time, cautiously, while keeping track of everyone present.

A whitetip passed through briefly, and the nurse sharks swayed and writhed in slow-motion all over the darkened site. Glammer and Annaloo came through while there was still light to see, and Samurai put in a rare appearance. Storm, unusually agitated, came just before I left. It was a good session. Everyone had fed and was happy, including me.

On the way home, I looked for Spooky. She was rarely with Kim and me on our outings any more, but sometimes she appeared late, winging homeward low over the bay from the ocean. She would circle, land on the boat, eat a fish, and fly on, drinking from the surface as she went. But that night, we missed each other—I found her asleep in her tree on arrival.

Once when Kim and I were out late in the morning, she suddenly appeared in wind that had become so strong that I could scarcely control the boat. I couldn't take my hands off the paddle to get out her fish. She balanced with difficulty on the bow, then glided beside my head until I was able to hand her the fish on the wing.

I caught Kim with difficulty, and the wind filled his wings so that it was hard for me to settle him. In the effort, I lost the paddle, and the kayak was blown swiftly away from it. I let go of Kim, who jumped overboard, slid into the sea, swam to the paddle, carried it back to the boat, climbed in, caught Kim again, and fought my way to shore. The good little bird ran up the beach and waited for me after our rocky landing in the waves, while I gave Spooky two more fish. Then I carried him to his pool. But when I went back to Spooky, she was gone. It wasn't until darkness fell and the stars were out, with just a bit of light in the western sky reflecting on the water, that she appeared, a dark flicker in the night air, and alighted on her perch.

One day it was perfectly calm, and began to pour with rain from a tiny cloud as Kim swam steadily seaward. Spooky was with us, and she sat playing with a fish, dropping it in the water, then ignoring it when I managed to retrieve it. The sun came out, and she shaded herself with her wing for a while, sitting bored with her mouth open, and finally she flew away home. As we neared the ocean, the wind began to blow. Blasted by wind and waves, and the brilliant glittering of the sea, and swept with spray, Kim bobbed solemnly onward, unperturbed. I wondered where he thought he was going and what passed through his mind during these excursions. Had he become used to not being able to fly? Had he accepted his new life as tranquilly as he seemed to? Was he, instead, lost in memories and dreaming of his former life, soaring the endless ocean with mate, family, or friends?

After we had been out there for an hour and twenty minutes, I decided it was time to go, and glided between him and the wind. It blew me toward him, and he found his way blocked; I scooped him up. It was a handy trick to catch him when the wind was high. On the way home, he didn't relax, but stood looking around in his glum way. As we approached the beach, he suddenly jumped back in the water. Spooky had soared out to meet us and was flying around my head, crying, "Ah! Ah! Ah! Ah! Ah!" I began handing her fish, and she dropped three out of four of them in the bay. So I swam out and managed to get one of them back, but she dropped it again. I rinsed the boat while letting Kim swim around a little longer, hoping he would come back in by himself, but after noticing that I wasn't watching him, he headed straight back out.

Rain began to pour down, and I was obliged to take the boat back out to get him. I didn't have any more fish for Spooky, so was taken by surprise when she landed on the boat close to me and started looking pointedly for more. When they weren't forthcoming, she started shouting, "Ah! Ah! Ah! Ah!" at me!

I reached Kim, and had a terrible time catching him in the pouring rain, while Spooky continuously yelled at us both. When I managed to get Kim into the boat, Spooky shouted at him, then as we approached the beach, she flew ahead to her perch. Kim jumped out, and this time, encouraged by yours truly, he ran up onto the sand.

Spooky was clamouring for more fish, which I prepared as soon as I put Kim in his pool, showered, and could defrost some. She waited on her feeding perch. When I came out with them, she flew at me, and gave a shriek at the moment she took one from my hand. Not only sharks felt they had a right to rudely demand food.

Kim's sores had slowly healed with long-term treatment, and I was trying to minimize the amount of walking he did so that more would not develop. He had days when he seemed unhappy. Sometimes, he didn't want to go into his pool, protesting frantically and trying to turn around in my hands as I took him outside. Moments later he would be waiting at the door to be let in, peering inside to see if I was coming, and relieved to rest on some cushions deep in the cool dark recesses of the house, pressing against me as I lovingly caressed his chin.

Our outings were not the same without Spooky, who accompanied us more and more rarely. As Kim paddled tirelessly onward or bathed, my mind was freed to muse upon the sharks. The last session had been a fluke with the young females off together somewhere, but that was not likely to be the case the next time. I was increasingly uneasy about the situation—what it meant, what it was leading to, and what I should do. Time after time, I rethought every detail, and each time, my only conclusion was that I would have to think it out again.

I couldn't forget the moment of dread I had felt about getting into the water after the second attack by the sharks on the boat. There seemed to be no escaping the conclusion that the next step in the long, long string of events leading up to the present moment was that it would be me instead of the boat who was attacked. Carrellina or Chevron would make the first move, and be instantaneously joined by everyone else. Surely, I was missing something. I would have to think it out again.

The next feeding session had been delayed a week because no scraps were available, and the evening before I was due to go back to the sharks, I asked myself what I had concluded. Kim had drifted so far out in the bay that if the water had all been removed I would have been able to see my sharks circling at the site; we were opposite it. The moon was waxing close to full, and the session promised to be an exciting one. I berated myself, after all this thinking, for having failed to come to any conclusions, or make any decision. And, as if in answer, the words came straight out of my subconscious as I gazed into the sapphire air above the site:

> *I'm reviewing the situation,*
> *I am in it now, and in it I shall stay.*
> *I shall simply take a shark stick,*
> *And teach them to behave the proper way.*

The idea of a shark stick is to place a rigid barrier between you and the shark. I visualized a short stick that I could easily manipulate, with a string looped through a hole in one end so that I could hang it from my wrist. If I felt threatened, I could take it swiftly into my hand, and point it at the problematic shark.

In hind sight, it was no solution to the situation I was in, since I would have needed thirty shark sticks and thirty arms to hold all my sharks off if they treated me as they had treated my kayak. But at the time, it seemed like the answer, a solution that I caught onto with intense relief and delight.

So on the following afternoon, I made a shark stick and set off.

Spooky flew after me and alighted on the boat; that day, she had stayed at home to rest. It was calm and sunny with a few light clouds, and in the lagoon, I found that the waves on the reef had risen. Spooky did not seem at peace as we progressed toward the site, and after I anchored and began to put on my gear, she stared downward as the sharks cruised by the boat, closer and closer to us. I was desperately worried for her. If she decided to play on the surface of the site, or anywhere in the area, with my sharks beneath—with Martha beneath—she would be happily inhaled at once. So I was relieved that she arose and, instead of circling the region, flew away in the direction of home.

Franck had taken pity on poor Kim, who had been put to bed early, and had taken him out for a swim on the bay. Spooky alighted on his kayak and she stayed with them for about twenty minutes, then flew home. Franck saw her sweep up into the arching trees where she liked to sleep.

I had decided that if the sharks attacked, I would simply wait, watch, and take photographs. I hadn't done so before because I had been so surprised and shocked that I had not been prepared. But this time I would. They would not be rewarded by having food raining down on them as a result of their bad behaviour. I would time the attack, and feed them when they stopped.

My worst-case scenario was that a lemon shark would be attracted to the scene by the submarine racket that my sharks were making, and come to join them. Lemon sharks were so enormous that they could easily wreak havoc with the kayak if they used their power on it. I would be dumped in the water. So I had fixed a string to the anchor rope. Should a lemon shark appear to be on the verge of battering the kayak, in a second, I could use the string to pull the anchor line to me, cut it, and depart. It was very unlikely, but I had been surprised often enough by lemon sharks. Their size, not to mention the size of their teeth, concerned me. I could count on no one for help, so had to prepare myself for all possibilities in advance. While I watched the sharks, there was a tiny utility knife, and the end of the string, attached to the kayak at my right hand.

But the sharks were slowly meandering around, and underneath the boat, and showed no signs of imminent attack, so I threw in the food. The wind was coming from different directions, and the boat was strongly affected by the streaming sea water. It restlessly waved on its line and eventually came to rest above the shark food.

Flora swept past at the surface to grab a piece of food dangling in the water, but she didn't touch the boat. I could see Madonna, Martha, Bratworst, Valentine, Cochita, and Brandy among the sharks. At times the water boiled as they took the pieces and coiled at high speed at the surface. I wanted to get in, but was afraid to jump in right there where the food was, and afraid to put my fins in the water to push the boat back. When I tried anyway, I couldn't get the boat to move, and became increasingly frantic. The sharks would be starting to leave! So when it seemed that not many were under the kayak, I hung my slate, camera, and shark stick from my wrist, slapped the surface hard with my fins to warn them that a truly powerful creature was about to descend, and jumped.

The shark stick rebounded like a Japanese fighting stick and hit me on the head so hard that I arrived amongst the sharks stunned and reeling with pain. The blow had caught me on the scar of an old but serious injury, and it had always been sensitive. I fought back waves of nausea and tried, with all the urgency of the situation, to continue to see the vast multitude of my sharks through the window of darkness that opened in my vision as I drifted against the large coral structure on the western end of the site.

The unbelievable pain coming on top of the extreme stress of the past minutes and days was overwhelming. It was so immediate, that I felt it as divine punishment for taking the stick, even for protection. I reached out and grasped the coral mountain, which was far from my usual lookout, and actually in the scent flow. Carrellina, Chevron, Sparkle, Sparkle Too, and all the others were passing before my eyes as I clung to consciousness. I lay watching, waiting for the pain to subside, but it dominated me for the session.

Nurse sharks were everywhere. There were three of more than three metres in length, five over two metres, and several between one and two metres long. Their presence in such numbers crowded and confused the scene. The visibility was bad with such current, and the clouds of sand raised by the nurse sharks caused my thirty sharks to appear and disappear mysteriously. Lying against the ancient coral structure, everything seemed unreal.

After a long time, I drifted down-current for my usual check on the sharks hiding behind the blue curtain, and found Twilight. Most of the resident sharks had left, and just the visitors were roaming the area and looking for scraps dropped in the coral. I began trying to feed the fish, molested by approaching nurse sharks and my own nervousness. Cinnamon suddenly appeared in the site, then swam away westward. She had passed by that way before, going somewhere, maybe to the border or out to sea. Darcy arrived as he always did, when the light began fading about ten or fifteen minutes before sunset. Samaria still roamed lazily through the region.

Sparkle came and made several aggressive approaches toward me, while I brandished the shark stick and splashed the surface in front of her. This had no effect on her and complicated my efforts to feed the fish.

The boat had steadied in the conflicting winds and current in the middle of the site. Usually, I made sure that it was well out of the way, and accessible up-current, but on this occasion, I just left it there. Whenever I went to it to get a fish head, many nurse sharks rose up toward me, and some investigated the boat. The crocodile needlefish was circling my head or swimming beside me. The scene was so confused with the opaque rivers of fish before my eyes, the dimly glimpsed writhings of the nurse sharks, and Sparkle's sudden dashes up to me out of the mêlée, that I often drifted on the periphery. When I tried to feed the crocodile needlefish, it seemed that each time I retrieved another fish head, the current got so bad that I couldn't maintain myself, nor could I see through the streaming sea, so I started to panic and threw the fish head away. When I did manage to extract a handful of food for the needlefish, I couldn't see her. Once when I threw some bits to her, where she was hiding behind me, the current carried them into my face, so she wasn't able to get any.

While I was feeding the fish, watching for approaching sharks, a nurse shark suddenly appeared over the coral edge and stuck his face into mine. He had clambered up the other side of my coral lookout. He couldn't maintain himself, though, and soon fell away. I leaned over and dropped the fish head after him so he would have something to eat, but someone else must have taken it from him, because soon he appeared again. In between his efforts to steal my scrap, Apricot repeatedly zoomed up to my nose, and ruined my efforts to photograph his funny face as he popped up hopefully, opposite me. But both of them were much too close to photograph anyway.

Throughout the session, there were sharks coming up to me in long lines at the surface, led by one of the bold sharks, especially Tamarack, who was developing the habit of persistently orbiting me. Fermat, the beautiful golden Antigone, Annaloo, and Fleur de Lis cruised through from time to time. When I had fed the fish, and the sun had set, I put the anchor in the boat and was swept away with the current before I could even get in.

The pain in my head was even worse once I was back to safety, drifting away to begin the long trip home in the falling night. I thought about the shark stick. Even after I had organized myself underwater, it had continuously entangled itself on its string with my slate and camera, which were also attached to my wrist, making the use of any of those things problematic, and quite impossible to use swiftly. It was like a bad joke. As I paddled, I mused about other sources of aggression for which it would be far more useful and was soon laughing at the image of myself proceeding through my day, my shark stick dangling easily from my wrist, and, whenever I detected a lack of respect, delivering a bonk on the nose.

I never used the shark stick again.

The restless lagoon heralded approaching storms. The next morning, Kim had to stay inside on his bed of netting while I was out buying food, and trying to get more small fish for him and Spooky. It was increasingly difficult since the Polynesians wanted to keep their little fish, and not sell them for seabirds. Just since I had begun rehabilitating seabirds a few years before, these small fish had been nearly fished out.

In the afternoon, Kim rested quietly on his pool, and I took him out at five o'clock as usual. There was a storm across the reef, and I kept an eye on it, but the surface was the image of tranquillity. Suddenly, tiny riffles appeared on the surface, fanning out as they spread around us. When I caught the force of the wind, I drifted in front of Kim, letting it push me toward him, and snatched him into the boat when he was close enough. Within a minute of that first ruffling of the water, the wind had become so violent that Kim was blown against my leg, and the water was whipped into foaming waves as I paddled at top speed to shore, which was luckily not too far away. By the time we reached it, torrential rain was driving horizontally down the bay. I rushed the boat up the beach at the last minute before I would have lost control of it. It was a sobering reminder of how fast conditions could get dangerous.

A few days later, we went out after a rainstorm when conditions cleared near sunset. Spooky was perched on the bow, while Kim drifted and preened on water the colour of a black pearl. Toward the ocean, curtains of rain were lit by a golden glow from the west, and over the mountain, rose-coloured clouds floated in a sky of ringing blue. We drifted silently, hardly moving in the surreal surroundings. Spooky asked for a fish, and just played with it on the bow, often dropping it in the water. Strangely, this one floated. I retrieved it and was handing it to her when I heard voices approaching through the mist. Two young Polynesian men were paddling their outrigger toward us. They were looking at us, and said hello.

The black petrel on the water stood out so dramatically that I was sure they had seen him. But they headed straight for Kim and ran over him! He started to fly across the water, but his wing was caught under the braces of the outrigger. With a desperate effort he freed himself, and swam toward me, as the speeding boat disappeared into the mist, the men looking back and laughing. Kim was all right, but badly frightened, and hurt a bit too. Spooky ate her fish and flew home. For the next few days she stayed home, and played with Kim when we went out, as she used to do.

Another evening when Kim and I were out near the ocean, Spooky materialized out of the distance and took a fish from me. She didn't seem too tired, and played with Kim as he swam and drifted, diving, and turning somersaults. He didn't want to go home, but once settled on the boat, watched the full moon rising above the mountain. Spooky perched on the tip of the bow like a little figurehead, watching the water passing. She ignored him. Back at home she ate, then asked for more fish, but didn't eat them. She kept looking at me. Not knowing what she wanted, I talked to her gently, and reached my hand toward her. She touched it with her beak and listened, looking intently into my face as if comforted. Sometimes birds have bad days.

At the end of the week I picked up the scraps from the fish shop, prepared, and set off for the lagoon. The weather was calm and sunny with a light east wind and a few clouds. Spooky was too tired to come with me.

I anchored correctly for once—the kayak came to rest over my lookout. Though no scent was leaking, many sharks had followed the boat from far across the lagoon. While I put on my gear, they zoomed around the site, obviously excited. I threw the food, upon which they pounced and soared away, chasing each other through the coral with their trailing scraps—I took advantage of their absence to jump in. It was a relief to find that the current had slowed somewhat.

Though the bad sharks, Carrellina, Sparkle, Chevron, and Sparkle Too, were there, along with Martha, Madonna and the other residents, the session proceeded tranquilly.

I was feeding the fish, and as the light faded, Twilight appeared. Fermat came, and the circling sharks were joined by Keeta, Marco, and Brandy. Fermat followed Keeta around the site as she circled the nurse sharks, who were crunching the fish heads I had thrown in for them. They munched the eyes first, creating startling sounds of underwater thunder, daunting proof of the power of their bites. This released a yummy odour of fat into the scent flow, which worked wonders in distracting the sharks from my activities. Keeta passed close to my face each time she circled, then she charged in and tried to take a fish head. Arcangela appeared and joined her in circling and badgering the nurse sharks.

Once, my leg touched something as I went to the boat for a fish head. It was a nurse shark swimming with me like a remora, having learned that when I went to the boat, it was to get food. I noticed that this same individual kept rising to sniff the boat. Watching them there, and reflecting upon their behaviour, I was convinced that the nurse sharks' poking and nosing around me was not something to worry about.

I received scraps unexpectedly in the middle of the week. Kim wanted to rest inside, and Spooky was off alone, so I felt free to leave. The day had been sunny with a strong east wind, which quietened slowly as I struggled out to the lagoon against it. A strong wind on the way out was never a good beginning to a session. The soft but unrelenting force that had to be fought for every metre, deprived me of pleasure on the trip out. Manoeuvring the kayak on arrival at the site and finding the precise place to put the anchor in the tossing water was hard. I dared not mark the site in any way, and in spite of the numbers of times I had gone there, there were occasions when the water level and waves were so high that I couldn't seem to find it. Sharks cruised beside the boat in greater and greater numbers while I turned the heavy kayak back and forth searching, becoming more and more frantic.

The large coral formation at the west end of the site was my main landmark, with the matching one at the east end just the right distance away. So as I crossed the lagoon, I took the right direction automatically and watched for the twin places where there was something just beneath the surface. When the weather was calm the two landmarks were easy to see; when the water level was low, they actually showed above the surface. When the sea was rough, waves broke over the westerly one, which made it easy to pick out, but since there were many other coral structures like these, sometimes I picked the wrong one, and ended up lost. Once I had missed them, they were very hard to relocate among all the others. But I could double-check my position by lining up the shipping marker at the border with a precise place on the opposite hillside to the west, at the same time that the red roof of a house on the nearer shore lined up with a particular mountain peak to the south. Then I could look for the oval of sand of the site, with my lookout perch providing an unusual configuration I could recognize, not far beneath the surface. If I could not see the mountains, I was running on instinct.

Madonna swam back and forth under the boat as I found the site and anchored, and many sharks quickly gathered, but with the strong wind, I couldn't see much through the surface, though briefly Apricot's dorsal fin appeared above it.

I threw in the food, and slid underwater. I was hoping things would get back to normal. The times that the sharks had attacked the boat now appeared to be just incidental, but still, I dared not return to my old habit of sliding in, and pushing the food into the water, with sharks thick around me.

The string attached to the anchor cord had turned out to be very useful for moving the boat after anchoring. I could pull it to move myself toward the site to throw the scraps in, then let go of it to move away, before going underwater. Each time, I carefully gauged the wind and current in order to anchor in the correct position, so that this would work out right. The crucial moment was my arrival with the food. Placing it uneventfully, and getting safely underwater, were the immediate priorities.

The usual sharks were there, with the exception of Martha, whom I guessed had gone off to mate. It was the full moon, and it was her time. Odyssey and Samurai soon soared in, followed by Arcangela. The session passed normally, and I cruised freely with the crocodile needlefish, watching the sharks and sketching new visitors.

As the scene darkened, Lillith came. She had developed a rash of white dots, as if she had been sprinkled with tiny snowflakes. These were most dense over her gills. On both sides, they formed an oval shape about twenty-five centimetres (10 in) long. There were a few larger snowflakes on her sides, too. In the past year, she had developed a diamond-shaped white mark about one centimetre (0.4 in) across, just above her left pectoral fin, and I wondered if it were a sign of age. Her colour appeared paler, too, though usually she remained dark, while the other sharks grew lighter in the reproductive season. Indeed, Lillith looked old.

The fish heads lay on the floor of the site, and as the nurse sharks began to devour them, the scent brought Carrellina, Valentine, and Lightning back. Christobel suddenly arrived, and started making close passes at me, even when I wasn't feeding the fish. Tamarack had been following me around for much of the session. Now he began making swift, close approaches, which felt like an attempt at intimidation. Bratworst and Glammer began to charge me! Glammer even made a swift circle close around me as I drifted down-current, yet she had always been so shy! I began to feel uncomfortable. I wanted to get a last meal to Spooky before nightfall, and left.

A few days later I went again with the usual weekend load of scraps. This time, the weather was calm and sunny with little current. Spooky alighted on the boat just before I got to the site, ate, and flew away. I never went anywhere on the kayak without taking two or three little fish for her. As I prepared, I kept an eye on the behaviour of the sharks. Mara'amu, Avogadro, Mordred, and Gwendolyn were roaming placidly; I tossed in some of the scraps, and slid in to find the sharks calmly eating. I wrote down the names of those present, and except for Gwendolyn, there were only males and juveniles. The females, it seemed, were off somewhere celebrating the full moon. I was even more certain that Martha had left to mate.

Everyone ate and left, just like in the good old times, and I found myself alone with Shilly. I followed her down-current with a scrap I had kept apart, and threw it for her. She startled. As she shot away, she performed several vertical jerks, as she had done when the nurse shark had frightened her. On her return, she had shivers running through her as if she were made of water and, from time to time, someone disturbed it.

Pippet had acted like that once before. It was an extraordinary performance—it had to be seen to be believed! At least I had an indication that this rare behaviour pattern was connected with a fear reaction. It made sense to consider that if the little shark had been frightened by a predator, this unpredictable twisting would be an excellent survival strategy. I wondered if the grey reef shark's famous agonistic display was also based on fear, rather than aggression, as had been assumed. Given the anti-shark bias in the literature, it was certainly possible. I threw Shilly another bit of food as she crossed the site, and this time, she curvetted up to take it as it fell.

It was time to feed the fish, and as I took a fish head from the boat, Apricot came. I did not like her. She was sneaky, always surprising me from behind, and there were times when she refused to leave me alone. I didn't like the way she dodged and accelerated when she saw that I had seen her. She had always been like that, and now that the change had come over the sharks and she was growing up, she was one of the worst of the bad sharks.

I fed the fish sporadically, because whenever I did so, Apricot came shooting through them. Marco followed behind her much of the time, but he paid no attention to me. Just to keep Apricot in view, I had to circle her as she circled me, because of her insistent efforts to come up behind me.

A big female came and began searching confidently for a scrap. I drew her fin on both sides, trying to remember where I had seen her before. Finally, it came to me—Cheyenne! It was the first time I had seen a female shark identified at Site Three, in the middle of the lagoon, in my study area.

The males, Blade and Fermat, arrived as dusk fell, and Christobel came charging boldly up. I put out my hand and pushed her away. I suspected that she had been involved in the attacks on the boat. She had never been a pest before then, but she was the same age as the other juvenile delinquents. I decided to add her to my list of bad sharks.

Cinnamon passed, ghostlike, to my right, around the big western coral formation, as I had seen her do time after time without returning. I followed her, and confirmed that she was heading straight to the border. When I got back, Cheyenne was still cruising around, and the site was a wild uproar, with wriggling, writhing, and undulating nurse sharks encircled by blackfins, and clouds of darting fish. I watched in a spell of wonderment until darkness occluded the scene.

As I paddled home, Spooky materialized in the air beside my ear. It was nearly dark. After checking to see if there were fish in my hands, she snatched one that I had put on the bow for her, and dropped it in the water as she flew on ahead of me. At home, she fluttered out to meet me, and I fed her on the beach when I landed.

By then, there was just enough light in the sky to see her. When she had eaten, she circled once over the bay to drink from the surface, then made repeated attempts to alight on her sleeping perch, fluttering around the beach foliage, apparently looking for the way in. A spectacular electrical storm out on the ocean began shooting fireworks. The clouds flickered continuously, and forked lightning shot from on high in crazy patterns into the sea I had just vacated, intermittently bathing the scene with cold, violet light.

Spooky managed to gain her sleeping perch deep in the shadows of the tree by using her routine approach over the water, flying up toward her branch, then hovering in about the right place until her feet touched it. I had the impression that she landed at night when she couldn't see, much the same way that we go around the house in the dark without the lights on—by memory and then by feel, using just the vaguest of visual clues to stay oriented.

CHAPTER THIRTY-FIVE

The Good and the Sad

The following week, I left for the lagoon feeling more confident about the shark situation than I had for many weeks. However, the waves on the reef had mounted to the dreaded level of a charge of white horses, rearing and leaping over the reef. The ocean swell surged through the lagoon, so it was out of the question to go to the usual site, and I anchored at the one I used when the current was strong. For once, I accurately found the tall, dead coral formation, which made such an excellent lookout. I didn't see many sharks through the streaming surface as I prepared, but recognized Bratworst when she rose to the boat.

When I was ready, I threw the food overboard, then slid underwater. The visibility was appalling, and swarms of sharks swirled around the food, vanishing and reappearing in the submarine fog. The first shark I saw was Marco. In order of appearance, I saw Chevron, Sparkle, Carrellina, Madonna, Ruffles, Shimmy, Sparkle Too, Pip, Darcy, Flora, Teardrop, Breezy, Gwendolyn, Shilly, Mara'amu, Lightning, Brandy, Christobel, Peri, Eden, Jem, Avogadro, Mordred, Samaria, Apricot, Muffin, Pippet, Tamarack, and Simmaron feeding. I was waiting for Martha.

Indeed, she soon appeared, having mated over the full moon as I had surmised. The lunar cycle had peaked the week before, and Martha's mating wounds suggested that she had mated a week before too—they were outlined in black, yet the smaller ones had not healed.

There was a new small female who had never appeared before; I drew her fin and named her Cygnet, then moved on to draw the fins of other new juveniles with her. The current was so powerful that I had trouble maintaining myself against it; lifting my hand to draw was a challenge. The extra effort required to navigate was evident in the rapid tail beats of the sharks as they slithered through the torrents between the coral formations. The fish sheltered inside. A tiny nurse shark had taken refuge behind my tall lookout. The fat little thing was the size and colour of a human baby with a frilly tail.

Martha approached me one time very lazily. I swam to the kayak and got the piece of fish I had brought for her, to welcome her back after mating, but when I returned to give it to her, she had left—she didn't come back! I was confident that everything was returning to normal, now that the period of reproduction was ending. Once Martha had mated, it was over. March had passed and April had come.

More sharks glided into view. Twilight and Arcangela were swimming together. My notes showed that an association seemed to exist between them, since Arcangela had appeared in the area that year. Whenever one of them came, the other did, too, and this appearance swimming side by side reinforced the impression.

Clementine appeared through the gloom. Her right pectoral fin was folded downward, near the centre, at more than a right angle—the outer part curved under her body. The injury had happened since I had last seen her during the past year. It was too dark to try for a photograph, particularly in such poor visibility in which everything was disappearing into shadows. She roamed around near me with Twilight and Arcangela after the others had gone, joined finally by Odyssey, who had come to the area while the moon was still waxing. Odyssey and Clementine had both visited at this time the year before, yet I had not seen either of them since. Cochita and Windy came just before I left. Windy was quite pale. She had grown to the size of Madonna and resembled her, being of the same build and colour.

I gave the usual treats to the crocodile needlefish, but even she was stressed by the current and did not remain beside me. By the time she left, I was too tired to hold on, and one leg was badly cramping from trying to keep myself balanced in the surge. So I caught up the anchor, and flew away.

During the week I was given another load of scraps, and though I wasn't expecting so many sharks, since they had already eaten, I returned in hopes of seeing Clementine again and photographing her with her injured fin. I hoped to see Kimberley, too, with Twilight in the area. I was becoming worried about her, since Twilight had been recently visiting with Arcangela, and Kimberley had been absent.

The tendency of sharks to roam with companions was one of the things I was trying to collect as much data on as possible—Arthur never failed to ask me how my graphs and tables were progressing when he wrote. The return of Odyssey and Clementine to the area precisely a year after their initial visit, was intriguing, and I wanted to investigate further.

It was sunny and calm, and mercifully, the current had slowed. Just one or two sharks were present when I arrived at the site, but by the time I got ready, there were more. One swam fast over my lookout, and another accelerated near, so I decided at the last minute to throw in the food, rather than returning to my former procedure of going underwater first, and greeting the waiting sharks before feeding them. Since they had calmed down so much, I was more and more determined to return to my former routine with them. But each time I went, there was some reason why I decided not to do it that time. And the longer I waited, the more they expected to eat immediately on my arrival, rather than waiting for me to feed them.

Underwater, the residents were eating, with some visitors among them. The waters were crystalline, and shafts of sunlight played upon my graceful sharks. When the food was gone, Martha, Chevron, Carrellina, Sparkle Too, and later, Christobel, kept circling around the site and me. Keeta arrived and joined them, and my bad sharks switched to slow charges in which they turned away just at my face. Martha was nosing around me. Samaria kept coming by me very closely and slowly, too, which she had never done before.

But later they left, and I was able to feed the fish sporadically from the fish heads I had, alternating with explorations down-current. Fleur de Lis came cruising unobtrusively, and the new little female called Cygnet did too. Eventually, Flora glided in up the scent flow and circled purposefully.

On returning from one of my tours down-current, I scattered some crumbs for the fish, and the site suddenly filled with sharks again. Carrellina was in the lead. Their bold approaches intensified. Mordred, and a slightly smaller male I had recently identified, Brambling, began scouting the site, which had become an alien spectacle of the lashings and writhings of nurse sharks, tearing fish heads apart.

Annaloo was passing down-current the next time I cruised around, after feeding the meat from another fish head, and as I returned to the kayak, Carrellina passed me. I watched her closely, and her eye looked strange. I was reminded of the way a horse will show the whites of its eyes in emotional moments. I didn't know if that meant anything or not, but I had never seen that look before. A moment later, she was at my elbow, approaching from the side, where I didn't expect her. She veered away, but after that she steadily harassed me, with swift approaches from behind, alternating with tight circling around my head. When Apricot joined her, dodging at light speed when I put out a hand to push her away, I decided to go.

Back in the kayak afloat on the magical lagoon, I felt a wave of relief, and drifted off, slowly arranging things on the boat, while the surface shimmered darkly with the colours of the west, and Carrellina circled, a mysterious shadow of the sea. She came repeatedly to the surface, and Apricot's fin appeared, too. Their behaviour was different than that of the other sharks. They habitually approached from behind, dodged when I avoided a charge, and charged me much too fast. Martha, Madonna, Flora, Valentine, Samaria, and the other sharks who came close to me, did so in a leisurely manner, swimming to me as I watched them, so even if they were behaving aggressively, as Madonna had done sometimes, it was an honest face-to-face gesture, not a sneaky approach.

Kim had stopped playing with things he found floating, and no longer did any diving, somersaulting, or vigorous bathing. It was more and more difficult to get him to stay in his pool. I was always alert for signs of trouble in this bird, so far from his true place. I had only managed to keep his legs free of sores by regularly wrapping them, and was constantly on the watch to see that he walked and jumped onto hard surfaces as little as possible. Though he was fat, with glossy feathers and no sores, apparently in excellent health, he had vomited a few times, and I finally consulted with a vet about a wormer for him, when this symptom got worse. I was concerned that this oceanic bird could be exposed to parasites that he was not equipped to handle, as a result of drinking tap water, and eating fish species he would not normally catch for himself. It was a relief when the vomiting stopped after I treated him.

In his pool, he watched the birds coming to the feeding table. When I appeared, the hundreds of birds flew up, and he gazed at them, and watched to see what I would do. Even when waiting at the door, he watched the birds, as they all flew up at once when I opened it for him. When they startled, he startled too.

One windy afternoon on the bay, when Spooky was not with us, Kim was preening as he drifted, becoming more and more energetic. Eventually, he began somersaulting, diving and leaping into the air. For the first time, he seemed to be having fun all by himself. The wind strengthened, and he started trying to fly, and managed to glide a bit, once going quite far. It was a long time since he had tried to fly. Spooky arrived and spent some time with us, overflying Kim and alighting on the water near him, so for a little while, it was like old times. I kept him out as long as possible, since he was having so much fun, and when I brought him into the boat, he clambered up me for the first time, and sat on my shoulder, to take a leap into the wind. I gave him a few more minutes to play before taking him home.

Spooky was at home less and less, but arrived at odd times to demand a fish. Once, I found her flying around my head when I got out of our pick-up truck at the end of the garden. Even though the vehicle and road were far from her trees, she had somehow learned that I could be in the truck!

Once, while Kim and I were floating on the bay, I saw three birds circling so high up that they were almost invisible in the blue. They wheeled together like eagles, but they weren't, and as I watched, one of them gradually descended and turned out to be Spooky, who alighted on the boat. It was the only time I saw her with friends, though other noddy terns I had looked after, had routinely brought friends home.

Another morning, we went on our usual jaunt, with a tired Spooky resting on the bow as Kim enjoyed himself. When we returned to the beach, I spent a few minutes with her, talking to her and feeding her until she flew up to her perch, while Kim waited patiently on the beach beside us for me to carry him to his pool. When he glided away on the surface, a piece of tapeworm drifted down from him. Instantly worried again, I gave him a dose of the worming medicine the vet had prescribed for tapeworms, and left him on his pool.

When I checked him again, he was very active, energetically biting up some pieces of aloe vera that I gave him to play with. Then he jumped off, and came inside to lie on his cushion. He seemed strangely agitated, and restlessly yanked on the books on a nearby shelf with increasing force, managing to badly damage one. Soon he began to stagger around, and tried to climb the shelves, yet his wings were falling loose from his sides. I realized that he was reacting to the wormer, and put him on his pool, where he could drink, but he couldn't maintain himself on the water and struggled frantically back out. So I returned him to the cushions and prepared a dose of medicinal charcoal, which I gave him with enough water to fill his crop. His agitation rapidly increased to a peak of intensity, and finally, he began to beat his wings hard where he lay.

I picked him up and felt a strange loose weakness in him, in spite of his muscular activity. When I put him down, he beat his wings on the floor, his legs out straight behind him. Throwing concern about his waterproof feathers to the winds, I put my arms around him and carried him to the couch, trying to comfort him. He crawled up under my body, rubbing his head against me as I stroked him, and tried to reassure him. He alternately flapped wildly and relaxed, seeming to be comforted by being held. I gave him more water, and when I pushed the tube down his throat, he opened it to the maximum, in a great spasm. His mouth was open and each breath came in fast, in two stages, one briefer than the other. The water seemed to calm him, and he lay quietly while I held him, his face against mine. I carried him outside, and he settled on my lap as I sat on the grass beneath Angel's big tree, stroking him. He was quieter; there was just the odd flickering of his wings. He lay in my lap, his head resting on my arm, and I could see his eye following the flight of some birds across the sky. The wind was blowing hard, and there was sun coming through the leaves. He could see the waves sparkling on the bay. He's calming down, I thought. It's passing. But then I noticed how awfully shallow his breathing was. I could scarcely see it. And then I couldn't see it. I held him up, and his head stayed in the same place, his eye open, his body and feathers so soft, yet frozen there. He was dead, incomprehensively dead.

I just couldn't have imagined that a wormer could kill him.

It was dusk when Spooky returned, and found me forgetful in the house, rather than on the kayak with her friend. She followed her usual routine the next day, but except to feed her, I did not go out.

The following morning, I decided I must make an effort for her—my anguish over Kim's death should not affect her. She was perched in the arching trees above, as I prepared the kayak to take her out briefly, and when I stood up, a noddy tern was flying straight into my face, veering away, and coming at me again. Still devastated by Kim's death, I was thinking very slowly, and thought that the bird was Spooky. But suddenly, there were two noddies flying at me. One took a fish from my hand, but the other continued to fly fast into my face, eerily reminding me of Carrellina's aggressive demands. She ignored the fish I held out to her. Finally I threw it to her. She deftly caught it, flew out over the bay, and dropped it into the water. Spooky came, and took another fish, then retreated to the rocks where she liked to sit in the morning sun.

I threw another fish to the newcomer when she came around again, and she easily plucked it from the air. Then she carried it over the water and dropped this one too! One by one, she caught and dropped all the precious fish I had brought for Spooky in the bay!

But why? She obviously wanted food. Could she not eat them? Mystified, I went inside to get some small fragments of fish meat, and the new bird caught and swallowed some of these. Then she kicked Spooky off the rocks, and stood there regally. Spooky flew to her sleeping perch.

I wanted to be with Spooky, and launched the kayak, hoping she would fly to it, but she did not. So, I drifted, watching the newcomer. She was thin and pale, and I realized, as the boat floated past her, giving me an all-around view of her poised on the rock, that this was Princess, a bird I had looked after for a few weeks two years before. I had not seen her in all that intervening time, yet here she was, sick. She had remembered me, and come to ask for help.

It was remarkable, almost impossible to believe. It seemed to be a sign that even though I had killed the petrel I had loved so dearly, goodness that I had done, love, concern, and care I had given in the past, was remembered by this bird. Maybe, my love and care were remembered by other birds too.

I put away the kayak, and went to sit with her. Her trust in me was unchanged. We sat side by side, and though wild birds will avoid being picked up, I was close enough to observe her closely, and finally to see that the inside of her mouth looked unhealthy and that her throat was swollen and red. No wonder she couldn't swallow fish. I prepared a dose of antibiotics for her, secreted in a bit of fish, went back to her, and gave her the medicine.

Later, Princess left for a while, and I gave Spooky another fish. Then she flew away. In the late afternoon, I saw the two birds flying in front of the house together and witnessed no conflict except for the moment when Princess took over the rocks. She had habitually used those rocks when she had been with me originally—ever since, I had privately thought of them as Princess' rocks.

Spooky returned at nightfall, ate as usual from my hand, and flew to her sleeping perch. Princess had already gone to sleep in her favourite coconut tree, so I was sure that she had not interfered with Spooky.

Spooky was gone when I got up at dawn and never returned. Noddy terns generally flew at the first lightening of the sky, before they were even visible from the house, so I did not know if Princess had driven her away. Princess was in her usual position, standing regally on her rocks.

I missed and worried about Spooky, looking for her each time I went out on the kayak. But I never saw her again, and there was nothing I could do. Princess resumed her life on the property, alternating between her rocks, and her coconut tree, while I gave her antibiotics and vitamins in her food. She soon recovered, and when the noddies left the island for the season on their yearly migration, she was strong enough to go, too.

CHAPTER THIRTY-SIX

The End of the Mating Season

Visits to the sharks had always soothed my grief over the latest dead bird, and I soon picked up some scraps and went to see them. It was impossible to be sad while watching sharks zooming around me—I wasn't responsible for them, and they were always there. Apart from the few frights they had given me, I habitually returned from seeing them feeling refreshed and happy.

The evening was calm, but clouds in the west obscured the sun. Many sharks joined the boat as I crossed the lagoon, including Carrellina, Cochita, Sparkle Too, Martha, and Bratworst. At the site, Flora began circling the kayak clockwise at the surface, as was her habit, and Apricot and Fawn repeatedly swam toward me as I prepared, as if they could see me above the surface. Then Apricot glided back and forth beneath, and I suspected that the little brat was thinking of slamming the kayak again.

I threw in the food and slid underwater, where I found the magnificent Odyssey soaring by. Cinnamon was near. Valentine had returned from her yearly excursions during the mating season, and looked as if she had mated over the full moon, as Martha had; their mating wounds were at the same stage of healing.

The food was quickly gone, and I was at the lookout to the left of my regular one, where the visibility was usually better, but which was harder to hold onto. Martha kept approaching me, time after time after time, circling, coming to the surface, approaching me, passing, swimming close underneath me, and doing it all over again. The other sharks were turning in the coral, just cruising around, nearly forty sharks altogether.

As twilight deepened, Lillith came. Her sprinkling of white marks seemed to have mostly disappeared, leaving just a line of snowflakes on the right side over her gills, and a bit of light mottling on her midline near her tail; the bright white diamond above her pectoral fin remained.

Dapples and Marco arrived as the rest of the sharks moved on. Their presence was an indication that Kimberley should be in the vicinity, and I was always pleased to see Marco, whom I had known since I started, and who came so rarely. The scene was gloomy due to the heavy clouds, but had settled into the tranquillity that I had used to enjoy after my resident sharks had finished before their strange change in behaviour—when they left, instead of roaming watchfully near me, when I could take pleasure in feeding the fish, while the sharks just roamed the region, poking through the coral for scraps. The usual horse-sized nurse sharks were there, but I was used to having a carpet of gigantic fish in the site by then. Indeed, I had come to look upon them as a blessing in spite of the murk they raised with their writhings. They created scent flows which continued to attract passing blackfins, while distracting the bad sharks as I fed the fish.

I cruised down-current for a look around. Indeed, the sharks appeared to have left. I found a head of a *saumon des dieux* that had been swept part-way beneath a coral ledge, and moved it thirty centimetres (1 ft) back out. The crocodile needlefish was swimming beside me like a little dog, and began circling my head. So I went to the boat for some food for her. She stayed so close that I just wafted the treats into her mouth.

Clementine soared in, grabbed the head of the *saumon des dieux*, and shook it. Instantaneously, several tons of nurse sharks converged on this tiny scrap, while my forty blackfins shot into the site, and dove onto the fish head, then orbited at top speed as the nurse sharks rose vertically, tails thrashing the surface, in their urgent efforts to scrape out a crumb. I drifted backward, up-current from the site, aghast at the turmoil caused by just that one movement of the fish head from under the coral! The scrap had been lying there ignored, all during the session!

It was impossible to feed the fish. Keeta circled fast, passing each time in front of my face. The shadowy green waters, half obscured with drifting clouds, were shot through with speeding sharks, orbiting the centre like stars around a black hole. There were so many present that there was not enough space; multitudes were zooming past me as if I weren't there, sharks known and unknown. The air of seriousness among them began to feel menacing, quite apart from the danger of collisions. Gazing, riveted, I felt I was looking upon a scene of utter madness, that I was an alien, fragile observer, hopelessly apart from it. Never, anywhere, even on television or in films, had I seen such a spectacle. As the shadows of night enveloped the scene, I drifted back to my boat, climbed in, threw in food for the squirrel fish and tossed the fish heads overboard. In scraping the meat from the heads, the nurse sharks would feed the fish for me. Luckily, I had already fed my crocodile needlefish; she had retreated and vanished in the tumult.

A stormy week followed, but on the usual day of my shark visit, the clouds began to lighten and clear away. I paddled out upon a placid sea, but once more, a thick bank of clouds in the west darkened the session.

As I put on my gear, I watched the sharks. Apricot was the first to come beneath the kayak, followed by the rest of the bad sharks, who coiled around the boat at the surface. Martha, Madonna, Flora, Valentine, Ruffles, Cochita, Windy, Teardrop, and the usual juveniles appeared with them. When I started throwing in the food, the surface boiled, then the sharks shot straight up towards me as if they could see me there above, peering in from the other side of the surface! They began slamming the boat with the same power as before, and leaping from the water as though they thought there must be more food I hadn't given them. But their food was already in the water! It was the evening of the full moon, a small point I had overlooked with all the things that had to be done before I could leave, and this time, the bad sharks had not gone off roaming, flirting, mating, or whatever it was that they liked to do when the moon was full.

When they lost interest in the boat, and descended to look for scraps below, I slid underwater. No food was left in the site, and everyone was roaming in confusion through the gloom. Undulating in mid-water, the nurse sharks looked far more imposing than when they were lying about on the sand. They filled the site with their volume, all in slowly rippling and chaotic motion, amidst the flashing fish. Gwendolyn, Rumcake, Shilly, Apricot, Mara'amu, Eden, and Trillium zoomed among them, and Carrellina flew to meet me.

She did not leave me alone throughout the session, and her repeated close approaches made it difficult to watch or follow any other shark. She was usually in the company of Bratworst and Sparkle Too, but they did not charge me repeatedly as she did. When I was on my usual perch, her harassment was the worst, and when she came to the end of her repertoire, she just circled around my head. What did she want? She came at me from different directions, and being smaller than the large females I had always been closest to, her speed was alarming. I tried holding the kayak between us and banging on it, but that had no effect on her. Neither did splashing my hand on the surface in front of her nose. I was afraid to push her physically because she was so swift, and she dodged faster than my eye could follow her.

On one of my tours down-current, Shilly came and closely circled around me, joined by Ondine, a shark of the same age who was her companion. This time, a third juvenile was with them whom I did not know. Carrellina had circled briefly away, but soon returned with Bratworst and Sparkle Too to see what I was doing. Strangely, they were not such a nuisance down-current.

Lillith began to cruise the region, and Cinnamon appeared and swam to the site with me. Darcy had come on schedule, and Keeta cruised in as the twilight surroundings darkened with the approach of sunset. Back on my perch, the bad sharks, as well as Cinnamon, Shilly, and her friends, were circling me tightly, as if they still thought I would produce more food. Long lines of sharks rose to the surface from time to time down-current from the boat, to see if they could smell anything. But I had no food left this time.

The crocodile needlefish had appeared, but I had to make her wait, given the mood of the sharks.

The biggest of the large nurse sharks was approaching, a pale blimp bobbing in slow-motion through the coral passageways, his pennant tail flowing far behind. He undulated into the site which was already filled with nurse sharks. I threw in a few of the fish heads. Everyone got excited again as the nurse sharks fell on them, and having succeeded in distracting the sharks, I was able to throw a few pieces of food to the needlefish who waited behind me.

She could tell from my body language the moment that I would throw her the food, and waited poised, her huge eyes fixed on me, anticipating me with the sensitivity of a bird. When the food splashed down near her, she accelerated straight to it, and carried it away up-current to eat it.

I held the kayak in front of my forehead so that the sharks had to approach me under it, though this tactic made little sense now that they had developed the habit of attacking it. The fish had come to swarm around me, which made the sharks suspicious, especially Carrellina and her buddies. I scattered a bit of food for the fish around my lookout, then swam around the site, and down-current, to avoid the sharks who soared up to me. There, Arcangela and Twilight were passing by.

As I fed the last of the fish heads, Marco passed, and I took a last excursion down-current in case more visitors, especially Kimberley, had come, though it was nearly too dark to see. Odyssey was roaming there, with Emerald, and a large female I didn't recognize. The crocodile needlefish drifted past my eyes again. I found one more piece of fish, tossed it to her, and climbed back in the kayak.

Heavy clouds filled the sky on the day of my next session, following violent storms the evening before. I watched and worried all day and finally left early, while it was possible to go at all. Heavy rain threatened to fall at any time and I was always afraid of the wind. It remained calm, but rain came while I was underwater, uniformly as far as I could see in all directions. It was dark, the visibility was poor, and I didn't have much food.

No sharks followed until I was almost at the site. Only then, when I looked back, did a dorsal fin show above the surface. I anchored, and Apricot again cruised slowly back and forth beneath the boat. Martha was circling, and I identified Cochita, Carrellina, and Mordred prior to throwing in the food.

Underwater, the usual group was feeding. Carrellina ate and left. The very young female I had watched in Section Two, Cotlet, joined them, coming to the site for the first time. Annaloo, Arcangela, and Storm cruised slowly in the vicinity. As the time passed, Kilmeny and Marco drifted in, and, a little later, Jessica. Sparkle Too stayed for the entire session, as had always been her habit, frequently passing me close by. Cygnet was there too. Teardrop came late, and looked anxiously for something to eat. Even though I had brought so little food, there was some for her. The sharks left, I fed the crocodile needlefish well, and was able to enjoy feeding my clouds of excited fish without harassment.

The nurse sharks filled the site and were restless because they got so little to eat. But generally, I felt things were going better, and I left feeling relaxed and pleased as I paddled home in the cold rain.

The next session was similar. There was some atmospheric turbulence blowing rain over the lagoon, but the waves coming over the reef were low, there was little current, and the visibility was all right. I threw in the food and got in with no problems. With the ruffled dark water it had not been possible to see any sharks through the surface, but underneath, the usual sharks were feeding calmly, with a few visitors, for a total of thirty-three sharks. The huge Odyssey, little Cotlet, then Darcy, Shilly, and Annaloo, came later as was usual for them. Martha cruised casually around with me. Once she spiralled vertically straight up past my mask to touch a floating piece of *turbinaria* with her nose.

The big coral structure on the opposite side of the site was especially good for watching for surprises coming up the scent flow, as well as being easier to use when the current was strong. From there, I faced into it to overlook the site. I could watch the two swim-ways approaching up the scent flow, and just by turning my head, could see the region down-current where shy sharks often passed.

When the crocodile needlefish came, I went to the boat to toss a bit of food to her, then returned to the down-current lookout.

I had a small bag of treats in the boat, so when Martha came to see if I had something for her, after most of the others had gone, I decided to get it the next time she was at the farthest arc of her circle. However, when I thought, *Okay, now I'll go*, she came zooming straight back to me, and slowly circled. I decided to wait. A few minutes later, however, the same thing happened, and when it had happened four times in fifteen minutes, I was mystified. She couldn't be reading my mind! So the next time that I decided to get the food, I watched closely—what she did, and what I did. First, I looked around, and moved my hands on my perch. Then I glanced above the surface, to see where the kayak was. Since it shifted constantly in the current and the wind, I automatically checked to see which direction to swim in. It was these signals, apparently, that brought Martha back. She was observing me!

Finally, I went and got the little plastic bag anyway and tried to get the pieces to drift out underwater. This was not automatic, because the pressure of the surrounding sea kept the plastic pressed tightly against the food. I had to hold the bag open, while fluids poured out, and shake it to try to get the water to circulate inside and waft the pieces of food out of the bag. Martha, who had been with me all along, came in without accelerating, and coiled up through the water column in front of me, taking the pieces one after another and paying no attention to the plastic or the movements of my hands.

Soon after that, Madeline swept in. She had been missing so long that I was beginning to worry about her, and her return reaffirmed that the season of reproduction had ended. She looked paler and browner, as all the females did. Just a few minutes after Madeline came, I was looking out from my perch on the down-current side of the site, when an enormous female shark went shooting by with a fish-tail in her mouth! It was months since this individual had first appeared, and just the fourth time she had come. I followed her, with an assortment of sharks whose attention she had taken too, and was rewarded with a brief view of the other side of her fin, which I sketched. The completion of the drawing and confirmation of her identity was worth all the trouble of going there. She was magnificent, and I called her Victoria.

When I got back to the site, Twilight was swimming into view, flanked by Samurai and Keeta. I kept the three big sharks in view as they roamed together through the coral surroundings in the strange, sullen light, and saw one of them turn nearly upside down to slam her side against a low bank of sand. The cloud of sand she raised drifted away in the water, evidence of her lightning-fast manoeuvre. Later, Twilight swam by with two large remoras flitting around her body—I wondered if she had tried to smack them off on the sand.

Carrellina was not a problem this time, in spite of intermittent close approaches and passes. Part of the reason I watched from the huge coral formation down-current was that it was harder for her to harass me there, so I could relax and watch the other sharks without always being on the alert for her.

The crocodile needlefish kept circling me, and sometimes swept closely by to get my attention. I went to the boat and threw her a piece of food, which she waited to receive. Each time I reached into the boat and looked back at her, she was watching, waiting, poised in the current. My arm was above the surface when I threw, yet she responded, and darted forward at the right moment, as if she could tell where the piece would plunge through the surface, just from my movements. But I spent only a few moments with her, and moved quickly away from the boat before Martha or Carrellina could notice what I was doing.

Back at my down-current lookout, I watched the many sharks passing and circling around, back in bliss. Some of them made close approaches from time to time, but nothing to cause me alarm. Shilly stayed around me all the time with Ondine and Cotlet, who had become very curious about me. I found a few more bits of food on the boat, and threw them to these young females. Carrellina and Martha finally left, and I relaxed enough to try to feed the fish. I was at the boat with Shilly and Ondine, when a little juvenile approached me from up-current, beyond the boat, to look at me. She came closer each time I went to the kayak, to look again.

There was a shark pup in the site, born in the past year, but it was impossible to get close enough to identify him. He appeared to be Spring, the baby I had kept track of during the past September in the local nursery, but in the awful visibility and darkened waters, I could not be sure. He was swimming near the surface, apparently attracted by the scent and activity. He came, left, came again, and disappeared. It was never possible to get close to the tiny pups; they were so nervous out in the open.

Back at the down-current perch, Cheyenne passed slowly; she was still in the area. She suddenly saw me and accelerated away—she did not return. I threw some fish heads to the waiting nurse sharks. The blackfins came back to soar around the flowers the nurse sharks formed as several ate from the same fish head, waving their pennant tails against the surface. With the sharks thus distracted, I was able to feed the fish at my regular perch for a moment or so at a time. It got very dark, and at the end, Jessica appeared, scouting through the shadows, and Cinnamon glided slowly past in her usual way, heading for the border.

But I waited in vain for Kimberley.

The end of April had come and the water was getting colder. I believed that the warm water of the southern summer had contributed to the change in the behaviour of the sharks during the third week of January, and was confident that soon I would be able to return to my usual routine. Madeline's return, and this ordinary and pleasant session, reinforced my certainty that all would soon be well.

Strangely, the fish shops had more scraps the next day, and phoned me to come and pick them up. So out I went again, through winds gusting from all four directions, blowing delicate curtains of falling rain before them. I had very little food, but went because there had been so many sharks the night before. Usually, if I brought food on successive nights, on the second one, many of the regular sharks were not there, so I would get a second chance to see the visitors. I was particularly concerned that all the signs had pointed to the possible presence of Kimberley in the area, but I still had not seen her, so welcomed the chance for another look. Others, including Droplet, had also been absent for too long, and I looked for them at each session.

In spite of the agitation of the air and sea, I managed, for once, to anchor perfectly, and watched the dark shadows appearing as I attached the paddle and began to put on my gear. Once again, I decided to throw the food in first and then go into the water. The sharks now expected the food right away, and what would they think if, instead of food, it was me who appeared underwater?

After the food was delivered to my excited protégés, I watched and waited until I could see no sharks, then smacked the water with my fins to indicate that I, huge and powerful, was about to descend. Then I slid in. I always turned around immediately to see my three-hundred-and-sixty degree environment as fast as possible, and there at ninety degrees, were Madonna, Martha, and two of the bad sharks gliding up beside me in a mass, with others beyond and coming behind. In that first second in which I saw them, they came on farther, grazing me as their motion took them past. This was too many sharks, too close, and it didn't take a three-digit IQ to figure out that they had assumed that I was more food descending!

I recoiled away, and they didn't respond to my sudden movement. They had an "Oh, it's you," reaction, and subtly changed their trajectory to descend once more. I was shaken, and cruised around the periphery to the coral structure on the other side of the site.

Most of the same sharks I had seen the night before were swirling around, to my surprise. Carrellina, Bratworst, Apricot, and Flora were present. And Jessica was there at the beginning, which was rare.

Though Shilly didn't come, Ondine stayed around me, often closely approaching. Carrellina, mercifully, left me alone, but Sparkle and Apricot pestered me sporadically. Finally, Droplet swept in. I was delighted to see her return. Victoria arrived with Keeta, and the two sharks casually roamed through, and circled in the site. It was a rare pleasure to have a new adult female visit repeatedly. Each time I saw Victoria, I perfected the initial sketch of her dorsal fin, to include all the subtle details on both sides. Madeline came at precisely the same time by my watch, to the minute, as she had the night before. A few minutes later, Storm cruised in, then Clementine! I watched all this from the structure down-current, looking around continuously. Grace drifted through like a dream, the three flares at the front of her white band unmistakeable.

As night fell, I began to feed the fish. The myriad nurse sharks were arriving, and searching for something to eat, so when Ondine shot up to me through the fish, I pushed her away, and threw the fish head into the site. There was an extra scrap saved in the boat, and I threw it for her. It was mobbed by a cloud of fish, which the young shark approached tentatively, then she took the bit of food, and after that, let me feed the fish in peace. I fed a bit to the fish, and threw the heads to the nurse sharks, who were undulating around in the image of confusion. A tiny juvenile skittered through, high above the action.

I fed the crocodile needlefish off and on throughout the session, and she left early. When the nurse sharks began tearing apart the fish heads, my blackfins returned, and zoomed around the site, passing centimetres in front of me. With them so distracted, I was able to sneak food to all of the fish who had waited, so often, so patiently, when the bad sharks had prevented me from giving them anything. From each fish head, I scooped out a handful of crumbs for the squirrel fish, one for the groupers, one for the needlefish and one for the excited wrasses, then tossed it into the site before the sharks realized what I was doing. While going to the boat one time, I met Chevron coming from the direction of the reef and looking very rounded—she paid no attention to me.

As it grew dark, I wanted to leave, but five nurse sharks were lying on the anchor chain with all the weight of their huge bodies, and some of their fins were pressing upon it. When waiting patiently for them to move and swimming near them didn't work, I was finally able to free it by picking up the anchor so that the chain moved beneath them. One at a time, bit by bit, they moved off it.

The following week on the day of the feeding session, I kept an eye on the wind, which had changed direction, and was blowing gently from the south. Sometimes, it let off a menacing blast, so I was not very comfortable about it. However, given various human-life factors, I chose not to cancel the session when it got worse as sunset approached.

I surfed out on the waves in record time, the wind behind me. Every time I looked back into the gale, the bay looked whiter, with more ominous white caps rushing toward me. The waves grew alarmingly, as the sea frolicked in the rising wind. By the time I reached the lagoon, big waves were breaking over the border, and violent gusts tried to blow the kayak broadside. I managed with held breath to surf across into the lagoon, and guide the kayak toward the site, but the force of the wind, pushing me like an angry power, was impossible to fight. The view beneath the surface was completely obscured, and as I lost control once more in the ever increasing blasts, I spontaneously threw out the anchor as I passed over the deeper site, the one I used in bad current. It was obvious I would have to swim, towing the kayak, to the shelter of the island, in order to get home.

It was the dark of the moon. I threw in the food as wild waters washed over the boat, and slid in, aware of only a few sharks around. As I fell through the water, several sharks appeared at my side with more soaring up between them. This time it was Madeline who brushed by me with Madonna against her, obviously assuming that I was more food falling through the surface. The water beyond and behind them was solid with sharks zooming toward me. In my surprise and anxiety over being mistakenly bitten, coming so soon after the battle with the wind, I failed to note the identity of the others. But again, when they saw it was me, they all subtly changed direction and descended again.

The scene was bedlam. I found myself in a fairly deep region, and in the pearly light beneath the waves, it took on an unreal quality. Two large nurse sharks circled and undulated around in mid-water, and my sharks were flying everywhere. One of the biggest nurse sharks was vertical, presenting a weird centrepiece, as it flung its enormous tail around for balance, and flailed its fins—usually the water was not deep enough for such huge fish to balance vertically. Everyone was unnaturally excited. There was no order, and nowhere to hang onto within view. I was appalled at the pandemonium.

My camera had just arrived back from America, supposedly repaired, and I had been looking forward to taking lots of pictures to make up for lost time. So I raised it, and drifted, watching, beginning to manoeuvre for a photograph, as several sharks tore a large scrap apart.

Suddenly, Bratworst, Madonna, and Martha left the centre of the feeding area, and swept up toward me. The gesture was so swift and unprecedented, so full of conviction, that I instinctively lowered the camera. They came in triangular formation, as they had in our first moment of meeting more than three years before. Bratworst was in front. Normally in these situations, I quietly faced the shark until she turned away, but Bratworst didn't turn. And the approach was far too fast. As she passed under my hands, she awakened some ancient instinct of self-protection, which told me that my hands were my circle of defence, and that on no account must a large predator be allowed to pass them. So I hit her on the back of her head. It was amazingly hard!

Like lightning she turned at right angles, and shot away, and Madonna was there. I was already raising my knees between her and my chest, and finned water into her nose, but she just dodged slightly, and kept coming! Leaning backward, I finned harder, and she turned. Martha was simultaneously arriving, and I pushed her away with my hand. She continued in the new direction in which I had pointed her, then rejoined Bratworst and Madonna. Back in triangle formation, the three zoomed away together and disappeared into the whirling sharks.

Watching their waving tails as they shot away, I told them in my mind, *I'm never coming back*. It wasn't even something I thought about—the words were spontaneous. I envisioned myself in that moment spending all my days playing with my birds and never thinking again of these sharks—forgetting them, and this awful moment, forever.

Badly shocked by their gesture, I wanted to climb back in the boat, but I knew that if I went toward it, the sharks would all come shooting along too, so I swam away. My heart was pounding so hard that I could hear it myself, and I realized that I couldn't leave in the boat because of the wind! I swam on until I was out of sight of the feeding, but soon, my need to see the sharks overcame my fear.

Sneakily, I moved closer. Martha came to me at once, and began a series of close approaches to my face. The second time she did, I snapped her picture, releasing the flash like a warning, but there was no response. She was so close that the resulting photograph was blurred, but her teeth were clear to see—her mouth was open! I dared not approach the feeding area, but still, Bratworst, Flora, Annaloo, and many others closely circled me. The large nurse sharks were undulating vertically in the water at the food, still at the vortex of a swirl of flying sharks.

I waited, poised, watching, and reviewing what had happened. Bratworst had a tendency to circle me, not charge. That was more typical of Martha and Madonna. Why had the three—these three—left the food in unison to charge in triangular formation like a tiny squadron of jet planes? I trembled as one shark after another swam up to my face or circled. I wanted to go home.

Waiting up-current, I began writing down the names of those present, photographing them if they came very close. Martha continued her close approaches, and many juveniles came to me just under the surface in waves so big that they were intermittently concealed from my view. Sharks continuously approached and circled me, and when they became too numerous, I swam away from them again. But soon, I came back. I couldn't help it. I was determined that my record be complete.

The sharks could not have been upset because there was not enough food. There was the usual load, as well as chunks of unsold fish meat, and due to the terrible conditions above the surface, I had simply thrown it all in. My resident sharks would have caught it as it fell from the surface. Martha must have been able to eat. Indeed, they had left the food to come to me!

I had brought half a small *saumon des dieux*, and had cut much of the meat off it for them. These good pieces had been among the scraps, and there was still flesh on the bones. Many sharks were hanging on to this large, bony frame, and chasing around the area with it, one after another grabbing hold and trying to shake it while the others changed places as they went. Lillith was soaring high like the monarch over all. Jessica was there. Shimmy, a male who had never paid attention to me, was repeatedly passing me closely, and Simmaron, just maturing, made a swift charge, which she had never done before. Trinket, who came so rarely, repeatedly circled me—he had never done so—and Annaloo, the mystery shark, did too. So did Valentine. As an older shark who had been there since the beginning, she had never harassed me, though she had been one of the ones who would swim up to me when I first entered the water, in the greeting or affirmation gesture, once she became familiar with me. There were many juveniles of all ages present, as well as the usual males who came when they were in the area, such as Avogadro and Ruffles. On the other hand, I did not see Carrellina, Sparkle Too, or Chevron.

Their behaviour was not returning to normal. This was obviously one of their special sessions that usually coincided with either the full moon or the dark of the moon, but these had always involved many visitors joining the residents. Here, I saw no visitors. I had written down only thirty-eight names, the names of all of my sharks.

I wanted to leave, but to do that, I had to extricate the anchor, which was much too close to the centre of the feeding area. I had been waiting for the sharks to move, but nothing changed. Finally, spiralled around continually by Tamarack, I was able, through repeated dives and manipulations, to drag it away. Grasping it as if it were a magical talisman, I swam away along the bottom without looking back, rose with difficulty to the surface, threw it into the boat, and just kept going. With my thoughts in a turmoil as dark as the waters, I swam the kilometre to safety in the lee of the island, as if all the devils of hell were after me.

On the shore, I hauled the kayak onto the sand, extricated my cell phone, and called Franck. The wind on the bay seemed to be nearing hurricane force—the mountains funnelled and accelerated the wind in the bay—it would be madness to go anywhere near it. It was the first time I had to be rescued by truck, and it was a problem because the kayak was too long to fit in the small pick-up. But Franck had made a structure so that it could lean against the roof, and arrived in record time to rescue me. All I had to do was sit in the back and hold my kayak steady as he drove, which was a great relief, because the session seemed to have stolen my strength.

I was badly disturbed about what had happened with my sharks, and during the days that followed, they circled continuously in my mind. My spontaneous decision never to return was irreversible. Never again did I want to be in the situation in which I had found myself, with my life depending on the whims of a fleet of emotional sharks.

Their new pattern of rising to the surface upon my descent into the water, thinking I was more food descending, in itself seemed very dangerous. Though my companion sharks were unlikely to take a swipe at me, an excited juvenile or visitor coming among them could easily do so under some unusual circumstance.

Secondly, while I had been sure that their behaviour would return to normal after their period of reproduction was over and when the water cooled, that had not happened. The reproductive season had ended more than a month before, and the water was markedly colder, but at this session, sharks who had never bothered me had suddenly begun to charge and harass me, and the level of excitement had been higher than I had ever seen it. The problem had not been restricted to one incidental move by Martha, Madonna, and Bratworst.

The behaviour of my three favourites was the most troubling question. Their charge had felt very much like an attack, just as the violent slamming of the boat had felt like an attack. While I was sure that they had not intended to bite me, it did appear that they would have rammed me if I had not defended myself. Bratworst, a large, high-powered missile driving straight into my solar plexus, would have been crippling. Then there were Madonna and Martha right behind. I had little doubt that their action could have become a general attack like the one made on my kayak, given the mood of the sharks, and the speed at which they can suddenly move.

But why had it been those three sharks? They had been in precisely the same formation in which they had approached me the very first time!

Day after day I pondered the question. In the years since I had first met them, I had become familiar with many other sharks who used the area; it was within their home range. Some of these had proved to be more aggressive than Martha, Madonna, and Bratworst. Bratworst had paid little attention to me once she got over her curious phase at the beginning of my study. Given the sequence of events that led up to this one, the logical sharks to be involved in an acceleration of the aggression would have been Carrellina, Sparkle Too, and Chevron, or any of these accompanied by more of the bad sharks.

I couldn't believe that it was just a coincidence that it was these three, the original residents of that precise area, who had made this final gesture. Given the presence of plenty of food, maybe their act was rooted in a more territorial motive. Perhaps they had been conveying the idea, "Look at what you've done! Look at all the sharks here now because of you! It used to be just us."

This possibility reinforced my decision not to go back. But it grieved me, because I loved them.

I asked all the other people I knew who had dive clubs and fed sharks on the north shore of the island if they had noticed a change in the behaviour of the sharks during the warm season this year. No one had, and some of them had been holding commercial feedings in the same places for twenty years, attended regularly not only by blackfins, but also grey reef sharks and lemon sharks; even the occasional tiger shark visited. The changes in behaviour had occurred in my group alone. Something in my relationship with them had brought about the evolving changes in behaviour. Like a tangle of string too big to trace all the strands, I just could not unravel the different influences, the different events involving different sharks that had brought me to this. There were some things I just couldn't figure out—difficult as it was, I had to face it, that I would never know why my sharks had behaved as they had.

The change in behaviour of a group of animals in a particular situation, was similar to the development of culture. My sharks had even developed their own method of eating—which they had learned, one from the other—of leaping from the water to eat from the back of a kayak! While I had read of different cultures having been observed in different populations of dolphins, I had never read of culture having been noted in communities of sharks.

It was also true that in spite of my long-term intimacy with so many sharks, under a wide variety of conditions and circumstances, never had I been bitten or slammed. Indeed, despite my mammalian instincts and emotional reactions, they remained the only species I had been in prolonged, intimate contact with, who had never hurt me. Even my dog often carelessly took my hand in her teeth along with a cookie, hard enough to hurt. I had once had a pet skunk, an animal lazy and cuddly as a stuffed toy. But he often lost his temper, sometimes for reasons known only to himself, and bit deeply. Other times, he would heave over the garbage can in a rage. My pet mynah bird had pecked my eye, and daily he pecked my hands, through impatience or irritation, though he lived free in the house and garden, and was very attached to me.

Yet all these wild, predatory animals, sharks of my own size, had never bitten, either through accident or a fit of pique, no matter what had happened. It had to be more than chance; they must have made a point of not biting me, as they did not bite each other.

Still, I could not go back to that group of sharks. Something had gone wrong, and it was time to stop.

CHAPTER THIRTY-SEVEN

The New Program

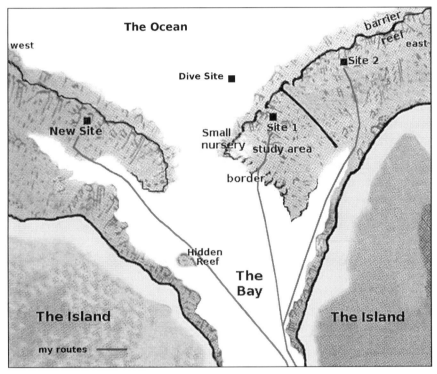

For convenience in keeping track of the sharks' ranges in my notes, I had divided the lagoon into five sections. Since I had already investigated the sharks in Section Three, I decided to hold my regular weekly feeding sessions in Section Two, and take some food to Section Five when I could.

At Site Two, I hoped to see most of the same visitors that I had seen at Site One. I would doubtless gain insight on how they travelled and where they were from. Thus, I would be able to continue to collect data about them and enjoy seeing them. It would be interesting to see how long it took my sharks to find out that the feeding site had moved.

I wanted to learn, too, whether my sharks crossed the bay to roam the lagoon on the other side. By doing a series of feeding sessions there, I would be able to compare an entirely new border region with my original one, and gain insight as to how much traffic there was across the bay. Did many of my rare or infrequent visitors come from there? It would be exciting to find out!

I would continue to visit my sharks during the day, but would let a few weeks go by first.

So the next time I had scraps, I divided them into two parts, stuffed half into the bottom of the fridge, packed up the boat, and launched the kayak for the far-off western point of the bay. It was a long haul through an unfamiliar area, and the point was unexpectedly turbulent, since the oceanic waves crossed the bay from the pass, and struck it head on. There was a sea mount in my path where waves broke, surrounded by treacherous waters—it had to be given a wide birth. I noted its location so that it wouldn't catch me unaware as I hurried home in the dark. Around the point of land, Polynesians lounged and fished. They waved to me from shore, and beamed their radiant smiles. I smiled and waved back.

Then the sinking sun shone straight into my eyes.

I had an idea where to go, having explored there before. For a long way past the end of the barrier reef, the lagoon was barren reef flats, so in order to meet the resident sharks, I would have to proceed well into the coral beyond. Unlike my original study area, much of the space between the reef and the shore consisted of a wide, deep channel. There was only a narrow strip of coral habitat between it and the reef.

The weather was calm and sunny with a light east wind, and since I was travelling west, the wind was on my side and hurried me onward. But as I approached the chosen area, the glitter between me and the sun was so brilliant that I could see nothing. I had always prided myself on travelling by kayak without ever touching the coral, but suddenly, I was being helplessly blown onto the coral formations, and becoming stuck. My eyes were aching from the blinding play of light, and eventually, though still virtually at the boat channel, I decided to anchor right there, and locate an appropriate site underwater. So I prepared to get in, and to my astonishment, one of my movements badly startled a large female shark. She had been just under the boat and splashed out through the surface in alarm as she shot away. At least there were sharks.

I descended into a region very different from the one I was used to. The coral was so thick that it was hard to find a way to swim through it trailing the boat, and the water was much less than a metre deep. Feeling cramped, I made my way with difficulty toward the reef, looking for deeper, more open strands, but there were none, and remarkably quickly, the water became shallower. I recognized the reef flats and *turbinaria* landscape typical of the inner edge of the barrier reef.

So I turned back, and explored until I found a little cove, where there was a sandy floor, and the water was deep enough to manoeuvre. I didn't want to waste any more time; it had taken nearly an hour to get there. So I put in the bit of food. A juvenile blackfin passed, and when I had retreated, he approached. I was expecting a very slow, boring session compared to what I had become used to. A session was a lot of work, I missed the sharks I knew, and based on my experience at my original study area and other locations, I would be lucky if a shark even came to eat. So it was nice to see a shark there. I carefully drew the curious animal's dorsal fin, and some patterns of light marks he had, and named him Randall.

The visibility was strangely poor, and soon a ghostly shark passed just beyond visual range. She was enormous, far larger than Madonna and Martha. I waited, watching, in hopes that she would approach. She did. I was able to draw her dorsal fin on both sides before she suddenly approached me at the same moment that a second very large female appeared from the direction of the reef. The two placidly converged on my nose. Mercy! There was nowhere to go! I was in half a metre (20 in) of water, and couldn't believe that here I was again nose to nose with big sharks, this time ones I didn't even know! But they turned away, visited the food, and came back to circle beside me.

I was mystified. And watching the first shark I had seen—the way she passed in the coral in front, then behind, then swam up to my nose—it seemed that I had experienced that before. I stared at her dorsal fin and realized—yes—this was Shenandoah! It had been a long time, but she was recognizable not only by her dorsal fin and her size, but by her unusual behaviour. I did not recognize her companion, and named her Mercy, as she continued to behave boldly, yet placidly, like Shenandoah.

The sharks began to eat and became excited. I wondered what had been wrong with me to set up in such a cramped space—I was stuck with a crowd of excited, strange sharks all around me, in a tiny space just half a metre deep, where I was crouched and pressed back against the coral. More and more sharks came. I drew their dorsal fins, noting rapidly which were females and juveniles. Everyone was new. There were quite a few with obvious distinguishing marks, though, so they would be recognizable even if I was unable to perfect the sketches of both sides of their dorsal fins. Soon, so many sharks were close around me that I was having trouble concentrating on drawing.

The sharks were the usual mixture of females and juveniles, though there was a much higher fraction of the very large females than I had seen at Site One. As night fell, I became uncomfortable with so many strange sharks cavorting around me as if I weren't there, yet ran out of space on my two slates.

It was much farther home than I was used to having to travel. Going back east, I was once more fighting against the wind, and had lost the beauty of the sunset because I had my back to it. I felt tired and sad, as I looked across the bay to where my beloved sharks were circling, thinking of me, as I was thinking of them.

Finding names for all these new characters was challenging. I kept a list of possible shark names taken from books I was reading, and even maps of my favourite countries. When I had identified a new shark, and was tired and pressed to finish quickly, if no likely name had come to mind, I could choose an appropriate one from the list. By this time I was watching out for close to five hundred sharks, not including those with incomplete identifications. I could never have remembered them all, had they not been named.

Franck continued to talk about how neural networks—one of his favourite intelligent software tools—would be sure to pin down the underlying rules governing their roaming, while I mentioned occasionally that, imperfect as it was, my mind was probably best for comprehending my sharks.

My sporadic searches on the Internet had revealed that animals under study were to be numbered instead of named. I couldn't imagine keeping track of five hundred sharks by number. I had already tried that, and they had all turned to a haze in my mind when I only had twenty. I would look at a shark, and wonder, but was that golden, young female numbered fourteen or sixteen? Fifteen or nineteen? I would never be able to remember if a shark was number 323 or 232, for example, particularly while watching them speeding by like grey torpedoes, each visible for less than a second. If I made a mistake and wrote 123 instead of 132, while cold and hurried or typing fast, I could overlook the mistake, and the wrong shark would have been recorded for the session. If mistakes were impossible to see, one could never be sure that one had not made any. If I spelt the name wrong, though, it was easy to notice and correct. Franck's computer program would spit out any incorrectly spelled name right away.

It seemed natural to associate a name with the animal. Everything around us has a name; I was sure that naming things has a genetic basis in our species. Nevertheless, I gave each shark a number, too. This was not hard, since they were all listed in chronological order on my computer, in tables for each site.

The night after my adventures in the lagoon to the west, I hauled the other half of the scraps out of the fridge, loaded up the kayak, and went to find an appropriate site in Section Two. Just within the reef was a deep region, in which large, widely spaced coral formations formed a cathedral-like area. I found a circular, sandy place with an appropriate lookout, and put in the food.

The water was crystal clear, and brilliant fish hung in the coral like humming-birds among flowers. It was a beautiful region. I waited, and when no sharks came, I began to feel peaceful.

After twelve minutes had passed, out of the thick coral, a group of little pups very tentatively appeared. They flitted toward me at the surface, retreated, and came again, five altogether. Finally, they grew bold enough to come close enough to draw. I was delighted, after all my encounters with large, bossy sharks, to dwell tranquilly in the company of these tiny nymphs. Sometimes, one picked up a bit of food, and carried it away into the coral beyond. With the passing time, one grew curious about me, and came close for a look.

After about twenty minutes, a young adult female appeared, probably a year from maturity, and I drew her fin as she circled the site and the food. She came and went, and the youngsters flitted about as darkness fell quickly, due to heavy clouds in the west. The little sharks grew bolder as it did so, so I was able to record the identities of them all. The next shark to arrive passed a few times off in the coral before she came close. When she did, she showed only one side, which was just like Cotlet. It was a while before I was able to see her other side, and confirm that it was Cotlet! I didn't think it was possible that this young shark, who had begun to come regularly to Site One sessions, could have been coming this far, and managed to be there each time on the right nights. But apparently, that was the case. Perhaps she had begun going to the border at dusk, and thus found my sessions.

It was getting dark when the first adult shark approached—Arcangela! So she was finally leaving Section One. Not long after that, Mordred came through, in confirmation of his wandering nature. Flannery came minutes later. By then, the site was full of tiny shark pups, and it had become confusing to identify the new ones. I was going to have to memorize them a few at a time until they all became familiar. Evidently, the thick coral through which the scent flow was passing, was one of the regions where they hid. It was a delightful finding.

Cotlet was the juvenile shark who had spent a long period passing back and forth in the coral during one of my two sessions there the year before. Her appearance in Section One may have been connected with her roaming more. She was at the age in which new juvenile sharks usually appeared in the region. Probably many had grown up in that area, considering that it was nearby, and was favoured by juveniles.

To enlarge my understanding of the lagoon's inhabitants, the next time I got some scraps, I took them to Section Five. Since it was too far to go by kayak, I drove there and parked near our former house. Then I found a place to enter the water, hid the car keys, tied the scraps onto one of the children's old body boards, and swam out to the reef. The little raft trailed restlessly behind me on a string tied around my waist.

It was an enchanted evening. There wasn't a wave or ripple to mar the reflection of a sky glowing in gold and rosy hues, as the sun plunged toward the horizon. I had explored this region well, since I had lived there when we first had come to the island, and as I progressed into what should have been the supernatural garden that had once filled me with awe, I was appalled at how much of it had died in just those few years. The visibility was dreadful. It was like swimming through a fog, and it didn't improve as I approached the reef. Weighed down by the trailing raft, it took me forty minutes to get there.

At the reef, the sharks appeared. First, a fat little female zoomed past, looking. A whitetip shark drifted in the vicinity. I started to look for a place to put the food. The coral formations were covered with a soft, brown, sponge-like animal, so there was nothing to hold onto. I had started shredding a scrap to create a scent flow to attract the sharks, while searching for a place to use, and fish quickly came to feed. It would have been very pretty, but the sun drifted behind a cloud, and everything turned grey. I found a place I could use as a lookout, put the food down-current from it, and the juveniles who had gathered began to roam around, looking the situation over.

A big female glided in. It was Amelia, whom I had identified at Site Three the year before. I watched her feeding appreciatively, very happy once again to have discovered a known shark so far away. I didn't expect to know the juveniles, but, so far on my travels in the lagoon, I had found few adults I didn't know! Quite a few younger sharks were circling, and I tried to keep them sorted out while sketching their dorsal fins. This was challenging in the appalling visibility. I was soon shaking with cold, and it was already getting dark, so I was obliged to leave. As I swam away down-current, I found many more sharks, some of which circled close enough that I was able to do quite detailed drawings of their dorsal fins. Uncomfortable though the session had been, I couldn't wait to go back. It was wonderful that my enthusiasm for sharks was returning.

At the end of the week, I once again picked up enough of a load to split in two, and returned to the unpleasantly shallow site on the other side of the bay, where Shenandoah lived. Once again the visibility was so awful it was like swimming in soup. I could see the particles in the water. I had left early enough to have time to look around for a better place, but there wasn't a good place anywhere in the region, so I set down the food in the little cove I had used the previous week. This time, there was no current to make a scent flow for passing sharks to intercept, and twenty minutes passed in which I slowly grew cold. Then Shenandoah glided silently behind me. She was more interested in me than the food. She swept in slow-motion though the misty surroundings, vanished and reappeared a number of times. I had never felt comfortable with her since that first meeting when she had come repeatedly up to me, showing me her dorsal fin for the first time from a few centimetres away.

She did not eat. She left, and for another fifteen minutes, there were no sharks. I wondered where all the juveniles who had come the week before had gone to. Eventually, a smaller adult female arrived. She had a bite out of her right pectoral fin, which was bent down in the way that Clementine's had been that year. She came straight to the food like a reasonable shark, and ate while I drew her dorsal fin. Shenandoah came back and ate with her, but no other sharks came to that session.

At the Section Two site the next night, I placed the food in a circular space lined with tall and pointed coral structures. It was a very pretty place, and gave me a choice of lookouts to use. Ali appeared in moments, and circled slowly around the area before she placidly came to eat. There was a large female off in the shadows, but she did not come forward.

I watched a whitetip shark who came to look at me, each time a bit closer, every few minutes. Then five large grey female blackfins sailed in, circled, and swept down to the scraps. Three of them had come to Site

Three—Cheyenne, Merrilee, and Tornado. They probably knew each other and lived nearer the centre of the lagoon. I did not know the two others, but one, whom I named Belladonna, turned out to have her home range right there in Section Two. She became a regular visitor to my sessions, though she had never appeared at Site One. The fifth shark had a distinctive, symmetrical wave on her tip line, and I later found her in my book at home. I had identified her the day that the big barracuda had been at Site Three, and called her Barracuda.

I was completely engrossed in sorting them out, and drawing their fins on both sides without mixing them up. Tornado I was fairly sure of, but sketched her again anyway for confirmation. I was rapidly drawing, when my view of my slate disappeared. The whitetip shark was passing my mask above my hands.

She had come when I was concentrating on something, just as Annaloo had chosen a moment when I was distracted, to drift across the site from left to right. I had seen sharks react to situations in which they understood that they could not be seen, or were not being watched, countless times. There was no doubt that this whitetip shark passed under my face when she saw that my attention was taken up by something else.

It followed that she understood what attention is. Defining attention as the mental light beam with which we focus on different things in our environment or minds, it appeared to me to be self evident, that if she could recognize it in me, she must have it herself. Here was another clue to the subjectivity of sharks—they know what attention is.

Another example was the crocodile needlefish, whose efforts to get my attention indicated that she knew whether or not she had it. When my eyes turned to her, hers were always looking straight into mine. How could she know when she had my attention if she did not experience it herself, and see it in other species, as well as in me?

The whitetip shark's act of swimming between my eyes and hands when I concentrated on drawing, after a series of increasingly close approaches, was my most striking example of the phenomenon yet.

Nurse sharks appeared as the day ended. Perhaps at least some of them lived in the lagoon, and were not just border visitors from the bay or ocean. Mordred, Arcangela, and Cotlet also attended this session. If Arcangela was moving up the lagoon, she must be travelling very slowly. Maybe she was doing a lot of socializing on the way. When I turned my attention to the identification of the pups, the one passing closest was Cygnet, the two year old who had recently come to the Site One sessions! Given how sedentary the adult females were, I found it remarkable that the tiny shark had been travelling so far. She must have found her way across the barrens and somehow found the sessions there near the border by chance. As time passed, it became clear that Section Two was her favoured region. By the time darkness fell, the site was full of little pups again, and Belladonna passed off in the shadows before I left.

It was another pretty day when I returned to Site Five. Though the dreaded south wind was blowing, I was sheltered there. It was such a pleasure not to have to care about it. Out by the reef where I placed the food, just enough of the wind wrapped around the mountain to balance the current. The breeze blew back my little raft, which kept me from drifting with the current so that, without effort, I stayed in one place.

While waiting for the blackfins to arrive, I watched a whitetip shark pup. He swam in and out and around a large coral structure not far away, and sometimes left it to come over to see what I was doing. I recalled the really big whitetip sharks Franck and me had seen at Fisherman's Point on Tahiti. They had lived in the holes in a gigantic coral structure, and it seemed that the maze of tunnels under and through this formation provided the little pup's home. Possibly, the juveniles of the species lived in the lagoon, too, and moved to deeper waters on maturity. It seemed strange that I had never found where the pups were born, but perhaps the adults just didn't put them all in one place like the blackfins did. The species was so secretive that it was possible that the little ones were around but rarely seen.

Dreamily speculating along these lines, I was awed to find myself gazing at a large and strangely familiar shark soaring past me at the surface. She had done that once before so dramatically and in such a beautiful setting that I had snapped a photograph of her, and since it was the best photograph I had of a reef shark, I had used it for wallpaper on my computer. Thus, there was no trouble recognizing Willow, identified the year before at Site Three. I held my breath as she turned to show the other side of her dorsal fin. I had glimpsed it and described it as being "just like Valentine's." Indeed, it was just like Valentine's, and this time I drew it.

It was curious that the only adults I had seen in Section Five were sharks I had identified at the centre of the lagoon. Yet none of the Section One sharks had come to Site Three, though the distance was similar.

A smaller female came with Willow, and while they joined some juveniles in looking over the food I had brought, I drew her dorsal fin. These two females stayed near until I left, while more juveniles, and finally Amelia, came. I was painfully cold by then, but decided to wait just another five minutes. Four minutes later, another very large female came winding slowly into view. I recognized her too. She was another older shark I had met at Site Three. Innisfree! Never had she come to Site One. I was thrilled to find her, and set off for the distant shore with a light heart, which soon began to palpitate as I travelled at top speed, frantic with cold. I had to slow down to keep from fainting. Finally, I had to stop, gasping to catch up with my oxygen supply until my heart began to beat normally. When I continued, I tried to pace myself. The physical difficulties with this remote site were a problem, but the results were wonderful.

With the scraps that came with the weekend, I returned to the site on the opposite side of the bay. The sun in the west was so blinding on the water that it was impossible to face forward at all, and with the wind ruffling the surface and pushing me from behind, I slammed again into a coral formation under the surface.

I was wondering if I was in the boat channel or had passed into the lagoon when a Polynesian man in an outrigger canoe drew up beside me. I had noticed him push off from shore as I passed the point and began the long crossing to the reef. But I had been so concentrated on my progress against the sun that I hadn't realized that he had come after me. There were such bad feelings about the commercial shark feedings, held for tourists, among fishermen that I was frightened immediately. The fish scraps were wrapped in a sugar sack, and it seemed that the smell of them filled the air. The pale red stains on the sack shouted out the shameful truth of my activities.

But he seemed focused on me being a female of the species, and commenced to make small talk. He was young, smiling, and full of charm. He didn't look at the sack of shark food, and I noted that he was upwind of it. He didn't even look at my boat. After we had talked for a while, I told him how nice it was to make his acquaintance but that now I was going to explore the lagoon. He failed to persuade me otherwise as I put on my gear and disappeared underwater. When I and my trailing boat moved away into the coral labyrinth, he waited, and then began fishing. Mercifully, he finally went away.

I explored farther into the lagoon, but the whole region seemed shallow, and I eventually put the food in a miniature barrens, like an underwater river bed, that was more open than the place I had used on the former occasions. The visibility was still very poor but the coral surroundings were thick and healthy. I waited.

After fifteen minutes a large shark passed down-current and did not return. It was probably that weird Shenandoah again. More time passed, then a little male appeared and commenced excitedly to eat. He was soon joined by Shenandoah and another juvenile, then the three left, and for fifteen minutes there were, again, no sharks. I had been there for forty-five minutes, and was beginning to think of leaving.

A small crocodile needlefish appeared, and throwing scraps of food to him amused me until a few more sharks arrived. There were the usual juveniles, a young female, and then Mercy, with a beautiful companion I called Carmen. They were well worth waiting for to watch, feeding close to me in the shallow water, and sometimes approaching for a look. Just as the shadows of the night obscured the surroundings, the area filled up with sharks again as it had during my first visit, but they were mostly males and juveniles. A large nurse shark arrived, too, the first I had seen in the area, and a larger crocodile needlefish appeared, taking the odd scrap from the water under the surface. I tried to leave unobtrusively, but badly frightened a nurse shark pup as I skirted the site to climb into the kayak.

The next night found me back in the familiar circumstances of fighting my way through a grey and turbulent atmosphere toward distant Site Two. I was beginning to dread going in bad weather. It was so hard.

But once underwater, all was calm, all was beauty, and Avogadro was with me before the food was in the water.

It was a pleasure to recognize most of the juveniles who came right away, and write down their names, instead of sketching their fins. Jessica was there in moments. The new site was just beyond her home range, and I had hoped to see her often. Belladonna came in behind her. She seemed agitated, twitching, making sudden rushes, then whipping back around in a tight circle. Her flighty behaviour attracted many sharks, all moving faster than usual. One of them was Twilight. Her companion, this time, was one I had seen only once, just when I had begun to hold the sessions. She had soared past at the surface, such a terrific sight that I had named her The Concorde. Here she was nearly three years later, soaring past my eyes just as she had the first time. Cotlet and Ann Boleyn came. Amidst the usual throng of cavorting juveniles, Droplet circled me slowly, time after time, as if wondering what I was doing there. In a period of relative calm, I recovered the scraps, put them out in the open, and added two fish heads for the nurse sharks.

Mordred glided over the scraps, and ate the one fillet, passing by me as he rose with it in his mouth. Good for him! He was rarely there at the beginning, so usually got little more than a crumb. Then Yorkey, a male identified at Site Three, soared in, snatched up a scrap, and passed over my shoulder as he left with it, followed by Raschelle, who had been so shy at the Site One sessions. It seemed that some of the sharks were more confident in their own ranges than they were when off visiting in strange waters.

Cheyenne came, with the magnificent Victoria, as darkness began falling. Then Teardrop came zooming in, and pounced on the food, thrilled. She was the first of my sharks to find the new site. But I was not pleased to see her, and wondered how long it would be before they all found me and recommenced their bad behaviour. I would have to play it by ear, just as I had been playing it, never certain, all along.

That week I returned to Section Five, and this time trailed scent all the way out from shore so that everyone in the area would know I had come. But I saw no sharks, though I went all the way to the reef. There was a shallow region just within the reef flats where shark pups were gliding. Since it was so close to the big nursery, it was likely one of the shallow refuges used by the little ones as they moved into the lagoon. Such places seemed to be found here and there all across it.

I found a good place to put the food within the edge of the reef where it was deeper, and began to feed the fish. The visibility there was fine for the first time, and excited juveniles began to accumulate. I sketched their dorsal fins, recognizing some from the times I had come before, while waiting in suspense for the resident females to appear. After a very long time, a female passed in the coral just within visual range, then came up the scent flow to circle and look at me. On both sides, her dorsal fin resembled a shark I had seen one time in Section Three, but not enough. It was a different shark, a first new sighting in the region. Shaking with cold by then, I could no longer control the pencil for drawing, but returned home with many new older juveniles identified to copy into my book; none of those already identified there had reappeared! But so close to the nursery where hundreds of shark pups entered the world each year, it was natural that there were a lot of kids.

The next time I struggled out past the turbulent point on the western side of the bay, the wind was high, the waves were wild, and the whole trip was an exhausting and alarming marathon. With difficulty, I manoeuvred through the shallows in the virulent current, and got set up to welcome the sharks. After waiting for a while, I started to feed the fish, keeping up a continuous strong scent flow for an hour until darkness filled the world with shadows. But no one came. Perhaps the entire population was simply farther along the lagoon, upstream from my scent flow. Such a thing had never happened before in a region known to be populated with sharks.

Surfing back toward the bay was easier with the wind coming broadside and pushing me homeward. But I passed the point where waves poured and crashed over the invisible underwater mount with held breath. Above, no stars penetrated the unfathomable blackness that had descended, and the strengthening wind howled through the jungles lining the bay. With each dip of the paddle in the flying waters, I questioned my sanity. Again.

My Sharks

It was time to go to see my sharks again. I missed them. So one tranquil morning in early June, I set off for Section One. Valentine was the first to find me feeding the fish at the site. She didn't see the piece of food I threw to her, and the fish carried it away into the coral. Martha came while she was trying to extract it. I tossed her a morsel, which she neatly caught. Eden and Pippet arrived. I threw them each a treat, then swam away with Martha. We went west on a winding path, gradually turning clockwise until we returned to the site. I grabbed my slate and camera, and we set off again, on a longer circle in the direction of the nursery and the reef this time.

When the light was just right on the meandering shark, I began to manoeuvre for a photograph, and was about to take it when I was rudely interrupted by Madonna zooming up to my face. I probably reeked of food. Martha turned and circled us about two metres away as Madonna swam up to me repeatedly, turning at the last moment. Finally accepting that I had no food with me, she glided away, whereon Martha resumed her former path, and I followed. Cinnamon, huge and pale, had come with Madonna, and remained within visual range.

Martha roamed placidly through the coral in her careless fashion for another twenty minutes. Sometimes she investigated a floating object, but otherwise appeared indifferent to her surroundings. But I knew how aware she really was. After wandering far through the inner edge of the barrier reef, she entered the region of thick coral where the juveniles sheltered. When the white sand floor fell away into the barrens, she accelerated, and left me.

Having renewed contact with my favourite shark, I felt better than I had for many weeks, and swam back toward the kayak. Madonna crossed my path when I was still far away from it, and followed me. At the site, she zoomed up as I got a piece of food to throw to her, but as usual she failed to locate it until long after it had landed on the sand. She never could manage to catch the pieces I threw to her, though I had spent much time trying to teach her. After eating, she left, and I fed the fish some crumbs before trailing the boat to the barrens to look for Shilly. There were sharks everywhere, grey sharks moving over grey reef flats, all running with light. But if I moved, they just weren't there. The only way to see them was to be so still that they appeared in visual range. Yet unless one looked intently, they remained virtually invisible.

I was also looking for the crocodile needlefish. I had felt badly about leaving her. She had behaved like a pet with me, which seemed remarkable in a wild fish. I thought it possible that she was elderly, and depended on the food I brought, after meeting me weekly for a year and a half, while the sharks did not. But I could not find her.

Suddenly, Madonna was charging me again, and then she cruised around me, darting from time to time at the fish. I went back to the site for one last look around before leaving, and the fish rushed to meet me. As I crumbled the rest of the food for them, Windy came, with a new companion. Then I drifted away across water so glassy that looking into the coral was like looking into an aquarium. Such calm was rare and to be savoured. The waves on the reef were low. The South Pacific, for once, was at rest.

It was rarely possible to cross the barrier reef, but that day I paddled carefully across the low lapping waves, and into the ocean. Farther out was a dive site where the sharks were brought food nearly daily by the dive clubs, so it seemed a likely place to have a look around. I rinsed the well of the boat, to provide a little scent that would bring any sharks who smelled it to the surface for a look, so I could see who they were. The seafloor was visible about nine metres (30 ft) down, where tiny bright fish pursued their affairs among the little corals.

There were some male sharks cruising there, and I dove down to see if I could recognize any. Amazingly, one was a large male who had come to a session just once, three years before, called Plato! His dorsal fin pattern was unmistakeable, and I was even able to get a photograph to prove it by zooming after him along the bottom. He circled, then accelerated away from this unexpected intruder. I nearly drowned flying back to the surface, it was so far away. Diving again, I recognized Brambling, a little male who had come to the last sessions at Site One.

The anchor was hanging about a metre above the seafloor, and a shark rose to touch it with his nose. I dove and recognized Trinket! Had he moved outside the reef as males did when they matured? What an unexpected discovery. It would be interesting to confirm if he actually made the region across the reef from the place I had known him, his home range. But apparently the males there didn't frequent the lagoon—I recognized no others.

I had been there for nearly an hour, and was feeling the effects of hypothermia, when I realized that my slate was no longer looped over my wrist. I couldn't even remember when I had last seen it. I searched back and forth, increasingly frantic, trying to remember every place I had been. Luckily I found it lying in only about five metres (16 ft) of water, so it was easy to retrieve—it could have been so much deeper.

When I returned to the kayak, several sharks were roaming nearby, and came to me as I swam up. The first was Cochita! She circled me several times as if to say, "What *are* you doing *here*?" and glided away into the depths. Then Bratworst came and orbited placidly at arm's length. The bronze shark in the blue looked so serene, after her shocking behaviour the previous time we had met. What a pleasure it was to find her again behaving so reasonably. Beyond her, Gwendolyn passed. I watched in wonder, then saw a female shark travelling along the reef just beneath the crashing waves. I swam toward her, trying to get close enough to see who she was. Flora! For years, I had wondered what she did during the day, because she was consistently absent from the lagoon, even the barrens, yet she was nearly always present in the evening.

The females did visit the ocean it seemed, more often than the males went to the lagoon. And here they were together. Perhaps this was an explanation for some of the times I had puzzled over an empty lagoon. My sharks had gone roaming in the ocean. I climbed back into the boat and headed home, swept along by the east wind wrapping around the mountain and joyful in the glittering light of midday.

Determined to keep up my data collection at other sites, two weeks passed before I returned to Site One. The dry season had come, and the sea was lying silently, flawlessly reflecting the blue above. It was nearly midday by the time I arrived, so I didn't expect to see many sharks. They were usually sleeping in the barrens—or the ocean—in the middle of the day. But I wanted to check, feed the fish, and look again for the crocodile needlefish.

I had a few scraps saved in the freezer, for favourites. Two sharks were waiting by the time I anchored, but they circled peacefully. Underwater, I found Valentine and Madonna, and threw in a handful of food, which they began to investigate. Muffin was sleepily cruising around, and Jem appeared. I was trying to get a good photograph of Valentine in clear water and good light when Martha soared in. I went to the boat, as she snagged one bit of food after another, then came to me, sank and passed beneath. I handed her the chunk of fish I had brought for her, and she coiled up vertically to snatch it. It was rare for Martha to get so excited.

She ignored the handful of bits I threw in, so I took the last bit of food, and waited while she circled, ready to throw it when she approached again, camera raised. Just as I threw, she coiled, and I snapped. Marco zoomed in behind her, and I took his picture too. Samaria entered the site, though normally at this time of day she was sleeping half a kilometre to the east. The scent had not moved at all in the still water. Martha was circling away and I went with her, angling slowly eastward. However, she slowly accelerated until she left me behind.

I returned to the site and found it full of sharks! Cinnamon circled, stunningly large and pale in the full light of day. What a magnificent shark she had become. Never had I seen one grow and change so fast. Just a year before, she had been a slender juvenile roaming around on the edge of the barrens. Had I not been underwater, my mouth would have fallen open in awe, as I turned with her orbiting me.

I hung in mid-water in the centre of the site, gazing at her and all the sharks swirling around me, too amazed to even go to my perch. Tornado sailed past with Cheyenne. Circling farther out was another superb female, heavy and grey with a distinctive wavy tip line—Barracuda! This was the group I had been drawing when the whitetip shark had glided between my eyes and slate at Site Two five days before. They were moving down the lagoon from as far away as Section Three or farther, causing excitement, it seemed, everywhere they went. Kilmeny was there again at the same time as Marco. Eclipse soared through. She usually travelled alone, or joined different companions as she travelled. It was the first time I had seen her in the light of day. Many residents were among them, the big females swirling in the site with the visitors, while farther off in the coral circled the males and juveniles. Beneath the surface, just at the edge, two small shark pups tentatively approached. They turned and vanished down-current only to reappear again, but always too far away to identify.

Still I drifted in the site as the big sharks orbited me like planets. So unprecedented was the event that I had forgotten all protocol. No doubt this unusual meeting created a lot of submarine noise, and had attracted everyone within a large radius. Even one of the nurse sharks came out to see what was going on. In all that time, I had never seen one out in the middle of the day. Gradually, the sharks drifted away. They hadn't wanted the food, which was still lying on the sand. They had only come to socialize.

And sliding forward beneath the silver surface came the crocodile needlefish! What a relief it was to see that she was still all right. I fed her the bits of food that I had kept separate for her. She waited, poised in the water for me to throw each piece, but stayed farther away than had been her habit at our former sunset rendezvous.

When another fifteen minutes had passed, the young friends, Shilly, Ondine, and Ali came circling around me. Of the adults, the only one remaining was Eclipse, who came and went many times, always at high speed. I never saw her circle the region in the pattern normally used by the sharks. She shot in and shot out again, making photographing her problematic. She too was getting old. She had a few of those odd snowflakes I had seen on Kimberley, Twilight, and Lillith. Her colour line, like theirs, was hardened and twisted, as though it had been outlined with a heavy black pen by a hand that favoured curlicues.

It was the day before the dark of the moon.

Subsequent visits to the lagoon during the day resulted in pleasant interactions with the residents, promenades with Martha, and sometimes Bratworst, Valentine, or one of the others. These, plus Madonna, Cochita, Eden, Windy, and Apricot, were the sharks I usually found in the vicinity of the site.

But there was not another visit like that, with the group of large visitors present. The extraordinary session emphasized the random nature of the sharks' appearances. Though they could follow a strict schedule, it was flexible, and could change from day to day or night to night, depending on conditions and events.

Sometimes, the barrens was filled with cruising, sleeping sharks, and other times it was empty. Shilly was often in the western part during the day, and she usually accompanied me when we met. She had grown a lot, and would likely mature the following year. Ondine, Ali, and Cinnamon frequented the same area. Samaria and Madeline were often resting in the barrens, or on the sand in the coral glades on its other side.

On one occasion, I fed Shilly when she came to me in her home range at the foot of the barrens, then cruised up to the barrier reef. After exploring there, I arced back down through the exquisite coral garden of Section Two. The water was much clearer there than it was in Section One. I swam slowly, looking around, hoping to see some of the local residents, looking behind from time to time in case anyone was following me. I had no food or scent with me. After swimming for a long time, I noticed a fish vertebra, and wondered who could have been holding a shark feeding there, until I realized that I had reached my own Site Two!

I started back, using the kayak as a sail to carry me homeward in the rising wind that had been piling up waves in my path. But an approaching shark appeared, so I let go of the boat. It was Bratworst, who closely circled me with a wondering look. She had followed me all that way, far beyond her usual range. Seeing that I had turned around, she circled nearby. I began to sail again, but came face to face with Shilly. I had fed her in the barrens an hour before! She swam to me calmly, then away, in her usual gesture.

This behaviour pattern of following out of visual range, when it was not associated with the search for food, suggested an unexpected level of curiosity—I had noted this in Bratworst before. Perhaps these reef sharks also followed other species of sharks or fish—one cannot expect an animal to treat one differently than it would treat other animals. Any clue to their way of thinking was intriguing.

Sailing back, trailing effortlessly behind the wafting boat, I made little noise, and far through the clear waters, could see sharks gliding, a spectacular sight across the vivid coral hung with watching fish. Sparkle Too, the rare male visitor Blade, and Tornado passed near enough to recognize. Knowing that they moved in loose contact, often out of visual range, made it impossible to tell whether Tornado was alone or still with her group.

When the weather was calm enough, I went back to the ocean outside the reef. This time I had a few scraps saved, and threw them in, in hopes of attracting the sharks from deeper down, who habitually attended the shark feeding dives there. The visibility in the ocean was very bad, though outside the reef the water was usually transparent. This time, it was Keeta who rose to see me, circling close and looking, her chocolate-coloured skin satiny in the sunlight. Then she angled down into the depths, but soon returned to circle the food, dimly visible below.

Some males were roving over the landscape, so I dove, and found the visibility below much better. The male sharks were all strangers, so I recorded their distinguishing features while catching enough breaths at the surface to stay comfortable. But this was challenging. From the surface the sharks were just dim shapes below, and drawing both sides of the same shark on one dive was rarely possible. But there were some with distinguishing features, which I would be able to recognize again. One was a young female missing the tip of her tail. Her fin tip line formed a very even wave. I named her Saragossa, after the Saragossa Sea, because I found her in the ocean.

I was thus occupied when Trinket zoomed by as if I weren't there. He still didn't look much bigger than he had when I had met him. Two sightings in two visits strongly suggested that I had found his new home range.

After an hour, I was cold and queasy in the rolling waves, so I decided to leave. I had seen between eighteen and twenty male blackfins. They were all unknown except for Trinket, and a large individual whom I had followed around until I had drawn both sides of his fin. It was Christian, identified the year before at Site Three.

Three large, silver remoras had joined me, so I fed them briefly. Then another female appeared far off in the blue and I swam down toward her. She paid no attention to me while I dove three times trying to see her as she scouted about in the coral, intently occupied. At last, the shark left what she was doing and came to the surface where she cruised around me briefly. It was Cochita. She was not interested in the food and pursued her business until she disappeared from sight. The three remoras flowed with me like water.

Again I was about to leave when I saw another large female in the distance. I dove down to look, using all my force to swim after her along the bottom, the remoras flitting over me as if the motion delighted them. I thought it was Madonna, but also thought I could be hallucinating. I was forced back to the surface, which seemed terribly high above, and it was many minutes before I was able, again, to dive down. I watched closely, gasping at the surface, for the form of the great, dark shark to return. When she did, I dove again, and this time was able to approach close enough to see the dark spot she had above her left pectoral fin. At the same moment, she became aware of me. As I flew to the surface, she came, zooming vertically after me, then spiralling with me as I flew toward the light. Rushing together upward through the vast blue space, we gazed eye to eye, locked together, just for that moment. Then I broke through the surface, and she dove straight down into the blue and disappeared.

I had such a feeling of joy to know that she was there and had always been, while I was doing other things—that they all were. How beautiful it was that they were all right, pursuing their mysterious lives, and that we could meet to our mutual pleasure.

O Madonna, Martha, Isis, Flora, and Valentine. . . There was us!

Far, far below, more sharks were coming, and suddenly three large tuna crossed in mid-water. After a few more dives to get a good look at the arriving sharks, I decided to get out of the water before I drowned. By then the remoras had attached themselves to me and appeared to have gone to sleep. When I scattered the last crumbs of food for them, they woke up and excitedly fed, chasing bits around me until one of them entangled in my hair. I freed it with difficulty, climbed aboard, and rinsed the rest of the crumbs into the sea. The silvery fish would have no trouble finding a shark. I put away my gear, took my paddle once again, and set off for the island, far off and shining like an emerald in a blue abyss of sea and sky.

In July I was feeding the fish in the middle of the day at Site One, and was spotted by some crested terns, who flew above crying out. In moments, they were joined by every other tern within at least a kilometre, if not two. I threw bits to the birds, and scattered handfuls for the fish, watching the birds underwater, as they dove into the shark site up to half a metre (20 in) to snatch up the pieces. Bratworst was gliding through the area, and finally came for a bit of food I threw for her. She ate and circled placidly. It would have been hard to imagine a more sleepy looking fish. Martha came, then Madonna and Valentine. I kept throwing some for the birds, and some for the sharks, and when I had finished, and looked underwater, there was no one. They had scarcely eaten anything.

That day, I found a net about a hundred metres (130 ft) long, strung across the lagoon from the reef toward the shore. It was all caught up in the coral, and though it wasn't nearly as deep as the one that had been strung along the border, it was very heavy for me. It took all my strength to remove half of it, untangling it from the coral first, cutting it, where convenient, with scissors I had begun carrying with me for such chores, and hauling it bit by bit into my kayak. Though it was an old and obviously abandoned net that had been there for some time, I felt obliged to keep an eye open for anyone coming into the area. I was too exhausted afterwards to continue, and had to return the next day to finish removing it.

However, throughout all these watchful meanderings, I failed to find my crocodile needlefish again.

CHAPTER THIRTY-NINE

Mysteries of the Sea

The sessions in the lagoon across the bay did not become easier, but I became familiar with the conditions. The winds accelerated there suddenly, pushing me into the breakers on the fringe reef. The ocean waves, invisible as they crossed the bay, suddenly grew mountainous, threatening to capsize my fragile craft upon the hidden sea mount. These natural forces had driven many ships to ruin. The remains of the Kersaint, the last French war sailing ship, still lay upon the reef a century after its destruction. I learned to judge whether I could pass between the shore and sea mount, or had to widely skirt around it, which was usually necessary. It was a marathon kayak trip, but fortunately I had no more encounters with fishermen.

The enigmatic Shenandoah was usually in the area and cruised endlessly, sensed rather than seen as she passed time after time through the vicinity. When she came to eat, there were usually some juveniles present. At times she never did eat, just drifted ghostlike, then vanished. How often did I believe that she had gone, only to find her close behind me twenty minutes later in the soup-like surroundings. Her behaviour was unlike any other shark I had known. The youngsters were Randall, the little male I had first encountered there, Tessa, Twinkling, Griffen, Kinnikinick, Pom, Coquille, who was a tiny pup, and many more. Two other resident females sometimes came near the beginning too, but the little community, though structurally similar to the one in Section One, was markedly different in that different adult females came at different times and different sessions. While this was similar to the pattern at Site Three, I speculated that here, it was because their lagoon home was so long and narrow—so were their home ranges. Perhaps only Shenandoah lived more permanently near the eastern end.

I was beginning to know and enjoy the females who came. They were often shy, and lingered in the coral, until darkness could hide them. Then they glided forth, and came over to look at me where I crouched under the low surface. Finally, with much coming and going, and approaching very close to me again, they chose morsels from the little pile of scraps, and ate while I drew their dorsal fins. Strangely, they were all very large females. Even the young ones were unusually large, as Cinnamon was.

There was Mercy, Carmen, Chimère, Tripoli, Tiffany, and Flicka, who were beautiful animals in the prime of life. Then there was Wrath, who looked very old. Ambrosia was pale gold and Nutmeg, her near twin, had the general look and colour of Cinnamon. There was a shark remarkably like Madeline, even to her swift and forthright way of moving. On both sides the two sharks' dorsal fins were nearly identical except for a few tiny details. She was also bigger than Madeline. It was the closest true lookalike I had found, and I named her Madelena.

When it got dark, there were times when the sharks became very bold, and alarmed me as they had at my first visit. They would shoot up behind me, or accelerate suddenly to nearly touch my nose when I was looking at something else. The visibility was always poor there, and in the dark, things became eerie. When I couldn't see and sharks were flitting invisibly around me like huge remoras I would leave.

Collectively they behaved more like Shenandoah than the sharks in my original study area, an odd phenomenon which brought again to mind the concept of a cultural difference. I looked for Janna, the highly aggressive shark who had appeared once with Shenandoah at Site One, and whose behaviour had been exceptionally bold—I had wondered if she and her companions came from a place where they habitually intimidated fishermen. But I never saw her there. Once again, in spite of the bold and sudden approaches that these unfamiliar sharks made toward me, no one bit!

It appeared from these forays, that the sharks did not habitually cross the bay. None of the sharks I found there had come to my sessions except Shenandoah. Living at the far eastern end of her lagoon, she had still only come twice in more than three years, and none of my sharks had appeared there. So even though they went out in the ocean on the other side of the barrier reef from their range, they did not seem to go too far to either side of that region, or cross the bay to the next lagoon.

I estimated that in general their home ranges were about half a kilometre (a third of a mile) across, depending on the shark, and that individuals could be more or less attached to their ranges. Due to strong individual differences in this facet of their lives, some, such as Martha, were nearly always at home, while others roamed more frequently, and some left for months at a time.

But my observations were made during the dry season when the residents were usually at home. During the period of reproduction, I might have seen some of my sharks there, given the numbers of travelling females passing westward through my study area, beginning in September, though the lagoon's main nursery was at its eastern end. The females in Shenandoah's lagoon might not come eastward to my study area because they, too, went westward for parturition. Perhaps as a result, the community tended to roam westward rather than eastward at other times, too. But I could only speculate.

Later I learned from Dr. Samuel Gruber, a close friend of Arthur's, that studies he had conducted with Atlantic lemon sharks had shown that a high fraction of the females returned to the place where they were born to give birth. His findings revealed a possible reason for the roaming patterns I had seen.

Indeed, the marine biologist, Johann Mourier, Ph.D., who began to study these sharks after I left the area, confirmed that this was true of the blackfins. He found a major nursery on the west coast, and evidence that a high fraction of the females on the north shore travelled there for parturition, because they had been born there.

Further, Johann found that a significant number of the females were going to another island when they left to birth! (Johann's findings are detailed in the epilogue.) This meant that the juveniles, once they left their nursery shelters, travelled far more widely in search of a home range than I could have suspected. Given the sedentary lives of the sharks I knew, it was hard to believe that some crossed wide reaches of ocean to other islands as juveniles, then returned years later, to their home island, and nursery, to birth. Yet this fact correlated with the unusual roaming patterns I had seen in certain individuals, including Valentine and Carrellina. If they spent part of their time on other islands, that would explain their long absences. Certain rare visitors could have come from other islands, too. The roaming of the pups had always been mysterious, because many appeared, but I saw few of them later. If they were roaming far afield, including among other islands, that would explain these observations.

In his book, *The Secret Life of Sharks*, Professor Peter Klimley described how he had found evidence that sharks may follow magnetic field lines while travelling through the ocean.

Further, the different islands in a region would surely be recognizable to sharks through the sense of smell. Each island could be visualized as a tall column in the ocean floor, with its top in the clouds, washed by rain and drained by rivers, each one home to a unique variety of species of plants and animals, terrestrial and marine. A long, slow scent trail presumably drifts through the ocean from each one, carried from the lagoons by the currents that flow into the sea, and by its rivers' waters. Such a scent trail could spread slowly and very far, yet, since we cannot smell underwater, we would remain unaware of it.

Whether sharks travelling through the ocean would detect these island scent flows, and use them for navigation, is not known, but it would seem possible. Salmon returning to their home rivers to spawn are said to be able to scent it from far off. Similarly, travelling sharks, using the sun and moon, and maybe magnetic field lines as guides, might easily sniff, from the soup of oceanic smells, the lost scent of their origin, and follow it home.

The sharks who are residents of those islands in the middle of the Pacific Ocean, are descended from individuals who initially traversed the ocean from the archipelagos of Asia, to find them. The ability to identify the presence of each of the many islands by its scent flow, would surely be even more advantageous to sharks roaming the vast distances between archipelagos. Given the individual differences between sharks, one could theorize that sharks descended from these successful travellers would have inherited the genes involved, and might tend to travel more themselves, than individuals of their species who were more sedentary. The routes they took to populate the volcanic islands that are now strung across the pacific are unknown, but DNA studies will eventually reveal those patterns.

One of the points I raised in my eventual article for the journal *Marine Biology*, was that sharks originating in the northern hemisphere would likely mate in the northern warm season—from April to September—while those from the southern hemisphere would follow the patterns displayed by my sharks. There could be places where sharks from both regions now live, and where, unless one kept track of the individuals, it would appear that they mated and birthed twice a year.

There were many unanswered questions, as I considered their roaming patterns. Carrellina's visits from December to April had begun the year before she was mature, so could not have been related solely to reproduction. I had no idea that she could have come from another island. She had come at the age in which a juvenile settles in a home range, but she had not stayed there as she matured. Then there was Valentine's case. She used the study area as a home range only outside the period of reproduction. And each of the others was different in her own way!

I hoped that my sessions in Section Five would shed some light on the travels of the sharks, so I went as often as I could. But it was hard to go. The water was cool during the dry season, and I became so cold I could scarcely think clearly enough to drive home on my return to shore. I couldn't use a wet suit because it made swimming such a distance impossible. The location was too far to go in my kayak. It was beyond my physical abilities to paddle the heavy kayak that far against the wind. The kayak was too large to be carried on our little truck, so taking it with me was out of the question. My art was not selling in the non-existent art market, so I lacked any funds to get a smaller kayak, or a little motor for the one I had. The sessions always had to be cut short because I was so cold I couldn't cope any longer. I had to get back to shore before darkness fell, too, so they were not as comprehensive as the sessions I held in the other sections.

But I was gaining a fair impression of who the residents were. By the time mid-July 2002 came, I had identified many juveniles, and determined that a few of the mature females I had met at the centre of the lagoon, actually had their home ranges at its eastern end. As it had seemed, they had just been passing by at Site Three.

Innisfree came regularly. I had met her so long before, it seemed miraculous to have found her again, with others whose location for years, had been mysterious. She was one of those individuals like Antigone, who was put together just right, in perfect proportion. Her skin was so radiant she glowed. Like Keeta, she was one of the biggest sharks I had found. She reminded me of the beautiful two-toned Carmelle, who had disappeared soon after I had met her. I enjoyed seeing Innisfree so much that I called the Section Five sessions the Innisfree sessions, and would say to myself as I prepared, *I must arise and go now, and go to Innisfree*, fondly remembering *The Lake Isle of Innisfree*, a poem by William Butler Yeats. It was worth freezing to death to see her.

But mid-July brought bad weather and the current increased, adding a new dimension to my difficulties. It took fifty minutes of flat-out battling through the streaming waters to get to the reef, trailing my little raft. I had very little food upon it, but they were decent pieces of fish meat, with no bones included. After I had found a good place, just within the reef flats, I threw in a few pieces of fish and crumbled one to get a scent flow going. With my raft relentlessly pulling me down-current, I was finning hard just to stay in place. Coralie, one of the youngsters arrived, and came swiftly up to my face, then passed, circled, and soared away, before she returned and began nibbling at the food.

Innisfree sailed into view. She swept down to the food then circled and came straight to me. I had to back up and fin rapidly between us. She dodged away. Luckily I was well set up. I just had to reach up to get the food, without taking time to extricate it from the bottom of a folded sack as was often the case. Innisfree curvetted away, and I threw in more food. A young male approached, displaying a fin I had not yet drawn, so I began to sketch it. When I looked up to double-check my drawing, there were many sharks speeding around, including a pup going too fast for the eye to follow. Everyone was excited, and strings of sharks were coming behind them.

I did not have enough food!

Innisfree dodged toward me again with frightening speed. A group of others moved with her, and I flew backward, raising my fins between us. She kept coming. Beyond, the coral was full of excited sharks all focused on me, all zooming forward. And still Innisfree didn't turn away! She shot toward me as I finned away from her faster and faster. Still backing up as the current pulled me toward her, I grabbed handfuls of food, and threw them in different directions until it was all gone.

Then I violently turned and swam away. No one followed. They would be busy locating the pieces I had thrown, and I didn't want to be around when they demanded more. I flew shoreward on the current, thankful that I had not become entangled in the line attached to my small raft, in my panic.

Obviously it was absurd to go there without my kayak to retreat to in case of trouble. I had not expected the sharks to become so pushy so fast, but then each one is different, and the behaviour of one affects the others. Innisfree's lightning-fast dodges were ones I associated only with my bad sharks. Yet she was older and massive, and she hadn't known me long to be so bold. At Site One, the big females had always been the shiest!

But the full moon period was upon us, which had not crossed my mind earlier. When I was at home, coordinating picking up the scraps, and arranging things so that I could get away for a shark session, was often challenging, and I didn't take such things into consideration. Yet, being alone at sunset a kilometre from shore with posturing, excited sharks, mentally slow from the effects of hypothermia, and lacking the protection a kayak could offer, seemed unwise. The situation had developed so unexpectedly and so fast, it had left me shaking. I trembled for the rest of the evening, and, needless to say, I did not go back.

Five days after the thrilling noon-time meeting at Site One, I held a session at Site Two. The group of female visitors I had seen there, which included Tornado, Eclipse and Barracuda, had moved back up the lagoon, and I was favoured with another session with them, at which they lingered, and for the first time with me, they ate. Twilight appeared while they were there, still without Kimberley. I felt a twinge of anguish every time I saw her, because by then I was sure that Kimberley had died.

Twilight came regularly to those sessions; her home range was nearby. Most female sharks living farther away than that, appeared to be in the once-a-year category of visitors to Site One, if they came at all. Flannery appeared more often than he had at Site One too, and I had frequently seen him at Site Three. I suspected that he was one of the unusual males who lived in the lagoon, and generally roamed through its centre. Mordred had always passed there, being the most widely-roaming shark I knew of.

Slowly I identified and named the sharks who came. Belladonna was the resident female. Though she was not far away, she had never come to a session at Site One. Solace was pale grey and very pregnant. I had met her the year before at Site Three. In the year of her maturation she had grown suddenly as Cinnamon had. Cotlet, Florin, Tuvalu and Megan were half grown sharks. Then there were all the tiny pups: Trill, Faroe, Cygnet, Kinnikin, Papillon, Coconut, and many more who came irregularly. Probably many circulated through the vast area of thick coral beyond, and came when they crossed the scent flow. Two had blotches of different sizes all over them. The only other shark I had seen with such colouring was Dapples. One of the pups had a deformed pectoral fin; it was too large, and an odd shape.

I knew most of the adult sharks visiting Site Two, but occasionally saw a new one. A magnificent female came one evening. She was very dark, with a narrow white band on her dorsal fin, enhancing the effect of a passing shadow. I called her Lochalsh, and enjoyed the breathtaking vision of her roaming near with Jessica and Belladonna. Kilmeny and Cochise, with Revel, another new and very dark male visitor, came to that session too. The rarely seen Raschelle, and even Storm came. They were travelling, and I wished I could determine where from and to, since the more I watched, the more it seemed that all my investigations had revealed more mysteries than they had solved.

Each time I went there, I stayed close to the shore until I was well past the barrens, and into Section Two, before crossing the lagoon to Site Two. That way I could be sure that Martha, Madonna, and Bratworst would not hear me going by. I did not want them to find out what I was doing.

Some of the younger females from Site One began to come to the sessions in Section Two. The haven for juveniles provided by the thick coral there made this piece of the puzzle fit in just right. Naturally, the juveniles would be the first of my sharks to cross the new scent flow drifting toward the barrens and discover my new site. They probably still spent much of their time in that familiar region, when they weren't exploring, prior to establishing a home range. Indeed, Mara'amu, Ali, and Teardrop began to come, and at one session, Rumcake met me.

The only outlets for the ocean waves surging across the barrier reef were the borders at the ends of the wide lagoon. Though the current was slower nearer its centre, it inexorably brought the news of my new feeding site westward toward the barrens. Since sharks can detect very faint scents, only a trace of scent would be enough.

On the day after the full moon at the end of June, a big group of residents from my study area, including Madeline and Christobel, cavorted into the site. Sparkle Too zoomed up to me, and circled closely. Shilly came straight to me too, circling tightly as she always did when we met. Cheyenne and Merrilee were with them—they may have been moving back to their own range in Section Three, joined by some of my sharks for a while.

I had often seen that residents will join travellers for awhile. Encountering a group of my sharks together with travellers known to live farther along their route, provided more evidence of this. It supported my impression that sharks from the same range knew each other, and at times travelled together too. There was a large gathering that evening, and the site was so full of little pups that it was impossible to keep track of them all.

I had known that my sharks would eventually learn of my change of site. They would be especially displeased on coming late after the Section Two residents had eaten all the food, when they were used to having the first pick. By early July, Jem, Rumcake, Teardrop, Mara'amu, and Mordred were usually cruising slowly in visual range when I slid underwater; all were juveniles, and males who travelled more widely.

Once when I slid underwater twenty metres (60 ft) too far to the northwest, Mordred appeared over the coral wall of my usual location, looked, and curved back down, as if to say, "We're over here." They were all circling around in the coral circle of Site Two, and Sparkle Too, near the top of my list of bad sharks, was there. But they waited while I placed the food, and didn't trouble me.

At the next session, I was happily watching the antics of the shark pups and noting the residents as it began to grow dark, when I was rudely jolted from my trance by the sight of a large group of sharks approaching up the scent flow as if on urgent business. They surged into the site, and orbited it and me—Madeline, Lillith, Glammer, Ondine, Ruffles, Darcy, Muffin, and Fawn. Christobel came a minute later.

Lillith was a dramatic sight. She was very dark, with prominent white mottling, and was swollen with pregnancy. Her behaviour was markedly changed at this site. The first thing she did was circle me swiftly. Then she went into search mode and ate the few scraps she ferreted out of the surroundings.

They all seemed thrilled, and flew around the area, looking for scraps and partying. Eclipse passed on one of her usual hit-and-run visits. By the time they dispersed, it was nearly dark. I went to the boat and got in. Exhausted, it took me a moment to realise that I had forgotten the anchor. So I slid back down. There was a big dark female shark in the site, and I drifted close to see who she was. It was Madonna—she had found the second site.

I drifted away, watching for dark shapes gliding under the gleams upon the glowing surface, as the shadows deepened, and the stars began to shine. It was a long way to go back now, so much farther along the lagoon, across the wide gulf to the island, and then all the way down the bay. I always had to fight the trade winds on the way out, except on those occasions when we were treated to a visitation by the southern wind. When I returned in the hush of night, the winds had fallen, and a different planetary effect was at work. The cool air from the island's summit flowed down its flanks and accelerated down the bay, fanning out at its entrance, and creating a breeze strong enough at times to raise white-caps against me, all the way back toward the island.

While paddling, I gazed at the perfect peak of the Shark's Tooth Mountain which was always in the centre of my view, a small but surely the most beautifully shaped mountain, pointing straight up to heaven, like a finger pointing to the divine. And suddenly it seemed as if there were a very bright star above me. Glancing upward, I saw that there *was* a bright star—unnaturally big and bright, brighter than all the others! As I stared in surprise, it very slowly dimmed during the following minutes, until it disappeared! I stared and stared into the blackness where it had been, and there was no trace. I had never seen anything like it before, nor could I imagine what it could be—no plane, no UFO, no supernova, no meteor. It was something seen but unexplainable, a reminder of the mysteries of the universe.

The next session at Site Two was on Saturday, July 20, 2002, at the usual time. The weather had been mildly stormy but began to clear in the afternoon. As I struggled with the heavily loaded kayak out to the distant site, squalls were passing across the reef, the sea was disturbed, and the wind gusted against me all the way. When I slid into the water, Lochalsh was passing, her unusually dark appearance and dorsal fin unmistakeable. She appeared to be travelling with a friend who was swimming on her other side. The two big sharks were moving steadily eastward in mid-water. I put the food in place right away but they did not come to eat. I was disappointed, because I wanted to see her again and wondered why. It was unusual for a shark to be present when I arrived, and not come to eat when I had been trailing scent, but she did not know me.

Ruffles was the first shark there, suggesting that he was travelling too. He was with Avogadro, and Tuvalu joined the two older males. They circled and ate alone for a while. Eventually a large, dark, male appeared and began to circle the area. I had never seen him before and called him Darwin. Of all the sharks who had come so far, only one, Tuvalu, who was a juvenile, was a resident, and all were males. This was unprecedented.

Finally Poppet came, and eventually Belladonna. Then Shilly came straight over to me in greeting, before circling to eat, and Florin, another juvenile resident male, followed her in. Papillon, Coconut, and Cygnet, all just two to three years old, arrived and ate with them, so by then I had a little group. But this was nothing like the usual sessions there. There were too many males and too few pups, juveniles, and females.

Sparkle Too swept in and right up to me, orbiting at arm's length and making some feinting, fast approaches. Then she went to the boat, sniffed it, and finally circled down to the food and ate. She stayed in the area until I left at nightfall, which had been her unique pattern at Site One, too, to stay until the end. More juveniles and males drifted in. Shimmy, Mordred, and Caro, a rare visitor, swept purposefully in together. Caro passed close by many times, looking. I returned his cool gaze while rapidly sketching the fine detail on his dorsal fin—he had never come close before. Merrilee appeared, and after a slow survey of me, she went into search mode, looking for the best scraps remaining.

As time wore on, the site remained strangely quiet. Sparkle Too and a few others passed slowly in and out of visual range as they cruised the area. Rachelle joined them, and an unknown young female came to look closely at me several times as the twilight darkened.

Suddenly a big, dark male with a deformed dorsal fin shot into the area. He was twitching in that odd way I had seen in very small juveniles. The shark jerked and instantaneously changed direction repeatedly and rapidly as if he were made of water with ripples and waves running through him. The effect was bizarre and startling. It was as if the animal had gone mad. Thrasher, also a rare visitor, was close behind him with Flannery. The disturbed shark remained in the area until I left, and his continual jerking and unpredictable flights added a note of alarm to the aura of strangeness over the scene.

The juveniles Pip and Fawn arrived from Section One, and completed the list of sharks who came that night.

This session took place four days before the full moon, so it was already the period of light at night. Though I considered the lunar phase in searching for an explanation for the bizarre session, I had never seen any change in the activities of the pups or juveniles due to the phase of the moon before.

The session had been very different from the pattern established at that site. Starting with the initial purposeful passage of Lochalsh and her companion, it had been dominated by travelling males and strangers. These males and two of the females had never appeared before outside of the mating season. Few juveniles, and only one baby, Poppet, had been there, whereas usually the sessions there were alive with tiny pups. Only two thirds of the usual number of sharks had come, and a high fraction had been rare visitors, or sharks I had never seen before.

Since sessions in the dry season tended to be predictable, and often weeks passed in which no visitors appeared, the change was even more peculiar. After such sessions I always looked forward more ardently to the next one, in order to gain a greater understanding of what was going on, but this time, I was unable to return to the lagoon again until another week had passed. My three step-children were visiting and there was simply too much to do to indulge myself with a mid-week trip.

On Saturday, July 27, 2002, I returned to Site Two, slid into the water, and put the food in the centre of the coral circle. No sharks were visible and I waited. Every so often I checked my watch. Five minutes, then ten minutes passed, and no group of thrilled shark pups flitted from the coral beyond, no residents came circling from the veiling light, no one came. Half an hour passed. The still lagoon began to slowly darken. I drifted down-current but the view was empty in all directions, where no shark glided. I returned to my lookout. Over and over my eyes travelled from the little pile of food on the sand, nibbled by a few fish, up the plume of scent that flowed away, and into the empty distance. The shadows of night began to deepen, and I drifted again down-current. After another long wait, a small shark passed in the distance, but it neither returned nor came to eat.

As the curtains of night drew over the empty landscape, I had no choice but to put the scraps back in the boat, climb in, and paddle home. I could not sleep, rose early, and before anyone would need my services, I put the scraps back in the kayak, and paddled at top speed to Site One. I had to see if the sharks I loved were there, safely living in peace in their home ranges. They were not. Trailing scent, I swam from the border to the site, where I waited, creating a strong scent flow and feeding the fish. No shark came.

After twenty minutes, I began slowly trailing scent toward the barrens, where I did an intensive search as far as the scent flow from the previous night would have reached, in the hour I had spent waiting. No sharks, not even one at the limits of visibility, appeared.

So I went home, got through the day, and flew back to Site One as the sun began sinking. I still had the scraps, guarded between searches in the refrigerator, and used them one by one to create a scent flow throughout the hour before sunset. Countless fish came to eat, unharassed by sharks or nurse sharks. How strange it was to feed them endlessly with no sharks appearing. As darkness gathered, I could not give up and stayed long after the fish went to sleep, long past the time I would normally have left. Just as I was telling myself there was no use waiting, a little shark drifted by at the visual limit. Then it moved slowly through the site, and I was finally able to recognize Tamarack. But he neither approached nor ate. The scent flow held no interest for him. I waited in case his appearance heralded the coming of other residents, but darkness fell upon an empty coral garden.

Deeply frightened, I hurried the heavy kayak home, rinsed my gear, showered, dressed, rushed to the phone, and called Bou Bou. He answered right away with the news that he had been intending to call me because the divers hadn't seen any sharks outside the reef for nearly a week. Starting on the Tuesday of that week, few had appeared, and since Thursday, there had been none. He was concerned to hear that the lagoon had emptied of sharks too, even the tiny pups who sheltered in its thickest hidden regions of coral.

Yet the weather had been settled all week. The trades had been blowing just right, the sky had been its usual abyss of blue, sometimes with faint high cloud, and nothing in nature was disturbed. The only thing I could think of that could have caused the sharks to flee, was a massacre somewhere. On Monday morning, I began to make discrete enquiries and talk to the other dive clubs.

The sharks farther along the lagoon had disappeared, too, I learned from those who held the hotel shark feedings at Site Three. None had appeared, either, at another commercial feeding site in the next lagoon, far west of Shenandoah's range. All dive clubs reported that the blackfin sharks had vanished from the outer slope of the reef along the north shore, at the sites where they were usually seen.

Dr. Michael Poole, the dolphin and whale biologist, told me that the marine mammal population had been behaving normally. His tours, and others, had circled the island all week watching dolphins leaping, and visiting dive sites and ray feeding sites, while buoyed by tranquil seas. No dead whale floated offshore attracting all the sharks to a banquet, and no one had noticed anything unusual.

I contacted the military and asked if submarine activities had been carried out which might have created such a submarine racket that the sharks had fled from it. None had.

Further, had some submarine noise been the cause of the event, the sensitive marine mammals would likely have been affected. Yet there had been humpback whales in the bay the day before, two hours before I left for my evening search for the sharks. Whatever had affected the sharks had not troubled the whales, or they would not have been resting there in the shelter of the bay.

I enquired whether there had been any earth-quake activity which could have produced submarine vibrations of a frequency that could have frightened the sharks over a wide area. None had been detected. It is something that is closely watched on the islands, due to the danger of resulting tidal waves.

Had there been some toxin in the water, other species would have shown some sign of reaction too. It was impossible to know about the nurse sharks since I never saw them as much at that time of year as I did in the wet season. I had not seen whitetip reef sharks during my search, but then I saw them infrequently, so could infer nothing from that. The fish were behaving normally. Had I been more organized or had help, I might have thought of searching for them, too, but I was focused on the sharks I loved.

No construction work had been under way, there had been no oil or pollution spills, and no human activities had affected the region in any way that I was able to discover.

And I was reassured over and over again that no sharks had been fished. Even the dive club owners were sure that such a thing could not happen without the news of it being spread. Someone would have found out. They agreed that in the twenty years that shark feeding dives had been held, on only one occasion had no sharks come, and on the following day a hurricane had struck the island. It seemed that the sharks had sensed the approaching riot of the storm, and had moved to deep waters. Indeed, due to the disappearance of the sharks, the dive clubs were expecting a terrible storm, but none came.

I contacted dive clubs on other islands, and learned that the sharks there were in their usual ranges, apparently unaffected by whatever had caused the evacuation of our local sharks.

No cause could be found to account for the disappearance.

I sent out word over the Internet, enquiring whether anyone had seen a similar event elsewhere, and received responses from across the planet that no such disappearance of reef sharks had ever been noticed.

Since July 23, three days after my bizarre Site Two session, the entire population of blackfin reef sharks, including the pups, had vanished from the lagoon, and the outer slope of the barrier reef. Everyone who was observing sharks reported their disappearance, but since dives were held on the north shore of the island only, the event could have been much wider in scope. Perhaps all the sharks on the island had evacuated, but no one knew.

Due to my studies in different places, I knew that the affected region included separate populations of sharks who had almost no contact with each other. I had been able to establish myself, that there was little contact between the sharks on the two sides of the bay, and even between the border region and the centre of the lagoon, too, so I could be sure on that point. The sharks had failed, so far, to develop cellular phone technology, yet the separate communities had departed from a wide region of the island at the same time.

The agitated dark male shark swam through my mind time after time, twitching, feinting, and appearing utterly unnerved. What had been wrong? What had I really been seeing? The population had already been in dramatic flux, and the shark had been in obvious distress, roaming through the gloom beneath a restless surface in a place far from his home. Juveniles who had briefly behaved like that had been recently frightened. Had the sharks become alarmed by something only they could sense, and fled into the heart of the sea?

But the babies! Why would they leave their sheltered safe places? Their disappearance was the strangest part.

Some dive clubs said they had seen blackfins at the northwest corner of the island, but Bou Bou assured me that there had only been grey reef sharks there, only female ones, and no blackfins, anywhere. Two of the dive clubs reported having seen a small shark or two pass, but affirmed that each one had stayed far away from the divers and had not approached to eat. Yet at the traditional shark feeding dives, competition was always high for the small amounts of food brought. Since the shark had been a small one it must have been a small male or juvenile. I too had seen Tamarack pass in the distance—a small shark—and once at Site Two, one small shark appeared just within visual range, but neither of them had come to eat.

I anxiously roamed the lagoon, looking and waiting. Martha had only left her home range once in three years, apart from brief absences for mating and birthing. But in spite of my concern, I secretly knew that their evacuation corresponded with the full moon. It was the only sign I had that the event could be a mysterious, yet natural one. They had shown signs of disturbance on July 20, vanished completely on July 23, and the moon had become full on July 24. Now that the moon was waning, would they come back?

On Tuesday, July 30, I battled against strong trade winds, and anchored within the border of the lagoon, southwest of Site One at 4:25 in the afternoon. Trailing the boat and scent, I swam toward it. After about two minutes Bratworst passed just close enough for me to see her dorsal fin.

I reached the site at 4:34, and the fish streamed out to greet me. But I had nothing for them and continued very slowly, trailing scent. By 4:50, I was crossing the region of shallow water where the coral grew thickly and pups and juveniles normally appeared. There, a small male passed me but paid no attention. I didn't recognize him. As I reached the barrens a tiny pup passed slowly, looking, very close beside me. I had never seen him before, and wondered if that was because he had been displaced from his usual hiding place, and was on his way back. Soon after that, another small pup passed whom I did not recognize.

I drifted across the barrens where no sharks glided. In the coral on the far side of this vast grey expanse, Bratworst passed me again. Good shark! It was 5:18, and the light was fading.

Suddenly, a lot of sharks came into view. Teardrop passed and disappeared, Barracuda appeared briefly in front of me as she vanished into the twilight ahead, and Lightning flew by. The youngster was followed by three more juvenile males who zinged past as if I weren't there and vanished into the gloom ahead. A little pup passed to my right, and at 5:30, Flannery crossed very slowly in front of me. The sharks all appeared to be travelling deeper into the lagoon from the direction of the border.

Fifteen minutes later, I was nearing Site Two and an unknown female moved away as I approached. There was another in the distance who resembled Victoria, and another who looked like Belladonna, but they were too far away to be sure. Farther on, two juveniles appeared off in the coral, then Cotlet circled me.

I drifted through the Site Two region until it became too dark to see, and at 6:00 p.m. Ruffles appeared nearby with an unknown male companion, and Faroe, one of the local juvenile males. It was interesting that Ruffles had been present at the session I held there before the sharks had disappeared, and was there again when the first sharks seemed to be returning to the lagoon. Yet otherwise, I had seen him only once in the Site Two region, and he had been with a group of other Section One residents. There had apparently been an odd displacement of male sharks around the island in concert with the disappearance, given my observations of strange males in the lagoon prior to their evacuation. If the sharks had fled into the ocean depths, why had the males who lived in the ocean, come into the lagoon? Some females, such as Lochalsh and Raschelle, had been far from their ranges, too, while most of the residents had already left.

The next day, I slid underwater just inside the border at 4:45 p.m., and followed the same path as I had the evening before. It was 4:53 when I reached Site One. Fish swarmed toward me and I gave them a few crumbs before continuing. No sharks appeared. At 5:08 as I passed through the coral garden where the pups had lived, a large female shark moved away. She seemed to be another of the strangers moving up the lagoon.

As I traversed the dull plain of the barrens, Bratworst swam slowly by, apparently staying with me beyond visual range, as she had the night before. The visibility became worse and worse until it was an impenetrable fog. Someone or something up ahead was stirring up the sand. It was very hard to see anything. A shark materialized out of the gloom. It seemed to be Glammer, but in the thick grey soup I couldn't be sure.

A tiny shark approaching to look at me, panicked and took flight as a small motorboat full of Polynesians buzzed over. This was an unprecedented occurrence—no one took a motor boat into the lagoon!

The clouded water limited the visibility to two or three metres all the way past Site Two, then the lagoon suddenly became crystal clear, with no sign to indicate the cause of the disturbance. Not far off, the shallow boat channel turned from the reef toward the island. Perhaps the Polynesians in the boat had left the boat channel to scavenge for shellfish, then took the motorboat on along the reef, looking out for coral structures just under the surface, instead of retracing their path to the boat channel.

As I swam through the thick coral of Section Two, Tamarack came with me, joined by Bratworst. They stayed near for five minutes, then both disappeared. Shimmy passed. I arrived at Site Two having seen only two more shark pups, and no one else. Though a few sharks had returned, the failure of the rest to come, day after day, deepened my concern. The only messages I had received from scientists in other parts of the world in response to my requests for information about disappearing sharks, had suggested that a massacre was the obvious reason. It was hard to fight back panic over this possibility. Just the fact that I had seen only Bratworst on successive evenings suggested that of the resident females, only she was present. Most of the sharks I had seen were young males of assorted ages, which was strange too. Tamarack was in that category.

I let a day pass without going out. I was exhausted, and had responsibilities to my step-children to honour. They were anxious about their dinner each time I returned after dark, and would come into the bathroom while I showered and prepared for the evening, to ask what we were going to eat.

On Friday, August 2, I returned to the lagoon in the evening. It had been calm and sunny for the past days with irregular winds, and a front was moving in. The calm, cloudy atmosphere deepened as I paddled out to sea, and the ceiling lowered. By the time I arrived, it was raining.

A spear fisherman was moving slowly past the site as I arrived, so I slid underwater to the northeast, close to the reef, to avoid being seen. It was 4:50 p.m.

I had acquired some scraps, and took one to shred as I swam toward the site, which I reached five minutes later. There, I fed the fish. The visibility was bad, and after a while it was obvious that there were no sharks around. Restless and worried, I decided to move on, trailing copious scent. My sharks would follow if they crossed the odour trail. I gave the fish a few last treats. They were all eating from my hands, delighted to see me, and I left them reluctantly. As I drifted away, a shark passed down-current, and paused. Apricot approached, and just then a whitetip shark passed by me. It was 5:10 p.m. Beyond, Breezy was flitting forward, so I returned to the site.

Sparkle glided in with Bratworst, and Windy was in her wake. I threw in a few scraps, which they engulfed with no hesitation. This was quite different behaviour from their normal endless circling and fussiness over the scraps. Bratworst in particular inhaled entire *mahi mahi* dorsal slices without shaking them, as if she were starving to death. Normally she wouldn't touch *mahi mahi* at all. My beautiful Madonna joined them. Then they all swept on their way.

Eden, Samaria, Kilmeny, Peri, Storm, Pip, Cinnamon, Lillith, Flora, Madeline, Avogadro, and Keeta were following. They gulped down some food and zoomed on. I was delighted. I even saw a pup shoot by, that I thought was one that I had identified at the nursery the previous year.

The sharks were on their way back.

He jerked and instantaneously changed direction repeatedly…

Chapter Forty

The Aftermath

On Sunday, August 4, I began my search at Site Two, so that I could be at Site One when darkness came. It was 4:15 p.m. when I slid underwater. The visibility was normal, but the light was cloudy from a heavily overcast sky. I trailed a bit of scent and drifted with the current through the thick coral. Tuvalu, Poppet, a second baby, then a male shark appeared gliding in the vicinity. As I travelled, more sharks came into view, and Darwin, the big male who had come to my last session before the shark evacuation, drifted by looking at me. This was the only other time I ever saw him, and I wondered if, like Ruffles, I had seen him on his way to and from the unknown place they had fled to. That strange, dark session that began with the two big females moving steadily away along the reef, often came to mind, sparking different speculations each time.

At 4:30, I anchored and put down a few scraps, so that the sharks in the vicinity would come close enough to be identified. When all was quiet, Teardrop hurried to the food, followed by Cotlet, Sparkle Too, Muffin, Christobel, Jessica, Droplet, and several of the local little ones. I waited until they ate and the flow of sharks ceased, then moved on. Avogadro arrived as I left. They must have all been in the vicinity to come so fast.

As I continued, I saw sharks off in the coral briefly, then no one. Approaching the barrens, Teardrop glided casually by in front of me, then turned, and passed me again slowly. Apparently she had come with me through Section Two and this was the point at which she was going back. I didn't see her again, and she didn't appear at Site One, providing another clue that this young female had made her home range on the Section Two side of the barrens, in spite of all her visits to Site One when she was growing up. I had been keeping track of her since soon after I had begun the sessions, and named her when she was very small for the drop-shaped mark on her side. She had always been a beautiful pale bronze, and though darker now in her first year of reproduction, she was still a light-coloured shark.

In the barrens, I poured in a lot of scent and drifted slowly, watching. I wanted to make sure that all sharks in the area were aware of my coming, especially since I was approaching from the unexpected direction. But all across that barren landscape, no sharks moved. Just as I reached the other side, Mara'amu passed closely, and at 5:00 p.m., when I traversed the shallow region, Lillith suddenly crossed my path, a metre in front of me, slowly drifting. She did not follow me to Site One, which I reached ten minutes later.

I threw in the scraps and waited. No one came for seven minutes, a long time when one is impatient, and not used to being the one doing the waiting! Finally Bratworst came, and in the next fifty minutes Madeline, Flora, Windy, Cochita, Shilly, Keeta, Ruffles, Apricot, Ondine, Tamarack, Fawn, Pip, Jem, Breezy, Shimmy, and at the very end, Mara'amu came, ate, and left. Their behaviour was still very different to what it had been before their mysterious evacuation, and they had not all returned yet.

At home I wrote everything down, entered it in the computer, and made a list of the sharks that had not returned—the *List of Lost Sharks*. I was mostly concerned about Martha, and there were several others who had not returned including Annaloo, Valentine, Chevron, and Gwendolyn. The only newcomers I had seen at this session were Mara'amu and Cochita. I may have missed the sharks leaving, but it seemed that I had caught them coming back. On the Friday night I had seen nearly all who had returned, zooming up the lagoon, many apparently going to their home ranges further on. They had come back nearly all at the same time, with a few exceptions.

All of the females were pregnant. Some of the scientists I had contacted on the Internet had suggested that this must be a mating phenomenon. But I couldn't see how that could be possible when it wasn't the mating season, the females were pregnant, and the whole population had gone, including the pups. The temperature was normal, and in the middle of the ocean, its condition didn't change fast enough to account for a week to ten-day evacuation. A pollution event would have affected other species and couldn't have spread throughout such a large region uniformly, particularly without anyone being aware of it. No event like it had ever been reported.

The wisest observation was provided, as usual, by Arthur. He wrote:

Dear Ila,

Your observation, dealing with the peculiar antics of the population, has no ready answer. Individual actions sometimes can be explained. But, the cause of a common response by a large segment of a population, often remains a mystery. Juvenile sharks, in general, often do not react in ways shown by adults and adult females sometimes do not do what males do. Adult females often segregate themselves from males over a given geographic range or prefer to move away from males.

There are only two apparent reasons why a population, as a whole, would move away for a period. One would be that something happened in the water prior to their move. You believe that nothing appeared to happen as regards a unique pollution or similar event. This appears reasonable since no other species appeared to be affected and also, the animals returned in a short time. A pollution event, involving a chemical or similar agent, would not be sufficiently dispersed from such a large area in only a week (time before return) and if it had been a point source, animals would likely have shown movements reasonably correlated with dispersal from the point source. Laws of chemical diffusion would not predict large numbers of individuals leaving rapidly from such a large region without some indication of differential spread.

The second reason is that of migration for purposes usually (but not always) to do with reproduction. Usually, juveniles of many animals follow such migrations and so juvenile sharks may follow as well. The problem here is that migrations often involve long distances over a lengthy time-frame. A week's absence does not indicate a migration.

So. . . here lies a mystery, especially when it happened this year and not before. The only other reason one could hypothesize is that a food source had been discovered and this unique event might somehow have been communicated. This is highly doubtful as only blackfin reef sharks moved and other species did not. Since sharks feed reasonably alike, regardless of species, a unique food-source drawing only one species seems unlikely.

Animals often do remarkable things that have no answer. Perhaps we'll understand someday, when we can ask them and they can provide us with an answer that makes sense.

Sincerely,
Art

It was a relief to have help in understanding the phenomenon, and to know that in spite of our human curiosity, it was all right to acknowledge that there were things that we just could not find out. I was starting to realise how little really was known about sharks, through my searches for information in books, on the Internet and by asking shark research scientists.

I asked Arthur if he considered it possible that the sharks had left due to the deterioration of their coral habitat, under the dictates of an instinct evolved on continents. I had wondered, on my frantic searches and musings during the event, whether, over the long aeons of their evolution, sharks had developed this population-level behaviour on occasions when their environment was no longer satisfactory—they went exploring to see if another habitat in better health was available. Since these sharks were on an island, they quickly found that they were already in the best place, so they came back.

Arthur answered:

Your idea has a slight probability of being true. If habitat destruction were the reason, other animals would likely have responded as well, particularly other species of sharks. If blackfins had responded, why not others? Perhaps blackfins are more sensitive. But if so, moving away and then returning after so short a time, seems hard to understand as it would not have been long enough to explore other regions. Even an island like yours is large enough to have taken more than a week to explore for better regions, with individuals returning, erratically, from regions miles apart. But many individuals returned in an extremely short time. If so, the degree of deterioration that drove them out must not have been severe enough to explore other regions before returning. The returning females are apparently in good health, since they are pregnant and look well and that condition had to have been present before they left en masse only a short time previously. One week away would seem to be very short for pregnant females to return since they should be the ones least likely to return rapidly. Pregnant females are often the most leery members of a given species, to problems occurring. However, that may not hold for sharks.

I thought that it did hold for sharks, given how much shyer the big pregnant females were than the others.

So many sharks of all ages, male and female, leaving their homes in concert over such a wide area, did seem to defy human explanation. But they had done so. The pups, females and juveniles had begun leaving first, apparently starting some days before the males. But the males, too, were on the move, and had not fled straight into the ocean depths. If they had done so, so many males from unknown areas would not have been passing through Site Two when I went there. Some who had been there were males who presumably lived in the ocean, and usually never entered the lagoon except during the mating season.

Then, between July 23 and July 29, 2002, no sharks were seen anywhere, except for the occasional small one who appeared in the distance at shark feeding dives, but did not come close or eat. Since juvenile sharks normally shelter in shallow water, this was plausible. Perhaps the rest of the population was much deeper, while the juveniles were drawn to shallower regions, even in the ocean.

When the sharks began to return, with the exception of Bratworst, it was the juvenile males who came in greatest numbers first, and a few days later the adults arrived, almost all at once. Did this mean that these sharks from our community had been all together somewhere, and travelled in a large scattered group back to their places, preceded by a flock of excited boys? It seemed possible.

When I learned that Peter Klimley had found that hammerhead sharks use magnetic lines on the ocean floor as routes, I wondered if unusual magnetic fluctuations within the earth's crust, something not monitored by humanity, were a possible explanation for the reef sharks' flight. I could speculate forever, but never be sure.

Sometimes I thought of the unnaturally brilliant star that had appeared above me after my last trip to the sharks before they had vanished, a dramatic celestial reminder of the walls of awareness of the human mind.

Arthur wrote to me often about his work and mine. He was obviously brilliant, and intellectually committed to the true understanding of any given phenomenon. He had worked for years with Konrad Lorenz, who, he wrote, applied his natural genius in ethology all the time, in whatever situation he was in.

I learned that one of ethology's major principles is to "know your animal," by observing its behaviour closely over an extended period of time. Since such long-term observations take so much time, Arthur told me that it's easier to place tags on animals and keep track of the actions of many individuals at once, rather than only a few. But tagging methods cannot explain, except in the broadest sense, how it is that some animals do what they do, since some individuals perform differently from others. That, he told me, was why my observations were important and why ethologists will always be around.

He disagreed that the "selfish gene" is responsible for events in nature, stressing that when an animal survives to reproduce, it is an individual animal, using all of its faculties, that has done so. Likewise, the things in our civilization were made by individuals. I had adopted this view early on myself, always focusing on an individual in a situation in my paintings, and studying the actions of individuals among the wild animals I had known.

With each message, he asked for news about whether I had managed to get my data into the form of graphs and tables (eliciting a startled *eek!* from yours truly). I had made some progress, and had planned most of a paper on the sharks' daily lives, but the voluminous data about their roaming was an organizational puzzle.

At night, I sat by the sea racking my brains over how precisely, to put the information I knew so well into graphs. There was so much of it, and each shark's pattern was different. There was no fine line between residents and the different visitors. For example, how could Carrellina and Valentine be placed in a category? There were rare visitors and very rare visitors, but there were also just the sharks who lived in Sections One, Two and Three, each of whom roamed around with a different schedule. How could they be grouped into categories? How should I decide what the categories should be? And how broad should each category be, since it seemed to me that each shark was in her own category?

Slipping away over the lagoon after each visit, I pondered my latest revelations and how they fitted into the greater pattern. Suspended upon the film of silver sea between sky and sharks, I felt as if I were roaming over a painting with a microscope, delighting in every detail, but unable to ever stand back and take in the entire work.

As I sat worriedly scribbling at the table, Franck would wander by, lamenting my inability to mentally analyse ten thousand shark sightings, and his lack of time to write a neural network program that could organize my sharks. He had not had much time to mathematically order them, and his sole conclusion had been that if a shark was seen in the study area, the probability was highest that it was Bratworst. In other words she had been seen the most often. (Gwendolyn was second, followed by Martha and Madonna). I didn't see how such a fact could be either of interest or of use, but was happy for him anyway.

Franck laughed at my tables, so neatly formatted on my word processor, and asked why my data wasn't on a spreadsheet. Since spreadsheets were for balancing one's accounts that made no sense to me. One didn't put sharks on them. Alternately watching the sharks, and going over their records, I concluded that my husband's preposterous mathematical ideas did not apply to them. Their behaviour was clearly in the non-computational category.

What was important was seeing the sharks arriving together when they came, the inter-individual distance at different times, whether two sharks were always in the area at the same time, whether or not they had appeared together reliably at my sessions during that period, the date, light level, time, and many other details of their behaviour that no numbers could convey.

The next time Franck sneered at my absurd methods, I suggested that his reasoning inferred that it would be best to send the kayak out with the food by itself, programmed to navigate to the site and stop there. A special camera could protrude underwater and film the sharks when the food was mechanically tossed in, and some visual recognition program behind the lens, that he could easily program, would identify the sharks. This information would be recorded, then transferred, upon the automatic return of the kayak, to my computer, and analysed by his silly program. I could sit around and eat chocolates.

"Yes," he said. "That's about right."

"What about the anchor?" I demanded. "It would get stuck right away."

He had an answer for that too.

Nevertheless, in serious moments he urged me to read his book on postgreSQL—a computer language and method for creating a database and use it. A proper database for so much material could be questioned, and the answers would be computed for me. But this involved much study, and then the use of the given methods to create the design, which then had to be transformed into a computer language. It was time-consuming and required deep concentration, especially given the stumbling block of how the data should best be organized for the database, which presumably I was supposed to know. Though I read much of the book and made some progress, most of the time, the project was on hold.

I went again to Site One when I acquired some fish scraps. Sharks came to circle the kayak, and underwater I found Madonna, Bratworst, and Cochita. We all swam together to the site, and when I stopped to put their food into place, they zoomed up to me, even Cochita, who had always kept her distance. Windy, Sparkle, Ruffles, Jem, and Brandy soon came, and Madeline, Breezy, Eden, and Ondine drifted in during the hour of the session. They were all the same sharks. My worry about Martha's obvious absence went up a notch. Annaloo, Valentine, Chevron, and Gwendolyn were also still missing.

Later, as the shadows of night descended and I had been alone for a while, I picked up the remains of the fish spines from the site in preparation for leaving. Simmaron and Glammer passed down-current, and a pup swam into the site, but too far to identify. Jessica began roaming the surrounding darkness, and I swam around with her for a few minutes. No one else was there. Just as I picked up the anchor, Christobel drifted by. This time the only new shark had been Brandy, who was still a young juvenile.

I let a week go by. Then, on a sunny evening after a flawless day, I went again to look for Martha. A spell of silence lay upon the sea. Anchoring southwest of the site, I slid underwater and found Madonna cutting through the scent flowing from the back of the boat that I thought I had plugged up. I swam and she followed, never coming to me but undulating behind, turning away, then circling, in typical shark fashion, to follow up to the kayak again at the surface. I kept an eye on her because dear Madonna could be pushy. As we drew near the site, I found Apricot lurking behind a coral, and by the time I got there, she and Madonna, with Windy, Bratworst, and Sparkle, were orbiting me. They knew where I laid the table.

I put their food in place, and was pleased that among the sharks who came were Trillium, Peri and Droplet, new sightings since the shark evacuation. Fawn had a long, deep, and swollen cut on his chin, which was new. Teardrop had come all the way from her range farther east. The big sharks Samaria, Glammer, and Keeta drifted in, and roamed through the surroundings in formation. At dusk, Cochita and Flora came together. Glammer's presence without Annaloo concerned me. They were almost always together, yet Glammer had returned early.

The sharks' behaviour was more normal at this session than it had been since their dramatic reappearance. They ate in their usual picky fashion and circled around the area socializing. They seemed to disperse, then when I looked down-current, I found them all roaming just beyond the visual limit. When I rinsed out the back of the kayak they all came back. Little Lightning went to sniff the strap hanging in the water—he remembered. I saw many of them, once in a while, quietly go to the boat and sniff it. But none of the other missing sharks appeared.

I went out again one quiet morning a week later, and entered the water southwest of the site. Bratworst circled far off at the visual limit watching, as she often did when she was alone on my arrival. Madonna glided from the veiling light and swam lazily up behind the boat, the golden light of the rising sun playing over her steel-coloured skin. Bratworst appeared behind her. I had only a few pieces of food, sufficient for a scent trail and Martha, so threw them just one small piece each, then swam toward the site. The two sharks came, circling me from time to time, especially when I crumbled another bit of food to renew the scent trail.

I didn't linger at the site, partly because of their presence. In the region of thick coral farther on were some new juveniles, and I sketched their dorsal fins as I went. Bratworst shot past, and swiftly turned around me. When I went on, the two sharks swam in formation with me and the boat.

At the barrens, Windy appeared and began scouting around us. I was trailing scent again, hoping desperately to find Martha, and she soon lost patience, and charged. I had to throw her the scrap. The three sharks went into fast search mode, each trying to find the tiny scrap before the others, and getting more and more agitated about it. Unfortunately, they couldn't find it, and Windy began shooting up to me much too fast. I threw another piece of food, but she didn't realize that I had done so, and soon her and Madonna were darting and gesticulating around me very closely. Bratworst joined them. I grew increasingly concerned, as Windy kept up her high speed approaches.

Finally I tried to smack her on the nose with my slate, which broke from being moved swiftly broadside underwater. Then it fell onto the barren floor. I was afraid to dive down with them spiralling and soaring around me. Finally they went into search mode, bumping into each other, and moving in quick, impatient jerks. I dove down, picked up my slate, and flew straight up through the surface and into the kayak.

Martha was obviously not around, and I had lost my taste for being the object of interest of irritated sharks. Drifting away, I remembered that other morning, when it was Isis and Martha who had met me. They had come along beside me while I had handed each a piece of food. Though they too, had missed them initially, they had suddenly turned, and each had gone straight to her treat and inhaled it. Watching Bratworst and Windy dart and whip around the boat as I drifted across the border and into the indigo waters of the bay, I wondered what alchemy had made them seem so wise, and these so manic.

Not long after that, I fell out of a tree trying to get a fish to a depressed seabird, and injured my shoulder. Since I could no longer paddle the kayak, I swam out to Site Two in an effort to keep up my records. It was hard to swim so far against strong current, and I spent the first part of the session leaning on a dead coral shelf with one fin off, trying to massage painful cramps out of my calf and foot. The sharks ate and Sparkle Too orbited me.

Most of the little pups and juveniles who had once enlivened and filled the site were missing. Belladonna was cruising the region with an unknown female visitor. Shimmy was there with Thrasher. Otherwise, the only interesting event in an exceptionally quiet session was the passage of Twilight. She swept in, picked up a scrap, and exited left stage, closely pursued by an unknown male with swollen claspers. Mating was not due to begin for another two months. At least I had confirmed that Belladonna was back.

The following week I paddled out to Site One in spite of the pain in my shoulder. The first week of September was already passing, and I was increasingly worried and sad about Martha's failure to return. This time it was Sparkle Too, Trillium, and Eden who were waiting for me. Sparkle Too was beginning to seem endowed with special powers, to be waiting for me, no matter which site I chose. When their food was in place, the rest of the same resident sharks who attended before, swirled in to eat. In spite of the advancing season there were no visitors. My sharks were excited and pushy, repeatedly checking the kayak, sniffing its straps, and coming to me.

Darkness was falling with the sun, when Gwendolyn came slowly up the scent trail. It was a month after the others had returned, in the same lunar phase. The surge of joy and hope was wonderful after my weeks of worry, and I waited at the heights of anticipation for Martha, Valentine, Chevron, or Annaloo. But no one else came.

It was mid-October, and six weeks had passed before I came again to Site One. There was an east wind, and the sky was clouded and lowering. No one was expecting me, so I slid underwater and looked around. It was dark. I paused to gaze again over the familiar, gloomy surroundings. The fish were all streaming toward me, but no shark was visible, so for a while I fed and enjoyed the fish. They had not forgotten me, though six months had passed since I had stopped the regular feeding sessions there.

Then suddenly more than a dozen sharks whipped into the site. They had not forgotten me either!

I pulled the boat over to my usual lookout and started firing the food overboard, displeased to find myself back in the unwelcome situation of being in water solid with excited sharks. When I looked underwater, it was bedlam. Gwendolyn was soaring past, sinuous after birthing. My sharks were flying around, feeding and chasing each other through the coral. What a pleasure it was to see them all again enjoying a party, to behold once more the enchanting spectacle I had left behind so long ago. Tuvalu from Site Two was among them! Indeed, the juveniles seemed to travel farther than any other segment of the population.

I was still waiting for Martha. Valentine was one who often roamed, so I still thought that there was a chance that she, and Chevron too, would some day appear. Annaloo had also been less regular in her visits. Given that I had stopped the feeding sessions, it was possible that they had relocated for a while, and would eventually turn up.

But Martha. . . Martha should have come straight home with the others. Her failure to return filled me with foreboding. Where had they gone and why had she not returned? What had happened to her? Drifting away across the lagoon in the fading light, leaving my sharks once again, a deep sorrow grew with the conviction that Martha, my strange and beloved underwater companion, was irrevocably lost. Could it be a coincidence that Bratworst and Glammer, the close companions of two of the lost sharks, had returned early? Sharks were known to flee when one of their number was killed. . . Had they fled home early after a disaster? There were long-liners on the ocean. . . But speculation was useless—I would never know.

I returned time after time to the lagoon to search for Martha, though paddling the kayak was painful, and my arm and shoulder ached for days after. I wondered if anyone had ever searched with such desperation, longing, and pain, through the sea for a fish. I always followed the same routine that had drawn her to me in the past, anchoring southwest of the site, and trailing some scent as I swam slowly through the familiar landmarks of our favourite part of the lagoon.

On one of these excursions, I was swimming alone, having just entered the water, when a whitetip shark approached me curiously. This was more like blackfin behaviour. Whitetips usually drifted by at a distance of two or three metres or more, but this one came directly to me, then turned, looking at me closely with a catlike gaze. She drifted by at arm's length. Her markings revealed that she was the one who had glided between my eyes and my slate as I was drawing, at that long ago session at Site Two! Wonders never ceased.

My reverie was interrupted by Windy, who suddenly shot underneath me, thrilled to have found me again. It seemed that she was using the same range that had been occupied by Martha. Her near twin, Madonna, was coming behind her, and as I threw the matched pair a handful of scraps, she swept up to my face in the greeting gesture, then the two shot around me. I watched for awhile as they searched for the scraps and ate, waiting to see if Martha or any other sharks would come to join them, then swam on before they could assault me for more. I didn't trail scent. Martha would come if she were in the area. Once I had fed some sharks, others would hear or cross the scent flow, and come looking for me.

I paused just south of the site, and watched for a minute. Windy and Madonna appeared on cue and swam past, but when their waving noses failed to pick up any tasty odours, they moved on. I followed, dripping a bit of scent in the lagoon from time to time. Lightning passed without pausing. He was nearly adult-size by then. At the barrens I trailed scent, watching, and sketching the dorsal fins of new sharks. Each year there was a new assortment of two to three year olds who appeared during the dry season, and it was useful to straighten out who they were as soon as possible. Each time I went home, there were more little fins to draw in my book.

I often saw Samaria, Madeline, or Shilly, too, and sometimes unusual visitors appeared in the barrens. During the day, they were often quiet enough to see closely, but usually they disappeared into the veiling light long before I could draw their fins in detail. Then, my only hope of seeing them on both sides was to throw in a handful of scraps and retreat, hoping a little group would form to feed, and the visitor would come close.

I had many interesting meetings during this period, and was surprised by how often I was able to tour through the lagoon with a rare guest whom I had met only once or twice before. It was always a dramatic experience to roam with one of the very large females from deep in the lagoon. Each one behaved and responded to me in a different way as we promenaded together.

When Shilly appeared, it was always from the same area, and I found that I could go and get her by swimming into her range. She would hear me, come, and then accompany me for a while. She was often with Ondine or Ali, both her age. Ondine, like Shilly, would circle me often for long periods, but Ali remained aloof, and would neither come near, nor let me approach her.

I continued to hope that one of these visits would bring Martha, Valentine, Chevron, or Annaloo back to me, but the reproductive season passed and none of them appeared. They, and many of the pups and small juveniles, as well as unknown numbers of sharks from other locations, never came back from their strange evacuation. I had seen them for the last time, unknowing, long before.

The lagoon was never the same without Martha. I had put three years into cultivating her companionship, noting all that she did, and pushing myself to get out and see her regularly no matter what else had to be done too. No other shark showed promise of becoming a companion in the same way. No other could take her place on my daytime excursions. Shilly was growing up, and as the time passed and Martha never appeared, I went to find her instead. But her home range was much farther away, and I wasn't able to devote the necessary time to the project. It was her unique nature, that when I found her, she came with me of her own free will, but wandered nearby rather than staying beside me as Martha had. Like a dearly beloved pet, Martha could not be replaced.

Cochita and Flora were the most responsive among the sharks in the original study area, but Flora wasn't around during the day. Madonna and Windy were always delighted to see me, but they inspired me to avoid them. Bratworst usually accompanied me on my travels on her own initiative, but she was aloof, following out of sight, or roaming within earshot, just coming near from time to time. Sometimes one of the bad sharks, such as Sparkle, hounded me around the barrens.

The bizarre shark disappearance, and my resulting preoccupation with the loss of Martha, effectively interrupted my sessions at the lagoon on the other side of the bay. If something is hard enough to do, it can become just too difficult, and I never did go back to see if Shenandoah and the other sharks had returned home or not. I had essentially gained the information I was seeking there. Section Two was more interesting, since there I could see the lagoon population I had studied from the beginning. So I held weekly feeding sessions at Site Two, balanced by an outing during the day in Section One when I had time. After so much invested effort to know the female sharks there, I wanted to stay in touch with them.

Chapter Forty-One

The Coming of Carrellina

It was hard to keep up my enthusiasm for Site Two, but in time I found new things of interest. The fishing industry had expanded, and I was often given far more scraps than I could use. It took time, and a supreme effort, to pack the heavy fish scraps home, sort and repack them in the kayak, look after the birds and the family, get ready, and paddle the loaded kayak down the bay and out to near the centre of the lagoon. It was always a fight against the wind, and as the wet season advanced, the current grew worse, creating the usual extra difficulties.

Belladonna was the only resident female there. The lagoon males most often there were Mordred and Flannery, whom I knew from both Sites One and Three. Flannery was a large male, and always on the alert. Yet on the few occasions I met him alone, he was very relaxed with me, letting me come near enough to photograph him and swim along with him for a while, which was rare for males. Any time I met him during the day was a special event, since he had come so rarely to Site One. He was one of those sharks so well designed that everything about him seemed just right, even to the graceful pattern of his dorsal fin. Flannery was magnificent.

The biggest male I knew was Danny, whom I had first met at Site Three. He appeared often at Site Two, but had never come to the Site One sessions. He always travelled with a large group, so his passage was especially interesting because of the sharks who accompanied him. These were often males I knew as rare Site One visitors, or passers-by from Site Three.

As the reproductive season advanced, many of the same female visitors I had seen in Section One appeared. I would often identify a group at a session at Site Two, then find them a few days later, resting in the barrens, or hunting along the border. They typically stayed in Section One for about two weeks, then moved back up the lagoon and attended the Site Two feeding session on the way back. Thus I was able to gain an idea of how they travelled—slowly! But after passing Section Two, they did not come again for many months, or until the next reproductive season, depending on who they were.

These groups moved in a similar time frame, which often brought the travellers to the border during the period of the full moon, possibly to hunt there. Many of these groups remained there, in my study area, during the period of the dark of the moon, and I speculated that they were waiting for the light to return before continuing their journey. They usually left as the moon waxed and grew bright again, so some stayed for close to a month.

Storm, Droplet, and Jessica, who had come regularly to Site One, now came regularly to Site Two; their ranges were in between. But though I was still seeing many of my familiar sharks, the dynamics of the sessions and the group had been altered by the change of location. Site Two sessions continued to be dominated by shark pups and juveniles. Occasionally no adult females came at all. Some of the youngsters were ones I had identified for the first time in the thick coral beside the barrens while looking for Martha the previous season. It was interesting how each different region of the lagoon had its own pattern.

Shilly came more regularly than before, and continued to give me special treatment, yet never pestered nor charged me. I had grown very fond of her. She often stayed near me looking around in the coral until the end of the session. Since I had lost my crocodile needlefish and did not feed the fish at Site Two, I kept a piece of food separate only for her. She took it on cue.

In November there was an odd incident when it was very dark, while I was watching an excited male visitor who had followed Storm into the site. Suddenly, he soared straight to me, while at the same time Sparkle flew in from the left, to perform one of her high speed close approaches. Shilly happened to be coming too. The stranger may have felt threatened by the others coalescing around him at high speed, and he charged me with the darting and feinting motion typical of the bad sharks. I finned swiftly backward to get out of his way and he changed trajectory at the last moment and flew past my shoulder. Cotlet was roaming around with two large remoras attached to her, and the three males, Jem, Avogadro, and Thrasher had also just arrived together. It looked as if they had all come from the ocean, as males had often done at Site One as night gathered. Apricot zoomed straight across the site at the surface to nearly touch my nose, ignoring the food still lying on the sand. She seemed excited to have found me again. Christobel and Sparkle were behind her, travelling as usual together. The three may have accompanied the males as they passed through Section One from the ocean beyond.

Sparkle Too, Cinnamon, and Jessica all birthed during the full moon of November, and as the height of the warm season approached, nurse sharks attended more regularly.

As December began, I acquired just a few scraps, and took them to the barrens to see who was around one morning. Bratworst had been missing, and she was on my mind as I paddled out. I still hoped to find Martha, Valentine, and Annaloo, too, unexpectedly, one magic morning. And it was as beautiful as it could be, so beautiful it was almost unbelievable, with the rising sun blazing upon the mountain, and flaring into a sea of ringing blue. The lagoon was a silver mirror as I approached it. The waves crashing over the reef were high, as they had been for some time, but they were not breaking along the border and bay, and there was not excessive turbulence as I entered the shallow water of the border, which were the signals of difficult conditions ahead, that I had learned to note.

I was so tired I wished I could just go to sleep, and yawned repeatedly. I planned on a languid time circling around with the sharks, maybe snap a few photos in the sunshine, then try to find time to rest for a while in the afternoon—maybe even sleep. Passing Site One, I dripped scent into the water. The size of the waves there were surprising. They lifted the kayak high in the air, and I worried that they would drop it on the coral. But being so sleepy, I didn't think it out. Obviously the day was tranquil. I was preoccupied with determining whether sharks were following the boat, and if so, who. There had been few sharks around the last twice I had come.

I anchored in the barrens, and many sharks converged at the surface, so I threw in the food and got in, holding onto the kayak. The visibility was awful because of the waves. When the current didn't sweep me away, I let go and flew sideways and down-current to a coral formation I could hold onto near where the sharks were feeding.

The first shark I noted cavorting excitedly around the food was Carrellina. She was back right on schedule, and had survived the great shark disappearance. Year after year in December she came, after a disappearance of eight months since April! Madeline was there—she had birthed—and Windy, Gwendolyn, Storm, and Glammer were swirling around with the usual juveniles. Thus happily focused on the sharks, I found it increasingly irritating that I couldn't seem to hold my slate against the current, while trying to write. Letting go of my lookout was out of the question. I became more and more interiorized on this little problem which seemed to get worse and worse as I alternately tried to write and watch the sharks.

I grabbed the coral with my other hand just to keep from being swept away. The sea surface was deceptively smooth, but it had huge undulations in it, each of which poured over me with daunting power. Each one seemed more difficult to cope with than the one before it, and I was becoming breathless. Soon the power of each wave forced me underwater and breached the seals on my mask and snorkel so that I kept getting water in my mouth. When I was able to put my head above the surface to gasp for breath, I saw another even bigger wave towering above. Under the surface I went again, and once more the power of the water took all my force to resist, and filled my mask and mouth again with water. The tremendous, long waves poured off the ocean without a break, and while I was underwater feeling as if I were drowning, I was watching Carrellina, who was directly beneath me, swimming into the surge. She had a knack for appearing like a bad dream at the worst possible times.

I wondered desperately if it wouldn't be better to try to swim back to the boat along the bottom, until I remembered that I wouldn't be able to breathe down there, and already I couldn't seem to get in all the oxygen that I needed. So there I was, stuck, clinging to the coral for dear life while the sea laughed and pushed me under, each wave filling my snorkel as my desperation grew. I was using all my power just to keep from being swept away, yet could not breathe.

I quickly went through the scenario which would ensue if I couldn't get back to the boat. It seemed obvious that I wouldn't—I couldn't even breathe. I could easily swim to shore then walk or get a ride home. That would be embarrassing, given my appearance in my underwater outfit, and the fact that I would be soaking wet with sea water. But it was possible. The problem would arise over how to get back out to get my kayak. The children had lost the paddle for the other one, and Franck hadn't bothered to replace it. There were still some scraps in the kayak, so anyone who found it would know that I had been feeding the sharks, which would reinforce my bad reputation with fishermen. Luckily I had the camera and the slate with me, one looped around each wrist.

Finally, the sea swept me away. As I began to fight my way toward the boat, I was already at the end of my strength, and I hadn't had very much to begin with that morning. I made no headway against the surging torrent, and began losing ground. Pushing the loops of the camera and slate up my arms, I began to swim with all of my power, with the camera banging around my elbow, frantically gasping for oxygen at the very limit of my physical reserves. At first I barely managed to stay in place, but then very slowly I found I was getting closer to the kayak.

Unfortunately, I had already used up all my energy. Even much later, when I was only a metre away from the boat, I was sure I would never be able to keep up the battle long enough. I was advancing too slowly and was at the utmost extremity of oxygen deprivation and muscular exhaustion, not to mention the sheer pain of the continued desperate struggle. But I kept on, arm over arm, seeing myself, in disbelief, centimetre by centimetre getting nearer to my precious kayak. Finally I caught hold of the tip of its tail, amazed that I had managed something that had seemed so utterly impossible. Hard as I could breathe, I couldn't get all the oxygen I needed, and went on hanging in the water, trailed by the sea, gasping at my limit for a very long time. I was so depleted I didn't think I would be able to get back into the boat, but was so incredulously relieved to have gained safety and self respect, that I even managed that. By then, the lagoon looked relatively calm again. Had the series of big waves not passed when it did, I wouldn't have made it.

Stunned at my own stupidity, I sat there upon the rising and falling waves looking for Carrellina, and around at the sea, too lifeless to do anything. How could I have gotten into such a situation after years of going out there in all sorts of conditions? I had looked for the usual signs signifying bad conditions in the lagoon, but they had not been present, and the sea was still strangely unruffled, except for the huge, long-wavelength undulations in it.

After a long time, feeling slightly re-energized by my rest in the sunshine, I slid back underwater, crawled along the kayak, and down the anchor line, picked up the anchor, and got it and myself back in without once letting go. Then I drifted much of the way home on the current and a slowly rising wind.

As the day passed, those long waves reached down the bay as far as our house, and soon, the atmosphere began to stir as a great storm enveloped the island. The ocean swell had turned, and was coming from the north straight into our shore—by the next day the waves were close to five metres (16 ft) high, toppling over the reef like mountain ranges on the move. I had been caught at what was just the beginning of a major oceanic change. For the whole of December the storm raged and the waves poured in from the north.

On December 15, 2002, the wind and rain temporarily abated and the waves suddenly dropped by half their height, so I paddled out to look around. The current was daunting, and the barrens were foggy with suspended particles. Yet, I found Bratworst, recorded a rare daytime sighting of Flora, and confirmed that everyone was still there pursuing their affairs and having babies.

It wasn't until January 11, 2003, that I was able to return to Site Two. There was a strong east wind to fight, but sharks began circling the boat as I arrived, though it was a month since I had been there. It was the first time at that location that they had come to the surface to meet me. I anchored and was preparing, when a shark swerved along the kayak beside me. I knew a bad shark when I saw one! No scent had leaked, and I slid underwater to see who it was. Sparkle swam up to me, and her friend, Christobel, was visible gliding beyond. Many were young males who knew me from Site One—Tamarack, Breezy, and Muffin—with the adults Ruffles and Jem. They must have all been in the area when they heard me and came. I took the anchor, and swam to Site Two, which I had missed in the river-like flow of the lagoon. The sharks followed at a distance. A whitetip was with them. They glided near as I pushed the food in. Droplet and Jessica were cruising around the periphery.

When I had moved the kayak out of the way and returned, Mephistopheles glided in over the white sand. He was back once more. I may have missed a lot in December due to the storm—it was the month in which he usually appeared—but at least I had not missed him completely. It was the fourth year in a row that he had come through during the mating season, and the first time I had seen him at Site Two.

He was even more of a mystery by then, because I had shown his photograph to dive club owners, who attended shark dives daily at various locations along the north shore. No one could see that shocking face and forget him, yet no one else had ever seen him! This seemed remarkable. Could he be travelling from a home range off some other shore? While some individuals roamed far—I had established beyond a doubt the vast difference in individual habits—I still believed that most did not. Yet, in the mating season, was it possible that roaming males could partly or completely circumnavigate the island? One could speculate that if Mephistopheles did so, staying mostly within the lagoons where the females were, he would remain unseen by divers attending shark dives outside the reef. Though he had appeared at Site One with Shenandoah on both of her visits, I had never sighted him in the lagoon across the bay, so was unsure that his range was truly linked with hers. But there was still the possibility that he came from that lagoon, as Flannery, Ruffles and Avogadro ranged through mine. I could have missed seeing him there due to the short series of sessions I had held, the tendencies of the sharks' ranges there to be long and narrow, and the greater roaming distances of the males.

It was also possible that he visited during the mating season from another island. How far the males actually roam is unknown, but now that it has been established that they travel between islands, more possibilities appear.

The current was at the limit of my tolerance but at least I was there. Shilly came on schedule and roamed near me. The whitetip shark who had been there on my arrival was the one who had met me near the border on one of my searches for Martha, the same individual who had glided between my eyes and slate while I was drawing. She remained for much of the session, often passing near, and gazing at me with those big soft eyes. She was the first whitetip I had seen repeatedly in different locations over time, and like the blackfins who knew me, she behaved as if I was familiar to her, even though I had seen her rarely.

It stormed for another two weeks before a day came that seemed calm enough to go to see the sharks. But as I approached the lagoon, a strengthening east wind fanned the surface, and was soon whipping up half-metre waves against me. With the surging sea, the flying water, and the brilliant light, it took more than an hour to gain Site Two. Unable to locate it in the seething waters, I threw in the anchor when a gust turned me broadside.

I waited for the boat to line up with the wind, but it didn't. Waves were being generated from three different directions, some with very long wavelengths. They smashed together, forming an extraordinary pattern of splashing peaks. Getting into the water was unimaginable. At home, I had considered myself negligent in failing to get out to the sharks. Now the question became, instead, why had I done so this time?

Slowly, I secured the paddle to the wildly plunging kayak, put on my gear, and threw in the food. When I slid in, the current caught me so strongly that all I wanted to do was to get out. In the impenetrable green water, the sharks flying around me appeared and vanished. Clinging to the boat, I was jerked back and forth as the kayak yanked on the anchor, and pushed three ways by waves and wind. With difficulty, I climbed down the anchor rope, dislodged the anchor, and flew away through the feeding sharks. Shilly swept up over a coral outcropping toward me, then Twilight and a friend scattered in surprise as I soared overhead, taking it all in, as my multitude of sharks flew away in every direction through their emerald world. Shilly stayed with me briefly, but I was soon so far that she turned back. For a moment, the whitetip shark coalesced before me, to flash out of the way as I swept on. Unwilling to leave such otherworldly beauty, I flew, thrilled, until I saw a region of such turbulence that I leaped back in the kayak. But safe again, I was happy to have briefly seen my sharks.

In the following weeks the conditions were little better. I was unable to let go of the kayak, even though I found a deeper place where the current was slower. The anchor dragged with my weight on it until it caught on something, and as I dangled unbalanced on the rope, when I turned my head, the torrent surged through the seal on my mask. It was not the best of circumstances in which to enjoy the company of sharks.

Once, after fighting my way to Site Two, I was resting for a few minutes after throwing in the anchor, and was surprised to see a large shark circling the kayak clockwise at the surface, just as Flora used to do. The holes in the well of the boat had been plugged so that no scent was in the water, yet it was her. She had been in the vicinity, and had come to the sound of my kayak—after all that time, she remembered.

At that session the visibility was spectacular, though the current was just as torrential. Flora had mated about three weeks before, and her pale body was covered with light scars. Jessica and Sparkle Too, who had reappeared after an absence, seemed to have mated at the same time, and little Ali suddenly appeared looking bigger and fat—was she pregnant? Perhaps Shilly, too, would mate that year.

It was impossible to look freely all around while bracing myself against the heavy flow, and I had trouble breathing with the pressure of the surge. Once when I turned, barely able to maintain myself against the latest wave, I found Carrellina's nose at my ear, and the sea water surged into my mask in the split second in which she whipped around and vanished. She had sneaked up behind me, and fled when I caught her, the brat.

Oh, Carrellina—always there at the wrong moments. It grew dark early that night, as clouds piled up in the western sky, and it was a relief to go. My muscles were cramping from maintaining my balance in the torrent.

Chapter Forty-Two

Ordering my Sharks

At the end of February, a calm fell silently upon the sea. After my morning chores were done, I went to see the sharks, revelling in a sky shining blue to the horizon, with just a few fluffy clouds above the island. It was the first calm day since December began, when the sea had begun to rise and nearly swept me away.

At the spawning site at the border, fish were jumping and blackfin sharks were surfacing in several places, often just before lunging forward. I went underwater. The black surgeons were close to the peak of their ritual, but instead of the full-sized fish spawning at dusk, these were little ones. They were just five to seven centimetres (2 to 3 in) long, yet they were spawning with the great flood-light shining straight down upon them!

Windy was approaching the clusters of fish at the surface, and she came over to me, then circled back to the spawning fish. There was no scent coming from the boat—my few bits of shark food were in two pots.

As I watched the spawning fish, Windy left them repeatedly and came to me, then circled away. Ondine found me about ten minutes later, and also returned sporadically. The rest of the time she cruised through the milling fish, usually out of visual range. After another ten minutes I got back in the kayak and went on toward the barrens. I was paddling fast, and did not think of looking back. But once I was well into the lagoon, I paused for a drink of water, and as the kayak slowed, a shark swiftly circled. When I went on, two little fins followed me. A little later Madonna glided by, and continued beside the kayak, her large dark side just under my hand, so that I could have stroked her. Often, she bumped the boat slightly.

I went to the barrens, because in the middle of the day that was the best place to see a lot of sharks, and I was more than half a kilometre (a third of a mile) from the spawning site when I anchored. Underwater were Windy and Ondine—they had followed me from the spawning site. Several months had passed since Windy had come to a session. She was a border shark, and had never come to Site Two, so was using long-term memory. But more surprising, she and Ondine had left a place where there was scent and visible prey, and where she had probably eaten, in order to follow me! Must she not have thought about the situation, and made a decision based on an old memory to leave the visible food? It was rare to catch a circumstance in which an animal makes a choice between two possible options, and here a wild shark had done so!

I had worked to know the sharks as animals and individuals, and form an idea of their intelligence. Yet, having read extensively on the subject, I had found that very little is known even about human intelligence, and how it manifests in different ways. The sorts of intelligence, quite different from ours, that animals express is not well understood. I had been collecting examples of incidents in which it seemed that an animal had displayed intelligence, spirit, or thought, for as long as I had been observing them, and it was harder to assess such incidents in sharks, but here was one! Here Windy and Ondine had made an unexpected choice—they had made a decision based on an old memory rather than visible prey and odour.

Pleased with this unexpected display, I began identifying the sharks present. Carrellina soon came, and in the bright light, I played at photographing her, diving down and waiting until she came over to me. Though there was still enough current to make the session difficult, many sharks came, including visiting males I usually only had a chance to sort out at dusk. Here it was easier, and I covered my slates with drawings.

The pieces of fish I had were old fillets, which had been given to me frozen by the fish store when they had cleaned out their freezer. I broke them up and threw in a few pieces every now and then. In the middle of the day, the sharks came alone, in pairs, or in small groups, and all behaved so sleepily that there was no trouble feeding them this way. When they had eaten, they left, and I waited for more to come.

When Fleur de Lis glided in, I threw her a piece and it floated. She zoomed up and took a swipe from the scrap. It was perfectly bitten through, showing the size and shape of her bite. Then Madeline was at my elbow, with Twilight and The Concorde behind her. It was only the third time I had beheld The Concorde, and the magnificent shark passed me from right to left at the same height relative to my eyes as she had on the first two occasions, glorious in the light of day. Several interesting visitors arrived at the same time, including a beautiful grey satin creature I had never seen before, whom I called Mandolin.

After that session, stormy weather plagued my visits to the sharks. Swimming with them was impossible in the relentless current. I had to hold onto the boat or a dead coral formation near it, so that I could be sure to reach safety when I needed to. As well as threats of wind and rain, darkness fell early due to thick cloud layers blocking the light from the low sun. I looked out across a dull grey landscape in which sharks and fish wriggled against the torrential flow to navigate the coral, and sheltered in the lee of the formations.

Yet in spite of the difficulties of these sessions, I was able to track in a limited way, the females' reproductive patterns for another year, and record the passage of some of the visitors. With the information gained at different sites and the many observations along the border at different times of the day and evening, I had added an enormous volume of data about the movements of the sharks in the past year.

During the rainy afternoons I perused my notes, jotting down when this and that shark had come, when she had returned, when she had been alone, when she had been with a companion, which one she had been with, and everything else that seemed relevant. I printed the table listing who had come to each session, dozens of pages of shark names and numbers, spread it across the table, and daily, I scrutinized it closely.

Eventually I copied the tables onto a spread sheet. More time passed while I figured out how to make the spread sheet work for me, but finally I managed to get the different columns to do a calculation, and generated a graph. As the days passed, I tried different ways of manipulating the data, until it became familiar to mathematically manipulate long columns of data at once, and generate a graph at the press of a button.

One of the long-term mysteries was that there were special sessions now and again. Noticing that in one session, three pairs of friends had reappeared after a long absence, several rare visitors had attended, and that everyone had been especially excited, I decided to analyse this phenomenon. The first special session was the eighth one, in which Bratworst had acquired her name by frightening me. There had been four times the usual number of sharks present at that session.

I went over the records of each session, counted up the fractions of the different segments of the population that had attended, analysed the relative movements of known companions, the conditions, and other more detailed information, and added the information in new columns in my expanding tables.

Then I fitfully filled multiple columns of computer spreadsheets with a variety of types of data, trying different categories, and mathematically manipulating them according to whim, to see what the results were. I got so good at this novel practise that I would re-analyse the sharks every few days, following some bright idea that had come to me during one of my long meditations on the sharks at night by the sea. And finally I had it. The graph was generated and the pattern was clear to see. I had found the secret and had mathematical proof after all!

Gleefully imagining going down in history for having discovered something, I rechecked the data, graphing it in different ways. Since a graph shows one type of data on the x axis, and another on the y axis, there were different possibilities for arranging information in a graph, and I tried different ideas. By the time I had perfected the way to graph my data, I had also realized how to present it in a scientific paper.

Firstly, I described my method. This involved the identification of the individual sharks. Concerned that my method of drawing their fins could be questioned by the scientific community, I prepared an argument to justify drawing, as opposed to relying on photographs. Anyone reading it would have no choice but to agree, that whereas photographs were useful, when it came to recording the dorsal fin patterns of multitudes of grey torpedoes shooting past the eyes, only the important human tool of drawing, would do.

Then I set out the section for results. Before I could address what the sharks were doing, I had to describe who was there. So I analysed the population, year by year. This information was easy to enter into tabular form. Having established that, I described the sharks' patterns of movements, short-term and long-term. The annual pattern was dependent to a large extent on the mating season, which meant that I had to mention the reproductive cycle. As I worked, I slowly realized the need for references for all the things I was referring to, and assuming to be common knowledge, such as the gestation period, the general reproductive pattern, the size of the sharks, and so forth.

So when I could access the Internet, I focused on searching for the needed references, and was increasingly disturbed to find that almost no information came up, except the usual hits shouting "shark attacks!"

This emphasis troubled me. Wasn't it self evident that a shark meeting a human underwater would consider us to have the same importance as any other animal which nature had made available to eat and enjoy? When eating a steak, no person considered that they were attacking a cow. Why should a feeding shark be seen as attacking? How could sharks single us out and grant us the exalted position that we assume for ourselves? Given our relative ineptness underwater, it seemed more likely, if they had thought about it at all, that sharks would consider themselves to be the superior ones.

The information was already available that even the great white shark (*Carcharodon carcharias*) did not consider humans to be food. Thanks to the movie *JAWS*, this species seemed to hold many people in thrall as the very image of a sea monster. Yet studies of shark bites indicated that injuries from great whites had resulted from the shark mouthing a person, then letting go, since humans are not fat enough for them. The mouth of a shark is their only body part selected for contact with solids, so sharks use their mouths to feel and sense things. I had watched the reef sharks do it often, sometimes not leaving a mark on the item they had picked up. Of course the results from such a mouthing by a great white shark would indeed produce a horrible wound, and likely a fatal one. But it should come as no surprise that sharks do bite sometimes, especially those species who eat mammals. We all are part of nature. We eat animals and we are animals.

Yet the anti-shark manic attitude I was finding on the Internet suggested a deep phobia and bias against these unusual and intelligent animals which disturbed me increasingly the more I encountered it. My searches were necessarily brief and always left me puzzled. It was as if the creatures I was reading about on the Internet were an entirely different class of animals from the sharks I knew so well.

I sat frowning at all the hyped up, slogan-type shark descriptions. The endless circling around the shark food and the inept attempts to extract bites from it compared very poorly with the ruthless fashion with which our dogs tore up the same scraps. This impression was later confirmed by the information that the bite of a shark is relatively weak. Their bites are significant only because of the shape of their teeth.

While one could, indeed, find arguments to support the idea of the evolutionary success of sharks when considering their anterior end, at the posterior one, they clearly had a lot of trouble. Even with two male organs, and with their females precisely the right size for them, the manner in which the males bit the females so badly during copulation suggested that in this department, they were far from the perfect predators ranted about, and very close to being failures.

Other information I found on sharks was sketchy or false. As well as looking on the Internet, I visited two local research facilities on the island that had small scientific libraries. One of them had no information about this species of shark. At the other one, the manager refused to even let me touch the books, and searched the archives for me. He pronounced that the males were two metres long, and the gestation period—and this was from a very special paper written locally—was sixteen months. "But that's not true," I said, in profound frustration. I wanted to look and read for myself, but he would not permit it. The inability to find any support for my findings after such careful and hard won observations was troubling.

On the Internet I was eventually able to make indirect contact with the author of the information that the males of the species were two metres long. The reply I received was that the record would be updated in the next edition of the reference book, having been based on just one fisherman's report. I was stunned. For years the two-metre males had puzzled me, and now, because of an e-mail sent, they had vanished!

Nor had I been able to measure the sharks to give a precise description their size. I took a metre stick to a session and watched closely as the sharks passed over it, but that was not very helpful.

But worst of all, no amount of searching revealed any supporting data about the sharks' gestation period and reproductive cycle, which was important to supplement my information on their travels.

Finally I told Arthur about my difficulty in finding references for my ten month gestation period. I told him that every web-site I had seen, listed the gestation period as being sixteen months long, with the exception of a few which stated that it was about nine months long.

He told me he would check into the matter, and his next e-mail gave me the names of the references for the two different figures for the gestation period. The finding of sixteen months was fifty years old! Arthur added that he had seen no good evidence for sixteen month gestation periods. Having already searched, I knew that the articles he named were not locally available, and was relieved when later he sent word that his friend, Dr. José Castro, had the papers and would send me copies. I hadn't wanted to ask for such a favour, so was deeply grateful for such kindness from a stranger.

When they arrived, I went over them carefully. I had understood that evidence submitted in a scientific journal had to be supported by facts and figures, in the form of graphs and tables. But the sixteen month gestation period was simply asserted, in the introduction to the paper, which actually described the development of the foetus. There was no supporting evidence, no mention of the number of female sharks examined and at what stages of pregnancy, or any reasoning given to allow the reader of the paper to understand why the conclusion of a sixteen month gestation period had been reached.

The paper asserting an eight to nine month gestation was more recent. The information had also been established by examining dead sharks. In this one, sample sizes were provided and these were of less than five sharks per month. For the important month of December, only one shark had been examined! Given the dramatic differences between the timing of the females' individual cycles, such methods would not reveal the full picture. Some sharks mated as early as November to birth in September, the others mating throughout the warm season until March, each birthing ten months later. Though the younger sharks were more sporadic, once fully matured, a female shark tended to mate and birth during the same lunar period of the same month each year, nearly always during the full moon.

Surely, the best way to destroy the evidence of the gestation period was to kill mama!

The truth sank in over the following days—no one but me knew what the gestation period of these sharks was. It seemed incredible, impossible. They were considered common reef sharks, regularly seen on dives from Polynesia to the Middle East. How was it possible that no one had checked their gestation period for fifty years? How could the unsupported figure of a sixteen month gestation period, baseless and arbitrary, have been accepted for all of that time?

I felt cold every time I thought about it. My investigation of the sharks had been purely for my own interest. I had never doubted that there were other people all over the world who knew everything there was to know about them. Now it seemed that I had gained some sort of responsibility. More than that, I was losing my faith in science. How could billions of dollars of tax-payers' money be spent to search for life on other planets when so little was known about life on this one? I had thought that everything accessible on our planet had been investigated, but apparently the truth was far from that. Any housewife in need of something to watch could discover something.

Reluctantly, I cut everything out of my manuscript except the information on the reproductive cycle, expanded it, added graphs to show the gestation periods and reproductive cycles of the sharks, and sent it to the scientific journal *Marine Biology*.

After the middle of March, conditions began to improve, and for the first time since December the prevailing ocean swell changed direction. Again I was on the protected lee side of the island, the current slowed, and for a while the water was crystalline—one could look out across the coral landscape, and see sharks moving far away. Carrellina came to the kayak as I arrived at Site Two and for the first time there, I threw the food overboard before going underwater. The session proceeded as usual, but there were deep cuts into Cotlet's back behind her dorsal fin. She had mated about three weeks before, but these cuts were fresh, and looked as if they had been made with a knife—stab wounds straight into her back from above.

Cinnamon, who rarely came to Site Two, glided in after sunset. She had also mated about three weeks before. She came and circled around me in a leisurely way while I drifted with her, looking her over. There were many sharks in the area, and in the clear water, the surroundings were a beautiful deep violet twilight.

Suddenly Sparkle shocked me from my reverie by zooming up to my nose with a male shark just behind her. It was a replay of the earlier occasion—the same darting, feinting male shark with the same unusual marking showing momentarily as he shot by. As before, I flew backward to avoid being slammed. Christobel appeared behind him, then Bratworst and Tamarack. Beyond them came a long ragged line of sharks flying toward me at the surface. All at once there were sharks everywhere—the Section One sharks were just thrilled to find me. I retreated up-current as they soared around expectantly, but once they had come to look at me, they drifted away, and scouted through the coral. I moved farther away, and drew the dorsal fin of a large pale female like Amaranth, who had joined in. It was Clara. She saw that I was focused on her and came tranquilly to look at me. She had appeared in my first year at Site One, and once in Section Three.

I had a small sack of extra scraps in the kayak, and went to get it. The sharks were flying freely through the vast clear space in a revel of socializing. As I swam, several shot up behind me, Simmaron swept up to my face, and a juvenile momentarily entangled herself with my legs. I was still some distance from the kayak when Bratworst made a close-up, straight on approach, reminding me of the session she had charged me with Madonna and Martha. It was shocking how fast the touch of fear escalated as I faltered. Finally, the sharks went sweeping away, after one of them grabbed a fish head and took off with it. I got the sack of food, swam to the site and emptied it. No one harassed me, and the sharks were thrilled.

But fear is bizarre. I suddenly found myself so frightened that I couldn't catch my breath. It took several minutes to calm myself while I clung to the boat, savouring the extraordinary grace of the sharks as they sailed around together. There was a wonderful moment when they all soared past me at eye level, one with a little scrap. But it was too dark and they were moving far too fast to take a photograph.

Drifting away over a surface shimmering and glowing under the pale green evening sky, I took off my gear, drank water, and rested, then detached my paddle for the long trip home. It was only then that I saw Shilly circling the boat. I had already travelled quite far, and she accompanied me for several more minutes.

I often thought of changing sites again, but Section Two seemed to be the limit I could go to regularly in terms of distance, given the conditions I often had to cope with, so no alternate feeding site was evident. I was not willing to stop the sessions altogether, and since I had not had any real trouble, I just kept on, week after week, taking my precautions, and keeping the kayak where I could easily escape into it if necessary.

I was still trying to find others on the Internet who were also watching the sharks, so we could compare notes. It's a big world, and there are many islands where sharks can be seen. I was convinced there were others doing the same thing. Sharks were so interesting that it was impossible that no one else had discovered the joy of spending time in their company, and had done what I had done. Besides, I loved the sharks. I loved to watch them. I loved their beauty, and the surprising and intelligent things they did.

In May I was drifting in darkened waters, thinking of leaving. No sharks had appeared for a long time when a dark female swept into view and circled me clockwise. I drew her dorsal fin. She left, and I waited again for a while, then the dark shark returned, and this time circled counter-clockwise. It was almost too good to be true—I drew the other side of her fin and looked her over appreciatively. I was turning to leave when Dapples appeared and I waited—Hurricane flickered into view with a group of sharks on the move. As they went into search mode, I sorted out their identities, wrote down their names, and put in the last of the food. Storm swept up and ate, and as they soared through the wide expanses socializing, a three-metre nurse shark undulated all around, his pennant tail slowly waving, pale in the violet gloom, to settle at last on a fish head in a flurry of frills and clouds. The scene deepened to shades of charcoals and greys, so I left, and called the new shark Carmeline.

Chapter Forty-Three

Thoughtful Sharks?

At home I received an e-mail from Arthur. He wanted to know if I had ever seen evidence of cognition—thinking in non-human animals—in sharks. He had been invited to speak on cognition in sharks at an international symposium in Germany, so had written to the leading shark behaviour researchers around the world—fifteen scientists in all—asking for their input. But no one had seen anything that could even be speculated to indicate cognition, except for the possibility of using cognitive maps. In other words, no one believed that this ancient line of animals was capable of thought, even very simple thought.

Arthur went on to describe cognition as the process of knowing through thinking, and mentioned a number of indications of cognition. The most important one was decision making, a choice reached through thinking about two likely avenues of action, and deciding on one of them, rather than using "trial and error."

He wrote:

Evidence of cognitive reasoning does have a tough set of rules, which requires that thought must have been involved to decide which alternative, amongst a set thereof, would be the best one to execute, to obtain the best benefit.

Cognitive-maps, for example, are exceedingly difficult to establish. They are inferred when an animal shows established routes when moving throughout a given region. Evidence for a cognitive map must show that the animal can use a route to a given spot without the use of specific landmarks or by sensory orientation (e.g. electromagnetic reception), alone. In other words, it must be able to use a short-cut never used before or use a detour never used before because of route-blockage or significant displacement. In essence, it must travel over unfamiliar ground. The implication is that it possesses a mental map that it has constructed and the animal is, thus, capable of making short-cuts or detours, since it can read the mental map as we can read a map in our cars.

I remembered Martha, and how she navigated around the lagoon, that she had turned four huge circles to the four points of the compass after the mirror experiment, passing in front of the mirror at the centre each time with no last-minute adjustment in her trajectory. She always knew where the kayak was, no matter where I had left it. And no matter where she was, she could turn and go straight to the barrens when she wished.

"I'll be awaiting your answer. . ." Arthur finished.

I gleefully wrote out a long missive citing possible signs of cognition that I had witnessed in sharks, and finished with Windy and Ondine's decision to follow me from the spawning grounds to the barrens.

Arthur answered:

Thank you so much for getting back to me so soon. What you've done is exemplary, since the so called "pros" rarely ever take the time, as you have done, to carefully test situations and try to think what might have caused the behavior shown by your pets. I realize that I shouldn't call them such, as you treated them as wild animals and have given them the respect due them.

It's uncanny that your individuals could learn to associate the boat above with you, and move toward the boat with you still in the water. Surely, if they learned anything, it had to be that you were the important factor associated with food, and to move away from you and move to the boat in your presence, seems contrary to logic. Even humans would have remained near you since you were the source of the food, not the boat.

Yet, there was no doubt about the sharks having treated the human-kayak combination as if they understood that it was an inanimate object that I used—when it arrived, they knew that I was coming too. If there was food in the kayak, they knew where it would be. There was no doubt that they knew that when I went to the boat, it might be to get food for them. Over the years, I had seen dozens of different individuals change direction to swim several metres to sniff the boat to learn if there was food in it or not.

I answered Arthur's many queries about my observations, and we exchanged several letters discussing each of the ways the sharks appeared to demonstrate cognition.

Arthur described some interesting findings he had made on placing a mirror in aquaria to see the effects on the fishes' behaviour:

In passing, I also have presented mirrors to fish. In one case, I read that aggressive fish in aquaria fight their image and get physically tired, thus, reducing their aggression on tank mates. So, I tried it with convict cichlids, a Central American species of highly aggressive reputation! Males in tanks with prospective mates often turn on such mates and do serious harm. So, I placed a mirror in a tank with a pair, the male of which was constantly taking it out on the female. Once the male encountered the mirror, he began to fight his image. Rapidly, the female learned to run behind the mirror whenever the male moved in her direction since often the male would pass his image and turn to fight the image and leave the female alone. The scene rapidly changed in about an hour, however. Instead of continually fighting the image, habituation to the image began. But the male's aggression, instead of becoming reduced, had become heightened, resulting in the increased level of aggression being directed at the female once the male stopped fighting his image. He soon passed by the image, ignoring it, but not ignoring the female. The presence of the mirror image may well have, in fact, heightened his aggression, which he, then, directed at the poor female. Soon, I had to separate the pair or the female would have been killed. I tried the mirror experiment several times during that week but soon left the mirrors out of all tanks.

Reducing aggression to tank-mates by such means did not work and the publication that I had read, which had predicted what I had hoped for, soon went into the trash-can. That's one of my experiences, using mirrors with animals.

Arthur stayed in contact with me as he prepared his talk, and he included four of my examples of cognition as well as some of my photographs. I was delighted with this unexpected opportunity to contribute something toward raising the status of sharks in people's minds. It was time they gained respect. I had formed the conclusion that they are intelligent animals who think about the events in their lives and form decisions about what to do. It was time to stop considering them as cold-blooded eating machines who act on instinct alone.

CHAPTER FORTY-FOUR

More on Thinking in Animals

My initial efforts in observing sharks were to learn what they were like, mostly because of their obvious intelligence. In observing wild animals, following the precepts of the field of zoology known as *cognitive ethology*, I had loosely organized intelligent behaviour suggestive of cognition or a subjective state, in one of the following categories:

- Emotional reactions
- Knowing others as individuals
- Self-awareness
- Categorization of events
- Memory and learning
- Problem solving
- Communication
- Attention, curiosity, and observation
- Courage, spirit, and intentionality
- Planning and predicting
- Decision making
- Deception
- Manipulation of the environment
- Eye gaze
- Imagination
- Insight
- New activities
- Cognitive maps
- Mirror reactions
- Appreciation of beauty

All my examples could be placed in one of these categories. Manipulation of the environment, for example, included tool use, but was more general. Though sharks were harder to observe than mammals and birds, I had seen them in so many different situations that I had collected examples in some of these categories.

Memory and Learning

While some learning could be classed as conditioning, other times learning had been displayed by the sharks in such a flexible fashion that it appeared to be cognitive. The slow evolution of their behaviour after my effort to feed Meadowes, in which they seemed to have lost their trust in me, was too complex to be due to simple conditioning. It had involved learning from many varied incidents over a long period, requiring

296

long-term memory. Researchers have found evidence showing that learning in fish cannot take place in the total absence of cognition and consciousness. *Declarative memories*, which are memories of the facts we can call on consciously, are considered to require some level of consciousness. In many circumstances the sharks had shown evidence of using declarative memories.

Breaching the surface to seek food is not naturally done by blackfin reef sharks. The instances in which the sharks did this brought up the question of their subjective concept of the surface and the space above it. Under normal circumstances, the space above the surface is not something that these sharks would have reason to consider. But they were presented with an artificial situation in which I came from above the surface and returned there, and so did the food in which they were interested. They would have stored memories about the surface from the occasions, particularly when they were small, when they swam through it or up against it while chasing a fish, though its unlikely they could have gained much of an impression about it, from such brief events. Yet, their behaviour suggested that they were aware of a volume above the surface in which things could exist, and from which I came and went.

Different sharks had different learning rates. Martha and Shilly had easily learned to target a treat I tossed them, for example, while, in spite of my efforts to help her, Madonna had not.

Communication

I could not see evidence of communication between sharks except through body language. If you have ever met an aggressive shark, you will know how well body language communicates at a physical level. The response arises deep within us without any interference from the frontal lobes of the mammalian brain.

I suspected that the close passes that sharks often made when encountering each other were a form of communication, in which each shark affirmed, "We are here." There had been moments when companions had acted in concert, leaving the other sharks, and swimming in formation to perform a specific act together. How they had communicated the decision to do this was unknown, as was the role if any, that communication had played and what form it had taken, in their mass evacuation.

The gesture they made of swimming slowly up to my face and turning away just as they reached me, defied reasoning. At times, it seemed to be their way of assuring themselves of my identity, yet other times, that was not it. What were they really doing? The gesture seemed ritualized, a ritualized affirmation. Perhaps it was linked, too, to a positive emotional response on seeing me. After many years, I concluded that it was analogous to the way one's dog will come excitedly when one returns home after an absence.

When done swiftly, as to a fisherman whose fish was of interest, it was likely intended to intimidate. My gardeners told me that sharks approached their faces fast, then turned away, when they wanted their fish. Yet sharks who had initially approached me at top speed, had not bitten, and had often not repeated the gesture. Mephistopheles, for example, had approached me like that only the first time, and on future visits, had behaved as if he knew me. So even when swiftly done, it seemed that the gesture could be the result of curiosity in an excited shark. Sharks also speed up when they are just excited.

In *The Secret Life of Sharks*, Peter Klimley described how great white sharks ritualize their conflict when a seal that one of them has killed comes under dispute. Each slaps the water at an angle with its tail, and the shark who raises the most water and blasts it farthest wins the prey. For this ritual to be effective, each shark must read the gesture as a communication, and the loser must acknowledge the winner to avoid a physical battle for the seal, which would badly hurt both sharks.

My best over-all example of communication involved a young red vented bul bul (*Pycnonotus cafer*), an ordinary little brown and black insectivorous garden bird, whom I called Pookie. She had been brought in after having fallen from her nest as a chick, and had been with us for nearly a year. Though as a young adult she pursued a natural life in the garden, she still visited me daily, and would fly to the window closest to me in the house. She was the fifth bul bul I had raised and studied, and my particular interest was the development of the bul bul language. I could whistle it so well that I could call wild bul buls down from the trees to hover in front of my face while we exchanged.

Our old house was full of geckos, small lizards of the family *Gekkonidae*, and when one was accidentally injured, Franck cared for him. At each meal-time he brought him to the table and took him out of his box. The lizard would look up and extend tiny arms toward Franck's fingers, grasp on, leap up, and run up his arm to lie on his warm skin. Franck would offer him banana and papaya, stroke him, and he would happily remain there throughout the meal. One morning Franck left his little lizard on the table after breakfast. Pookie arrived as I was going out, and when I returned, Franck emerged from his office very upset, and said he had found the lizard dead under Pookie's perch. He had held the tiny corpse up to show her, shouting the French equivalent of, "What the hell is this?" Pookie looked at the lizard, looked at Franck, and flew to the top of the wall, where she began shouting at him. Possibly she thought it was her lizard, not his. They shouted until Franck opened the door, and she shot outside. It may have been the first time that an animal had shouted back at Franck. Indeed, many people did not.

Pookie did not return for two weeks, and after that she would not come inside. I fed her at the kitchen window when she occasionally appeared, but soon our three children came to visit, and on the only occasion that she came into the house, their noise and agitation frightened her away. When they left, we went on vacation for a month, and I feared that I had lost her for good. But one morning soon after we returned, she was at the kitchen window at dawn. I had none of her special food made up, so offered her a bit of buttered bread. She ate a bit and left, but returned at 4:00 p.m. For the first time in more than three months she came inside, flying neatly onto the bar at my eye level, in front of the kitchen counter where I was working.

I offered her food, but she wasn't interested. Instead, she chattered pointedly away to me, as I offered her one favourite treat after another with no response. Finally I stood looking at her in puzzlement, while her voice took on increasing urgency. She hopped back and forth from one foot to the other, coming to the edge of the bar so that she was talking right into my face. I still didn't catch on, and she flew down to the counter top and peered over the edge. Looking down together, we could just see the edge of the toaster.

"You want toast?" I asked her, and pulled the toaster from its place on the shelf below.

The tone of her voice changed immediately, and she waited expectantly as I plugged it in, put in a piece of bread, and took a knife from the drawer and the butter from the fridge. She followed all these motions with apparent approval, and we waited together for her toast to pop up. When it did, she hopped onto my hand, chittering as I buttered the toast. Then she ate.

Her efforts to make herself understood, and frustration in not being able to initially, had been transparently clear. Prior to this incident, she had only ever had toast at the table when she happened to be in the house at breakfast time, which was not often. It was not one of her usual foods, and I could not remember her having a chance to learn about the toaster and the difference it made to bread. I had given her bread when she had come at dawn that morning, and it seemed that getting some hot buttered toast, instead of bread, was what this visit was all about. And she knew what was needed for toast—the toaster! She even knew where I kept it.

After that we always respected the toast ritual. Though Pookie had been increasingly independent and wild, and no longer let me touch her, for the toast ritual only, she would fly to my wrist, and ride upon it around the kitchen as I got out the butter, the bread, a plate, and a knife. Then she would sit on my arm looking down into the toaster at her bread being toasted!

Attention, Curiosity and Observation

A capacity for paying attention, and observation, was most clearly demonstrated by Martha. If a coconut floated across the surface it was she who noticed and moved vertically to sniff it, followed by everyone else. When she circled me slowly, moving her nose closer with each tail beat, to touch my slate, hands, arms, and side, her behaviour was not associated with searching for food, or any other biological directive. It suggested that she was curious for mental reasons. Her habit of flying to my side the moment I decided to go to the boat to get her a treat, was also my best example of directed attention and observation in a shark.

But there were many more subtle examples seen in the sharks generally. Their habit of waiting out of sight for some signal, indicated that they were comfortable observing the object of their attention using their sense of hearing and their lateral line sense.

The lateral line detects vibrations made by others moving in the water all along their sides, and apparently coordinates that information, vibrations coming in from near and far all around them, into a reasonable facsimile of reality that the shark can act on. Humans have no idea what it would be like to perceive reality like that, but the lateral line sense of sharks has been evolving just as long as our ears have evolved, and could be quite refined. I soon believed that the reason that sharks looked so uninterested when they passed, was that they were paying less attention to their sense of sight, than their other senses.

The length of time they were willing to devote to their hiding and listening was an indication of their capacity to remain focused on something for long periods. The sharks were using their sensory information in expectation of a future event, which put it into the category of "planning and predicting." Cognition is indicated because the shark must have referred to declarative memories in synchrony with sensory input, and thought about them in the present context, to make moment-to-moment decisions.

Shilly's habit of following me around since she had been tiny, was an exceptional example of this pattern of shark behaviour.

Another illustration of attention and observation was the way shy sharks reacted when I, or someone else, looked above the surface. They noticed immediately, and chose that moment to approach for a closer look, or to eat. While swimming in the lagoon with my stepson Romain, on one occasion, he climbed on a coral head to look around, and Eden, who had been loosely accompanying us, came swiftly to sniff his legs at the very instant his face disappeared above the surface. She knew that she could approach unobserved when his head was out of the water, and spontaneously used the opportunity to her advantage.

Though I did not find it discussed in my searches for more information on cognition, the factor of attention seemed to be a vital one in determining a mental life in animals. To me, the placement of the attention is the focus of our moment-to-moment mental experience. Sensory input, especially visual input, has priority in taking my attention, but it moves swiftly among many associated thoughts as well. These related considerations seem to be stimulated by the current scene, and are examined one by one as they appear, in a logical train of thoughts.

Analogously, the placement of the attention could be likened to being in a darkened place full of multitudes of varied things. When you cast the light of your attention on something, it appears like a pop-up window in movie form, pulling related subjects with it, one after another.

I tend to think in images, but have discussed this subject with others who can't imagine having a thought unless it's in words. So you might think in words, have a screen with writing that comes up instead of a movie, or perceive ideas like a voice speaking, as movements in a volume, or in some quite different way.

Dangerous things will grab one's attention and painful, threatening, and emotional subjects will tend to keep snatching at the attention when one tries to concentrate on daily life affairs. Dwelling on such thoughts will draw up related subjects deeper toward the roots of the mind, in the realm of the subconscious.

To give an example: I might be musing about my sharks while driving, with a small fraction of my attention monitoring the visual image coming in. Should a dog begin to cross the road in front of my vehicle, my attention instantaneously leaves the sharks and becomes riveted exclusively on the incoming image of the dog. I watch the progress of the dog in relation to the speed at which my car is approaching it, subsequent to having hit the brakes. The action of hitting the brakes was automatic. It was a subconscious act as a result of conclusions firmly in place, arrived at long ago, regarding the appropriate way to respond when a dog runs in front of my car. I pass the dog, and accelerate as my thoughts return to the sharks. The day goes on in this way. My attention is directed partly toward subjects of current importance, and partly by the situation I'm in, and its demands on my attention.

When I want to concentrate deeply on some project or idea, I arrange to free myself from environmental interruptions and distractions, which is why we have offices with doors!

Though doubtless we all have our own version of this subjective life, probably this is not too different from what most people experience. As an analogy, we all look different, but have the same anatomy. Commonly used phrases, such as "Pay attention!" or "She's just trying to get attention," suggest that the subject of the placement of the attention is generally understood. Since the focus of the attention can be consciously changed, its manipulation is voluntary to a degree, and seems to be under the direction of an overseer, the self. But, since this self has never been scientifically found, though it is acknowledged by the legal profession, we are now in controversial territory. While it may seem to be self-evident that the individual directs his or her attention, and is consciously aware of doing that, proving it scientifically is another matter.

Much of our train of thought is automatic. We tend to think all the time just as our hearts beat all the time. The predominant nature of thought in our subjective experience suggests that its roots are deep in our nervous system, and argues against the position that thinking is a recent evolutionary development. The way this automatic flow of thought tends to centre on the interests of the self, further supports the hypothesis that self-awareness and self-interest were selected for, and are likely primary in other species as well.

So, in considering cognition in animals, it seemed logical to look for signs of a directed mental focus. Indications that animals had noted the placement of my attention seemed equally significant. If an animal understood where my attention was directed, it must understand what attention is—it experiences it too.

The most impressive example of this in sharks was the occasion when the curious whitetip shark had made several head-on approaches, then glided between my eyes and slate when I became focused on drawing. The animal had been aware that my attention was concentrated on something else.

The habitual approach of shy sharks from behind, indicated their awareness of frontal views, and that they were escaping my attention by staying out of my sight. Familiar sharks always approached my face when they came to me; they didn't approach my head on all sides. Then there was Annaloo, the mystery shark, who for six months, drifted by me from left to right, only when I was looking the other way.

The sharks showed that they were aware of whether they could be seen or not, and used this awareness to their advantage. This capability is also associated with self-awareness, which I will discuss further on.

The crocodile needlefish who had come so faithfully to Site One, had clearly been aware of my focus of attention. On arrival, she orbited my head, gazing at me until I began looking at her. Then, once we had made eye contact, she waited behind me up-current from the sharks until I had a chance to bring her a piece of food. It was her spontaneous way of getting my attention, and she knew when she had it.

The fish I had fed at Site One also showed signs of comprehending my placement of attention. Those little fish had to wait, and while they were waiting, they were excited. They would flutter around my hands, and look into my mask. That these fish would make an effort to attract my attention while I was watching the sharks, provided evidence that they had mental lives of their own. They understood what attention was, and knew that while I was watching the sharks, they didn't have mine. They seemed to understand, in spite of the mask and my different body form, that inside the mask was *where I was*. Eye contact is an important part of exchanges between people, and is the strongest indicator of the direction of attention. Months after I changed sites, many of these fish were still in the vicinity, and remembered. Once when I had brought no food, I held my hands out to them, and they swam into them—I caressed them.

A different example of an animal's effort to get my attention is provided by a Stellar's jay (*Cyanocitta stelleri*). Jays are members of the crow family of birds, known to be highly intelligent. In the mountain valley where I lived in Canada, these jays sailed through the sub-alpine forests that surrounded my house. Since I painted outside when the temperature was above freezing, I was easily available to any passing bird, and the jays soon discovered that I could be coaxed into producing a peanut.

One female especially liked to sit with me, on the table or the railing around the deck, preening and resting. Sometimes after I had painted for a very long time, ignoring her entirely, she flew off and returned with one of the thousands, no doubt, of my peanuts that she had stashed around the place. This she placed upon the table in front of me. Of course, I looked. She looked at me, and flipped the peanut around a bit as if to say, "Now, *this* is the subject of consideration here!" Naturally I went inside and returned with some peanuts.

She had found a way to tell me what was on her mind, indicating insight, thought and planning. She clearly knew, too, that my attention needed to be diverted from my painting to her situation, as she sat waiting there, with lessening patience, for peanuts!

Other birds in my care also showed increased impatience if they saw me settle with my attention on papers, as if they realized that once I began going through them, I was in another world, and they would not get my attention for a long time. Sometimes they chose that moment to fly to me, and indicate vocally that they wanted to go outside.

Intentionality

When an animal's behaviour correlates with a mental image, which is held in the mind as a reference, it is said to be in an *intentional state*. For example, when I decide to get a drink of water, I get up, walk to the fridge, take the bottle, and pour the water, with the intention of drinking. The desire to drink may come from my biological condition, but the action required to get the water involves some intentional and mental acts, including deciding what to drink, what glass to use, remembering where the drink is kept, and so forth. In the same way, an animal who goes through a series of actions in order to get food, such as a predator who has to go to some trouble to hunt something, is acting intentionally. The behaviour suggests that the animal has a mental image of the intended result—eating—and is making some relevant decisions along the way. This is closely associated with *planning and predicting*.

Intentionality was discussed by Arthur's friend, Donald Griffin, in his book *Animal Minds* (1992, 2001), and by John Searle, Daniel Dennet and many others. According to this concept, the sharks were in an intentional state whenever they were following my boat across the lagoon, and in countless other situations when they showed the intention to do something, including when they decided to snap up a particular fish. They felt hungry, decided to eat a fish, chose one, and did so. A curious shark would also fall in the category of being in an intentional state.

Intentionality was linked with artificial (computer) intelligence through the idea that it was an epiphenomenon arising spontaneously from complex systems such as thermostats. Professor Roger Penrose, in his books *The Emperor's New Mind*, and *Shadows of the Mind*, wrote beautifully crafted mathematical arguments to nullify assertions that machines, such as thermostats and computers, could ever achieve intentionality and consciousness. He made the point that many things in nature, even simple things, could not be computed by any algorithm or computer program, however complex, and concluded that consciousness is in that category. For example, a computer can compute, but not understand the mathematical reasoning behind the decision to make a particular computation, which is obvious to the person considering the problem.

So because consciousness is non-computable, no computer, however fast, will ever be conscious. Penrose felt that biologists were not taking the true nature of matter, as understood by physicists, into account in their studies of living systems, and provided evidence that quantum mechanical phenomena could be involved in thought. If he is right, the common argument that a large brain is significant in determining cognitive potential would be meaningless, as studies suggest that it is.

It was remarkable that science could be prophesying that human-constructed machines either were, or soon would be conscious, while denying consciousness to animals! Animals are very poorly understood living systems. They are far more like us than any inanimate object, they evolved in concert with us, and, like us, are the only known living and acting children of eternity.

Planning and Predicting

The best illustration I witnessed of sharks preparing for a potential event in the future, was their method of hiding outside visual range, listening, unseen, to what I was doing, and planning to come shooting in, the very moment that some signal told them that I had produced the predicted food. The best examples of this occurred when I was waiting for Meadowes, when he was ill and I was trying to medicate him. Each night,

I had to think up some new tactic to get his medicine to him, because the other sharks always seemed to be a step ahead of me. I sometimes put scent, only, in the water, and waited with a chunk of medicated fish for him. Some of the sharks waited outside of visual range for me to throw the food, then zoomed in and snatched it before those who were actually present noticed what I had done. These sharks were waiting for a *predicted* future event with the *plan* to act, when they received the awaited signal.

Martha's focus on me, with the intention of joining me, should I show signs of going to the kayak, is another example. She *planned* to swiftly come, should I show signs of doing what she *predicted* I would do. To accomplish this, she was holding in her mind the idea of what she expected to happen.

It seemed that the sharks tried to be one step ahead of the object of their interest. The way they leaped out of the water to snatch the food for themselves from the back of the boat fitted precisely with this theory. In long-evolved predators who catch swift and evasive fish for a living, the strategy of watching and waiting, and trying to predict from past experience what the prey would do next, could well have been selected for.

Decision Making

My best example of sharks making a decision in a way that could be understood and documented, has already been described. Windy and Ondine decided to follow me from the spawning ground to the barrens, even though they had not seen me for several months, and they had not been fed in the barrens before. They appeared to have chosen the action that was based on a mental reference—a thought—in spite of being in a situation in which they could see, hear, and smell food, moving in a stimulating way. This is an example of sharks using declarative memories, which is considered an indication of consciousness.

There were also more subtle indications of decision-making in the sharks' daily lives. For example, when one of them went outside the reef one morning, though most days she stayed in the lagoon, did that not require a decision? Similarly, when I was swimming with Martha, and Madonna appeared, and began circling me, looking for food, Martha orbited us instead of going on by herself. She went on with me when Madonna left. This flexible change in behaviour due to Madonna's arrival, apparently indicated a decision to stop with me, instead of continuing on her way. Such incidents happened regularly when I was with the sharks, suggesting that they were responding to their environment through thinking about it, rather than through automatic, instinctual responses.

Manipulation of the Environment

I had seen blackfins and whitetip sharks flip over on their backs to wriggle in the sand, presumably to scratch, or to free themselves of parasites. On other occasions, a shark whipped the side of its body against a sandbank. Sometimes a remora was dislodged; the little fish flew swiftly after the shark as it swam away. Though the floor of the lagoon was made up of sand interspersed with reef flats and coral, the sharks invariably chose only sandy places for such manoeuvres.

When a shark positioned himself to use a smooth, flat surface of dead coral to scrape a remora from his ventral surface, he had chosen that structure from the others available in the environment. Therefore, the shark must have had a mental image of the required object, and referred to it while looking around in search of something that would serve his intended purpose. Intentionality is involved, as well as decision making.

Though this may not seem to be very impressive in terms of thinking in sharks, the availability of surfaces to use in this way doesn't mean that the animal will realize how they can be of benefit. For example, mynah birds (*Acridotheres tristis*), and junglefowl (*Gallus gallus*), the wild ancestor of domestic chickens, both spend much of their time foraging for insects on the ground, and both have strong feet for walking. However, mynah birds haven't discovered that they can use their feet to help them uncover these insects, while junglefowl do so instinctively.

Categorizing Events

The sharks' ability to place different sorts of events in categories was suggested by their changed behaviour when Bou Bou came with me. They vanished initially, then after many minutes they swam straight to him in long lines, led by Carrellina and Martha. Their actions indicated that they could categorize people, too—I was the familiar one, and Bou Bou was the stranger.

Particularly in the first two years of my study, whenever anything was different about my visit, whether it was in a different place or at a different time, the sharks' behaviour became more cautious. This implied that they had the event categorized, and immediately noticed anything different enough to take it out of the expected category.

Fishermen who complained that dive clubs bringing scraps for the sharks caused them to harass spear fishermen had failed to understand this crucial point—it was the fishermen themselves who were attracting the sharks by holding a dying fish underwater and trailing scent. Fishermen have always attracted sharks in this way. Sharks easily discern the difference between fishermen and shark feeding dives.

Deception

I had not seen deception in sharks, but there are famous examples of it in birds, including the well known one in which a bird pretends to have a broken wing in order to lure a potential predator away from her nest. She flexibly changes her behaviour depending on the circumstances.

My best example of deception was also provided by a bird, a junglefowl hen. She had been dropped off at the house as a tiny chick with only half a wing on one side, and had become a pet. Her name was Babalu, and her mate, who was also dropped off as a very sick young bird, was called Kubilai Khan.

One day Babalu was mauled by a dog. With parts of her looking like the remains of a chicken dinner, large areas of her body had to be kept cleaned, disinfected, and bandaged. Without continuous intensive care, she would have died. When she was able to walk again, she found it difficult to keep her balance, with her body and wings bandaged, and tended to stumble and fall. I helped her stay on her feet, was always there to support her, adjusted her bandages if necessary, and took her outside for a few minutes each day to see Kubilai. She received lots of attention and sympathy over her predicament.

Sickened by what had happened to her, Franck and I shared her care. Daily, I removed her bandages, cleaned and disinfected her injuries, applied healing balms, ointments, and honey, and bandaged and wrapped the struggling, shrieking, and wildly kicking bird. Then I gave her antibiotics and took her to Franck to recover while I cleaned up. He comforted her, and let her rest beside his computer. To Babalu, Franck was her saviour and hero!

She spent her time delicately picking, with undying patience, at her bandages, and often dislodged them so that I had to remove them, and re-wrap her. Sometimes I spent the whole day wrapping and re-wrapping her, frequently changing my approach as I improved my method and found new bandaging products to try.

As she began to feel better, she was permitted outside, but the first thing she did was take a dust-bath, filling her bandages and injuries with dirt. So I stopped letting her go out, explaining to Franck that until she had healed, she must stay clean to avoid infection. Sympathetic to her as he was though, often Franck insisted that she be allowed to go outside anyway. Babalu, therefore, had daily experience with our reactions to her evolving situation, and to each other. Repeatedly, she saw that Franck could overrule me. Though this may sound implausible, hierarchy is a concept that junglefowl know very well. In their society, too, males and females look different, and the male is more powerful than the female. He's the one who fights, for example, and takes charge of matters as the leader, just as human males do.

After many weeks had passed, Babalu seemed to be healed up, and we thought she was out of danger. However, one afternoon I noticed that her injured wing was red and swollen and she seemed sick. Close examination revealed that the wing injury was not actually closed. Where the two sides of the dog-bite slash had grown together, there was a long crevice all the way to the bone. It had filled with dirt, which was not apparent, since a thin scab concealed it. I spent the afternoon cleaning the dirt out, holding a light and magnifying glass with one hand, and removing each particle of dirt with the other.

Kubilai was also hospitalized by then. Since his mate was stuck in the house, he had gone looking for another hen in the jungle on the other side of the road, and had collided with a car. With a badly smashed wing, he was also clad in the latest in netting bandages, and was installed on a low perch in the bird corner at the other end of the house.

During my extensive cleaning of her infected injury, Babalu cried repeatedly to Kubilai, who returned each of her calls. The house echoed with their cries, which increased in urgency as the afternoon passed. By the time I finished, they were shrieking at each other, and Franck emerged from his office to find out what the hell was going on. I explained the situation as I gently set Babalu down. I had put a light, loose bandage around her wing, just to cover the injury for the night, since she had already suffered such an ordeal.

But she staggered across the room, and fell gasping against the wall. Franck turned to her to watch, and she intensified her efforts. She stumbled, fell, and careened around the room, leaning drunkenly against walls and furniture, as if she had lost all muscular control. He watched her in rising alarm, telling me to get the bandage off her, while I insisted that she was faking it. He didn't believe me, and if I hadn't put the bandage on myself, I would have considered it impossible that the bird did not have a serious neurological problem. But since she was behaving as she had done when she had been fully wrapped with a tight bandage, though in a much more exaggerated way, I stuck to my position. We argued for some time.

And Babalu won! I was compelled to remove her bandage, whereon her behaviour became perfectly normal, affirming Franck's conviction that it was the bandage that had been too tight.

Then I put Babalu on the low perch beside Kubilai, and she fell off. Repeatedly I placed her on the perch, reassuring her and steadying her carefully so that she was balanced, yet as soon as I let go, she fell off, in an apparent and most successful effort to elicit Franck and Kubilai's pity! Her deceptive act had succeeded in manipulating the feelings of another species, and she was rewarded by having her wishes satisfied. Further, she seemed to have an understanding of the emotion of pity, or how could she wilfully try to arouse it in Franck and in her mate?

This example is even more remarkable because it could not be argued that Babalu was running on instinct, due to the artificial situation in which she was functioning.

Eye Gaze

I had been considering the importance of the gaze for as long as I had been observing wild animals. If one watches one's own subjective state, one finds that some emotion colours nearly all of our thoughts, though faintly. These feelings express themselves to others through body language, and the nuances of facial expression and speech. Whether we approach another person with confidence or anxiety, suspicion, anger, joy, or perfect serenity, the attitude is seen in subconscious signals and gestures. Thus, our emotions communicate, without our deliberate intention or awareness, and anyone who has a pet will have noticed that such communications pass between species as well.

The most compelling physical communication passes through eye contact.

Once as I walked in the wilderness, I came face to face with a grizzly bear (*Ursus arctos horribilis*), tales of whose ferocity are often featured in the news media, the way they are with sharks. The bear and I stared at each other, and in those questing eyes I read only calm curiosity. After a long, uncertain moment, in which I stood tranquilly gazing, the bear began to pace toward me with that heavy, uneven gait unique to bears. The eye contact did not waver as he approached, and neither did his expression. I searched his eyes more and more deeply as he came up to me, never reading in them anything but interest, with an underlying wariness. He was locked as deeply into reading me as I was locked into his gaze and all it held. The result was a mutual mesmerization as he came very close. He did not pause nor hesitate, but sniffed my nose as if nothing could be more natural. The deep communion through eye contact was broken when he glanced off, turned away, and walked on into the trees.

Having first become familiar with bears while exploring the forests around my home as a child, I had noticed the importance of eye contact with them in countless meetings. My quiet presence resulted in some intriguing interactions as it had with the sharks. As a young woman I sketched their attitudes and gestures beneath the full moons of summer at the town dump where they congregated at night. The bears would pass close by me but ignored me, other than to avoid stepping on me.

It was my first experience of how reports of wild animals could be incompatible with what I knew of them first hand. As in the case of sharks, society really did not know them at all. Stories of their nastiness and guile were recounted by hunters who left out the part wherein they had shot the bear and were in the act of killing it when it fought back, in a desperate effort to live. Unfortunately these stories were accepted at face value. The result is that it is generally believed that the way a bear behaves when someone is killing it is normal bear behaviour.

Another example of the power of eye contact and inter-specific communication involved a skunk I found lying at the roadside as though dead. A certain tension in its posture betrayed suffering, so I stopped, wrapped it up, put it in my truck, and took it to the closest vet. She refused to treat it on the grounds that it was illegal for a veterinarian to treat a wild animal in Canada. It was surprising, since at the time, skunks were considered trash animals and anyone could go out and shoot as many of them as they wished. How could it be illegal to try to save one? They did no harm, as secretive, nocturnal animals who live on such things as insects, frogs and spider webs. I was obliged to take the unconscious creature home, unexamined and untreated. I made her comfortable on a bed in the kitchen, and as I did my housework, from time to time I squeezed a sweet, nutritive liquid into her mouth. Finally, at midnight, as I was encouraging her again to swallow liquids, and talking to her softly, she opened her eyes.

Her expression was easy to read as she found herself nose to nose with a human. Her eyes grew bigger and bigger until they bulged, while her muscles tensed and her tail rose. I murmured gently, and used the hand with which I had been stroking her to stroke down the rising tail. The skunk continued to stare at me, and I at her, our eyes centimetres apart, while with exaggerated gentleness I encouraged her to have some of the delicious liquid melting now upon her tongue. Gradually she relaxed and lowered her tail—I saw the moment of acceptance in her eyes. She began to lick up the nutrient.

I made up a bed for her at the high end of the bathtub where her spray would be easiest to clean up if she did let loose. My house was for sale, and the reek of skunk would render it unsaleable at any price. But she remained quiet, and visitors had no idea that there was a wild skunk in the house. She meticulously, painfully lifted herself from her bed to do her business at the other end of the bathtub to keep her bed clean. I was always quiet around her when I brought her food and water and cleaned up. Even though I had to move her when my son, Peter, or I wanted to have a shower, she had decided to trust me, and never did she spray. After two weeks she was well enough to be released.

She had perceived my intentions well enough to overcome her instinctive responses, and the moment of understanding had come through deep eye contact.

Observations of the power of eye contact extended beyond the more sophisticated animals classified as vertebrates, to a mollusc I met upon my porch as I painted in my mountain valley one evening.

A garden slug, one of the fat, yellow, fifteen centimetre (6 in) sort, was hunched up and dry on the wood of the deck, its slime turned to traces of silver around it. Moved to pity, I sprinkled a handful of water over it. Infinitely slowly, it moved. Gradually its body lengthened as it lifted its head, as minutes passed, toward me. Its eye-stalks grew longer as its eyes turned, not toward my hand now dripping water into a miniature pool in front of it, but toward my own, and focused on my face directly. It reached up toward me and its eye-stalks actually twisted around each other as though the animal was possessed by an urgency to see who or what had brought the lifesaving balm of water. I was in communion with a slug.

After a very long time of gazing up at me, as I stared back in astonishment, the slug slowly bent its head to drink from the layer of water in front of it. Awed by my clear perception of the little being within the slug's common and despised form, I made sure I did no harm to slugs after that. Still moving at a snails' pace, the invertebrate descended into the grass and, plucking a tendril, swept gradually away.

Earlier I described the moment a moray eel, a creature normally considered impenetrably alien, slowly took a large scrap of food from my hands while maintaining eye contact. And there was the strange event in the school of perch, in which one fish after another took the position in front of my eyes to gaze back at me. The needlefish had also often sought eye contact, apparently for renewed affirmation of our connection. This search for eye contact, and responsiveness to it among such different classes of animals, suggests an affinity across the vast gulfs of evolutionary time between us.

Though sharks don't make eye contact in the same way, when I was with them, their attitude at times seemed to reflect my own subjective state. For that reason, it seemed important not to become frightened of them. Though their eyes look out on each side—they cannot look straight ahead—they tend to look closely with one eye, and there is no doubt that they know when they are observed. On approaching me closely, with each undulation they would look at me first with one eye, and then the other.

Self-awareness

The subject of eye gaze naturally leads to considerations of self-awareness, since the animals in eye contact seem aware of each being an entity separate from the other. Donald Griffin pointed out that when an animal hid itself from human view, it was demonstrating its awareness of itself. He described how Lance A. Olsen had reported that grizzly bears sought places from which they could watch hunters while remaining hidden. Other observers had reported too, that bears tried to avoid leaving tracks. The researchers concluded that these bears were aware of being present and observable, as well as creating effects—their tracks—through their movements, which could be seen by others. The sharks' habitual way of remaining concealed behind the veiling light until an opportunistic moment, or approaching from behind to avoid being seen, is in the same category.

A prominent example was the lemon shark who came early to a session, and cruised around while I clung to the kayak. He came and went as if I weren't there, and it was easy to imagine that he had not seen me. He had been out of sight for fifteen to twenty minutes when I drifted away from the boat, whereon he appeared immediately, passing just within visual range to investigate the kayak. He had been aware of me, keeping track of my location, and came to the boat when he could do so while remaining virtually out of sight.

Sharks are self aware at least to the degree of being aware of being present and observable. Since any animal is a self-serving entity, seeking food for the self, protecting the self, saving the self, and so forth, it is logical that to be aware of the self, as distinct from others and the environment, would result in survival benefits. Thus, evolution, through natural selection, would favour self-awareness.

New Activities and Social Learning

When some of the sharks realized that they could get the food from the kayak without waiting for me to feed them, a cognitive leap was required on the part of the sharks who initiated it. This activity was performed by this group of blackfins, and no other. It seemed that leading sharks Carrellina and Chevron began, and were spontaneously joined by the others present. This appears to be an example of social learning, since the sharks learned something new from the leaders.

Cognitive Maps

Often when I went looking for Martha, I swam trailing the kayak until I found her, then anchored it and went with her. The enormous circles she performed in figure of eight or cloverleaf shaped pathways, returned her to the central point, and thus on some of our promenades we often passed the kayak. Her circles were so precise that we passed the boat when we completed the circle, no matter where in the landscape I had left it.

Sometimes she accelerated away from me when we passed it, as if this was her way of ridding herself of her human companion. Why did she take the trouble to accompany me to my boat when she could accelerate away from me at any time? I found this curious. It was oddly reminiscent of the way Merlin had taken me back to the beach when he wanted to be rid of me, indicated by body language that he wanted to eat (deception), and then went flying off at top speed when I went to get his food. He too had an excellent cognitive map in his mind.

Martha's action provided further evidence of her awareness of my connection with the kayak. Since reef sharks lack a home, their home range being just a large familiar area, they must have little sense of ownership. This emphasized the curious nature of the sharks' understanding of my relationship with the kayak.

Whenever Martha decided to go to the barrens, she stopped her sinuous roaming through the coral and went straight there, no matter where we were, as if she were always aware of her location relative to it. I often looked above the surface to check my location relative to the kayak, reef, or island, but the sharks could not. They seemed to have excellent cognitive maps, of their own ranges, and of the surrounding regions too. To roam as far as they did, a cognitive map would doubtless be selected for.

Mirror Recognition

The best known test for self-awareness is called the 'mark' test. It consists of putting a mark on an animal's face while he is unconscious, and showing him his reflection in a mirror when he wakes up. If he tries to remove the mark from his face, that is considered proof that he recognizes himself, and is self aware. However, is there really a connection between recognition of the self in a mirror and self-awareness?

Only great apes and some dolphins have passed the mark test, and the leap of judgement has thus been made that only these animals are self aware. However, doubt has been thrown on this conclusion in recent years, and it has been suggested that there are many levels of self-awareness, and that the mark test may be a misinterpretation.

Many animals see a conspecific in the mirror, as Arthur's fish did. Birds often treat their image as another bird, though not after they have seen it more than once or twice, I have found. Roosters, renowned fighting birds, sometimes attack their image at first, but soon appear to enjoy looking at it. I provided a very sick junglefowl chick with a mirror for company. He was ill for months and lost all his feathers. After he had grown into a weak and crippled rooster, he caught sight of himself in the mirror. His appearance was so changed with his fire-coloured mantle and iridescent green and blue accents, that he thought the image was another rooster, and shrieked. Then he just stared. He had realized that this was not a different bird, though whether he understood that it was him, I cannot know. Yet it appeared that he did understand. Subsequent views of himself in the mirror resulted in him just looking; his image didn't startle him again.

Though I rarely gave mirrors to seabirds, Angel had used hers to watch me when I was behind her. Often when she saw me enter the room in the mirror, she turned her head to look at me, indicating that she understood that it was a reflection.

The sharks paid a surprising amount of attention to the mirror when they saw it. Their level of interest, and the physical effort they devoted to investigating it, suggested curiosity and thought. When I showed the mirror to our dogs on my return from the lagoon, placing food in front of it as I had for the sharks, the dogs showed almost no interest in it—since they lived outside, none of them had seen a mirror before, either.

The sharks' investigation of the unusual object they found in their environment suggested problem solving, too. Though it was impossible to tell if they recognized themselves or not in the reflection, they clearly did not think that it was another shark. Their actions suggested that they realized that nothing was really there. Having reached this conclusion, which infers that they had thought about their investigations of the mirror, they ignored it.

In other words, they came to the right conclusion, not bad for animals who had to circle in front of it to see it. The mirror experiment was my first concrete evidence of just how intelligent sharks are.

Knowing Others as Individuals

Relating to conspecifics as individuals is the prerequisite for a complex social life. In travelling with preferred companions, sharks demonstrated that they knew each other as individuals. Their habit of swimming with a companion may have facilitated their acceptance of me. Arthur had noted that bonnethead sharks knew others as individuals, and many species of fish have been shown to do so as well.

Imagination

My accounts suggesting imagination are based on the actions of animals while dreaming in a state of unconsciousness or semi-coma. I saw no evidence of imagination in sharks.

Insight

A clue to the sharks' capacity for insight was their apparent understanding of my use of the kayak. Their discovery that they could take the food from it without waiting for me to serve it to them suggested that they were capable of insight.

Kim's effort to cope with Spooky's harassment by grabbing her out of the air and trying to drown her, was my most striking example of insight. In *Animal Minds*, Donald Griffin mentioned a few reported cases of a hawk drowning a smaller bird by standing on it in shallow water until it quieted, prior to eating it, but mentioned that such reports are rare. He wrote, "We can only speculate about how such behaviour might have arisen," and called the action an example of "versatile inventiveness." He wondered whether a hawk who killed both fish and birds would know enough to kill birds, but not fish, by holding them underwater, and had no reports of birds other than hawks drowning another bird.

Kim was a fish-eating bird. His strategy of standing on Spooky, was precisely like the descriptions given by the witnesses to the hawks' method. However, Kim and Spooky were in deep water, he lacked talons, and she was able to get out from beneath him by herself. The hawks drowned birds in shallow water where they could pin the victim down against the ground.

Related to this subject was the behaviour of one of Franck's dogs, Cesar. Long before, we had rescued a German shepherd dying of starvation at the roadside—there are many such dogs in Polynesia. When, after a long struggle she returned to health, she became mysteriously pregnant and gave birth to a large litter of black puppies with brown accents who became huge dogs. They were handsome, calm and friendly, and during the following years they protected us from thieves, and our birds from cats. People stopped their cars just to look at them. Cesar was born into this family, and was possessed of great sensitivity and spirit. He and Franck were inseparable and he was so handsome, loyal, and gracious, that he became known all over the island. He was also the biggest dog on the island.

Each evening, Franck and the dogs went for a swim off our beach, and played uproariously. Occasionally, Franck would play a trick on them, and pretend to have drowned. None of the dogs had ever done anything about it. However when Cesar saw his beloved partner half-dead in the water, he grabbed him by the hair and hauled him to the beach. He did this whenever Franck tested him, which wasn't often. He understood that Franck needed to breathe, and shouldn't be underwater. So Cesar lost no time in saving his life!

Humans are the champions of insight. Our countless discoveries and inventions form the history of science, and have been largely the result of this capacity.

Plato first described a world we could access only by the intellect. He showed that it has an independent existence outside of space and time, where the transcendent laws of mathematics, physics, chemistry, music, and so forth, exist. Only by going there can we understand the world and the universe.

An example of something that exists only in Plato's world is the square root of minus one. This is the number which, when multiplied by itself, will give minus one. While at first glance this could seem like nothing more than a mathematical joke, since all numbers when squared are positive, the square root of minus one has proved indispensable for working out some of the details of the functioning of the universe.

The behaviour of subatomic particles cannot be understood without it. The mathematical phenomenon known as the Mandelbrot set is the solution to an equation invoking the square root of minus one.

The illustration shows it graphed, and then a part of the set magnified over a million times. The intricate boundary does not change on magnification, which is one of the qualities of a fractal. The remarkable beauty of the graphed Mandelbrot set, named for Benoît Mandelbrot who found it, was inaccessible until we developed the power of the computer. Yet it was always there, in a virtual world of its own.

Music, too, leads us into Plato's world. Some birds sing using the humanly defined scale, so it appears to be accessible to other species.

The history of Easter Island is a startling illustration of how human insight can fail. The isolated island in the eastern part of the Pacific Ocean was originally populated by people crossing from Asia many hundreds of years ago. Since it lacked a barrier reef, boats were necessary for fishing. As time passed, the people got into a competition over who could build the largest statue (the instinct of territoriality). Later, crowns of red granite were added to make the huge faces looking out to sea even more impressive. The forests were cut to provide logs for rolling the enormous stones into place, and to make ropes to pull them. Finally all the trees were cut for the project, but work was still ongoing on the statues when the last rope broke. Without boats for fishing, everything edible on the island was devoured, every lizard, rat and snail. Finally the communities became militant and turned to cannibalism. By the time the Europeans arrived, there were no wild animals on the island but insects. The survivors had boats stitched together that could scarcely stay afloat, even with continual bailing. There were four boats for about two thousand people.

History reveals many examples where individuals and societies showed dangerous blind spots in terms of the real situation they were in, to their future detriment.

Humans seem to operate more on instinct, than reasoning, than we assume. War, for example, evolved with us out of our animal state. How else could the mass killing of other people be so easily accepted? The tendency to view people from other countries, other cultures, other races, and other religions as inferior, opening the door to war, slavery, oppression and torture, probably has genetic roots. While the force of instinct in other species has been exaggerated, in ours, it appears to have been underestimated. We distanced ourselves from nature, but unfortunately, we took our instincts with us.

Emotional Reactions

Emotion in sharks was apparent in their body language and actions. Basic ones, such as fear and excitement, were clear to see, but the most impressive one was their slamming of the boat, an intense emotional response to a particular series of circumstances. They have positive emotions too, as we will see later.

Appreciation of Beauty

The appreciation of beauty is another category that is strong in humans, and there is evidence that some animals are aware of beauty, too. Two examples are the creation of bowers by bower birds, which in every way resembles art, and the perfection of songs by birds, some of whom compose hundreds, and perfect them single-mindedly. I observed several wild species of birds, including junglefowl, who strove to perfect their calls and songs, and occasionally one achieved an aria. As in humans, the artistic ability seemed to be partly hereditary, and partly due to the efforts of the bird to improve his or her own vocalizations, in comparison with an inner concept of the musically beautiful, which it found, presumably, in Plato's world.

There have also been reports of wild mammals who were seen to go to a lookout to watch the sunset. I had one bird, a junglefowl named Diamond, who spent each evening on the beach, and often stood gazing across the bay, shimmering with the glory of the sunset.

I could not perceive appreciation of beauty in sharks. But so much of nature is beautiful—so many animals have evolved a surpassing beauty—that one could postulate that beauty was selected for. That is, the most beautiful individuals had the most offspring. If that is the case, the appreciation of beauty could be very wide-spread throughout nature, while imperceptible to human eyes.

That the lion fish had evolved its extraordinary beauty under the influence of crustaceans seemed hard to believe, though it was written in my reference book. But if intricacy, rather than beauty, was selected for in order to produce a disguise, why was it of such loveliness? Just the natural workings of physical law have a tendency to produce beauty, though how this applies to biological appearances is one of science's neglected facets. It would invoke the physical laws at work in living things, all the way down-scale to quantum mechanical behaviour.

Sharks have exquisitely coordinated sensory equipment, and their behaviour suggested that their mental discrimination was of a comparable fineness. They used their sensory input to make moment-to-moment decisions, and appeared to use deductive thought to respond flexibly and appropriately to their evolving situation. They remembered the events in their lives, and formed concepts based on memories, to which they referred in response to changing circumstances. They were curious, but cautious, and learned quickly. In their daily-life repertoire of behaviour patterns, were many indications that they were using cognition. The accuracy with which they followed their own schedules and invisible paths at different times of the day, lunar cycle, and year, suggested a mind capable of fine discretion. Their versatile behaviour, differences in cognitive abilities, and ways of approaching different situations, was not indicative of a stereotyped set of stimulus-response reactions. Especially illustrative, was the way each shark roamed according to his or her own schedule, guided by both the planet's luminary bodies.

Yet, so far away across the expanse of evolutionary time that separates us, the true states of subjectivity experienced by sharks must remain a mystery.

As a lone individual faced with a such a different species, I found that my spontaneous interpretation of the sharks' body language was accurate, and they responded rationally to mine. That this species barrier was penetrable suggests a higher level of affinity between us, than is usually thought possible. Our subjective states, with their corresponding body language, were enough alike to communicate, so that we could function as companions. The expression of emotion through body language is likely the original medium of pre-vocal communication, an important factor in multi-species biological communities, and the reason why we can enjoy pets.

Though fish may seem primitive when looking down on them from the altitude of *Homo sapiens*, in fact they are highly complex and evolved life forms. No brain is simple, as anyone who has observed the activities of a spider will appreciate, and nothing else in the discernible universe is so close to us in type as other living things on our planet. As vertebrates are organized physically in an analogous fashion, it follows, due to evolutionary continuity, that other species experience their own versions of subjective states, too.

Logically, how can we believe that consciousness occurs in us, and nowhere else in all the universe?

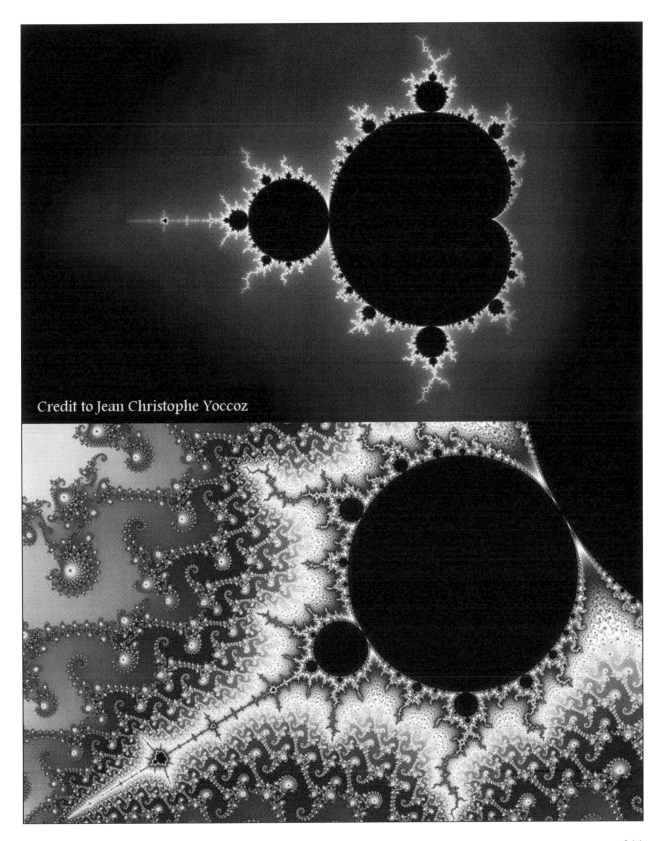

Credit to Jean Christophe Yoccoz

CHAPTER FORTY-FIVE

Christobel and Flannery

On the evening of May 3, 2003, the wind went wild as I passed the protection of the mountain, and crossed toward the lagoon. Swept broadside, it took all my power, paddling on the lee side of the kayak, to advance through rushes of confused waves. It was a great relief to throw in the anchor as I was blown backward past Site One—gaining Site Two was impossible.

The sharks met me as if my visits there had never been interrupted. Apricot swam so closely around the boat, that it seemed she would swim out of the water and into my lap. I reached out and stroked her, and was amazed by the softness of her dorsal fin. It felt so alive, and like water—water just a bit more solid than the liquid surroundings. Shilly was there, though her home range was so far away. She and Apricot had been greeting the boat at Site Two, so how they both managed to meet me there instead, was a mystery.

Underwater, Madonna, Trillium, and Ali came to me. Ali was nearly unrecognisable, she had grown so rotund—the little thing was pregnant! The passing of the many months since I had come in the evening to Site One was marked by big changes in the appearance of the maturing females. Trillium was growing up from the tiny wisp that she had been, into the stage of perfect beauty that Christobel had shown, when I had first met her—slender and graceful, with glowing bronze skin—before the heavy maternal build and scars of mating changed her. In her year before maturity, Trillium was making the border region her home range.

Windy, Cochita, and Brandy, who roamed along the border, had never come to Site Two. Flora's mouth had been badly scraped along the left side, and her pale skin was mottled from her last mating. A large, unknown female soared through, and I was sketching her dorsal fin when Odyssey and Clementine, still with a bent pectoral fin, followed. It was a year to a day since they had come, on May 4, 2002, and I blessed the wind for bringing me. Equally surprising was that Cheyenne was with them, as she had been the year before! With the appearance of Keeta, Lillith, Hurricane, Madeline, and Columbine, eleven very large females were swirling in the site. Flannery, looking lean after the mating season, joined in too. As the light faded, the sharks spread out, socializing in the coral. I drifted with them, and they treated me like another shark.

Two weeks later, I returned and slid into a tumult of excited residents. They surrounded me even before my bubbles cleared my view. Trill, a small pup from Site Two flitted by, surprisingly far from home! Christobel and Sparkle, together as always, came to see me before eating, and Lillith flew through in a huff, shot vertically in an effort to bite her remora, and made a dart at me for the first time in all those years. The Concorde drifted by high in the water, as she had done each time she had come. Samurai, Victoria, Eclipse, and Jessica came in the gathering night, roaming together in search mode, snapping up the odd bit of skin that remained, and trying to extricate the pieces that had been swept under the coral. I drifted with them, and The Concorde let me swim with her down-current until it was too dark to see.

These sessions were such a pleasure that I began returning to Site One for an occasional evening feeding, though I still held the regular sessions at Site Two. Madonna and Flora still swam to my face in their gesture of greeting when I first slid underwater, and in the absence of Carrellina, and the passage of so much time, the sharks behaved calmly, and did not try to push me around.

On Saturday, June 7th, Christobel appeared with a shark bite thirty-eight centimetres (15 in) across, on her back at her second dorsal fin. I swam just above her looking closely at the shocking injury. It was fresh. There was no sign that healing had begun, or that exposure to the salt water had whitened the flesh. The upper jaw of the big shark had raked across her back from her right side, just missed her spine, and removed the flesh above it. Each tooth had left a deep line, three centimetres (1.25 in) apart, and parallel across her back. Below, on her side, was a line of punctures where the lower teeth of the shark had grasped her.

Christobel had become a big animal. She would have provided a lot of food, yet perhaps her heavy build had helped save her. The size of the bite suggested that the predator had been a tiger shark, a nocturnal species. Here was another clue that the sharks went out into the ocean at night.

Given the danger of being eaten by larger sharks in the ocean, it seemed surprising that it was the first time I had seen a shark bite on one of my sharks. Perhaps big sharks rarely missed their prey. Perhaps there were no shark bites on the reef sharks because the large oceanic sharks had been so heavily fished for the Asian market for shark fin soup, which was emptying the oceans of the pelagic species.

When I had taken tourists to see wild dolphins, one of the problems with recognizing the different individuals had been that they no longer had any distinguishing marks, whereas, in earlier years, a large fraction of the dolphins had recognizable scars from shark bites. (The shark would go for the baby dolphin, and its mother would be raked by the shark when she defended it). This, and the increase in the dolphin population, seemed to be the result of the disappearance of the oceanic sharks.

I was still following Christobel around, stunned at the sight of her injury, when Flannery swam by. A month had passed since I had seen him. He had appeared lean then, but now he looked as Meadowes had done two years before, emaciated and reminiscent of a tadpole. His skin was dull and his colour line had hardened, as if someone had outlined it in black above and whitened it below. But he did not look blotchy and muddied. I noted uneasily that he was ill at precisely the same time of year as Meadowes.

I inquired on the Internet in hopes of finding someone who had seen a similar illness in a shark, but no one had. Dr. George Benz, with whom I discussed the possibility of a parasitic involvement, told me that few people had observed wild sharks over long periods so there were few records to draw from. He advised me that he did not treat a shark if he did not know what he was dealing with, and that seemed wise. After my troubles treating Meadowes, who had been relatively easy to find, to try to treat Flannery, who roamed widely and had never regularly attended my sessions, seemed an impossible proposition.

I returned on June 13, six days later, with some food for him. The usual sharks came, and as night fell, Christobel glided in. It was too dark to see her wound in detail, but the lower row of teeth punctures, and the edges of the bite, had turned black as healing began. The exposed flesh was darker and smoothed out.

But Flannery did not come. I had his food in a plastic box, and as I took off the lid to distribute it at the end, the cube of excellent fish chunks accidentally fell into the lagoon. Sparkle Too grabbed the whole thing, and bombed off with it, doubtless incredulous over her unprecedented find so late in the session. The rest shot after her, and they all got something good to eat that night.

For two weeks storms perturbed the ocean, and I returned to the sharks with the calm. Golden sunlight illuminated the coral, and flickered over the graceful creatures as the clouds blew apart in the west. The session was interesting as well as beautiful. Merrilee reappeared. She was the one I had identified at Site One during my first year, who then disappeared until I found her two years later at Site Three. It had been nearly a year since I had seen her, though she had visited Site Two regularly when I had first changed feeding sites. Her habits would take some time to unravel. Yet, could it be that she was only now returning from the great shark disappearance? Could I still hope for the return of Valentine, Annaloo, and Chevron, if not the long lost Martha?

Christobel glided in. It was now a month since she had been bitten, and the injury had turned black. Deep parallel lines across her back showed where each tooth of the shark had raked her. Given the resources, I could have determined the size of the shark from the spacing of its teeth! The leading spine of her second dorsal fin, that had remained in place due to having been between the teeth of the predator, lay back over the wound, and appeared to be cemented to it.

I waited until the sharks were only motions in the darkness, but Flannery did not come.

But the following week he appeared late in the session, skirting the site on the up-current side. He was shockingly thin; the curve of his ventral surface was concave. Though I never saw him eat, he was adjusting his jaws, so had already found a scrap. He swam slowly away toward the reef, and I flew after him before he could disappear from my view. I followed him far along the reef, with the rays of the setting sun flaring through the water towards us. I knew he was leaving this world, and was filled, again, with the grief of facing the death of an animal I had known and cherished, one who pursued his life as I did mine, and who would do so no longer. He vanished ahead into the veiling light, and it was the last time I laid eyes on Flannery. He had sickened and died exactly two years after Meadowes, with the same symptoms.

When I returned to the site, Bratworst and Ruffles came to meet me, perhaps perplexed to find me arriving from up-current, and returned to Site Two with me. There, I found a pale and blimp-like Solace trying to eat. I watched her in wonderment as she tried, time after time, to pick up a bit of food, overshooting it in the current and trying again, her fat body manoeuvring almost vertical, her fins extended down towards the sand to help her balance. She would be one of the first of the females to birth. Christobel had come too, and I tried to photograph her bite. It was still reasonably light for once. In the two weeks since I had last seen her, the black lines where each tooth of the shark had scraped deep into her back were even more prominent, as the advancing healing blackened the edges. The spine on the leading edge of her fin was still lying back across the injury, and the space below had filled in, to make a diminished second dorsal fin.

As I drifted away after the session, the sharks circled with me. Cheyenne, who had seemed so much older, Faroe, Papillon, Fawn, Tuvalu, a slender Bratworst, Hurricane. . . They came with the kayak for a long way as I drifted, paddling sporadically, but mostly just watching my beautiful sharks and imagining how nice it would be if we didn't live in such different worlds.

In June I went again to Site One. As was her habit, Cochita circled the boat clockwise. Since Martha's disappearance she had become the first shark to find me as I crossed from the border to the site. This time as she passed, I reached down and stroked her. The next time she passed, she turned before reaching the place where she had been touched. This was why I had rarely touched them. They didn't seem to like it. And it was a matter of respect too, not to be too familiar unless invited. The sharks did not caress each other, and they did not caress me, so I did not caress them.

Madonna and Bratworst were circling the boat too, still there in the same place all these years later, joined now by Windy. As usual in the dry season, more and more two to three year olds were appearing, and I was busy much of the time sketching their fins, and trying to sort them all out. Flora passed beneath me as she went to the food. Madeline, Brandy, Trillium, and Storm arrived, then Dapples, my faithful herald, so I waited in anticipation to see who would come next.

Samurai, huge, dark and scarred, Twilight, Columbine, Fleur de Lis, Droplet, and Cinnamon soared in, with Ruffles for company. He still didn't show the slightest sign of age, but he had been mature when I had first met him. As night fell, only Ruffles and Columbine were passing from time to time as they circled through the region. I got some bits of skin from amongst the corals and put them out in the open for them. Ruffles tasted them and spat them out. Several more juveniles came through as I picked up the bones that remained in the site, to obliterate all sign of the session, and as I returned to the kayak, Fawn drifted by.

Arthur continued to write to me, telling me of the many conferences and papers he was working on. In one letter he described a major conference at The University of Maryland, College Park: *The First International Conference on Acoustical Communication in Animals*. This was his pet subject, on which he was one of the world's authorities. He had focused on the concept of *Interception*, which he had developed on his own. For me, his correspondence provided fascinating glimpses into the world of a leading ethologist.

He told me:

Many experts came from Europe, Asia, and Australia, and I renewed innumerable friendships with individuals, whom I hadn't seen in years. My lecture was at peak-time and it went beautifully! The talk covered Interception by predators, by prey, by competitors and by prospective mates in numerous groups: bony fishes, sharks, frogs, many birds and mammals, including bats, whales, dolphins, and terrestrial forms and even crickets and parasitic flies. I only had time for acoustical examples, but had ready numerous cases of electro-receptive, chemical and visual interception. The talk, fortunately, ended on time and many had nice things to say about it during the conference. At least 300 experts from around the world now know of the process and, now, I've got to put together the final manuscript, dealing with the subject on a grand scale and not limit it, as I've previously done, to fishes.

Don Griffin was my room-mate at the conference in College Park; but the day before he was to speak, he became extremely nauseous due to his chemotherapy (he has cancer). So, they moved him to his own room at a nearby hotel; but he couldn't remain at the conference. Unfortunately, he didn't speak and left for his home on Cape Cod the day following his time-slot. We were all greatly saddened when he left. I've known Don for more than thirty years and I can only hope that he pulls through his problems.

Your knowledge of cognition surpasses my own. Your latest letter has innumerable points and I shall print it and carefully examine each and every point. I have invited talks to give, around Miami and Ft. Lauderdale, tonight, as well as several during next week. I can't find time to even unpack the stuff that I brought to my office from the meeting.

So. . . After I closely read your wonderful letter, I'll get back to you and let you know my thoughts. Again, thank you for your continued help. Don got ill so rapidly, I didn't have the chance to discuss cognition and animal-awareness with him. . . one of the world's most eminent scholars on the subject. . . my poor luck.

It was a shocking revelation to learn that the pioneer of cognitive ethology, Donald Griffin, was dying. Arthur also wrote about Konrad Lorenz, with whom he had worked years before. Since I had started reading his books when still a child, it was exciting to gain Arthur's insight into the great ethologist:

You asked about Konrad Lorenz. He was probably the brightest thinker that I ever met. He was never at a loss for words in almost any language. He spoke seven languages fluently: German, English, French, Italian, Spanish and even Russian and Hungarian. The latter two, he learned as a POW for about five years in a Russian Stalag, working as a MD, which he also was, besides having a Ph.D.

He visited me twice in the U.S., both times here at the Marine School. I took him and his wife, Gretel, on a ten-day cruise through the Florida Reef Tract with two other behavioral scientists and six assistants, most of whom were graduate students. It was for all, the trip of a lifetime, except probably for Konrad and Gretel, although I'm sure that they enjoyed the cruise. In his book, Die Sogenannate Böse *(which you are reading), he wrote about the Florida reef fishes, which he thought he saw on the reefs. He saw the fishes, but they really weren't on coral reefs as he had assumed. But, he finally saw such fishes on actual coral reefs during the cruise and for that he was very happy.*

He and I talked together many times, while at his Institute, as he often visited the fish-aquarium labs, where I usually was during the day and I truly learned what it meant to be an ethologist from him, i.e., always observing behavior of animals (including humans) and trying to explain the behavior in terms of biologically significant functions. That is what he did before an aquarium, while watching ducks on a lake, or humans at a sporting event. Wherever he was, his lectures held the audience spellbound with stories about science and scientists, whom he had known, both famous and not-so famous, but after his lecture(s), the latter became famous!

In passing, I've sought out lectures by Nobel Laureates to find out something about what kind of person gets such a prize. I've heard about twenty to twenty-five such lectures over a period of twenty to thirty years and in every case except for one (which proves the case), each lecture was brilliantly accomplished. The persons were spell-binding in their presentations.

And Konrad was such a person.

Chapter Forty-Six

Crisis

On August 10, 2003, I returned to Site One for an evening session. No sharks followed the boat, and none appeared as I anchored and put on my gear. Underwater there was no one, and as I waited longer and longer, no one came. Sharks coming from farther along the lagoon to hunt on the border in the evening should have been crossing the scent flow and arriving, but there seemed to be no one in the area. Something was wrong. After nearly fifteen minutes Flora came. She circled, picked up a scrap half-heartedly and moved on. My anxiety increased. Shimmy passed through. And again for a long time, the beautiful coral garden was empty. Could it be another shark disappearance? It was the full moon corresponding roughly to the lunar period in which they had left the year before.

Night was falling when a shark emerged from the gloom, and swam toward me very slowly. She appeared to be dying. Her movements were far weaker than Meadowes' had been before he died. She came straight to me, and passed underneath as I drifted sideways, looking her over.

It was Trillium. A large hook was embedded in the angle of her jaw, and there was a long tear through her flesh beside it. She was swimming with her mouth open. Moving painfully slowly, she picked up a piece of food and carried it onward, out the other side of the site. I swam beside her while in slow-motion she swallowed it, gulping now and again over a period of a few minutes. She glided listlessly away, became a shadow, then only a movement in the coral, and vanished.

My blood ran cold. Sharks fled when some among them had been killed, and the disappearance of most of the population seemed evidence of a massacre.

I had met Trillium one morning three years before, when she was very small, velvet bronze, and radiant with life. She stood out among the others as such a pretty shark that she had become a favourite, and for a long time she had attended infrequently. Now that she was older, she had been spending her time in the border region, and probably had been fished while out in the pass or ocean. Obviously she had been fought nearly to death, and had barely escaped when the line broke. I waited in mounting panic while the night fell, and no one else came, except for two juveniles who spent their time between Site One and the barrens. Probably whoever was slaughtering the sharks was doing so out in the ocean, so the small juveniles who stayed in the lagoon would be unaffected.

By the time I had paddled the kayak all the way home in the light of the rising full moon, I was terrified that Trillium was the only shark who had escaped. Never before had I seen a hook in one of my sharks, except on two occasions when a tiny hook had snagged in their bodies. This was different.

I called Bou Bou, who told me that he had noticed no change in the patterns of the sharks outside the reef, but that he had started seeing hooks in them. I was in an anguish of waiting while a storm careened over the island, frolicking with the sea in such an alarming fashion that I dared not venture out, so I contacted the editor of a local news magazine. He invited me to come for an interview, and after we had talked for a while, he told me that a man had been going around the island handing out a poster offering a good price to fishermen for shark fins. He searched through his piles of papers until he found a copy.

"Good news!" the poster read. "I am the representative of a company in Singapore. This company collects all the dried shark fins. Go and fish any time, since it is a pleasure to make money by fishing for shark fins." Blackfin sharks, and the other reef sharks I knew were specifically mentioned in the poster as valu-

Trillium

able species to target. The company was very organized. It had distributed the poster, and set up a network to collect shark fins from fishermen on hundreds of islands in the archipelagos of Polynesia. The local men had never fished for sharks before, but the offer of easy cash had dramatic results. Divers began finding the bodies of finned sharks on the slopes of the reefs and in the passes. One was a three-metre lemon shark in Rangiroa, which was brought ashore as evidence and photographed by one of my contacts, Peter Schneider.

The editor of the magazine published an article about it all, and included my observations. The dive clubs were lobbying the government to protect the sharks, and I sent a report about the unprecedented impact at my location to their spokesman and to government representatives. A few days later the daily newspaper, *La Dépêche*, featured a long article about it, describing the massacre in all island groups in French Polynesia. Stories from the different dive clubs, my report, many large photographs of finned sharks, and commentaries by biologists on what it would mean for the coral habitat if sharks were eliminated, presented the situation as a nationwide crisis. The article demanded that the government make Polynesia a sanctuary for sharks, as it was already a sanctuary for whales.

Shark finning had only been in wide-spread practice since the 1980s, yet it had already resulted in a worldwide loss of sharks, comparable with the slaughter of the buffalo on the North American plains in the 1800s, but on a global scale. The depletion of the oceanic sharks around Polynesia had been due in the past to rusty little boats from Asia such as the one we had visited in the port, who hunted oceanic sharks to supply the growing demand for shark fin soup. Until now, the reef sharks had been left alone.

The sharks were caught in nets or on long-lines, then hauled on board, where fishermen sliced off their fins with sharp knives, and threw the wingless sharks back into the sea. They slowly sank into the crushing densities of the abyss, frantically wriggling and writhing in an anguish of pain and suffocation, or landed on the reef's outer slope, to lie gasping, still gazing out through their blue world, until they died.

China's growing middle class provides the biggest market for shark fin soup, with millions more newly rich Chinese being able to afford the luxury dish each year. China, Malaysia, Indonesia, Japan, Taiwan, Singapore, South Korea, and Thailand are the biggest consumers of the dish, which sells for about a hundred U.S. dollars a bowl. While the body of the shark is worth little, dried shark fins sell for eleven hundred U.S. dollars per kilogram, more than seventy times the value of tuna, making them one of the world's most expensive sea foods. The party tradition of shark fin soup is responsible for the cruel and wasteful death of an estimated minimum of seventy-three million sharks yearly while millions more are caught by sportsmen or as by-catch. But the true numbers are impossible to learn, since the slaughter takes place globally, hidden on the high seas.

Sharks are susceptible to over-fishing because they have long life cycles, mature slowly, and produce small litters of live pups. There are approximately four hundred twenty shark species, with more being discovered all the time, and they are top and middle predators in marine habitats. Oceanic ecological systems have evolved through the aeons under their influence, and their decimation is thought to threaten the very health of the oceans. The slaughter goes on continuously, three sharks per second, out of sight, and out of mind, as we go about our busy daily lives.

With the market for shark fins growing fast, the sharks of the South Pacific were being hunted with increasing intensity, and as their population sank, they were more and more difficult to find. Highly motivated modern companies such as this one from Singapore used every trick, such as making use of the know-how of the local people, to get as many shark fins as possible wherever they could find them. Since the numbers of sharks were so depleted, and the efforts to catch those remaining was so intense, the fraction of the sharks left alive plummeted annually.

I was witnessing only one of the latest moves of these voracious killers: the targeting of dive sites. Some regions, of which the Marshall Islands was an example, had their shark population nearly eradicated in just a few days. The Polynesian Islands represented a unique Pacific habitat, attracting tourists worldwide, many for the incomparable shark dives. Without protection, it could soon be devastated.

I believed that due to the islanders' traditional attitude to sharks, the reef sharks of Polynesia would provide the one exception to the rule. But I was wrong, and even more in error in the belief that some public outcry of horror would go up to match my own dismay. Only the divers cared. Most people were in sympathy with the fishermen. In the agony of knowing that my sharks were being slaughtered so cruelly, I sent out pleas over the Internet to write to the government to ask that Polynesian sharks be protected by law.

By then, via the Internet shark and bird discussion groups, I had developed a substantial network of contacts who were receptive to my concerns. Letters poured in to the government, to the President, the Vice-president, and the Ministries of Environment, Economics, Fisheries, Tourism, and the Protection of the Sea.

As soon as the weather calmed, I hurried to the lagoon across a restless sea. Once over the border, I searched in anguish through the surface, flowing with furls of violet and green, where the enchanted coral garden appeared and vanished like a lost dream. Finally Cochita appeared, her little fin slicing through the surface right behind me. At least one more was alive. But underwater, I found that she, too, had a large hook driven though her jaw, and trailed a length of fishing line. Bratworst arrived sometime later, also hooked through the side of her mouth and trailing line! Sparkle Too was in the same condition, but her hook was bigger, and there was a deep cut up into her face from the place it was embedded. It was fresh—she must have been hooked in the past twenty-four hours. I saw Windy pass, but was unable to tell if she was all right or not—she would not come near. The only other sharks who came, besides a few young juveniles, were Shilly, who looked fine, and Storm, much later, passing far away in the gloom.

As darkness fell, Trillium appeared briefly, moving more strongly, but her mouth was pried open on the side where the big hook was jammed, as if the structure of her jaw was damaged. Lillith drifted down-current, and she approached when the others had left, but when she saw me, she turned and disappeared. She did not enter the site. The sharks' behaviour was very different, and most were absent. Appalled by the high fraction of them with hooks in their mouths, I continued to spread the news on the Internet.

One of the messages I received was from the owner of a dive club, Bertil Venzo of *Shark Dive*, on the neighbouring island of Tahaa. He told me that more than ten years before, he had been the dive monitor at *Club Med*, on Moorea Island. Three times a week, they held a shark feeding at a popular dive site called *The Tiki Site*. There were always many reef sharks there, grey reef sharks, blackfin reef sharks, and lemon sharks. The site was considered to be one of the top ten shark diving sites in the world. One day a foreign fisherman who worked in the kitchen, killed two female whitetip sharks, and used them to decorate the buffet in the dining room. When the eight scuba monitors arrived for lunch, they raised a scandal, removed the dead sharks, and placed them in the restaurant's fridge. The next day they took the two cadavers with them to *The Tiki Site* to observe the reactions of the other sharks. The boat was met by many circling sharks as usual, but seconds after the immersion of the dead sharks, they vanished. No one saw a shark for the rest of the day. Even the lemon sharks had gone.

He wrote: "We understood that they don't eat each other."

It was more than two weeks before the first shark returned to the site, but over a month before they were plentiful there again. Since then, Bertil told me that whenever he found a dead shark, he buried it.

It was the only report that supported my hope that the sharks could have fled on first being fished, and were taking their time to return.

The following evening I found Madonna, and circled her as she circled me, to find, with deep relief, that she was not only still alive, but that she had no hook in her mouth. But there were still very few sharks in the area. Drifting through the familiar scenery, from time to time I glimpsed one of the very large nurse sharks roaming the area, so far away he was blurry as a cloud in the dimness, appearing and disappearing with his pennant tail waving in slow-motion above him. The emptiness of the coral garden, and the silence in which he moved, highlighted his eerie passage, a reminder of all that was mysterious and unknown in the sea.

As night fell, Lillith passed, once more avoiding me, and I was able to add Marco, Eden, Ondine, and Ruffles to my list of sharks still living. Jessica passed as the gloom deepened, and circled me briefly. Then the nurse shark reappeared closer, and I drifted into position to try for a photograph of the enigmatic creature. As I focused on him, he responded by swimming over to me, which had never happened before with a nurse shark, even during feeding sessions.

Letters poured in by e-mail from all over the world, from dive clubs, shark divers, scientists and conservationists, asking the government to make French Polynesia, which is nearly the size of Europe in oceanic area, a sanctuary for sharks.

As August passed, I alternated between searching the lagoon, and writing e-mails. The sharks continued to appear with hooks in their mouths, or not to appear. Trillium's mouth remained disfigured, and she lost weight. Her ability to catch food was apparently compromised since she couldn't close her mouth. Other sharks who had escaped from inept shark fishermen trailed metres of fishing line which was slowly covered with a thick growth of algae, creating a continuous heavy drag on their mouths.

I wondered if people who sport fished for fun would still do it if they could see the suffering they caused, and the harm that they did. But soon fishermen wrote to assure me that science had proven that fish, including sharks, don't feel pain! With the memory of Trillium's behaviour after nearly being fought to death still sharp in my mind, I couldn't believe what I was reading.

How could a living creature manage its life without feeling pain? Only a small percentage of fish who come into the world live to adulthood, and any weakness would doom them. An inability to feel pain would result in inappropriate behaviour, and the fish would go straight into evolution's garbage can. Pain is a warning sensation that is important for survival. The sharks who had escaped being finned, as well as female sharks with mating wounds, showed the same signs of being in pain that I had seen in birds and mammals. They seemed mentally interiorized, were less alert, less reactive, and slower moving. If fish couldn't feel pain, observations should support that. But fish and sharks were shy and cautious and appeared irritated at times by just the soft touch of a remora.

Fishermen argued with me that fish don't feel pain because one will sometimes bite a baited hook a second time, after being unhooked and thrown back into the sea. But while it may be obvious to the fisherman what he is doing, how could it be obvious to the fish? These men assumed that the fish understood much more than it possibly could about its situation. It could have no basis among its experiences in life for comprehending the fisherman's practice of deception—the possibility of a hook hidden in the bit of food it had found. There was no sign of a dangerous predator underwater. How could the fish possibly imagine that above the surface a man was waiting, hoping to trick and kill it? Even a human walking by the sea would never suspect that there was a creature waiting for him beneath the surface with a plan to trap and kill him.

Even if the fish had already bitten a bit of food with a hook in it, why should it assume that the next bit it finds will also hide a hook? This is true even if it had been reeled in, the hook had been roughly yanked out, and the fish tossed back into the water. How could the little animal understand what had happened to it? It probably sensed something like enormous relief, then ignored its sore mouth as it continued to forage.

Another fisherman's argument was that sharks continue to swim with mortal wounds. But humans, too, will continue to function with terrible wounds in heightened states, such as in fear or in battle. Fear is generated from deep in the brain, and over-rides pain at times, with obvious survival benefits.

Investigation revealed that a fisherman, James D. Rose, had published the notion that fish don't feel pain in a fishing journal, as scientific fact. While acknowledging that fish may act as if they feel pain when hooked and yanked out of the water, he claimed that they lack the brains to be aware of it. Without having done a study to confirm it, he claimed that the neocortex, the outer folded layer of the mammalian brain that is so highly developed in humans, is the seat of all higher mental functions, including the consciousness of pain. Since fish lack a neocortex, he argued, they cannot feel pain. But in comparing only the brains of humans and fish, with no mention of the straightforward evolution of the vertebrate brain, Rose's article seemed biased and anthropocentric.

I knew from extensive reading on cognition and the structure of the brain, that from fish to man, the brain has the same general structures, arranged in the same way. The only exception is the neocortex, which evolved in mammals, and was considered to have taken over certain higher functions, which were *already present in* fish, amphibians, reptiles, and birds. The expansion of the forebrain has occurred many times in different species, including in some fish, whose brain structures would fall into Rose's category. All teleost fishes (bony fishes) have elaborate forebrains, and the degree of forebrain development has been correlated with social behaviour, and communication abilities, which are considered to be integrated with cognition.

Fish continue to develop neurons throughout their lives, and do so at a faster rate when confronted with a stimulating environment, indicating a link between experience and neural development. For example, triggerfish have advanced foraging techniques, and have a relatively larger telencephalon—the front part of the forebrain—than most other families of fish investigated.

Rose provided no evidence that consciousness depends on the neocortex, either. It is well established that birds feel pain and some of their cognitive abilities surpass those of humans. *The Human Nature of Birds*, by Theodore Xenophon Barber (1993), and Donald Griffin's *Animal Minds* (2001), review a wide range of modern studies that have found birds to be on the level of humans in some capacities, though their little pea-brains lack a neocortex. Here are a few examples.

The most common experimental subjects (pigeons, the crow family, and parrots), were found to be able to put together a series of elements creatively to solve a problem, learn a general principle to use to solve different problems, distinguish man-made things from natural ones, and identify humans and the different parts of human bodies of all races, no matter how they were clothed. They can distinguish the work of one human artist from another, and identify paintings by that artist, even those they hadn't seen before.

Bower birds show every sign of creating art when decorating their bowers. They stand back and look at their work from time to time, then go and make a small adjustment and stand back for another look, just as human artists do. In Australia, I had the pleasure of watching bower birds perfecting their bowers.

When played at one-quarter speed, a hermit thrush's song is like a human composition that has between forty-five to one hundred notes and twenty-five to fifty pitch changes. It approximates a pentatonic scale with all the harmonic intervals. Each bird sings its own songs and may compose as many as a hundred.

Birds construct secure and durable homes protected from predators, sometimes with multiple rooms, and when their nests are damaged, repair just the damage, working flexibly as needed.

Migrating birds demonstrate navigational skills that far surpass what the average human is capable of. They use a set of sophisticated guidance systems, and a surprising array of strategies. They decide sensibly when and if to migrate, and use the best weather patterns and altitude to conserve energy, constantly changing their angle of flight to allow for the sun's movement. When flying at night, they use the stars. They are equipped with excellent senses compared with ours, have an accelerated perception of time, and are more alert than people. So are sharks. The miniaturization of the animal did not affect mental capability.

Treating a variety of bird species with many sorts of illnesses and wounds, it was impossible for me to ignore their sensitivity to pain. To remind me of this daily, one of my patients had been given so many shots, that when she saw me coming with a hypodermic needle she tried, and often managed, to kick it out of my hand with a shriek, before I could get it near her. Birds are naturally vocal, and their expressive voices convey the subtle nuances of a wide range of feelings. That's why bird veterinarians are so concerned about using precisely the right medications for pain control.

It is accepted that birds feel pain, and have advanced cognitive abilities, though lacking a neocortex, so higher mental capabilities can be found in a brain that is wired differently from ours. Dolphins, too, show high cognitive capabilities, but their brains have a different form than primate brains, though both are mammals. There are people in which the expanded neocortex failed to develop, but who have normal psychology and IQs. So even in humans, it seems that the neocortex is not necessary for consciousness.

Other researchers had concluded that the neocortex was *not* central to higher functions. Arthur's friend, Donald Griffin, theorized that consciousness came with the centralisation of the nervous system, due to its survival value, while the expanded human brain was more likely to have allowed the emergence of the complex subconscious mind.

Scientists had traced in some detail how structures in avian brains and fish brains have developed differently, to handle different functions than those same structures do in the mammalian brain. The intricate neural network that forms our nervous system and brain, is thought to have evolved dynamically with cognition, over the aeons, as the myriad of animal forms strove for survival. These findings and others, suggested to some researchers, that it is the way the various regions of the brain are integrated, that generates consciousness.

The question of cognition in fish emerged as being vital in establishing whether they had the mental capacity to feel pain. In the same year as Rose's article was published, 2002, Redouan Bshary, Wolfgang Wickler, and Hans Fricke published a comprehensive review of the evidence of fish cognition, in the journal *Animal Cognition*, called *Fish cognition: A primate's eye view*. Here are some examples:

Recognition of others as individuals has long been established in many varieties of fish, both visually and acoustically. It forms the first step towards complex social lives, in which cognition is often most evident. My sharks, too, related to each other as individuals.

Social learning is illustrated by the migrations of the surgeon fish, *Acathurus Nigrofuscus*, described in detail by Arthur in 1998. These fish leave their territories all over the lagoon, and travel in single file through paths in the coral to their traditional spawning grounds. They go and return along the same paths each night at precisely the same time, as I had seen in the local lagoon, year after year. I had noted that the spawning ground was the only place along the lagoon's border where the outflowing current was exactly balanced by the incoming surge, so that the huge cloud of spawn that they left in the gathering night, stayed in place. These short term migrations had been shown to be the result of social learning; each generation of fish learned from its elders where to go to spawn, and when.

Triggerfish often feed on sea urchins. Usually, they try to 'blow' them onto their side to get access to the unprotected body parts beneath. Hans Fricke observed at Eilat how five different individuals successfully hunted sea urchins by first biting off the spines, which allowed them to grab the urchin and take it to the surface. They fed on the unprotected parts underneath, while the urchin slowly sank. In spite of decades of observations, Fricke never saw this behaviour anywhere else. It appeared to be the result of social learning.

Intertidal gobies (*Gobius soporator*), live in tide pools, and during low tide they can jump from one to another, without being able to see their target pool at the beginning of the jump. Experimentation showed that the fish had memorized the lay of the land around the home pool by swimming over it when the tide was in, so were referring to a three dimensional memory to navigate when the outgoing tide left them only a labyrinth of pools.

With the exception of humans, fish are more skillful than primates at nest building. At least nine thousand fish species build some sort of nest, either for egg laying or for protection.

The male minnow (*Exoglossum*), selects more than three hundred stones, all of the same size, from over five metres away, to build a spawning mound of thirty-five centimetres (14 in) across, and ten centimetres (4 in) high. Another fish builds dome-shaped nests from ten thousand pebbles.

The jawfish (*Opistognathus aurifrons*), collects stones of various sizes to build a wall, leaving a hole just big enough to pass through. This involves repeated rearrangement of the stones. In between, the fish searches for new stones that might better fit the available space than the ones it has already collected, using flexible behaviour depending on the circumstances.

Another example showing surprising flexibility of behaviour is the ability of the ten-spined stickleback (*Pygosteus pungitius*), to build his nest around the eggs if the female has already laid them, though he usually builds the nest first. Great care is required, and a different technique has to be used to avoid damaging the eggs. Since those eggs do hatch, the males achieve their goal.

Bshary described seeing cooperative hunting between Red Sea coral groupers (*Plectropomus pessuliferus*), or lunartail groupers (*Variola louti*), and the giant Javanese moray eels (*Gymnothorax javanicus*):

These two large species of groupers were observed regularly approaching the eels that were resting in a coral cave, and shaking their bodies in exaggerated movements, usually at less than 1 metre distance to the moray eel. In 7 of 14 observations, the moray eel left its cave and the two predators would swim next to each other, searching for prey. The groupers would often come so close that the two predators touched each other at their sides. While the moray eels sneaked through holes, the groupers waited above the corals for escaping fish.

Another unusual form of cooperation among different species is cleaning symbiosis. Cleaner fish come from many different fish families, and depend on cleaning for their diet to varying degrees. They clean the dead skin and ectoparasites from their clients in return for a meal. Full time cleaners may have about two thousand three hundred interactions per day, with clients belonging to one hundred different species!

According to the evidence, cleaners have their clients categorized as those who only come to their local cleaner, and those whose home ranges include the territories of other cleaners. For the latter, they have competition, so give them priority over the locals, who have no choice of cleaner.

Cleaners sometimes cheated by feeding off the client's healthy flesh, as well as doing the usual cleaning job, and the clients with no choice of cleaner punished the cleaner by aggressively chasing it, and inflicting a bite or two, as they saw fit. But these clients benefited in the future, because the cleaner fish were seen to give them, but not others who visited in the meantime, a better-than-average cleaning service on the next visit! Yet they were distinguishing more than one hundred individual clients belonging to various species.

Cleaners will hover above the client and touch it with their fins, in an effort to influence its decision to come for a cleaning. This touching tactic is also used to try to reconcile with a client whom they have cheated. Cleaners even exploit the presence of a third party in an attempt to make aggressive clients stop chasing them, by going to a nearby predator and caressing it, so that the client dares not continue the chase!

Cleaners will behave altruistically toward their clients, if they are being watched by potential new clients —but only those who could visit another cleaning station. The sight of another fish being treated very well is more likely to convince the new-comer to come for a cleaning, than if the prospective client sees another fish being chased. This tactic suggests a short term image, or social prestige, that determines their success in attracting new clients.

Such complex social behaviour—cheating, reconciliation, altruism, species recognition, individual recognition, punishment, social prestige, and bookkeeping, as displayed by full-time cleaners fifty to one hundred times per day, is generally considered to indicate consciousness when displayed in primates.

A similar example of social judgement is given by predator inspection, in which different individuals take turns to lead others away from the school to look over a predator. Fish who don't take their turn cooperatively, will not be trusted by their partners in the future. In other words, the fish make an evaluation of the behaviour of another individual, they remember it, and they take it into account in future decisions.

At the end of the review, Bshary wrote: "We are aware of only one experimentally shown qualitative difference in mechanisms between primates and fish, and this difference is the ability to imitate."

All of this easily available evidence that contradicted Rose's conclusions, was omitted from his review. He ascribed Pavlovian learning, only, to fish, denying any possibility that consciousness could be involved. Other researchers found evidence that showed otherwise, and denied that learning in fish takes place in the total absence of cognition and consciousness, stating that the learning processes seen in fish, "may require the formation of declarative memories."

In reviewing the relationship between learning in fish, memories, and conscious cognition, some researchers concluded that such fish behaviour is better explained within a theoretical framework that includes primary consciousness. Others wrote that cognition cannot take place in the absence of consciousness since there must be an overseer doing the thinking.

Continuing my quest for understanding, I asked veterinarian friends their opinion, and found a very different response. A leading bird specialist in Australia, Dr. Pat Macwhirter, wrote that she had assisted in surgery on a fish when a fish vet had come to work at her animal hospital. She described the fish being more sensitive than birds to electro-surgery, and said that the anaesthesia had to be deepened. There was no doubt, she told me, that the fish had felt pain.

Another bird specialist, Dr. Ross Perry, also in Australia, wrote me the following story:

I have befriended a wild Eastern Blue Groper (Achoerodus viridis), *that has a passion for sea urchins, the big ones with long sharp dark purplish red spines. She is very selective in how she approaches the sea urchin before striking it repeatedly to crack it open and suck out the contents. Groper have big fleshy lips and tiny teeth. She prefers me to uncover the underside of the urchin and to hold as many spines back out of the way as practical before striking. Even so, a large spine broke off in her lip that fortunately left enough sticking out for me to cuddle her on her return so I could grasp the spine with finger and thumb and pull it out. This has transformed my thinking about fish as sentient beings.*

With their training in healing, and experience with distressed animals, veterinarians are in a much better position than fishermen to judge whether or not an animal is in pain. I too had noted that the sharks who had escaped being landed for finning, as well as female sharks with extensive mating wounds, showed similar signs of pain as mammals and birds.

Temple Grandin and Mark Deesing at the *American Board of Veterinary Practitioners Symposium* of 2002 declared, "the ability of an animal to suffer from pain may be related to the amount of associative neural circuitry linking sub-cortical structures to higher levels of the nervous system." They considered that one could assume that an animal was in pain if it actively sought pain relief, protected injured parts, became less active when sick or injured, or self administered pain killing drugs, all of which are done by fish. Their bodies, too, release strong analgesics which relieve pain.

Then I found the work of Dr. Lynne Sneddon, at the University of Liverpool, in England. Her team had found fifty-eight receptors located on the faces and heads of trout, that responded to harmful stimuli. They resembled those in higher animals, including humans. A detailed map was created of pain receptors in fishes' mouths and all over their bodies.

Dr. Sneddon injected the lips of trout with acetic acid, bee venom, or saline solution as a control, and found that those injected with the noxious substances showed symptoms of pain, including an accelerated respiratory rate, rocking back and forth on their pectoral fins, rubbing the affected areas on the substrate, and taking longer to resume feeding than the control group, whose behaviour remained normal. A morphine injection significantly reduced these symptoms. The relief of the fishes' symptoms by the pain reliever shows the interconnection between the nociceptors, which sense the tissue damage, and the central nervous system. Here was proof that fish are aware of tissue damage as pain.

Other researchers published papers showing that fish vocalize when they feel threatened. I had often nearly drowned in underwater laughter on stroking a triggerfish hiding in a cavity in the coral over the squeaks it made at the moment of each caress.

Rebecca Dunlop of the Queens University of Belfast, found that fish learn to avoid pain. She said:

"Pain avoidance in fish doesn't seem to be a reflex response, rather one that is learned, remembered and is changed according to different circumstances. Therefore, if fish can perceive pain, then angling cannot continue to be considered a non-cruel sport."

When told of these findings, James Rose replied:

One consequence, at least where I live, is that all the revenues that support research on the habitat of fishes, that monitor the health of the fish populations here—all the biologists who do this work are funded by [fishing] *licence revenues. If there was no fishing there would be no one to do their job. That would be a catastrophe. That would be a colossal loss, and believe me, there would be no other funds from other sources to do the same job.* (James D. Rose speaking to science reporter Abbie Thomas, the producer of *All in the Mind*, at ABC)

This seemed a further indication that his review paper stating that fish don't feel pain was politically motivated, rather than an honest desire to find out the truth. He had simply declared that a difference between the human brain and the fish brain proved that fish can't feel pain. And it can be argued that research by and for sport fishermen is of questionable ultimate value, since it is done to prove a point that would convenience them, and focuses only on the target species, to the neglect of the others in the aquatic community. My experience with data from fishing studies on sharks, was that the information was all erroneous.

Sneddon's results have been found since by other researchers who have done further research, and pain relief is now systematically used by veterinarians who perform surgery on fish, in the full belief that they feel pain, and that the pain system in fish is virtually the same as in birds and mammals.

But Rose ignored these findings. His credibility seemed to depend on the the beliefs of fishermen that fishes' brains are so simple that they are well understood, and could not possibly support consciousness. Since his article was highly publicized, people remembered only that science had proven that fish can't feel pain, without looking into the subject in further detail. Indeed, with no knowledge of brain anatomy, people wrote me claiming that fish were "missing part of their brains!"

I was faced with the mindless acceptance of a barbaric tradition resulting from a discovery made in the stone age, and incredibly, still bragged about in the computer age. It was daunting how many people were proud of their efforts to outwit fish. They didn't seem to see any contradiction in claiming that fish were too simple-minded to feel pain, while being proud that they were smart enough to outwit them.

Shark finning had hurled my sharks into a black hole of suffering, and sharks all over the world were in danger of being sucked into it. The dangerous idea that they couldn't suffer stood directly in the way of their survival, and it was perpetuated by those who wanted to kill them.

I stopped the regular sessions in Section Two, and drifted instead through the regions inhabited by the sharks I loved, trying to keep track of who was still alive and who was missing. At home I encouraged others to write e-mails to the heads of the government about it, and investigated the shark finning operation. The only reason that so many sharks were escaping was that the new company from Singapore was buying the fins from local fishermen, rather than investing in the equipment to fish the sharks itself. Since the government didn't want to lose the support of the fishermen profiting from the new practice, they hesitated to do anything about it.

Fishermen are more important than fish. They can vote.

Chapter Forty-Seven

The Nurse Shark

The weather turned windy on the day I got some shark food as August drew to a close. I stood on the shore looking out to sea as it howled down the bay, and simply could not bring myself to go out upon such wild grey water. The wind screamed all night long, and the next morning brought sheets of rain. By afternoon, the atmosphere began to settle, and the wind had calmed from a gale to streaming gusts. The sky was a hazy grey, and confused waves beat against the kayak from all directions, as I struggled out to the sharks. By the time I arrived, my muscles were trembling and my hands stiff from clutching the paddle, which the relentless wind kept trying to tear from my grasp.

My plan was to go to the barrens, putting some scent into the water as I crossed Section One, and hold a little feeding, to see who was there. Then I would go by kayak to Site One to hold a normal session. I was looking especially for Gwendolyn, Madeline, Ruffles, Samaria, Apricot, Amaranth and Ali, whom I feared had been fished, and the best place to look for Madeline and Samaria was the eastern side of the barrens. There, I had a chance of seeing sharks from Section Two as well.

I released the scent as I anchored and a shark darted away. It must have been gliding under the kayak. As I put on my gear, sharks appeared and disappeared, barely glimpsed through the chaotic surface.

Underwater, Windy and Madonna with a crowd of my sharks, excitedly encircled me. So many more were zooming out of the deep blue distance, through strangely clear waters, that I threw in all of the rest of the food. Most were juveniles, who were temporarily safe, in the lagoon, from being fished. Cochita, Shilly, Avogadro, Jem, Kilmeny, and Lillith were among the adults. I tried to photograph those with hooks, particularly Cochita, who had two metres of fishing line trailing behind her. Sparkle had a second hook jabbed through her jaw. The first was still in place trailing fishing line. Gwendolyn swam in slowly and unsteadily as if infinitely tired. She too, had been hooked.

All of the females were barrel-like in pregnancy. Those who had been finned had died with their babies.

Christobel appeared. The area on her back where she had been bitten was now just a blackened region, and her second dorsal fin was a dark, flattened arc. But it was filled in, whereas at first only the spine at the leading edge of her fin had remained, lying back over the injury.

One of the three-metre nurse sharks swept down in slow-motion to munch on the scraps, while I roamed around looking at everyone, and trying to get photographs while the light lasted. Twilight passed down-current. She would not approach, and I only had a glimpse, but it was enough to know that she was alive and pregnant. Billows of sand drifted with the scent from the writhing nurse shark. I started writing down the names of the juveniles, which I had still not done, and sketching two or three new ones who were flitting through the region. Ruffles came to eat as night slowly fell, and I turned in the darkening waters, watching. More nurse sharks arrived, and the clouds of sand they raised lent an air of mystery to the twilight scene.

By the time night fell, I had been nearly motionless in the water for over an hour and was in that dream-like state that comes from bobbing around in cold water for too long. As I drifted, trying to see each shark on both sides to be sure I had noted everyone's condition, a large black shark came swimming straight toward me. My consciousness paused. I had forgotten the dream of the black shark, and suddenly here it was. It was real! I stopped breathing as she swam, not into my arms with her fins cut off, but up to my nose. Then she sank slowly as she turned, and sedately glided down over the food. It was Teardrop! How could this pale shark have suddenly turned black? Shaken by the apparent re-run of the dream that had set me on this path, with the actual finning taking place nearby, I hovered around her, and saw that my lost Madeline was with her! Terribly relieved that she was still alive, I watched her circle the lazing nurse sharks in her forthright fashion, and go into search mode. She was dark purple-grey in colour. Beside her, Teardrop was black. Teardrop was a completely different colour than the other sharks present, yet the last time I had seen her just two weeks before, she had been her usual light brown. I couldn't take my eyes off the black shark, and tried to plumb the meaning of this extreme colour change in a shark outside of the reproductive season.

So a long time passed before I realized that one of the very large nurse sharks seemed always to be drifting near—whenever I turned my head, I glimpsed him behind me at the periphery of my vision. I was so accustomed to being surrounded by sharks, and the big nurse sharks moving in the site, that I had paid no attention. But finally it became impossible to ignore the way this draft-horse sized fish was always with me, both of us circling in slow-motion.

The biggest nurse sharks had always been the most cautious and shy, arriving as darkness was falling, and withdrawing if I moved around, to float gracefully back later. Never had a nurse shark, or any shark, followed me persistently in this way, over a long period of time. It seemed at least fifteen or twenty minutes since I had first noticed him high in the water behind me, instead of on the sand.

I had never held a feeding there in the evening before, so assumed that the big nurse sharks present had never met me, or even attended a feeding. But it was possible that he had attended the feeding sessions at Site One. In such a case, he could be paying attention to me because of having rediscovered me so unexpectedly after a year and a half. There had also been a very big nurse shark who had approached me recently there. But this didn't ring quite true. Many sharks expressed curiosity, or even a sort of greeting, but this relentless, slow and peaceful following, with a daunting, steady, momentum was new.

Then Trillium appeared, dragging fishing line from another hook. She seemed weak. She must have been caught again. Over and over I dove down to look at her injured mouth before the night enveloped us, but it was just too dark to see in detail. And throughout these manoeuvrings, the huge nurse shark followed me like a pet chicken. Once, I grabbed onto the kayak as I passed it with him very close, to make sure, if he grabbed me, he wouldn't be able to pull me under. But of course, with its momentum alone, a fish that size would have broken my grip on the kayak. As I clutched it, he glided beneath my body, moving imperceptibly slowly. I held my breath to see such a huge animal drifting centimetres away. How strange it was that he did not startle with me and the kayak right above him. He seemed to know what he was doing!

It was a startling sight, when I looked back at him, flowing along behind me like a blimp with frills, his long tail slowly waving. Surely he had met me before, given how very shy the big ones usually were.

Eden and Simmaron glided in, and some male visitors began cruising the area. Tuvalu, the young male from Site Two, appeared with a hook in the side of his mouth, which was worrying, since it meant that either this very young male was going outside the reef, or that the sharks were being fished out of the lagoon. Bratworst suddenly swam excitedly up to me, and her supple motion seemed expressive of an animal feeling fine. Her hook didn't seem to be bothering her. I even saw Grace, for only the third time in all those years, that evening. She arrived with the male sharks from Section Three, Lavoisier and Marten.

But I spent most of the time cruising around with Teardrop, the great nurse shark in my wake, marvelling at the blackness of the black shark.

Whoever was fishing the sharks was obviously inept, for so many of them to have broken the line. The obvious thing for these fishermen to do was go to the store and buy some bigger hooks and thicker line. Then these sharks would all disappear. Our landlord had been a well-known shark fisherman, and the heavy iron shark hooks he had left lying around the property were more than thirty centimetres (1 ft) long, twelve centimetres (5 in) across, and spiked with barbs four centimetres (2 in) long. With such equipment, no shark would ever escape. The man was still deeply resented by local Polynesians for his shark massacres. They told me that wherever he had been, no sharks were seen for weeks afterwards. He sold their teeth in his little shop.

They told me about the *Black Shark*, a shark god who lived in the bay near the ocean. A sea mount just off the border of Section One was named after the deity and considered his abode. This sea mount is marked on the diagram at the beginning of Chapter Twenty-three; it's indicated by the shipping marker just outside the border region of my study area. Strange as it seemed, as well as existing in my dreams, the black shark had a mythological counterpart in the minds of the Polynesians, and was thought to dwell close-by. The Black Shark, they said, came to warn of danger.

Finally, I picked up the anchor with the great nurse shark still flowing behind me, and got back in the kayak. The enormous creature floated beneath, then slowly descended into the dark.

Back at home, I had no sooner landed when the wind strengthened again, and soon it was howling. I had barely managed to slip through the curtains of storm blowing over the Pacific ocean, to get to the sharks one more time.

By the time the ocean quietened, Franck had recognized the gravity of the sharks' problems, and had joined my efforts to elicit support for them. Since he wrote eloquently and powerfully in French, he was far better than I was at spreading the word to the French people, locally as well as in Europe. He wanted to see the sharks for himself, and came with me.

They met us at the surface, and Windy, Bratworst, Madonna, and Cochita surrounded me when I slid underwater, while many passed farther away. Due to the current, I took the sack of food to the bottom and began dumping it out, but when the sharks all converged on me, I let go, and returned to the surface. Windy grabbed the sack and shook it herself. But she didn't have any more success at freeing the prickly fish scraps than I did. The others roamed over it, sticking their heads inside, and trying to get a piece. Franck paddled his kayak overhead and fired the fish heads into the scene, startling everyone. Then he attached his boat to mine, and descended.

The fish shop had given me half a tuna which had become too old to sell. I had cut it into pieces, and began tossing these into the water in front of Franck, so he would have a good view. It was mostly the juveniles who rose to the surface to intercept them, while the big ones prowled around below. But here came a bigger shark, sweeping up, thrilled, to snatch a piece—Carrellina! What was she doing there in August? It was my only sighting of her outside of her usual visitation period in December to April. She put on a good show for Franck, and I was happy to know that she was still alive.

Franck felt sorry for all the sharks with hooks cutting through their cheeks, and when the pieces of tuna ran out, he swam down and dumped out the sack of scraps. There were a lot of sharks there from both sides of the barrens. This was a better place than Site One to keep track of them all, as long as the ones from the border were alerted when I crossed over the lagoon.

Sparkle arrived and swam up to my nose. She had a big hook in the right side of her mouth, and was trailing heavy line. She came repeatedly and swiftly up to us at the surface. It certainly looked as if she were irritated, and with that big hook in her mouth, one could see why. Franck swam down and found a tuna spine still loaded with meat, and held it out to feed her. She came and ate from his hand, taking bite after bite, bending and spiralling around him. Except for the stark ugliness of the hook in her, she was lovely—dynamic and radiant in the light at the surface.

I had always avoided doing things that might result in having dozens of sharks nosing around my hands looking for something to eat, but this one-time gesture on Franck's part was a pleasure for both of us.

After that he played with the sharks, feeding them with more scraps he found lying on the barren floor. They circled around him curiously, at all levels in the water column. I was checking on who had hooks, and diving down to see them close up, when Shilly arrived, and circled me. How beautiful it was to swim with her for a while, worried as I was for them all.

The big, older females, Cheyenne, Storm, Hurricane, and Lillith were there, and Franck was impressed by their size, not having seen any of the shy, very large females before. When Trillium came, I tried again to get a close look at her mouth. Swimming along the bottom beside her I could see that the hook was no longer there. Somehow she had rid herself of it. But there was a deep crevice along that side of her jaw, giving her an odd look, as if part of her face were missing. And the arc of her teeth in her lower jaw was uneven—there was a place where it was concave, dented inward. While repeatedly diving to swim with her, I saw her readjust her upper jaw, so the damage appeared to be in the main joint of the lower jaw. Her mouth was twisted open.

I hoped that she would be able to heal, now that she was free of the hook. The sharks had shown themselves capable of remarkable feats of healing. This fact was on display with Christobel circling, sporting a reformed second dorsal fin. It was more arced, and smaller than it had been, but it now appeared the normal width and colour, with a brand new black tip on it. The flesh and skin on her back showed no sign of abnormality that I could discern looking down at her from the surface in the dim light. Her body was smooth and rounded, the skin covering her former wound was the same deep rust-red-brown as the rest of her. No sign remained of the ragged hole in her back where a very big shark had taken a bite of her.

It was August 31. It had taken two months and three weeks for her to replace the flesh of her back and her second dorsal fin, and perfectly heal.

Teardrop, my black shark, circled at the end, having arrived with another who had been freshly hooked. But it was too dark to see who it was. As before when I had brought a second person with me, no nurse sharks appeared. It was uncanny, given how many had been there the previous session. What could it mean? That they knew me and not Franck? That two people were too many, but one was all right? They certainly seemed more perceptive than they looked. It was another unsolved mystery, that these languid sharks were so fussy about the presence of a second person that they would forgo an easy meal to remain invisible.

I continued to patrol the lagoon when I could get away, to check on who was there. I still felt acutely the absence of Martha, and now Samaria, Apricot, and Amaranth, as well as several others of which I could not yet be sure, were gone, too.

At night I sat looking out to sea, while the waves lapped gently upon the beach—a trill of sound, a pause of silence, another trill, measuring the time. Outside the reef, lights slowly passed. Always there was a tension. I never knew who would die next.

August gave way to September. On a rare blue day, without a cloud or breath of wind, I collected a load of scraps from the fish store, and when the sun was low, headed for the barrens. No sharks appeared as I anchored and prepared, and underwater I learned why. Only males and juveniles were present. But as time passed, Bratworst, Eden, and Droplet drifted in, followed by Cinnamon. More time passed, and Christobel came, then the rare visitor Vixen, from Section Three. She, like Cinnamon, had grown fast, and was now a big and very pregnant shark. One more was still alive. The sharks were excited, moving fast, and spread out over a wide area, which I surveyed from high above, my view unhindered by labyrinths of coral, as it always had been at the other sites.

From far off, a fighter plane formation of dark sharks came soaring together through the deep blue distance. They swept into the feeding site—Samurai, Hurricane, Cheyenne, and Emerald, loosely accompanied by Jessica and the male Caro, and they excited Gwendolyn to join them, her scars shining in the dying light.

Shilly arrived, and instead of joining the excited shark reunion, she came to me, the little dear. I wondered sadly if I would soon have to face her death too. So far she was one of the few who had remained untouched by the fishing efforts, and I hoped that suggested that she was not visiting the places that were being fished. I drifted, watching her, as the sharks flew through the vast grey space in the fading light. Lillith had joined the communal cavorting. Cheyenne looked as old as she did. Columbine and Victoria had come too, and for the first time, Victoria closely circled me.

As I watched, I realized for the first time that the great nurse shark was turning with me too.

Trillium arrived about then, and I tried to stay close beside her, to examine her mouth. There was nothing to give her to eat—the food had all been gone for some time. Every time I turned around, the nurse shark was right behind me, much closer this time than he had come before.

When Trillium left, I began following a new male visitor, concentrating on drawing his dorsal fin. The nurse shark moved with me. After so long in the water, possibly due to the cooling of the brain, it dawned on me but slowly that maybe my priorities were mixed up, and it was the nurse shark I should be paying attention to. He was following so closely, and I was unsure how long he had been doing it. It seemed that for at least fifteen minutes I had been catching glimpses of him behind me when I turned. He was usually within a metre now, whereas the first time he had followed me he had been twice that distance from me most of the time. When his face approached my right arm, I tried an evasive movement, doubling back past him. He flexibly turned with me at the same snail's pace, to resume his position in my wake.

As I saw the absolute dedication with which he was following my every motion, I became alarmed. It was already well after sunset, and shadows were swiftly enveloping the scene. I was quite far from my boat, and without changing my swimming pattern, drifted toward it. Once there, I flew into it like a dolphin. The huge shark moved on, just under the kayak, his frills undulating around him.

As I drifted away across the glimmering silken sea, the full moon came over the mountain.

The nurse shark's behaviour grew more unfathomable the more I mused it over. Two other big nurse sharks had been present on the occasions when the shark had followed me. One of them had been there for about half an hour before my follower had arrived. While I came and went, looking at the blackfins and swimming with some of them, he avoided me, and if I moved around very much, he took a tour down-current and returned later. That was the normal behaviour of three-metre nurse sharks.

To follow me for fifteen to twenty minutes, almost uninterruptedly, during each of the two sessions, seemed extraordinary for such a large individual of a species with a preference for lying languorously munching on the lagoon's floor. And he had remained hidden during the intervening session in which Franck had come with me. No matter how I thought about it, the basic fact that no other nurse shark had ever done such a thing returned to the fore. Finally, I described the situation to Arthur and he replied:

I've never encountered repeated following such as you describe. Young sharks are always more inquisitive than older (especially large), sharks. I've considered that such is based on the experience(s) gained by older individuals and such differences are typical for ontogenetic differences among vertebrate species that learn from previous experience.

The fact that the nurse shark following you is large rather than small, counters the above conclusion. Such a large nurse should be more leery of you than a small nurse.

The fact that his apparent attention to you, even when you're doing other things (inspecting a nearby shark), based on his close presence whenever you turned around, seems to mean that he keeps his distance from you, albeit a short distance. Such may well be the case since he could easily swim directly up to you and over you, while you are attending to other things. He clearly has not yet done that and so he must be aware of you (and your size?) and prevents himself from coming directly onto you.

The immediate question is why is he attending to you so closely? Since it has happened close to sundown, I wonder about his eyesight. Although sharks are thought to have relatively poor sight, such is untrue whenever it has been examined. But, as the sun lowers itself to the horizon, vision by diurnal fishes becomes poorer. I wonder if the nurse has cloudy eyes. If so, such indicates that its vision is impaired and it could not rely on such to maintain a safe distance from objects. Also, as you well know, certain odours bring sharks directly up to an odorous object. Nurses use klinotaxis, which brings it directly and rapidly onto the source, while at least the lemon shark (Negaprion brevirostris) *uses rheotaxis, which will eventually bring it to a source, but more slowly. If, indeed, this nurse shark is recognizing something chemically, on your person, he obviously is stopping shortly before reaching you, indicating that he recognizes something that prevents him from continuing his movement directly up onto you. If the nurse recognizes, chemically, something on your person, why aren't the blacktips also doing the same? That answer is not obvious, but no one has shown that species of sharks are, for the most part, equally adept at chemical recognition. It may be that nurses are simply more adept than blacktips.*

Klinotaxis refers to comparing chemical (or odour) concentrations to the right and left and moving to-

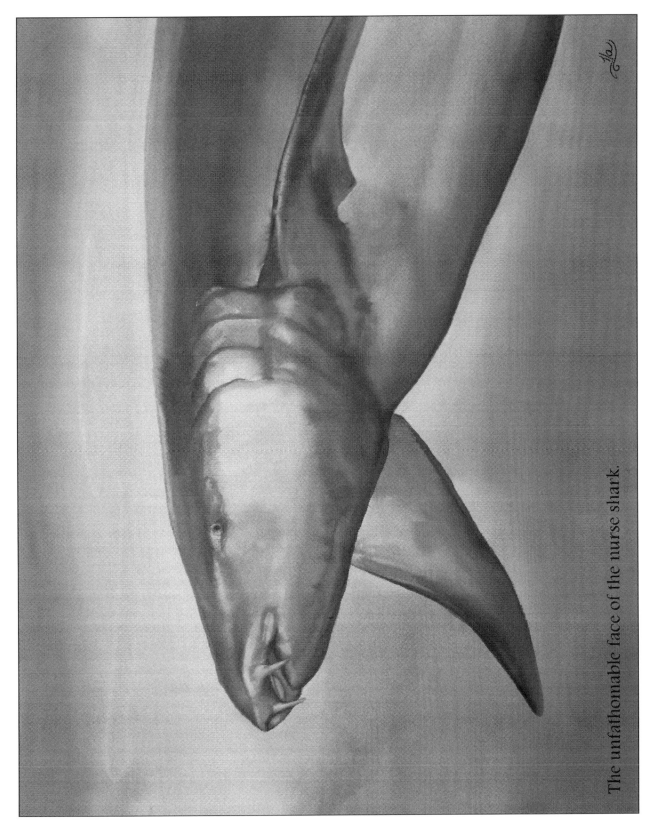

The unfathomable face of the nurse shark.

ward the side on which the concentration of scent is strongest.

Positive rheotaxis refers to moving upstream toward a chemical stimulus until the stimulus is no longer perceptible, then moving left and right until the stimulus is again perceived. This pattern is repeated until the source of the odour is found.

Negative rheotaxis is the same, except that the animal moves downstream.

Here was the explanation for the speed at which I had seen nurse sharks find scraps of food, compared to the blackfins, which probably use *rheotaxis*, since they are requiem sharks, like the Atlantic lemon shark Arthur had referred to. He continued:

So. . . a big shark follows you closely. Nurse sharks can do damage, especially large ones. They can bite and they do not let go. Their weight prevents one from dragging them, so a large nurse is extremely dangerous if it bites and hangs on!

If you choose to swim again in the area where the big one is, take a long rod (about four to five feet long), down with you and see if he, again, finds you. If he does, swim below your boat and see if he follows you there. If he does, push the end of the rod against the front of his head and prepare to surface and jump into the boat if the nurse shark swims against the rod directly toward you. He should not. He should reverse himself and swim off, or swim away to the immediate left or right. If the nurse moves away from you, don't do anything else; simply wait and watch in all directions. I'd only do such the entire time that I'd be in the water that time. I'd stay in the water for an hour or until shortly before sundown and watch to see if the nurse shark returns. And he may return from another direction!!!

If, indeed, he returns to you and comes closely to you, use the long rod again and hold tight to it. If the rod falls from your hand(s) watch the nurse before picking the rod up. The rod's length is such that it will not rapidly swing in your defence and so if the nurse moves toward you, face it and put something between your body and the shark (your leg-knife). Use the knife (object) only as leverage to move to one side and don't try to jab the nurse. If you decide not to pick up the rod, shoot up to the boat as rapidly as you can and jump over the gunwale.

I really think that the next time you go to the area where the big nurse shark is, take someone with you, who respects sharks, but does not have an intense fear of them, to help you with this problem. This shark has come up to you twice and he may well do the same thing a third time. And you must be ready for such an occurrence. Sharks are a peculiar bunch and your nurse fits in well with that premise.

I've taken time to present the above epistle as you have a situation which bothers me and obviously you also. As you correctly surmise, it calls for thought since it is exceptional and not noted by you or me before. Please be careful and seek to take someone with you the next time you visit that area, even remotely. Large nurses are dangerous.

I carefully thought out what Arthur had said. There was no point taking anyone with me if I wanted to see the shark again, because he only came if I was alone. I felt that Arthur was right about there being something wrong with the animal. Nearly always, when I had seen animals behaving strangely, it was due to a physical problem. I just hadn't thought of it that way until Arthur pointed it out. A blind shark would be feeling me solely with his other senses, in a way I could never know, hearing my heart, and my movements, sensing the living electricity I generated. This could make his behaviour different from that of a normal shark. As a warm blooded animal, the shark could simply have been intrigued by how unusual I was.

I was sure that he was not following due to an odour from me. The way I rushed around preparing, dragging heavy sacks I could scarcely move across the lawn in the tropical heat, swiftly sorting them out and loading the kayak, rushing back and forth bringing my gear, always racing against time to take care of the house, family, and birds before I left, the smell of hot human must have poured from me and streamed away as a separate scent flow the moment I slid into the water.

Other human scents I could have generated at different times due to emotions, pheromones, and the time of month had never attracted any shark interest. At first I had put coconut oil on my hair to protect it from the sea water, but when I suspected that its scent might have played a role in the frightening eighth session, I had stopped. Surely if any of my natural scents were interesting to sharks, I would have noticed that long, long before. Indeed, was it not strange that the part of me the sharks approached was my face? Not infrequently, mammals notable for their sense of smell, such as dogs, pushed their wet noses into other anatomical parts for a sniff. In comparison, sharks seemed so transcendent—they looked into my eyes.

In accordance with Arthur's suggestion, I planned a session to investigate the behaviour of the unusual nurse shark. Though I chose a stick to take, I did not feel comfortable about using it as Arthur had suggested, and of course had never taken a knife. I would observe the shark, try to understand his behaviour, and look at his eyes. Depending on how things went, I would decide whether to continue to feed the sharks there. Such a large animal was too much for me to cope with at feeding sessions, and I didn't know why he was following me. Since he was coming gradually closer, a certain danger seemed to be impending.

By the time I was ready that late Saturday afternoon in mid-September, I was already exhausted. I had picked up a heavy load of fish scraps, including half a *saumon des dieux*, which was so large I had to strap it to the front of the boat. It took a long time to sort everything out and pack. I flew around completing the last things necessary to leave the house and birds for the evening, then swiftly washed myself, my hair, and my gear with unscented soap, just to make sure I was covering the "interesting scent" theory. Then I hauled the heavy kayak into the water, flung myself in, and began the long voyage out to the lagoon. I forgot the stick I had put aside to take, and impatiently went back for it, shoving it carelessly up underneath the *saumon des dieux*. It had been windy all day, but each evening the wind had died down, so I expected it to die down as I paddled out to the site.

Instead, the wind rose. As I hastened toward the barrens, it beat me back with greater and greater force, as if it were a personal enemy engaged in an intensifying argument with me. As I pulled harder and harder to advance against its relentless attack, it yanked and jerked at the boat, blasted water into my face, and tried to flip the paddle from my hands. I could not rest, nor lessen my efforts without being swept broadside. It took all of my power just to keep the plunging kayak facing into it, and an endless time seemed to pass in which all my concentration was focused on just staying straight, while I seemed to make no headway at all, and my concern about my predicament grew. Each wave rushed over me in snowy gushes, and I was blinded by the glitters of light in the spray flying from the wave-tops. As I drew near the place where I thought that the barrens began, I was at the end of my strength. And quite without warning, an overwhelming new blast from the sky swept the boat broadside, then rapidly backward. I threw the anchor in, snatching at my mask and snorkel as the rope shot out of the boat, and nearly took them with it.

Weak and trembling, I sat there gasping and feeling done in, as the kayak leaped about, trying to find a balance in the forces pulling at it. But I was close to my goal, and could easily swim to it, trailing the boat, as I had done so many times. Piece by piece, I put on my gear, glancing around every few seconds. And suddenly, solid in the chaos of flying waters, a large, grey, triangular fin slowly and steadily passed, appearing and disappearing in the silvery uproar. Though the submarine world was invisible, and until then no sharks had appeared, at that moment my sharks arrived, and began swiftly circling the boat at the surface.

I sat, pounded and shaken by the elements, and watched, frozen with uncertainty. The battle with the wind had clutched hold of my mind, and the sight of the lemon shark, just drifting by, accompanied by the arrival of so many thrilled blackfins, caused the underlying sense of danger resulting from Arthur's warnings to leap to the fore. By chance, the fish scraps I had picked up had been unusually bloody, which no doubt explained the presence of the lemon shark, and the delighted cavorting of my favourites around me at the surface. I didn't know what to do. A dangerous nurse shark, uncontrollable winds, too many excited and disrespectful blackfins, bloody scraps, and now a lemon shark! And the sun still fifteen degrees above the sea! As I surveyed the wild scene in consternation, Cochita came to me and raised her head above the water in obvious expectation.

So I started throwing in the food, bracing the sack of scraps against the side of the boat, as I tossed handfuls to the sharks. On coming around again, Cochita reached up and grabbed the heavy plastic netting. I lowered it so that she could disengage her teeth, and the boat nearly capsized. Balanced precariously between the worlds of wind and water, I stilled myself, my mind and the boat, and went on pushing in their food. Soon the sharks glided down to look it over, and choose pieces to eat. The lemon shark had not reappeared, so I prepared to go in. But one of the very big nurse sharks was right under me! I paused in renewed uncertainty, but managed to slide slowly in, when the kayak was blown over a dead coral formation. I grasped it, slowly shifted my weight out of the boat, and sank beneath the surface, looking swiftly around for more surprises.

Underwater, many sharks were feeding, and they flew through the coral to the limits of visibility, obviously very happy with the situation. The lemon shark was not in sight, and the nurse shark's behaviour was normal. It was unusual for such a big nurse shark to have arrived so fast, and so early, too. Surely the animal knew me; a strange one would never come right under the kayak in full daylight, particularly when I was moving around in it. He lay placidly on the floor of the lagoon, munching his way into a fish head, while multitudes of bright fish flew around him, feeding on the scattering particles. He was not the one I awaited.

Drifting quietly away from the food, I put the anchor on an open stretch of sand, from which I could pick it up from the kayak, from above, if necessary. Then I watched, holding the gunwale, tossing scraps, writing down the names of the blackfins present, and looking around for the big sharks that concerned me, while balancing in the waves. The kayak often collided with my head, which did nothing to make the situation easier. The shark stick was still stuck under the heavy frame of the carcass of the *saumon des dieux*, and it would be impossible to extract it if I needed it. What I was going to do about this troubled me increasingly as the time passed, and sunset approached.

The group that had met me stuffed themselves and were satisfied, having found plenty of pieces of meat among the scraps, cut from the *saumon des dieux*. They still passed through periodically to socialize with the other sharks arriving. The bloody scent flow attracted a steady flow of sharks, and I was kept on the alert, throwing food, writing, and turning quickly to look behind me for the lemon shark, or even the big nurse shark, whom Arthur had warned me might come from another direction.

But I expected him to come openly into view. The lemon shark, however, could very possibly sneak up behind me.

It was time to have the shark stick ready, but I was afraid to detach the huge frame of the *saumon des dieux* with so many excited sharks present. Already nervous, I imagined that working at the kayak with my head above the surface, would precipitate a rush. In waters wild with sharks, I wouldn't be able to see if the lemon shark had joined them. The kayak could overturn in the wind and waves as I tried to dislodge and pull the heavy carcass off it, causing incalculable trouble.

The problem circled through my mind faster and faster, as darkness deepened and the probability shot up that at any moment the dangerous shark would appear. Paralysed with anxiety I stayed riveted in place, making and abandoning plans, while Arthur's words of warning gained alarmingly in significance.

Madonna suddenly dashed up to me while I was looking the other way, and startled me badly. She was accompanied by Madeline, and for a few moments I relaxed in the pleasure of throwing them some good bits of food. Madeline got the pieces I selected for Madonna, time after time. But in the end the two were satisfied and well fed, and I felt more relaxed. I simply went to the boat and started working on freeing the carcass. I took my time, looking underwater frequently, and in fact the huge, meaty frame was easy to detach. While the sharks were temporarily down-current it fell uneventfully to the bottom. It was a reminder of how paralysing and often irrational fear is.

I swam watchfully away from the scent flow, as Madonna glided toward me. Yellow fish surrounded her and the enormous fat nurse shark as she passed just above him. He was still lying in the same place, methodically scraping meat from a scrap. The tranquil and beautiful scene reminded me once more of the contrast between my fears and reality, as if I lived in a world of illusions. Why was I haunted by this recurring fear of the sharks?

The sun had set, and I picked up the anchor and hoisted myself back into the boat as the wild wind carried me away. At least I was travelling in the right direction, but it was hard to control the kayak as I tried to point it toward home while being swept across the bay. Except for the few moments with Madonna and Madeline, the outing had been an unforgettable nightmare.

It was one more reminder of how important it was to think everything out in advance, and stay home, if for some reason I couldn't bring my full attention and energy to the situation. One never knew what was going to happen.

Two days later on September 17, 2003, Bruno Sandros, the Minister of the Environment, announced that there would be a ban on "the commercialization of shark fins," though what we were hoping for was a sanctuary for sharks. But time passed, and the passage of the law was not confirmed. Enquiries were met with silence, and the sharks continued to vanish.

The only time I learned anything about what was going on, was while talking to one of the fishermen who supplied my fish shop. When he told me how worried he was about the poor fishing, I asked, "Well, are you at least still able to sell your shark fins?"

"Oh yes," he said, and told me he brought in about thirty cases a week. I was shocked. He was just a local fisher. The tuna long-liners would be taking countless times that number.

I pondered this, and at the next opportunity, got into conversation with the owner of the fish shop that this fisherman supplied. I had known him for a long time, and we had often talked easily about the sea, boats, and fishing. Since he was French, rather than Polynesian or Chinese, I hoped he would talk more freely to me than the others in the fishing business I knew.

While he helped me put the fish scraps from his cooler into my pail, we talked lightly as usual about the prevailing weather and fishing, and when the time was right, I mentioned the shark finning issue that was so commonly in the news.

"Is there really much of a business in fins in the islands?" I asked innocently, hoping to encourage him to talk. We had finished and were standing by my truck—he had just lifted in the pail for me. We were looking at each other, and his casual look faded. Slowly, his face went vacant, and his eyes lost their focus on me. It was like watching someone fall into guru meditation. Eventually, he stopped looking at me at all, and stood gazing far out to sea. I tried to think of something to say, but couldn't. Eventually, he said he didn't know anything, but I should ask at the other fish shop. They would probably know more.

But I didn't ask them. They were two Chinese partners, and in spite of the smiles we shared, there was an impenetrable barrier between us. It had taken me years to get their confidence to the point where they would even admit that they understood what I was saying. They would not share news of this highly sensitive and Asian subject with me if they had any. But I had wandered through their establishment so often, and so freely, that if they ever handled sharks or shark fins, I would have seen some sign of it. Never had I seen a shark, or parts of one, in their building, even though sometimes when handling the scraps, I asked in curiosity about what species of fish they were from. They sometimes had unusual creatures, and always pleasantly and openly talked about them to me.

One newspaper article had mentioned that the shark fin racket was based in Papeete, so it was possible that the tuna long-liners left their fins there, before going to the other islands. That would explain why there were never any at that shop. The other fish store was supplied by their own fishing boat and local free-lancers, so might receive fins from sharks caught and finned locally.

It seemed impossible to get any information about the business, so secretly was it run. Even the names of the companies responsible for the massacre were impossible to learn, though I spent hours on many days, phoning different government bureaus trying to get information, and being passed from one person to another, none of whom had any knowledge of the shark finning situation, any knowledge about the proposed law protecting sharks, or the name of the new company from Singapore who was responsible for the shark finning. For each ministry, the problem was the concern of someone else. If one wanted to find out who was selling pearls in town, the answer was straight and clear. Not so with shark fins.

Why was the massacre permitted to continue? The divers speculated that President Floss didn't want to do anything to lose the support of the fishermen before the election.

And no one weeps for sharks.

Due to the announcement of the imminent protection of the sharks, international concern had fallen away, but I kept broadcasting the need to continue to write to the government, asking that sharks be protected before the delicate coral ecosystems of the far-flung archipelagos making up the island nation were irreversibly damaged. Unable to gain any relief from the inner anguish I felt in the situation, I began writing down this story of my sharks, determined that somehow the world would find out what interesting animals they were, how misunderstood they were, and what had happened to them.

When conditions allowed, I tried again to find the enigmatic nurse shark. There had been a long period of rain and high wind, and as I paddled toward the lagoon, it changed direction and came lightly from behind me. I hoped it would not strengthen—I was afraid of the south wind.

The lagoon was still turbulent and clouded, so it was another nerve-racking session. Everything looked grey and shadowed under the ruffled surface and low sky.

The current was too strong to maintain myself against it, and I trailed in it, down-current from the food, with my back to the direction from which the nurse shark would approach. Thus positioned in the scent flow, sharks were flying past me from behind, and the need to glance backward while holding my mask in place was a continual distraction. I wondered what I was doing there, trying to investigate a potentially dangerous animal when I could maintain neither myself nor my mask in the current.

Vixen was there, swollen huge by her pregnancy. How she had changed from the fox-coloured, slender creature I had first seen three years before. She was with Fleet, a male from Site Three, who had been with her on her previous visit. A large hook was stuck in the corner of his mouth. Their presence suggested that more visitors were probably in the area, and suddenly Antigone glided in, the vision of loveliness I remembered, but more brown than gold this time. He widely circled the area. Soon two large females completed the visitation from some unknown region. These rare visitors were the highlight of the session, because the nurse shark I waited for never came.

Cochita arrived well past sunset, maybe late coming back from the ocean. I threw in the food saved for a last special visitor, and watched until the excited sharks flying through the shadows faded to flickering motion against the darkness.

In spite of the windy conditions, I managed to go again a few days later, drawn by anxiety to check on some of the border sharks whom I had not seen at the barrens sessions.

It was a long time since I had been to Site One. As I anchored in my usual place, Cochita circled close to me so slowly it was almost unbelievable. Each time, she passed underneath where I was sitting, instead of circling the boat. Watching her, it was impossible to feel anything but love and gladness that she, and the others there, were still alive, that they waited, and remembered our long ago sunset rendezvous. . . I slid in.

The sharks were all watching from the site and beyond, and turned to come when I appeared underwater. As had been my habit originally, I took the anchor line and trailed the kayak to my flat coral shelf, checked for stonefish, climbed on, and threw the food from the back of the boat to the sharks. As of old, they began feeding beneath my hands, splashing and careening all around me. How beautiful it was to feel relaxed, undisturbed by tremors of fear. They were alive, these sharks. I finished, grabbed my slate and camera from the boat, and ducked back underwater.

The site was a shark tornado as the excited animals swooped around feeding. I watched them critically, noting those who had hooks and whether they were new or not.

Sparkle still had two metres of line trailing from the big hook in the side of her mouth, which had grown a layer of algae two and a half centimetres (1 in) thick. That would add an enormous weight and drag to the line, always pulling on the delicate tissues of her face. Fleet was in the same predicament. A high fraction of the sharks present had hooks, or showed scars from hook damage. But Gwendolyn had lost the big hook I had seen at the session when the nurse shark was trailing me.

I spent a lot of time trying to photograph Trillium's mouth. In the poor light I couldn't see it well, just that it was strangely twisted, with a deep trench where the hook had been. She didn't like to come close to me straight on, always turning away while still too far for a close-up of her face. So I lay on the sand and photographed her as she sailed above. Her mouth was open at an odd angle in all the photos.

Lillith appeared. She showed no sign of having been hooked, and I wondered if the shy old shark no longer visited the ocean. I got the last small sack of scraps, and emptied it for her.

Fleur de Lis shot in like an arrow. She had birthed since I had seen her in the barrens a few days before. She might have put her young in the local nursery, and I made a note to go and check if the sea calmed down enough. Ali, of about the same age, still looked like a little blimp as she clumsily manoeuvred to search out scraps under the coral.

Droplet appeared with a companion from farther up the lagoon, and before I left, Avogadro and Tamarack came. Many juveniles still circled as the waters darkened, and I climbed back into the kayak. I sat a long time, watching above the site while the daylight dome of sky darkened, became transparent, and allowed the cold light from outer space to shine upon us.

Cotlet came and I could see the pale scar of the slash she had behind her dorsal fin. Had someone been trying to fin her the moment she had escaped? It looked that way. The sharks kept gliding around the boat, and beneath it, but there was no sense that they awaited more food. They were just coming to look. I watched, terribly concerned for them, and the ones who had not appeared.

Chapter Forty-Eight

Interrupted

Franck and I planned to go to France to visit his family for a month at the end of October, and as I prepared the house to leave it, bad weather kept me from returning to see the sharks. Finding a trustworthy house-sitter willing to look after the birds and dogs, who would spend each night there, had been challenging. It was too dangerous to leave a house empty overnight, but most people who already had a home were unlikely to commit themselves to actually living in the house for a month. One couple who were willing to come, backed out at the last minute due to other obligations, so the tension was mounting as we put up posters at the harbours, and an ad in the newspaper.

Only one person responded, but he seemed to be the one we were looking for. He had lived on a farm, so was used to looking after animals, and loved birds especially, having had several as pets. He seemed very sensitive to mine, when I introduced them to him, and described their needs. The arrangement seemed perfect because he lived on a boat, and needed somewhere to live while it was put in dry dock for repairs and upkeep. He was delighted from the first by the prospect of staying in our house, with its beautiful view over the bay, after living in the cramped quarters of his boat for so long. The setting, with the two white sand beaches and the long, tranquil garden surrounded by the volcanic ridge, seemed to please him very much.

The wild birds I had been rehabilitating were either dead or independent, so only my growing flock of handicapped junglefowl required care. Babalu and her mate, Kubilai, had hatched a family of chicks who had developed fowl pox when only a few days old. The ugly disease had grown with them and finally overwhelmed them. Growths had covered their faces, especially their mouths and eyes, for weeks before they turned black and fell off, leaving the faces of the young birds scarred and damaged, and their vitality compromised. They lived naturally in the garden, but needed to be fed and protected by a human presence.

The only bird who still required daily treatments was a very timid and gentle young hen, called Tricklet, because she had been just a little tricklet of life for so long. She had become a beautiful iridescent black bird, long legged, long necked, and elegant. The disease had destroyed her eye-lids on one side, so she was unable to close that eye. Each evening, it was important to check it to make sure that no dirt clung to the cornea, and administer eye-drops.

There was also a tiny handicapped hen called Larzac, who had one leg twisted behind her, probably by someone playing at swinging her around by her leg. She had been brought in starved and in poor condition. A hen with one good leg is in a bad position, but after a struggle, she had developed a life of her own in the garden with the help of our young rooster, Diamond, another of Babalu's offspring. Roosters have a natural concern for others in the flock, and helped me greatly in caring for handicapped hens and orphaned chicks.

Luckily, the timing worked out right for Larzac, who had accumulated enough eggs to want to brood them. If she spent the time we were away sitting on her nest, she would need no extra care. Her nesting box was near the food in the hen house, where she would be safe as long as the door was closed at night and opened in the morning. The eggs were boiled so that she would not break them, nor would they complicate matters by hatching. Diamond would help the crippled bird when she emerged from time to time, and hopefully everyone would manage until we came home.

I had treated Tricklet each evening since she had been a tiny chick just two weeks old, and helped her through countless infections and other troubles. Her cornea had gradually degenerated, and her eye went blind. In the long, long time of caring for her, I had grown to love her very dearly, and was concerned about leaving her welfare to someone else. She was the only pet I had. However, as long as the drops were put in her eye daily, she would probably be all right until we returned. The problem was that dirt got into her eye, and if it wasn't cleaned out, it became infected. The infection could spread to her brain if she did not get antibiotics right away. I therefore explained the danger signs to our house-sitter, and left directions for giving her antibiotics should the need arise. But, would he do it? That was the question. It was a lot to hope for.

However, during our house-sitter's visits to get to know the dogs and birds, my impression that he loved animals, and was happy to spend a month there with them, was continually reinforced. When he treated Tricklet, he always repeated while stroking her gently, "Elle est adorable!" (She is adorable.)

I scrubbed the house, and packed away our personal things to make it as pleasant as possible for him. There was so much to do that I was lucky to get a chance, between passing storms, to get out one last time to see the sharks before we left. I went to the barrens late one afternoon following two beautiful, calm days. But as I hurried out there, the wind grew in force, carrying clouds in from the east, and piling them upon the island. Again, I arrived trembling with exhaustion. As I prepared, glancing around at the tossing waters for sharks, the thickening clouds lowered, and the day darkened.

All of the same sharks met me, still with their hooks and trailing lines, with the usual crowd of juveniles. When Lillith came, I swam with her as she cruised the area, then curved westward. It was good to be with her for a while. After months of stress over the fishing of my sharks, I had a profound feeling of foreboding about leaving them—for a whole month—and savoured every moment with each one.

When I returned to the feeding site, a familiar pale shape swept around behind me, and followed me in a casual fashion a metre behind. Two other large nurse sharks were already lying there tranquilly crunching on the scraps. My faithful follower accelerated to rush one of them, and chased him out. I couldn't tell if he bit the other shark or not. I couldn't remember ever seeing a nurse shark do that before, and such behaviour didn't fit in with the theory of him being blind. It fitted in with the theory of him being aggressive.

At first I had not been concerned at his arrival, but after that, I drifted to the kayak. I had forgotten to take a stick with me. My former worries about this problematic creature had faded from my mind, in the far greater concerns over leaving my sharks and birds. With the sun sunk in heavy cloud in the west, the scene was shadowed, so when the nurse shark glided away from the food and infinitely slowly circled around me, just one time, it was impossible to see his small eyes in enough detail. It seemed that he was less interested in me because I remained still. It was when I was moving around that he was stirred to come with me.

He seemed more active than the other nurse sharks. He was constantly moving around and changing position; he trailed another shark, nose to tail. Perhaps he was simply a very unusual individual. He certainly didn't look handicapped. For such a big creature he seemed to be in excellent shape to be so active.

Night slowly fell again as I watched from the vicinity of the kayak. I needed to spend more time with this shark to understand his behaviour better, and decided to make it a priority when I returned from France.

While I was actually with the dreaded shark, watching, I didn't have a moment of anxiety, and considered again how strangely fear worked.

I left troubled, without seeing Apricot, Amaranth, Sparkle Too, Mara'amu, Peri, Lightning, Samaria, Droplet, Shimmy, or Keeta, all of whom had not appeared for weeks. There were many others, too, who had failed to come on schedule, but I couldn't be sure about the disappearance of non-residents. I hated to leave them in danger of being fished. So I decided that no matter what happened, I would be sure to get out to see my beloved sharks one more time, before I had to put on my dress and high heels, and head for the plane to Paris.

At least, I thought, paddling worriedly homeward, the magazine *Shark Diver* was featuring an article about the massacre of the sharks of Polynesia. The news of Trillium's fate, and those who had died, would leave Polynesia, lost down on the tail of the world, in a more significant way than through e-mails.

While I showered back at home, the storm that had appeared on my way out hit the island, and the wind howled down the bay all night long. By the end of the week, a seven metre (23 ft) ocean swell was predicted, and people like me were forbidden to go to sea in our small boats. Instead of visiting my sharks, I had to re-roof one of the two small out-buildings used by the junglefowl. The wind had blown its roof away.

Arthur was preparing for his talk at the conference on cognition in Germany, and we were writing back and forth over the final details. One of his last messages concerned fighting between sharks of the same species. He wrote:

I've one final question. Have you ever seen fighting among the sharks besides two grabbing the same piece of food and pulling it between them? I think that question may come up in discussion. During my many observations over about twenty-five years or so in the field, I've never seen a shark acting aggressively toward another shark other than males pushing or biting females during what appeared as reproductive tactics. I've observed lemons, tigers, bonnetheads, silkies, oceanic whitetips, and blacknoses for reasonably long periods and nurses and blacktips for very short periods of time. I've not seen what I'd call aggressive actions by the animals except a few of the fore-mentioned instances. How about blackfins?

I answered that I had not seen sharks fighting, either, and that they did not even seem concerned about preserving an inter-animal distance. This included different species, as when the lemon shark nosed the tiny nurse shark pup who was munching on a scrap.

It was one of the first things about them that had surprised me, the apparent camaraderie among so many individuals of many species, which I had never seen among terrestrial animals. Especially the sight of the two-metre nurse shark munching on a scrap in a crevice in the coral, with a moray eel of about its same size lying against it, had been memorable. The presence of whitetip sharks and visiting sharptooth lemon sharks had no effect on the harmony in the site. The only exception had been the sudden entries of the lemon sharks, which had startled everyone. However, after the startle reflex, the peaceful scene had re-formed.

Apart from mating wounds on females during the mating season, I had not seen sharks with bites on them, with the exception of the unique case of Christobel, which was an effort at predation, rather than an aggressive attack by a blackfin. It was interesting to have Arthur's testimony that he had seen no intra-specific aggression in sharks, either, among many more species than I had been able to observe. The dramatic contrast to this information was their appalling fashion of making love!

Arthur told me that he had finished preparing his power point presentation on cognition in sharks, which included some of my photographs, and said that it would be the first time that evidence of cognition in sharks had been scientifically presented. It was thrilling for me to be part of such a project. He concluded his message:

"At least, I don't think that I'll bore the audience although a few, who believe cognition only occurs in humans, may well snooze through it."

"I'm sure that they will all be waiting in suspense to hear what you are going to say!" I replied. "It will be interesting to learn if cognition in non-human animals is becoming more widely accepted now, and what the basis could be of a scientific belief that only one species, among so many, has evolved cognition. Sometimes the detection of cognition in *Homo sapiens* can be elusive, too."

He left for Germany and soon afterwards, I left for France. We crossed at the *Charles de Gaulle International Airport* in Paris, just four hours apart. But it was four hours too many, and it was the closest I came to ever meeting Arthur.

A few days before we left, Tricklet showed signs of illness. She was resting all the time, and her face was hot. I prepared all her doses of antibiotics for our house-sitter, and left them in the fridge; all he would have to do was empty the syringe into her mouth each day. Everything required to treat her was handy and I had simplified everything to a few basic steps, which were written down.

The morning we left, on October 23, at 5:00 a.m. the house-sitter came, and I showed him everything. He was kind, reassuring, and seemed concerned about Tricklet's well-being. I was worried about my little pet, knowing that she would feel abandoned without all the supportive care and affection I had always given her. As we left our beloved island, I had an overwhelming sense of foreboding that I would never see her again.

However, a week later we received a message from the house-sitter saying that she was much better. I was relieved, feeling that at least she would be alive when I got back, and then I would make up for her abandonment by giving her lots of special care. But no more messages came.

We returned very early on the morning of November 23, after a twenty-four hour trip from Paris. The garden was silent. No dogs barked, and no birds flew or called from the trees. After we parked, the dogs crawled stiffly from under the house, shockingly emaciated. No junglefowl were in the garden, except some wild ones from the jungle on the other side of the road, who instantly vanished. The house-sitter was absent.

My initial shock turned to panic as we roamed around, finding evidence. All Tricklet's doses of medicine were still in the fridge in the syringes I had prepared. The house-sitter had never treated her! The dogs were on the verge of death by starvation, there was no water available for the animals on the property, and the birds were gone. The house was filthy, and the key lay on the step outside the back door. There was as much dirt on the floors inside as there was outside. While gusts of anxiety swept me, I began to frantically wash the house with soap and water, just so we could function. We were exhausted and hungry, and had planned a leisurely breakfast and a nap. Instead, Franck left to confront the house-sitter. Sure that the birds had simply gone wild in the forest across the road, I tried to console myself, listening, as I cleaned, for their calls. I knew each voice by heart.

Sometimes I crossed the road and wandered through the forest, looking for my little flock. I couldn't find them, but saw two hens surrounded by newly-hatched chicks. Franck and I had initially thought that the starving dogs had begun eating the junglefowl, and that the others had fled to the safety of the jungle. But if our dogs had been eating the birds, the chicks would have been the easiest targets, so suddenly I rejected this theory. Further, the initial presence of the wild alpha male in the garden with his flock, suggested that our birds had disappeared, and that others in the vicinity were taking over an abandoned territory.

I examined the garden, looking for clues. Under the house, where the dogs would have carried the birds to eat them, there was not a feather. They couldn't have eaten the birds. But one of the windows to a storage building on the property had been forced open, and things inside had been thrown around. Further, Larzac's nesting box, which was heavy, had been moved across their little room. Its new location suggested that it had been placed where someone could climb on it to reach the birds on their sleeping perch. There had been a human intrusion, and one apparent goal had been to steal the birds. This impression grew as the day passed, while I cleaned the filthy house, searched the mountain-side, and heard no roosters crow.

Eventually Franck returned. The house-sitter said that he had left because a rat had come in. This, he said, he could not bear, and so he had abandoned our animals without food, water, or protection. That was the only explanation he gave. But rats were a common problem, an everyday part of life in Polynesia, even on a boat. So this made no sense for someone who had lived on a farm, where all sorts of unexpected things happen and need to be handled. Why had he made no effort to let us know that he did not want to stay? He sent no message, and did not phone the emergency number we had left him, to give us the possibility of contacting other people to try to make some other arrangement. A neighbour would have been willing at least to put out food and water for the dogs—everything necessary was on hand.

Franck argued and shouted, but there was nothing he could do. The house-sitter was perfectly contrite and agreed that he had done a terrible thing. Finally, Franck left with the parting words: "If you see Ila, you had better run. She will strangle you."

I was shocked to the point that I couldn't go on with my life. I could scarcely eat or sleep, and awoke from nightmares when I tried. I had been exhausted on arrival from the long flight and the extreme jet lag caused by going from one side of the planet to the other. Our holiday had been tiring, too. Meat was the only protein usually offered to eat, so as a vegetarian I quickly got protein deficiency, which exhausted me during a month in a winter climate, in which we had to function in high spirits all the time, and were constantly on the go. With the stress of the loss of my birds, and driven by a continual need to search for them,

I lived in a twilight zone. It was an intolerable anguish to think of Tricklet and all of the others in the hands of cruel people. Why the hens had been stolen, even handicapped ones, was incomprehensible. Cock fighting was wide-spread, so roosters were always in danger of being stolen, but why Tricklet? And the little crippled Larzac. . . Snatched from her nest! During the day, I hitch-hiked around the island, talking to people who might be able to find out about a big theft of birds.

It had taken me some years to learn that junglefowl were the most abused animals in Polynesia, but when I became known for rehabilitating birds, people had occasionally left them at the house. Whereas other species either died or flew away, the junglefowl knew a safe place when they found it, and stayed. Only a few crossed the road to live wild in the jungle.

The first case was Babalu, a tiny chick with half a wing on one side. Shortly afterwards, two older chicks were dropped off in a plastic bag, emaciated, nearly featherless, and suffering from a respiratory illness. Unlike other wild birds, junglefowl were prone to countless diseases, probably from having been kept, for the five thousand years since they were domesticated, in the poor and often filthy conditions humanity has traditionally provided for them. Diseases evolved.

Junglefowl are the ancestors of domestic chickens. They evolved to walk through the jungles of southeastern Asia, an environment as infested with predators as any on earth. A variety of cats, primates, raptors, and more, were always on the alert for a hot little snack. In the pools where they drank lurked eels, snakes, and crocodiles waiting to snatch them. Faced with this, they evolved into a unique and remarkable species.

Walking, instead of flying, and foraging in the leaf litter for insects, young plants, fruits, lizards, mice, spiders, and other edibles, their bodies became heavy, their legs strong. This is the reason humanity considers them excellent food. Since they were so voraciously hunted by predators, they evolved an accelerated reproductive cycle. This profits humanity too—imagine how many eggs are eaten around the world each day.

The result was a species in which the individuals stuck together in hierarchical flocks in a united effort to protect eggs and chicks. For the alpha males, that meant getting and keeping a good territory, warning the flock of approaching predators and fighting them, and helping the hens to find food and safe nests.

For this highly developed social life, an expressive vocabulary evolved. For example, a shriek of alarm announced a predator on the ground, whereon the flock flew into the trees. But when the approach of a swamp harrier was proclaimed with a growl, the birds ran for cover. The beta males supported the alpha, played, tried to mate the hens, and helped find food for the chicks. When a rooster found a special morsel, he called to the hens to come and partake of what he had found; the better the titbit, the more excitedly he called.

When the mother hen first led the newly hatched chicks from their nest, she moved slowly, finding tiny insects, and giving them to the chicks. They ate only what she took and gave them. Since she had brooded the eggs for three weeks, she needed food herself, but since mother and chicks needed foods of different sizes, she could feed herself while finding food for them. They learned swiftly what was good to eat, and soon actively hunted themselves. Within days, the chicks were chasing each other around, squealing in excitement, with insects too large for one alone, until their combined efforts tore it into pieces small enough for them all to eat. What they learned in the first days of their lives seemed to affect them forever after—a trauma at this time left a permanent mark upon the bird.

The newly hatched chicks imprinted not only on their mother or care-giver, but also upon foods, their environment (whether jungle, open forest, or garden), and their experiences. Hens appeared to adore not only each chick, for which they were willing to fight a predator to the death, but eggs as well. I often saw a hen unexpectedly find another's egg, and caress it with beak and breast for awhile before going on her way.

Though Polynesia lacked the great terrestrial predators, there were so many rats and cats who took the chicks from under their mothers at night, that often all the babies were quickly lost. Dogs roamed the jungle in packs, killing any hen without adequate cover with a bite through the thorax, just for the fun of it. Some hens succeeded in finding particularly safe nests, and managed to raise three or four little ones, but it was more common that a family of seven to thirteen chicks disappeared within a few days.

The mothers tried each night to force the tiny chicks to come up into the trees with them because of the dangers on the ground, but as long as the little ones were unable to flutter and climb up, she flew back down while it was still light enough, to lead them to their nest. After about six to eight weeks, they were left increasingly on their own, as she became reproductively active again.

The spectacular males of the species were renowned fighters, an ability that possibly evolved through defence against predators. Before humanity developed its sneering attitude toward chickens, roosters were considered symbols of courage and power, the only bird who could kill the tiger. This legend is not as far-fetched as it seems, since the birds aim with stunning speed, power, and needle-sharp spikes, at the eyes of their opponent, and a tiger with punctured eyes could die.

The junglefowl alpha male acquired an entourage of hens and juveniles, and was always on the alert for danger. He avoided fighting, while his sons and nephews attacked any intruder for him. Thus, once having achieved his position, he could maintain it for years, as long as he had the respect of the younger males. His sons helped him, as well as trying to mate the hens whenever his attention was distracted. These hens ran to him for protection, so his ability to maintain his leadership was constantly tested. Hens with chicks were respected, and young males tended to stay near them helping to feed the chicks. Later, these little ones often became the first members of his own flock. The species had a social life as supportive, complex and communicative as those found among social mammals.

Fights between roosters were generally brief squabbles between brothers or cousins; the weaker or more timid bird ran away. Young males chose a favoured area where they spent some of their time crowing, and which eventually became their territory if they could keep it. The alpha male and his entourage moved slowly, foraging, among the territories of his grown sons, and they, with their families, foraged with him until he moved on. He crowed from time to time as he moved, announcing his presence.

These territories were worked out by neighbours in a ritualized way. The roosters confronted each other on the border, each standing sideways at the place he believed it should be, head and tail raised, as if he were defining a fence. He crowed. If his neighbour chased him, he repeated the action a little closer to the heart of his own territory, or he turned, fought and chased the other bird, until the neighbour decided that he had gone far enough, and in turn, made the fence gesture and crowed. Finally the position of the border was agreed upon. Neighbours knew each other, and usually settled the position of the border without fighting. But if one of the birds detected that the other was weakened, he chose that opportunity to fight to extend his territory. So birds who were injured found a hidden refuge to rest, until they healed up enough to systematically win their territories back again. Until this was done, the territory was more important than their hens.

The older, powerful alpha males had known each other all their lives, and called back and forth daily, proclaiming their territories. Their calls echoed around the bay when they exchanged their proclamations, and some answering birds were so far away I could scarcely hear them. They occasionally visited in each others' territories without conflict. By that age, their spurs were needle-sharp and so long that their fights resulted in serious wounding—the head, eyes, breast and wing joints were targeted.

But, as among sharks, each rooster and his circumstances were different. Some lacked ambition and avoided fighting all their lives. Such birds often roamed vaster territories than the aggressive ones, by remaining quietly on the periphery, rarely crowing, and never answering a challenge.

From observing different flocks and their evolution, it seemed to me that the beta males were used by the alphas when a fight was necessary. They helped their sons if needed, but it was the beta males who tended to get injured, and once an eye was blinded, the life of the bird was essentially over. Thus the alpha male avoided injury, and his potential rivals, his sons and nephews, were killed off at the same time. The mechanics of human wars are similar. It is not the alpha males who are killed in war, at least not until the end. It is the young men of the country (the territory) who are sent to fight.

When the alpha male was hurt, grew old, or weakened for other reasons, he was often killed in the battle that ensued for the leadership of the flock, because he would never give up. I kept track of a few of these terrible battles, and never saw such long-term single-minded determination, courage and spirit as shown by some members of this species, not only by the alpha male, but by the beta male aspiring for leadership. Sometimes badly injured former alpha males became loners. They adopted orphaned chicks, and helped hens feed their babies, but avoided contact with other roosters, and did not crow.

Other alpha males retreated from action as they grew older, accompanied by the elderly hens who had always been their companions. When younger powerful males were in the region, they simply moved away, and thus were able to enjoy a peaceful old age. The roosters' famous aria declared "I am here!" and those who didn't want to attract opposition over that idea, remained silent.

Territories such as these differ from the reef sharks' home ranges, in that they are defined by the bird and defended. The sharks' home ranges are just preferred regions used by individual sharks, and they overlap.

But when my first cases of sick junglefowl were brought in, I had not learned this. I looked askance at the two pathetic, terrified, sick, and half naked young birds who arrived in a plastic bag. I had never owned chickens and knew nothing about them. I didn't even want them. I looked after wild birds, not chickens!

I started them on antibiotics and put them in a large cage which I kept beside me while I painted outside. Never had I seen such a look of strain and anxiety on the faces of birds. They were frightened of everything. But gradually, they relaxed as they became used to my presence and my manner as I talked to them intermittently, and offered them a variety of foods to encourage them to trust me. Soon they relaxed, and when they were stronger, and finished their antibiotics, I transferred them to an open outside enclosure where they could poke around in the grass and flowers. I visited them often, and put the little chick Babalu, with them, hoping that they would make friends. When they were well enough to escape, instead of going wild in the forest as I expected, they came to the house, and the three young birds foraged together nearby.

With good food and a healthy environment, these two little creatures flowered into beautiful, exotic looking birds. I was awed by the complexity of the patterning on each new feather, and the brilliance of the iridescent colours that unfurled like a flowering as their mantles grew. They were never apart, and cuddled together like dandies with their extravagant and frilly plumes in the shade in the middle of the day. Babalu was much smaller and turned out to be the only female. She was at the bottom of the pecking order.

In the evenings, the two cocks stalked regally into the house, such a sight that Franck would come out of his office just to look at them. Each of us had a favourite. Franck's grew into a huge, powerful, dark red and green bird. He stood tall and straight compared to the other junglefowl we saw on the island, but to me he was less alert, and not as well configured as my favourite.

Mine was the smaller one, the fast moving, nervous, more daring, and obviously the brilliant one, whose feathers were such a miracle I couldn't take my eyes off him when he was near. His mantle was palest gold, and the feathers on his back were triangular shaped, each rimmed with platinum, and filled with iridescent turquoise, glimmering with green, blue and purple lights. He was elegant and slender, with long legs and neck, and each movement was the very image of grace.

Standing together in the living room one evening, giving our favourites a cuddle, we considered names for these amazing creatures, who were obviously the most beautiful birds ever seen on the island, and we decided to call them Genghis and Kubilai Khan.

My bird was Genghis Khan. He pressed his face against my cheek and I stroked and spoke to him gently. His feathers were soft, and I smoothed them around him, and whispered to him. He nestled against me. It amazed me how affectionately this abused bird responded. When I encircled him in my arms, caressing him and talking to him gently, he rubbed his face against mine, murmuring softly. Treated with respect, the two had become as reasonable and intelligent as any other bird I had cared for. The nuanced vocabulary they used in their continuous conversation was just one sign of it.

We let them sleep on the seabird perch until they began crowing, whereon we carried them to a sheltered tree in the garden each night.

One evening, one of their play fights continued and grew deadly serious. They alternately faced each other, with beaks, necks, bodies and tails horizontal, then suddenly leaped, kicking straight at each other, and raced, whirled, and kicked again with stunning speed. I watched them fight as night descended, a pale swirl around a dark one, unsure what to do, or what it meant. The fight went on for a long time—even now, after all the fights I have seen, it stands out as one of a kind. For each bird it was the first fight, and they had no role models to guide them. Their fighting instincts, bred into them for countless generations, took over.

It was too dark to see much when I scooped up Genghis as he ran to shelter behind me. As I picked him up, he made the most blood curdling scream I had ever heard. I carried him into the house with Kubilai Khan stalking after me. Both of them were now alternating menacing low cries with startling howls of rage. In the house their incredible fight continued, accompanied by ear-splitting shrieks, and flying feathers. I was only able to settle them by creating a perch for Genghis Khan in another room, out of sight of Kubilai. Even then, it took nearly an hour to calm the birds, and I was very concerned about their future relationship.

But the next day, all was as it had been before. They walked together, cuddled together in the shade, and did not fight. However in the following days, there was a subtle change in Genghis.

He seemed distracted. There was no distinct sign—he just didn't seem as focused as he had been. Often he stood staring away toward the mountain with his back to me. I had developed many little activities for the birds, and often we circled the garden looking for insects under rocks and logs. There were traps for insects that we opened together. I used these activities more and more often to keep his interest.

One morning, when I began to worry that Genghis was actually going to start walking in the direction he was looking, I tried to get his attention back by uncovering insects. He came, chattering to me as usual, and participated with Kubilai for a few minutes, but then he turned again to look toward the jungle. I intensified my efforts, but he did not respond. As if drawn by a magnet, he walked a few steps. I called to him. He stood a moment, and then he just walked away, and vanished into the jungle on the other side of the road. I was heart-broken, sure that I had seen my beautiful pet for the last time.

But in two weeks he appeared at the far end of the garden with a small group of hens. Later I realized that their rooster must have been stolen for cock fighting, but at the time, I was unaware of this practice, which is enthusiastically pursued by nearly all boys and young men in Polynesia, behind a veil of secrecy.

Since it seemed clear that Genghis Khan had departed because he had lost his mortal battle with Kubilai, I waited to find out which of the two would be more successful, the big strong one, or the small smart one.

At first my beautiful pet lost weight, but he soon recovered his equilibrium. He would suddenly appear in the garden, and come rushing to me for protection from Kubilai. I would accompany him to the house, get him some food, and sit with him while he ate. Then we walked together down the garden and back to the place he crossed the road. He hid until the road was silent, then flew across, long plumes trailing behind. He seemed to understand its dangers very well. Genghis developed a pattern of coming to the garden each morning at dawn, and I put a bowl of food at the bottom of the stairs for him. Kubilai slept with his family in a little outbuilding, and I didn't open his door for another hour, so that my ethereal visitor would have lots of time to eat, and return safely to his jungle.

When his hens were with him I stayed inside to avoid startling the wild ones, and watching each morning from the window, I saw Genghis' first offspring grow up. They did not develop fowl pox as Kubilai and Babalu's chicks had done, and they became striking birds. One grew slowly speckled, then grey, then flowered, to my amazement, into a gorgeous white, cream and gold cockerel.

By the time we went to France, Genghis Khan's flock had grown. He was the alpha male in the jungle across the road, and his elegant white son was his lieutenant. The two were inseparable companions, who went everywhere together, chattering and crowing, allowing me to keep track of them. When Genghis brought his flock to eat at dawn, I noted who they were, kept records of when each of his hens had chicks, and other details. When the little flock returned to the jungle, they waited to cross the road until it was quiet. Then they flew over into the trees, the leaders first, and the chicks following, looking more like sparrows than chicks. Through them I began to learn about the natural social lives of this little-studied species.

CHAPTER FORTY-NINE

The Cock Thieves

Genghis Khan had been stolen as well as our own birds, and his spectacular son and heir had become the leader of the wild flock in the jungle. I circulated Genghis' photograph among known cock fighting circles, and promised a reward to whoever could return him.

On Saturday, November 29, 2003, a week after our return from France, three little boys appeared at the driveway, and Franck went out. They told him, "We know what happened to your birds."

Franck ushered them in. They told us that they had watched some boys they knew capture all of our birds, beating off the dogs with clubs and stones. One of the young birds, they said, had been so frightened it had run into the sea in an effort to escape. They also described a black hen with one white eye—Tricklet!

They gave us directions and names, and we gave them some money and time to get home. Then we descended upon the thieves.

At the first location the boys had described to us, we drove slowly down a long driveway, and parked in an open space with a few shacks scattered around the periphery. Though people had stared from their windows at us driving in, no one came out to meet us as we parked, got out of the car, and looked around. Many junglefowl of all ages watched us, half hidden in the shadows of the thick surrounding foliage.

Finally, a large Polynesian man approached and Franck began to talk to him in a reasonable voice. Slowly, others emerged, including a woman who seemed outraged by Franck's simply stated words. She angrily insisted that I come and look into their cages. I went. Indeed, the hideous little cages were empty. The hostile couple, with rising voices, insisted that their boys didn't steal cocks.

While Franck was talking, I looked around at the birds, and was struck for a moment by how much a little cockerel, walking by looking at us, resembled Tricklet. Then a large rooster walked out of the shadows under the trees and approached our truck. There was no mistaking Kubilai Khan, though he was black with filth, and his long tail trailed on the ground. As I rushed toward him, a gang of young boys suddenly appeared around me like a fleet of my sharks, and set about trying to catch him. Kubilai looked at me intently just as we surrounded him, and for a moment I thought he would come to me as I softly urged him, but at the last minute, he fled. He was soon cornered, and the boy who grabbed him handed him to me. He was feather-light, and his keel (breast bone) was sharp as a knife.

I carried him to Franck, who gathered Kubilai in his arms. It was the proof Franck needed that our birds were there. He warned the man that either our birds would be returned or the greatest scandal that had every struck the island would descend upon him. Suddenly the man admitted that though his family was not involved in cock thieving, other boys did leave stolen birds there, and he began to talk.

In the meantime, the boys who had caught Kubilai were gleefully chasing and catching a young cock, which they gave me, and then another. It was obvious that they knew exactly which birds were ours. The speed and practised motions with which they ran the frightened birds down and caught them was amazing. One boy flipped on his back at lightning speed and caught a cockerel by his feet as he flew over him. They had obviously been doing this much of their lives, and functioned together as a group in perfect harmony.

Then the boys declared that the other birds were at Brad's place. So we drove back up the driveway with the boys accompanying us on bicycles. On the way we met Brad, who begged Franck not to tell his father, then joined the rest of the gang as we drove slowly on. Not far down the road, we turned into another driveway, and I stayed in the car holding Kubilai and the three cockerels while the boys disappeared into the overgrown garden of another filthy shack.

Suddenly the most terrorized squealing I had ever heard began. One could not doubt that the creature making such a sound believed it was about to be killed. Diamond, pale, and tailless, was yanked out of a bush. He seemed to have been tormented to the limits of his endurance, and his high screaming continued as one of the boys dropped him unceremoniously onto me through the window. He didn't move after that.

Finally Franck returned. There were no more of our birds there, he told me, but the boys were going to look for their friend, who had all the rest of them, and we drove back along the road with them as they looked. They had admitted to having taken all of our birds, even describing the handicapped ones, and how they had caught them, laughing about the one who had tried to escape into the sea. The hens and Genghis Khan were all still missing.

When we found their accomplice, he tried to escape by swimming away, and Franck, whom I feared was attaining murder mode, pursued him. Left in the tropical sun, the car rapidly became intolerably hot. I was concerned about a general panic among the traumatized birds if I moved, but more worried that they would die of overheating if I didn't get them home.

Franck was out on a platform in the sea, where he was confronting an enormously fat boy of about sixteen. A large group of teenagers surrounded them. So carefully, still holding Kubilai, I managed to ease myself from beneath the pile of shocked birds, got out, walked around to the driver's side, got in, and drove home, with Kubilai unmoving on my lap the whole time. The others didn't move a muscle. Home was a hundred metres away; the families of thieves were our neighbours.

I put the birds in their little building, and hurriedly prepared plates of their favourite foods. When they saw me return, laden with bowls of banana, corn, papaya, shredded cheese, and rice, they poured off the perches to eat, obviously ravenous. But Diamond sat hunched at the back of the building, with closed eyes, and would neither eat nor respond.

I returned to where Franck was questioning the thief, whom he had caught and brought to shore. The obese child was sobbing piteously. A large circle of neighbours watched, while he managed to give us no information at all, crying that the other boy had my hens, while the cruel one who had tortured Diamond denied that he had ever seen them. I was frantic to find Tricklet, Babalu, and the poor crippled Larzac, as well as two other beloved sisters of Diamond. But these children yielded no information, and finally Franck went to speak with their parents, and look for the boy who had stolen Genghis Khan. I walked home.

Diamond screamed in that insane-with-fear way when I went near him, and this was his principle reaction to any movement around him for the rest of his life. Stroking, comforting, and caressing him with exaggerated gentleness had no effect on him. I took him to the corner of the kitchen I used for treatments, and examined him. He was so weak and emaciated that he would certainly have been dead had I found him a day later. His face was scabbed, dirty, blood-smeared, and white. His head was swollen, particularly around his eyes, and scraped all along his brow. His half-closed eyes were seeping. It seemed he had not only been used for cock fighting, but tortured as well. His tail feathers had been pulled out, and his legs were so painful he could scarcely walk. Cock fighters use the cords they tie around the birds' legs to jerk them around and this results in leg-pulling injuries—cracks and damage to the hip joints. I washed the crusted blood and filth from his face, and put Tricklet's ointment in his eyes. He cried as if he hurt all over. I started him on ciprofloxacine, a strong antibiotic suited to birds.

While I was working on him, Franck returned home, and I rushed out to see if he had Tricklet and the others. He handed me Babalu. The boys had put the rest of the hens up on the mountain, and would bring them the next day. Genghis was dead; several witnesses had confirmed that he had been fought to death by the thieves right away.

It was painful to imagine the fates of the others, the noble Genghis Khan, the delicate and timid Tricklet, and the helpless Larzac, in the hands of those cruel children, given the example of Diamond before our eyes. Perhaps Diamond had seen his hens die, too. His emotional state certainly revealed the depravity with which he had been treated. It had taken just two weeks for these people to bring my magnificent bird to the verge of death, and destroy his calm and courageous temperament.

I took Babalu to Kubilai, who was so weak I had to lift him down from the perch so that he could be with her. As they were reunited, he vocalized for the first time. Kubilai Khan had been stolen once before, and I had followed the thieves by hitch-hiking. I had learned where they lived, but was unable to get Kubilai back, so waited for Franck so that we could go together to confront the thieves. A white woman in Polynesia was about as significant as a street mutt, but the people could be intimidated by white men.

Rushing around to look after the rest of the birds, so I would be free to leave the moment Franck arrived, I had found Babalu lying on the grass. She didn't move when I picked her up. She had not a feather out of place, nor did she show signs of poor co-ordination, which might have indicated poisoning, or botulism. She just didn't move. Convinced from her behaviour that she was hurt, I examined her closely three times, leaving her on the emergency perch in the bird corner while doing other things. The next day she was about the same, but when Kubilai was returned to us in the early afternoon by the thief's father, and they were reunited, she suddenly recovered. Her reaction showed the depth of her emotions, that she became catatonic on seeing her mate kidnapped. This time the two birds had been separated after they were stolen. She must have been so frightened, and she was the only hen who had survived the ordeal. At least the two were reunited, and their love affair could continue.

Kubilai had always been the benevolent king of his family, warning of danger, keeping order, not letting anyone harass anyone else, not letting the chicks fight, finding food for the little ones, and even protecting unrelated handicapped birds given into my care. I had developed the habit of taking very sad cases to Kubilai, knowing that he would watch over them, encourage them to eat, and protect them. He had gone to a lot of trouble to look after the half blind Tricklet, for example, running after her, calming her and bringing her back to the flock when she got lost and panicked. So it would have been a terrible thing for him to have lost the care over his flock, and seen his family broken up and killed. Perhaps he had even been forced to fight with Genghis or Diamond. He was emaciated and distressed, but recognized that he was back under my protection in his familiar home. It was because of him that we had recovered them—he had emerged from the shadows and shown himself when we arrived, though he was weak, and in pain.

A large cage I had used for transporting wild birds had also been stolen, and Franck had brought it back. He said he would talk to the boys' fathers after the weekend.

A little later, the three boys who had told us where the birds were, returned to let us know that the black hen with the blind eye—Tricklet—was dead, and assured us that Genghis was too. Sick with regret and grief, I had to rush out to do some grocery shopping, since friends were due to arrive for a visit. It was a challenge to stay focused to prepare everything and get ready on time. I was giving Babalu her first cuddle when the visitors came. She clucked softly as I stroked her neck, leaning out from my arms to look around at her familiar garden. After she had spent some time with us being fussed over, I put her with Kubilai, and left the birds to rest, eat, and realize that they were back, safe in their own home. Some progress had been made, at least. I found a glass of wine and let the welcome company of our friends distract me from sorrow.

The next morning the birds came out into the garden. They formed an uneasy little group resting close to the house—Kubilai and Babalu, with the three young cocks. Babalu and a cockerel had become ill, so I put them on ciprofloxacine, and treated them for parasites. Never had they had a parasite problem before, but now they were practically overwhelmed with them. It was another sign of the shocking conditions in which Polynesians kept birds. There were no signs of joy at being home; they were all suffering too deeply.

I waited all that Sunday for the boys to bring back the other hens, intolerably frustrated over the situation, and when no one came, on Monday I walked down the road with the intention of visiting Brad's mother, and asking about them. I met her walking along the road with Brad. A big, hard-faced woman, she was hostile at first, but after a while, as others gathered around us, she warmed up. Nevertheless, I could get no information about the hens. Brad claimed that he had already given Tricklet to me, and then said that he had put her and the other hens up on the mountain.

The others drawn to the discussion group that formed there on the roadside described the scene in which a crowd of teenagers had been running down the street with birds in their arms, carrying the big wire cage, and a variety of other things, some of which was extra material I had set aside when re-roofing the bird's building before we left for France. The story of the terrified bird trying to escape into the sea was repeated.

One young man who had been standing watching, said he would try to find out more for me, and we talked for a while. He asked me if the red cock had come from a Chinese man living in the next valley. I said I thought so. A Chinese friend had dropped off the two chicks, saying that they were from a family member in that valley. It seemed that Kubilai had been noticed a long time before by the local cock fighters as being from the special line of fighting birds bred by this professional. He was well known all over the island, and his birds were greatly admired and very valuable. Even as we were standing there talking, a youngster came with a pathetic but colourful cock in his arms, and asked if I would trade for the red cock.

"No," I said. He went on pleading with me, and made quite a sales pitch. He informed me that this bird was the only one of that colour on the island, and I would be very happy to have him! I looked at the sad rooster, and realized the enormous difference between our splendid birds and the junglefowl the Polynesians called *mountain cocks*. Finally the truth sank in—my birds were highly bred fighting cocks. That was why they were so much bigger and stood taller than the native junglefowl.

Some of the boys involved in the theft joined us, and told me that they had come many times to get all the birds, including by night to raid the buildings. Tricklet, her baby, and their adopted son, slept cuddled up to Diamond on the upper shelf in the corner of their house. Larzac's nesting box had been moved beneath it so that the boys could stand on it to reach them. This had been my first clue that they had been stolen. Tricklet had died in the hands of the people who had tortured Diamond. The knowledge was intolerable.

Franck stopped on his way home that evening, to talk to the parents of Nano and Brad, the two ringleaders of the affair. He told me that Brad and his mother had confirmed that Tricklet had been eaten by a dog! Yet they had both assured me, in the presence of their friends and neighbours three hours earlier, that they knew nothing of her or the other hens. It was impossible to trust anything they said. Even approaching them with the most tranquil kindness, and honest appeal to compassion, had no effect. They lied. I was uncertain as to whether I could believe the younger boys' claim that Tricklet was dead, or whether I should intensify my efforts to get her back. If she were alive somewhere, I had to find her! Then there were the others, three other hens who had counted on me, and Larzac, about whom I could get no information.

Nano told Franck that Genghis had been unable to walk when he had caught him, and that he had died in the cage they had stolen from our property. They were evasive about the other birds, though they seemed to know exactly which ones they were—they could describe them accurately.

Later, one of the boys told me that every morning when Nano passed on the school bus, he had seen my magnificent pale Genghis there at the end of our garden, and had targeted him. He had caught him and fought him to death right there in our garden. At least it seemed to be clear that Genghis had died, and that there was nothing I could do about that. With the lost hens, it was another matter. I never found out.

It was unbearable to think about my bright and beautiful Genghis, and to know that his visits, which I had loved so much, had led to his capture and ignoble, painful death. Other people unexpectedly supplied reports of our property being full of not just children, but adults as well. It seemed that the whole neighbourhood had come to ravage the place, party and use the beaches. While abandoned and starving, our dogs had suddenly been descended on by shrieking, whooping boys, who repulsed them with rocks, and beat them until they retreated. Their ordeal had been terrible too, and they had witnessed the fate of the birds they considered to be part of our family and under their protection. The support of the humans they had counted on had vanished like smoke. As long as we lived in that location, whenever the cock thieves, or any of the people involved in this attack on them passed, we could tell who it was because of the intense hostility displayed by the dogs. It was a useful warning that danger was near, and I would run out and watch, to make sure that the young criminals passed by our property.

After Genghis Khan had disappeared into the jungle, the reason for the fight between the brothers became clear. It had been Babalu. The little hen was always at Kubilai's side. He found food for her, and the two birds talked endlessly, always involved together. Soon she was occupied with a nest, and in only three weeks, Babalu emerged with seven tiny chicks. It seemed like a miracle when one actually witnessed it, that a common egg, without interaction with the environment, could in twenty-one days become an intelligent little bird. Diamond was one of these chicks. Like Tricklet, he had been overwhelmed with fowl pox when just a few days old. The first growths that appeared on Diamond's face were shaped like horns, growing in front of his eyes and on his beak in a diamond formation for which I named him.

While some of the chicks were completely blind, Diamond was still able to see backward through a tiny slit from one eye. He managed to eat by himself by locating a bit of food behind him, then turning and pecking the place that he remembered it to have been. In this difficult way, he filled his crop long after his siblings had given up trying to eat. Even as a baby, he distinguished himself as a bird of courage and will.

His determination helped him recover more rapidly than most of his siblings, and in a few weeks he resumed a relatively normal life outside. But one eye was bigger than the other due to damage to his eyelids caused by the disease, and he was quiet and rather slow. He moved around the periphery of the flock, and eventually as he matured, he established one corner of the property as his own little place. But as time passed, it was obvious that he was lonely. When I sat outside looking out to sea, with my mind on the sharks, he would often stand near me, for company.

It was during this period that the young crippled hen, Larzac, was brought in. She had to be kept in a large outdoor enclosure for protection, and Diamond often rested near her. Since she could not walk, she could not be a proper mate for him, but his company made a big difference to her state of mind. Partly due to his solicitous attitude to her, and his regular company, she gathered the courage to fly from her enclosure and make herself a life free in the garden. When she began to lay eggs, Diamond helped her look for a nest and eased her through the excruciating anxiety she seemed to feel much of the time. When she became distressed, I could count on him to help her.

Tricklet was in and out of intensive care due to her exposed eye. Since she could never close it, the cornea slowly dried. She suffered from recurring eye infections and required daily treatments, as well as help just to get her through each day. When she went outside after her treatment, often she could not find the main flock. I recall, so many times, standing on the steps calling to Kubilai, and urging him to go after her, pointing to where she was running wildly down the garden, when bushes obstructed his view of her. And he understood, rushing off immediately, to patiently guide his littlest hen back to the flock.

But she was often frightened, because she could see nothing on one side, and thus felt that danger could land on her out of thin air. Once when Diamond surprised her, she began cock fighting with him. How lovely she looked with her feathers raised, her wings curved out around her, her black iridescent feathers glowing in the sunshine, her head high. She was just beautiful.

So I began carrying her to Kubilai after her treatments to make things easier for her, and he would run to meet us, and pounce on her the moment I put her down. But soon he grew impatient with her troubles and abandoned her. She became afraid to go outside at all, and for two weeks she hid in the darker recesses of the house. Efforts to persuade her to go outside, resulted in panic-stricken flights to the roof and no progress. So when I had a few minutes, each day I took her out to a secluded place, set her on the grass, and picked a few blades to encourage her to forage a little, safely, with me beside her.

Even these attempts were rarely successful, but finally, in a little shadowed glade, she consented to step off my arm and poke around in the grass. Diamond had seen us and watched with interest from a distance, slowly moving closer as if he just happened to be coming our way. She raised her feathers and chittered nervously as he approached, then darted behind me. Diamond seemed to sense a need for delicacy. He foraged tranquilly around us until he found something especially good to eat, whereon he called to her. She straightened and looked at him. Her feathers raised a bit and, clucking inquisitively, she very slowly and

Tricklet

tentatively took a few steps toward him. But then she changed her mind, and took refuge behind me. I sat with them as the sun slowly set, and gradually they relaxed together. When I rose to go back inside, however, the little hen hurried to follow. But the next day Tricklet chittered softly when Diamond appeared foraging near us. And this time when I returned to the house, she stayed with him. As the day passed, the two birds remained inseparable, foraging and resting in the shadows on the periphery, avoiding Kubilai's flock.

In the weeks following, Tricklet came to the door in her tentative way, and when I opened it, proceeded to search through the house for a nesting site. I showed her some possible places, and she examined them, and worked on the bedding I put down for her. Then she re-emerged, and continued her search. Finally Diamond came to the door looking for her. He saw the food I had put down in the bird corner, and came eagerly in to eat. But Tricklet walked slowly toward him making very soft sounds, soft as only this most delicate of hens could make. And Diamond switched his attention to her, as if she had spoken a sentence he understood. He forgot all about the food and began right there to look for a place to make a nest. Starting in the nearest corner, and commenting continuously to Tricklet, he tried each possible site. Then she tried it too. Together they examined each nook and cranny in turn, as they toured through the house. Finally, Diamond flew up onto a high shelf in the hall, knocking off books and decorative seashells, whereon Franck came out of his office, and threw them both out.

Outside, the exploration and examination of all likely places continued as they circled the house. When they passed the doorway, Diamond stopped in front of it and turned as though he was going to come in, then crouched and chittered, making nesting movements, as if to say that this door leads to good nesting sites. His gesture was close to being a symbolic communication, since he was transmitting a thought about something not actually present—the remembered sites in the house. Finally they settled on the laundry basket, and there, in the following days, Tricklet laid ten eggs, and settled herself upon them.

Three weeks later, a small round chick appeared, peeking out from Tricklet's cosy feathers. I offered crushed grain and papaya to give him strength while we waited for the other eggs to hatch, but twenty-four hours later, there had been no further developments. So, while gently reassuring Tricklet, I removed each egg and shone a light through it. When no shadow appeared in any of them, I carried Tricklet and her baby into the light, put out a variety of foods, and opened the door for Diamond. He began encouraging Tricklet and baby to eat, and soon the little family was busily searching for insects outside in the morning sunlight.

This was the golden age of Diamond's life. Tricklet's baby was just three months old when we left for France. He was very black, with the large eyes and beautifully formed face both his parents had inherited from Babalu. By then he had an adopted brother who had been badly injured when tiny. Diamond looked after him as he recovered and grew, and the two chicks were inseparable.

Tricklet's baby had been the first one to come to us when we arrived at the place where he was being held with Kubilai Khan. I just hadn't recognized him right away. During the weeks since I had seen him, he had doubled in size, and his proportions had changed. He had grown green tail feathers, a fire-coloured mantle and saddle, and his face had elongated and acquired a remarkable resemblance to Tricklet's.

The first night when I went to say goodnight to him, he closed his eyes as I stroked his face and murmured in the way he had always done, his voice now more mature. When I met him in the garden, he stayed close beside me. The three young cockerels seemed the least affected by their terrible experience. They had been too young to fight, so had been free and relatively unmolested. But their traumatic experience showed as stress lines in the iridescent decorative feathers they were growing.

Babalu and the sick cockerel recovered on ciprofloxacine, but the problems of weakness, strain, emaciation, and parasite infestation lingered longer, in spite of intensive care for all.

Kubilai had pain in his legs and walked with difficulty. It took a week for him to recover enough strength to hold his tail up off the ground. I gave him a bath, then, which he seemed to appreciate, lying for a long time unsupervised in the warm water while I made an extra rainwater rinse. Usually junglefowl did not like to

Tricklet's Baby

be in hot water. I dried the magnificent bird carefully with the hair dryer, while he showed increasing signs of pleasure. Back outside, he shook himself, and his body language suggested that he felt good for the first time since his return. With his clean feathers, and more natural posture, he was gorgeous. His iridescent dark green and red feathers were radiant, and he began preening again, which he had not done before.

Diamond continued to scream in abject terror when there was a sudden movement near him, which is a symptom of *Post Traumatic Stress Disorder* in people. Though for several days I kept him near me in the house, and handled him only with extreme gentleness, he no longer accepted comfort from me as he had before. His personality had changed. In spite of intensive treatment for injuries, starvation, dehydration, parasites, infection, anaemia, septicaemia, fungal infections, coccidia, protozoa, and severe leg pulling injuries, Diamond's recovery was slow, and at times imperceptible.

I was determined to save his life, but couldn't seem to get him out of danger. I continued to give him intensive care for weeks, yet he seemed always dehydrated, and due to recurring infections, I had to change his antibiotic more than once. He had a multivitamin formulated for anaemic birds, shots of vitamin B12, and an iron supplement. Yet his face remained stark white, and his skin looked old, and kept breaking out in little scabs. He was rapidly losing feathers—his tail and much of his mantle fell out. His limp became worse, and his legs oddly swollen. Walking was painful, and he easily lost his balance, so had to be on an anti-inflammatory medication just to keep him on his feet and functioning minimally. More and more he lay on his breast or stood on one leg. The scales were coming off the less painful one. I had tried one treatment after another to relieve his diarrhoea, but nothing worked, and it seemed to me that this pointed to some fundamental problem that my treatments failed to touch. I wondered if another bird pounding on his back in a fight, or some unimagined cruelty, could have damaged him internally, but there was no way to find out.

I had covered all the main medical problems that could be indicated, without recourse to lab work, mercifully guided by the renowned avian veterinarian Jaime Samour M.V.Z., Ph.D., Dip ECAMS, Medical Director of the *Falcon Specialist Hospital and Research Institute* at the *Fahad bin Sultan Falcon Centre* in Riyadh, Saudi Arabia. I had always been able to count on him to help me with puzzling cases, and he had sent me a copy of his book on avian medicine, a marvellous source of help on treating a broad spectrum of bird illnesses, parasites, and injuries. He was one of the best known and respected avian veterinarians in the world, and I was very grateful to have his concerned guidance, particularly in such painful cases. He agreed that psychological trauma was a strong influence behind Diamond's physical problems and told me that there was nothing like time and love to heal. So I waited.

Slowly Diamond began to gain back the weight he had lost—nearly half his body weight. On the day after our arrival home from France, during searches for my birds on the mountainside, I had found two of Genghis' last offspring—very small chicks—trapped in the roadside culvert. Unable to find their mother, I had installed them in the bird corner. As they became old enough to go outside, Diamond took an interest in them. He adopted and raised them, keeping them with him and clucking softly to call them when he found something for them to eat. It helped him to begin to make another start.

Genghis's beautiful white son continued to come to our garden in the mornings to eat as he had with his father. He had apparently been aware of terrible danger, because when we returned from France, no bird called from the mountain. Perhaps he had seen what had happened to his father, and the other events in the garden. Like sharks, junglefowl are smart and sneaky, and watch curious things from hiding.

But once Kubilai resumed calling, he did too, in a hoarse voice that slowly wound up to his usual melodic proclamation. His call was as unusual as he was, a pure, flute-like note rising melodically to a high trill, and dying away in a cascade of minor tones. Genghis's, on the contrary, had been a truncated croak, so where his son got this riveting call was a mystery. I kept track of him as I went about my day, by the intermittent soaring of his aria from the mountainside above. One morning I rushed out to meet him across the road, and put his bowl of food in a little glade within the cover of the trees so that no one would see him visiting our garden. While I was sorry not to see him any more, at least no one else was seeing him, either.

Once when I went to leave his food, he was standing in the dawn light in an open space above, looking ethereal with his green and pale frills laid over each other in complex layers, shadowed in violet. I looked away, not to appear rude; when I looked back, he had vanished. When that exotic vision stepped into a sunbeam from the tangled jungle, it was enough to make you believe in God. And after I had seen him once sunning himself after a storm, on the grass in view of the road, with cock fighters slowing down to stare at him, I vowed I would never let him be victimized like his father. Genghis Khan's legacy became my determination to protect him and his extensive family, much to the mystification of everyone who knew me.

To steal birds, the cock thief brought his own trained and experienced rooster with a cord attached to his leg, and tossed him as close as possible to the targeted bird. With his territory thus invaded by a strange male, the resident rooster challenged him, and a fight began. As the birds leaped, kicked and twirled, the victim became hopelessly entangled in the cord, whereon the thieves pounced on him, untangled him, and put him in a sack. This method of capture often resulted in the death of one of the birds. The stolen bird was not well treated, and soon died in a fight or through injuries, stress, starvation, and neglect. One bird I later rescued from the thieving neighbours, a descendent of Genghis, was being kept in mud, underneath a laundry basket so small he couldn't stand up in it. It had rained all night and he was covered with mud and shaking with cold. He had not been fed or given water, yet had been in custody for two days.

All his life, I kept Tricklet's baby near me, in memory of his mother, as my only pet. When Kubilai died, he became the alpha male of our flock. But later in his life he was blinded first in one eye, and then in the other, during his squabbles with challengers. Finally, I built a special shelter for him, where he could live surrounded by handicapped hens and orphaned chicks. Though this was less than ideal, he still crowed, had a small flock, and a life in nature. But one day he disappeared, and a machete and cord left beside the place where the thief had grabbed him, indicated that he had been stolen. I talked to all those I knew who were involved in cock fighting and stealing, about this theft of a blind bird, and offered rewards for his return. His theft was reported to the *gendarmes*, and described in the daily newspaper *La Dépêche*. I put up posters, and went door to door five kilometres (three miles) on either side of our house, talking to everyone I met about it, and spreading the word that the boy who could find him for me would receive a large reward.

When I still was unable to find the slightest trace of him, the local television station RFO, did a story about his theft, and interviewed me. I was able to say publicly that it was wrong to steal other people's birds, and if men had to fight cocks, they should raise their own and look after them responsibly.

But he was gone. Tricklet's baby, at the age of three, suffered the same fate as his parents, but being blind, it must have been infinitely worse for him. As long as I rescued junglefowl I could never feel they were safe. Even with the dogs in the garden, and locks on the doors to their buildings, they could be stolen at any time.

I learned that while children played at cock fighting, there were big cock fights held by serious players, in which large amounts of money changed hands. Large stadiums were devoted to cock fighting, which indicated that people high up in the government approved and supported the practice. Many of the cock fighters frequenting these places were there all the time; their lives were devoted to fighting birds to death. Others just attended at the weekends for sport. Professionals owned hundreds of fighting birds, each kept in a small cage; they were given steroids instead of exercise. I infiltrated one such cock fighting arena, and documented the experience in my junglefowl blog at http://junglefowltrust.blogspot.com/

Though cock fighting was illegal, the position of the *gendarmes* was that it was "tolerated," though it was clearly written in the law that the penalty for cruelty to animals was two years in prison. But like so many other laws, it was not enforced. Since they shared the prejudice against 'chickens,' the French did not care about cock fighting and the black hole of cruelty it created, which further influenced the people to have no respect for these magnificent and intelligent birds.

I remembered wistfully wondering what the people who lived in paradise would be like before we moved there. Surely they were the people most blessed on Earth. How would they be different? Would it show in their faces? And now I knew—they were no different than spoilt people everywhere.

CHAPTER FIFTY

Back to the Sharks

After I had found the birds, and taken care of them, I finally checked my messages. After five weeks, there were hundreds to go through, most on the subject of sharks.

Arthur had sent a message about the conference on cognition. He wrote:

The cognition symposium dealt largely with "ecological intelligence" a term that seemed less offensive to "humanists" than using the term cognition, but many speakers continued to use the term cognition, when speaking about the process of thinking by an animal. One person even summarized activities of invertebrates and their likely cognitive abilities. Numerous individuals spoke about mammals, including primates. Others discussed birds, reptiles, amphibians and one person, from England, spoke on fishes, largely restricting himself to symbiotic activities of anemone fishes.

*My talk involved differential brain size re: body size among elasmobranchs, operant and respondent conditioning in sharks, complex social organization among bonnethead sharks and finally several of your anecdotes. I showed several slides of your blackfins and, of course, your picture as well. I ended the talk with a graph, showing the close correlation between the thresholds of hearing of different pure tones as shown by behavioral responses and thresholds of action potentials from the eighth cranial nerve (CNS), i.e. a coupling of behavioral actions and brain action to do with sensory processes in lemon sharks (*Negaprion brevirostris*), reminding the audience that the evidence that cognition (= thinking) requires a similar correlation between a behavioral action and action potentials directly reflecting a thought-process. Such a correlation will surely be found by someone, likely a neurophysiologist, in the near future.*

The Chair, Dr. Hans Fricke, was apparently pleased with the presentation, as was the audience. Three days of talks and discussions resulted in agreement among those present that cognition can be openly discussed regarding animals and that term and its processes need not be treated as a non-scientific entity any longer.

It was thrilling to know that cognition had been scientifically acknowledged in sharks. Arthur told me that the scientific journal *Animal Cognition* was going to devote a whole issue to the papers resulting from the conference, and he was preparing his.

The news from *Shark Diver* was disappointing. They didn't publish the article on the Polynesian shark massacre they had asked me to write. The story of the tragedy, and terrible waste of the Polynesian sharks for Asian parties, was not going to be reported after all.

But in my absence, other e-mails had come from the media. One was from *Rolling Stone Magazine*, and one was from the *British Broadcasting Corporation*. The BBC was looking for solidly scientific stories on the smaller varieties of sharks, with an emphasis on their behaviour. With Arthur's triumph, and my work on cognition fresh in my mind, I dashed off a few sentences to them, on the way through my messages.

Rolling Stone Magazine did not respond to the information I sent them, but the BBC researcher asked to know more. I didn't know what to say. I had scarcely eaten nor slept since I had returned from France, and I had already been exhausted past my limits when I had arrived and had to start washing the house and searching, panicked, for my birds. This emotional and physical exhaustion had resulted in a depression so deep that I couldn't seem to throw it off. I was drowning in sadness. I felt as if I had been trying to climb up a cliff, concentrating and making some progress, and someone had come along and knocked me off with a smash in the face with a rock. I had fallen all the way down to the rocks below, and couldn't get up again.

The continuous tragedy of my sharks being fished had plunged me into a nightmare, which had not been relieved by the voyage to France. It had utterly exhausted me, and then there had been the horror that had struck my birds. I was still suffering from jet lag, and was troubled by nightmares if I had the misfortune to fall asleep. All I wanted was to lie down. I couldn't figure out what to say to the BBC, and finally suggested that the researcher, who was called Stephen, look on the Internet.

I wanted to see the sharks. I was sure that if I could only get out there to see them, I would start to feel that I was coming back to life. So when Saturday dawned brilliant and calm, I picked up the available fish scraps, got through the chores in record time, packed up the kayak, and left.

After crossing the border, I paused to have a drink of water, rest my arms, and just gaze at the sea again. As usual during the wet season, storms were passing, and the ocean lay pale as a pearl to the horizon, intermittently ruffling as if thinking of dancing. It was December 13, two months since my last visit. What would I find? I was so afraid of more tragic surprises. Except that they weren't surprises any more. They were expected, predicted, known about in advance. I sighed, drifted, and drank water. I was only just within the border of the lagoon.

Suddenly there was a shark gliding slowly past under my elbow. Windy's fin appeared beyond, with Cochita's. Eden was closely following. Flora was there. As I sat looking in surprise, they came alongside the kayak, and undulated against it. One slid against the paddle, and all around me the sharks placidly glided, dorsal fins above the surface, pushing the curves of their bodies against the boat as they swam, going underneath and pressing against it, then again against the paddle. Never before had they displayed this slow, almost sensual undulating against the boat, and the paddle too, where it was trailing in the water. I couldn't believe my eyes, and spontaneously reached down to stroke them as they passed, instinctively responding to what I felt as an affectionate gesture. I was amazed that after all these years they were showing yet another behaviour pattern, and comforted by this greeting I hadn't known was in their repertoire.

Who could imagine, that it would be my sharks who would offer comfort?

I didn't tell anyone but my closest friend, because due to the bias against sharks, no scientist would believe me. No one would believe me. But I write it now because it is part of the truth of what happened. I realized that it was important to my sharks that I come, and not just because of the food I brought. From then on, I had no doubt that my love for them was returned, that among the emotions that they experienced was something analogous to affection in sharks.

They swam with me as I went on to the barrens. As I anchored, attached the paddle, and began to put on my gear, they rose toward me in the boat. Cochita swam to me, and lifted her head toward me as she reached me. The others, too, came, looking up at me, and lifting their faces above the surface, side by side, another indication that they knew that I was in the kayak. I fed them right off the boat, then slid in.

Shilly was beside me with Bratworst and Gwendolyn. Others were placidly feeding beyond. The usual array of juveniles and males who used the barrens, were zooming around too. Sparkle came shooting up to me in her habitual way. She still had the big hook trailing two metres of algae-covered line—I winced at the thought of the drag of it on her mouth all those months. As she ate, she kept biting repeatedly in the direction of the hook, which seemed to be a severe irritation to her. Mercifully, Cochita had lost hers.

Fleet was there, and he was still trailing his long algae-covered line too. Never had he remained in my study area for so long, and it seemed possible that he had not left due to this handicap. Many more sharks from both sides of the barrens were sailing through the grey landscape, feeding spread out. How beautiful it was to be back with them.

Glamorgan, Columbine, and the large grey female I had identified at dusk, months before, glided in as darkness fell. Carmeline! What a treat to see her again. I redrew her dorsal fin, checking it closely to make sure all the details were accurate. It was only my second sighting of the shark, but the way she approached me was identical to the first time she had come, and I recognized her face.

Another grey visitor came soon after, and I was able, with difficulty, to draw both sides of her fin, too. Teardrop swam into view. My black shark was pale once again. Her back was the colour of yellow autumn leaves, shading into a redder gold along her sides—she was now a rare two-toned shark.

Trillium arrived on schedule at nightfall. Happy to see her, I dove down, and this time we met nose to nose, and she did not turn away. Her mouth was in the same condition it had been before I left.

But Madonna, Madeline, Lillith, Sparkle Too, and the missing sharks I longed to see, did not come. The familiar surge of anxiety returned as I got back in the boat, and began the long trip homeward. I felt I had reached the end of my rope.

The next day was Sunday, and after the chores were done, I retreated to my bed with a book, hoping to fall asleep. Lying there stunned, reading and thinking in a sludge of hopelessness, I realized that I had better pull out of my torpor and give a proper answer to the BBC. My messages had been very brief, and I couldn't expect people on the other side of the world to understand what my work was about unless I sent them some details. So I climbed back up, went to the computer, and wrote down a proper answer to Stephen's questions. Then I returned to my book.

In the next days I continued to try to escape from exhaustion by resting when I wasn't looking after the birds. But every time I checked the computer there was another message from Stephen at the BBC asking for more information. So I spent more and more time analysing and answering the questions, providing requested photographs, and considering the needs of a film crew with respect to my sharks.

By Wednesday, my torpor had turned to extreme, irrational agitation. It seemed that the BBC crew were very interested in coming and filming my sharks! The final decision was still to be made, but it didn't take a three digit IQ to figure out that they wouldn't be writing to me every day if they weren't interested. Since my sharks were wild, my relation to them was unique, the information about cognition had just been presented to the scientific world, and there were so many other species there in a supernaturally beautiful South Seas paradise, it seemed that I might actually have to cope with the reality of a film crew in the site with the sharks.

I would have to get the sharks to return to the original site, since the barrens was so grey and desolate, and the sharks were so spread out there. If the film crew came, they would need to film the sharks as they had been in the good old days, with the sandy circle of the original site in the coral garden full of fish, and the banner of the huge nurse shark waving in slow-motion off in the purple twilight. . . I would have to buy a mirror, since they would like to repeat my mirror experiment. I would have to get my diving gear checked in case I needed it. I had suggested, when they asked about other shark species, and specifically the cookie cutter shark *(Isistius brasiliensis)*, that we could do a shark feeding out in the ocean at night.

Whatever had I been thinking of? It had seemed like a good idea at the time—it wasn't something I could do alone, but with a professional team and a plan, shark sticks, and lights, it would be feasible. One thing I had learned was that if you go, watch, and wait, there will be something to see. It was just speculation that such an event could attract cookie cutter sharks, but I had found that lots of animals will come to look at something unusual. Cookie cutter sharks are tiny sharks of about forty to fifty centimetres long (15 to 20 in), who, I had learned from my dolphin tours, were reputed to stay deep down in the ocean during the day, as deep as three and a quarter kilometres (12,000 ft), and come to the surface at nightfall, looking for something to bite. Nobody ever went and looked out in the ocean at night, which was a good reason to do so. What would be more likely to attract these mosquitoes of the sea but us, with food, in the ocean at night?

I would have to get Franck to bring his motor boat over from Tahiti. You couldn't expect the BBC film crew to go out in a kayak! But so far I hadn't found the right way to tell him about the matter. When I had sent them a series of photographs that they had requested, he complained that it took half an hour for them to go out over the Internet, and what did I think I was doing sending so many photos?

"Well, the BBC. . ." I began, and he cut me off, saying that wasn't the BBC. I began to cry, and he told me that what had happened was very bad, but I really must try to put it behind me. When he left for Papeete, I looked at the address I had sent the photos to. Indeed, it said, "bbc.uk."

The word soon arrived that the whole team had consulted, and were in agreement with the producer that they should come and film my sharks. They had already planned their schedule. I read the message, and though it was mid-afternoon, flew back out to the sharks, grabbing a couple of pieces of fish I had saved in the freezer and plunging them into warm water for the trip to the lagoon. On the way I broke them up to make sure they thawed out. Crossing over the border, I slid in and trailed the boat as I swam toward the site.

I had to cultivate the individuals more intensely, as I had done with Martha, so that they would swim with me during the day. Since the shark finning had begun I had neglected to do that with the sharks. I was hoping for Cochita's company, remembering the slow undulations with which she had caressed the boat with her body. Maybe she would caress me.

But there was an energetic current. I was soon gasping for breath, yet making little progress. Time passed as I wriggled through the torrents pouring through the coral canyons, increasingly frustrated and irritated once again at having made a stupid plan. I was at the end of my strength when a shark appeared, swimming straight toward me from the east. She had not been following, but was coming to meet me. I waited with interest to see who it was.

Carrellina! Why did she always appear at such unwelcome moments? Indeed, I had forgotten that it was December and the bad shark was due back. She cruised past me very slowly looking me over, as I surveyed her, and then she glided away. But soon she came again and stayed within view as I struggled onward.

I soon realized that Carrellina could be my best bet. The main problem was going to be that the film crew would frighten the sharks badly, and they would stay beyond their blue curtain instead of coming forward to be filmed. I needed a courageous shark to lead them now that Martha was gone. Carrellina was smart, and always came to meet me when she was in the area. If I started spending as much time as possible swimming with her during the day, I could have another shark companion as dependable as Martha had been by the time the film crew came. I had been told that there would be seven people! My sharks had become suspicious and shy when it was just Bou Bou who had accompanied me, so I was going to need some help, and Carrellina could be it.

Cochita appeared and sniffed at the boat. There was no odour escaping since the food was in a basin of water, yet approaching the site, the school of yellow perch that had always attended my sunset rendezvous, was swimming all around me. They still remembered.

I threw in the anchor and tossed a piece of fish to the sharks. Carrellina noticed first, turned as it fell, snatched it sideways and swam placidly away, with Cochita following, nosing the trailing piece.

Eden swam up, passing with her eye to mine, as more sharks moved in quickly behind her. Cochita and Carrellina circled as Brandy snatched the next scrap I threw. She sped away. I struggled to maintain my balance in the current without flailing, while tossing pieces of food to Tamarack and Bratworst as they approached, then rapidly sketching the fins of two new juveniles who swam in after Ondine.

They devoured everything as it fell, and as Windy arrived, I threw in the last pieces. Then I circled with her, and eventually tried to swim with her, since Cochita and Carrellina had vanished. But Windy changed her direction and speed continually, often turning to face me, then returning to the boat. It was obvious that she would not go anywhere with me as long as she thought that I might give her more food. When I glimpsed Teardrop in the distance, I left Windy and went to join her. She came circling curiously, and I drifted near her a while. Gwendolyn was also circling widely through the region, and for a while I did too, staying for a few minutes with one shark after another. None were going anywhere—we just cruised the area observing each other.

Flora came. She had birthed, and her skin was pale platinum. I swam along with her, looking at her closely, marvelling once again at the amazing difference in a shark's appearance before and after parturition. How long had it been since I had seen her in the full light of day? She had been hooked—there was a red streak extending back from the corner of her mouth, and her once radiant golden skin was dull. She circled me as I studied her, and slowly turned with her as she orbited, her eye fixed on mine. Was she ill? She looked terrible. Teardrop, Windy, Tamarack, and Bratworst joined us, circling while I finned gently to keep myself upright in the current, turning with them, closely examining each one. My sharks. I felt overcome with sadness for them. These had all been youngsters when I had met them; with the exception of Gwendolyn, not one of the older sharks remained. Madonna again was absent. I tried to be happy because the BBC would soon be there with me to film them and tell their story. The world was going to find out! But sorrow for my poor sharks so overwhelmed me that the current was sweeping us all, still circling, over the border by the time I recovered my senses and swam back to the boat.

I returned to the sharks just two days later, when the fish store called and asked that I pick up their scraps. The BBC's scientific researcher, Stephen, had emphasized that I should feed the sharks at the place I had described to them—Site One—so that there would be as large a group as possible for filming. I therefore cancelled my investigation of the nurse shark who had followed me around, and began a program of preparation. The goal was to have as many sharks as possible understand that checking for food at Site One could be profitable.

There, the coral surroundings would create a background of incomparable visual beauty, with the sharks gliding through it. Since most of the sharks there now had known me since they were small, I was sure that it would not take long to re-establish the original sunset rendezvous with the next generation attending.

But it was hard to get myself moving. I was so overwhelmed with exhaustion and grief that all I wanted to do was play with my birds or lie down, whenever I could sneak a spare minute. The marathon of preparing and going out to the sharks was hard to confront, but I took it one step at a time, picked up the scraps on Saturday, did the chores, packed the shark food in the boat, and paddled out to the lagoon.

On December 20, 2003, not yet a month since we had returned from Paris, I paddled out under a milky sky, across a pale, calm sea. The current had quietened. I prepared and slid underwater.

Flora was passing in front of me on one of her habitual clockwise circles, and she didn't change direction when I appeared beside her. I turned swiftly and saw many more sharks arriving in excited disarray rather than circling, calmly waiting, as they used to.

Windy charged straight to my face much too fast, and when she did not turn, I placed a gentle hand on her head, ready to push. But she shot away at right angles at the speed of light with shocking power. The sharks were expecting the food, and here I was splashing down amongst them. Flora just circled, but the others came straight toward me, which was exactly why I hadn't wanted, I had never wanted, to be in that situation again!

With difficulty, since the wind was blowing the boat against my head, I grabbed my lookout perch and threw in the food, using all my force to hold the kayak steady. The delighted sharks pounced upon it, as Carrellina soared in. She had stuffed herself two days before, yet she stayed with me for the whole session, often swimming around my head at eye-level instead of circling through the coral like a normal shark. This time, she was welcome. I was becoming excited by the prospect of visiting her often, as I had Martha, while she was in the vicinity. It was good to have something to look forward to—she would be fascinating to get to know outside the framework of the evening feeding sessions. Once I got rested, it would probably be easy to find where she spent her days and spend time with her.

Teardrop and Sparkle came together, accompanied by a larger female who had recently mated. She charged me repeatedly. Her second dorsal fin and half of her pectoral fin on her right side had been cut off cleanly. Another botched finning attempt? It was sickening.

Sparkle was eating in her forthright fashion, and came across the site gulping down a fish spine. Slowly it descended into her wide open maw, as she repeatedly swallowed, her whole body convulsing with the effort. Finally only the tail was sticking out of the corners of her mouth, too wide to swallow! She drifted on into the coral, followed by me and the visiting female, and ejected the backbone. All the meat on it had already been eaten by the ever present perch, but the two big sharks were grabbing and dropping it and passing it back and forth as they returned to the site.

I threw them some more food as a group of males swam in—Breezy, Mordred, Dapples, Jem, Marten, Avogadro, Tod, Marco, Ruffles, and Fleet, still towing the two metres of algae-covered line. Mephistopheles, the shark with the bite out of his nose who came through each December, was with them, and when he turned, he revealed a large hook stuck through his cheek. At least I had confirmed that he was still alive. It was the fourth year he had come in December.

Lillith, Madonna, and Madeline were still noticeably absent. I had just received my last roll of film back from developing in America, and was disappointed that only a quarter of it had been exposed and developed. This had happened sporadically with the camera, causing me to lose about two hundred irreplaceable photographs—twenty-eight on each roll of thirty-six. Yet it had supposedly been repaired during all those months in America. The camera had been nothing but trouble since I had bought it, and now I had lost the photographs of the rashes of snowflakes evolving on Lillith, an important record. Lillith was now missing and may have been finned. There were no other old sharks left in the immediate area. I could not be sure if some still remained in Sections Three to Five, but I had not seen any pass through.

Late in the session I tried to feed the waiting fish, who remembered our parties there two years before. When I began by thrusting a handful of food into the grotto where the squirrel fish, butterflies, groupers, moray eels, and rock cod waited beside me, Carrellina nearly swam down the hole after it. Again, I felt a surge of hope that she would be an exciting replacement for Martha.

If the film crew could get out to meet the sharks more than twice, they would get used to them. But it would be important for them to meet the new people, strangely clad and equipped, at a regular nightly feeding session. The sharks would have the most confidence in the new situation if they met the crew at their familiar event. Then those who came when we introduced the mirror the following morning, would not be so shy of the strangers.

As the waters darkened, Hurricane and Christobel joined us, and just before I left, the beautiful dark female I had identified, my new mystery shark who only came in the twilight, glided slowly in, and came to circle me. Carmeline. Watching her circle, then search out a scrap to eat, I sketched the details of her fin. But when I got home, I found that the drawing was not accurate. One side seemed to be different. I made a note to double check her identity. It was odd, because I had been so sure of who she was, not only because of her colour, appearance and the details of her fin, but because of her behaviour, each time circling me one way, then the other.

Chapter Fifty-One

More Trouble

I was still in such a state of exhaustion that I was scarcely functional, and couldn't seem to throw the feeling of overwhelming disappointment and sadness off. If only the birds would get better, I could get on with other things. If only the shark finning would be stopped, I could rest tranquilly. If only I could find Madonna. If only.

The week went by in a daze, as the previous weeks had. I busily circulated among the birds and other chores. A sick bird requires constant attention, care, special foods, messages to avian vets around the world asking for advice for one thing after another, calculating and mixing drugs, and keeping them and their beds and housing clean. Diamond had to be tube-fed regularly. At the end of the week, I had done all the chores, picked up the shark food, and was finishing up some work on the computer prior to leaving. Then I went out on the deck to look toward where my sharks were, estimate the strength of the wind, and see whether the white-caps on the bay were beginning to die down.

I had run out the door and sat on the deck railing to gaze out to sea a zillion times since we had moved into that house years before, but for some reason, instead of sitting on the railing, I sailed right over it. I landed flat on my back, on the bowl of a coconut tree there that leaned out over the water. Luckily I didn't put out my arm to break my fall, or I would have broken it, which would have kept me from paddling the kayak. As it was, it was the sort of fall that could have broken my back had I fallen farther. But instead, it felt as if I had just broken a couple of ribs. The problem was that I couldn't seem to get my breath.

Franck helped me into the house, where I lay on my back on the couch gasping for oxygen, in a panic over the sharks. How was I going to get their food to them? I moved my arms, and they seemed to work fine, so I planned to rest for a little while, then take my time and paddle out there to them anyway. If hell yawned open in front of me, it would not keep me from my sharks, I assured myself. But on finding that his wife was unable to catch her breath, never mind stand straight, Franck announced that I was going nowhere.

I argued and cried but he was adamant. Finally I begged him to take the food out to the sharks instead. He laughed! I considered telling him that I had to get the food to the sharks in order to prepare them for a visiting film crew, and thought better of it. Instead, I intensified my efforts to persuade him with my "poor me" angle. But nothing worked. I had to lie around as the sun sank and the flies laid eggs on the shark food, gasping and frantic, regretting that I had thought things were bad before, because they had just become a lot worse. Remembering how I had flown over the bench, I became sure that my subconscious mind had covertly arranged the situation, though why was a mystery.

I didn't go to my doctor, because I was afraid that she and Franck would forbid me to go to the sharks for a very long time, if the ribs were confirmed to be broken. Instead, I planned to heal up enough by the following week to go, pain or no pain. But the problem was that I couldn't expand my rib cage. I just couldn't breathe in. Walking around, or doing anything other than staying absolutely still, caused distress, because I couldn't get enough oxygen. Slow walks up and down the garden were a marathon of agony as I fed and treated the birds. I was riveted on the fact that the BBC could be coming as early as January, and I couldn't do anything, never mind get out to the sharks.

The next afternoon I lay down when I finished the chores, to try to beat the pain and exhaustion in a period of calm. My hip, side, and rib-cage all appeared to be seriously damaged, and getting comfortable

Genghis's Son

was impossible. I was no sooner settled, after five minutes of trying to get into a tolerable position, when I heard an alarm call. Luckily, I went painfully outside and found three Polynesian teenagers trying to steal Kubilai. I was trying to convince them that they should not steal or fight cocks, when one of their mothers drove by, and stopped to talk. She was one of the women I had met when Kubilai had been stolen before. She and the friends with her in the car quite agreed with me that the boys should not be there stealing Kubilai again, and told me the identities of all of them. Thus I was able to talk to their parents.

But though that time it had worked out, it was a dismal reminder that even in such dire straights, I could not rest. Always, someone was after my birds. By then there was an attempt nearly every week to steal one of them.

Each night I listened for the wild cry of the white bird on the mountainside. When he flew into the trees to sleep at nightfall, he called, and I thought, *he's all right tonight*.

The cockerels who had been stolen had developed crushes on me since I had saved them, and their attentions were blatantly sexual. The three exotic creatures had taken to chasing me around and pouncing on me when they saw me. The result was that they were easy to catch and bring inside. I had them sleep temporarily on an emergency seabird perch in the large shower stall. They were afraid to return to the building I had gone to so much trouble to furnish and re-roof for them before we left, and I didn't want them to go wild.

Diamond's symptoms did not abate, and as long as I was afraid he was going to die, I couldn't escape emotionally from the pain of the theft and killing of the birds. The event continued to cast a shadow, just as the workload of having the seriously ill birds to look after had become a painful burden after my fall.

A week after my injury, on January 3, 2004, determined to fulfil my promise to myself, I got the fish scraps, and loaded just the best of them into the kayak. My luck was with me that day—it was calm. The sky was hazy and pale, and beneath it the atmosphere meditated, above a sleeping sea.

I could feel each broken place in my side, and tried to use muscles that didn't touch them, twisting slightly to dip the paddle through the silver surface, on one side, then the other, pulling just a little bit. It worked! Very slowly the shore began to slide by, and the lagoon, imperceptibly, to come closer. I forgot all else but the glow of the lagoon, and the gentlest movements between me and the sea. What a change after all of the battles I had fought with it. Everything seemed quiet and gentle—the honeyed air, and the soft, soft sea, shimmering and making little rivulets charmingly. Even the pain was tolerable.

Eventually I drew close to the site, and drifted, watching, unable to turn to look, but aware of the movements of the gathering sharks. It would be impossible for me to function in the water at all, and especially to pick up the anchor, so I threw it into the middle of the site where it wouldn't catch on anything, and I would be able to draw it easily up, using the string I had attached to the anchor line.

Cochita appeared to my right at the surface, and swam slowly under my hand. I stroked her. She slid against the kayak, and disappeared below. Then several sharks soared straight up, undulated against the boat, and went straight down again, so that their tails flashed above the surface around me momentarily like the wings of birds.

Since the brains of fish and birds scarcely differ structurally, there is no reason to assume that fish can't feel positive emotions. Birds are intensely affectionate, and Diamond was a constant and sad reminder of just how intense their feelings could be. Further, Arthur had pointed out that the brains of sharks are much larger for their body size than those of fish, and more complex—they are as large, comparatively, as the brains of birds and mammals.

Watching the sharks sliding against the boat, then the paddle, and undulating from one to the other, it was impossible to see it as anything but their version of affection. I would never have imagined that they would express this physically as they did—I had never seen them undulate against each other. But in the perfect calm of the evening, through the flawless clarity of the still water, everything they did was clear to see. When Flora arrived, she came swiftly up to the kayak, undulated once against it, then slid off the paddle on her way back down.

The only time that sharks would undulate against each other that I could imagine was during sexual play—copulating sharks were described undulating together as one shark. The fact that the sharks undulated against the boat, fitted in with the idea of affectionate feelings that were usually only expressed towards other sharks. The hull of my kayak was curved, and shaped more like another shark than I was. Did that inspire them to wriggle against it? I could never know for sure, but in the birds I was caring for, the correlation between their affection for me and their sexual whims was clear to see. There had to be some natural basis for my sharks' behaviour, and no other ever came to light.

Another surprising thing about what they were doing was that they were doing it in unison! It was very rare for them to act in unison, and each time they did so, the reason for it—the trigger for it—in that particular instance, remained obscure.

Windy, Cochita, who had mated, Simmaron, who had mated, Christobel, Gwendolyn, Pip, and Carrellina were there, with many deeper, whom I couldn't identify from above the surface. I slid the food overboard, jarring myself as little as possible, watched for a while, drew in the anchor, and drifted away. What a relief to feel back on track!

I vowed to do everything in my power to make the BBC filming as good for the sharks and their image as responsive, intelligent creatures, as possible. That included piling as much food as possible into the site in the time I had, to attract as much life as possible. I wanted the place opaque with fish and sharks, carpeted with nurse sharks. And since the nurse sharks would probably hide when the film crew was present, the scene would have to be good in all other ways. Since I had changed my feeding patterns and had used only the barrens for the past many months, I knew it would take time to recreate the amazing scene I had marvelled over at Site One for so long. I was riveted on this necessity, constantly planning and estimating. I still had no idea how long I might have to prepare.

But that week I heard from my BBC contact, Stephen. He had become doubtful that they would be able to come, due to the extreme cost, and other practical difficulties. It was particularly necessary for there to be a reason to film the sharks, a way to move from the cookie cutter sharks they planned to film, to the reef sharks in the documentary. He said that he would like to call me to discuss the entire situation.

I replied that the reef sharks had evolved with the coral reef environment, just as the cookie cutter sharks had evolved a remarkable ocean niche. The importance of the reef sharks' eye-sight, as demonstrated by the mirror experiment, was an illustration of their adaptation to an environment full of obstacles. I suggested that interesting examples of how the individuals used their habitat would emerge during the filming.

Stephen soon called, and we talked for nearly two hours, though it was well after midnight in England. He seemed fascinated because I had developed a personal relationship with the different individuals, and said he found the idea that a shark could recognize and relate to a human surprising. We seemed to be in perfect harmony over what we wanted to film, and I gained much insight on the complex nature of the project he was putting together. He had read some of the anecdotes I had written about the sharks, that had been posted on different web sites on the Internet, and we talked about the mirror experiment. I carefully described how the sharks had reacted to it, and he was intrigued by the idea that the tight circling of a shark in front of a mirror was an entirely new behaviour, one which could be filmed.

He was concerned about whether or not we could be sure of seeing sharks. I assured him without any doubt that there would be sharks. But how many sharks? Well, I thought that between twenty and thirty was a conservative estimate. I mentioned that they were shy but that I could get them used to coming to the proposed place—a clear circle of white sand deep in the coral landscape of the lagoon, where the water was just a bit less than two metres deep. Then, as long as the crew would be able to come back a few times, the sharks would be sure to get over their initial timidity, because of course, they were curious!

Stephen seemed surprised and impressed that they could expect so many sharks. He hadn't appreciated the drama of my sessions, and of the mirror experiment, from the accounts on the Internet. But he continued to stress the importance of the BBC having a large group of sharks to film there, because the other stories that they planned to film were of isolated sharks. The tentative date for the filming was April 19 and 20, but they were having difficulty budgeting for the extra cost of the trip to Polynesia, and were running short of time, given the stories they wanted to film in Australia. The decision would be made in the next few days.

If they came, he told me they would film only the story of my sharks, so I no longer had to worry about locating a motor boat, and taking them out in the ocean at night in hopes of filming a cookie cutter shark. They would be doing that story in Hawaii, where no doubt they had a crew with sophisticated equipment to take them. That greatly simplified my problems—I had no resources at all. I couldn't even breathe!

I had assumed that the crew would just film the sharks, and that a narrator would give details about each individual while the shark was on the screen. But Stephen told me that they planned to film me talking about my research. I had no intention of actually being in the film myself, so this brought a new and riveting worry. I looked like a witch who hadn't left her castle in three months. How would I ever speak to a TV camera, knowing that millions of people would be watching me close up? At least I had more than three months to heal, and get the sharks used to coming to Site One, so I felt reassured after the phone call.

I had spoken to a neighbour who made outrigger canoes, and he had agreed to rent two to us, so I had a way to get the film crew to the site. It looked like things could still fall properly into place if they came, and if they didn't, it would be an enormous relief to me. I would devote myself to completing the story of my sharks, which had been interrupted by the trip to France and the theft of my birds, and try to get to know the unusual nurse shark. Inwardly, I was convinced that dealing with a film crew was not one of the things that someone like me could ever do, and to consider it while chronically depressed and in such awful pain and breathlessness was ridiculous. More and more, I relaxed in the certainty that they wouldn't be able to come.

The weather was stormy when the weekend came, so I applied myself to healing up, and spent unprecedented lengths of time lying around reading. Arthur wrote and asked me if I would agree to be named as the co-author of his article on cognition in sharks, which he had finished and was preparing to submit for publication. The news cheered me greatly. But he also told me that Donald Griffin had died, and Arthur's deep sorrow over the loss of his friend communicated intensely.

A magnificent silver patterned tropicbird, who had been found drowning far out at sea, was brought in later that week. She was in pristine condition and surprisingly fat, but one leg had been severed close to her body. It had healed cleanly, and though a bit of bone was exposed, it was not infected or ulcerated.

I installed her at first on a bed of pillows, and since she was calm, put her in a pool to see if she was waterproof. She liked that, and rested tranquilly floating on the water.

The next day, she managed to squeeze out through a hole under the bathroom door, wriggled down the beach to the sea by pushing herself along with her one leg, and went sailing along the surface with the breeze, bathing, preening, splashing and trying to fly. I was lucky to be able to intercept her before she got so far that I would have had to use the kayak. She would be easy to lose if she managed to get out in a wind, travelling along as she did like a little sailing boat, with my inability to make any exertion. At least I had learned that she could not fly.

Since her feathers were not waterproof, and this promised to be a long-term case, I was affectionate with her, gently rubbing her chin, and the back of her head, always washing my skin first with detergent and water to remove its natural oils. She responded enthusiastically, wriggling all over, and cuddling into my face and neck. When she seemed ready, I gave her a short version of the seabird water-proofing bath, then kept her in her pool as much as possible, especially after feeding, so she could keep her face perfectly clean.

But after three days, she became restless, then agitated, and her wings hung slackly away from her body. Soon she was twitching, crying, beating her wings, racked by spasms, and trembling violently, with rapid breathing and speeding heart. I calmed her with a sedative, made her comfortable, and treated her for poisoning, which was suggested by the sudden onset of these symptoms. Since the only things she had ingested were the small live fish I had trapped, ciguatera was a possibility. This is a form of poisoning that comes from eating fish that have fed on algae growing on dead coral, and it is a serious problem for humans too.

Her condition deteriorated close to death, with intensifying seizures, as night deepened, but the next day she was still alive and calmer. I tube-fed her with a thick mixture of egg yolk, yoghurt, fish meat, electrolytes, and vitamins, adding a treatment aimed at ciguatera poisoning. But disturbing her to feed her triggered more seizures. Her only leg was thrown backward, she stretched forward, and spasms moved through her body like a series of electric shocks, accompanied by rapid beating of her wings. Simultaneously, violent vomiting racked her, and the fishy mixture I had so carefully instilled far down in her proventriculus (a seabird's crop), came up in waves which she inhaled. Each feeding was an emergency. I held her head up high, and cleaned the muck out of her trachea with one q-tip after another while spasms continued to convulse her.

In this desperate struggle for her life, I found that I could arrest the spasm reaction by stopping up her oesophagus with a finger pressed against it on the outside of her throat as I pulled out the tube, then by massaging the paste deeper in with that finger, with which I continued to block her meal from coming back up. Without pausing in these gentle, stroking movements, I smoothed down her feathers, tucked her wings close to her body, then turned her head, and hooked her beak behind her wing. Nearly every time, the spasms and vomiting stopped, her head remained elevated, and she slept.

Mid-January came. Six weeks had passed since Diamond was rescued. He had developed a scab on his breast from lying on the ground, and still walked as if he were in great pain. He seemed to have suffered permanent damage to the joints in his legs, and it was not possible to take him off the anti-inflammatory medication which enabled him to function. Sores appeared on his face and joints. Any movement around him still caused him to startle and cry out, and often he shrieked without warning or apparent reason.

But gradually, there were slight signs of improvement, indications that he wanted to continue his life. Kubilai normally circulated through the garden in front of the house, while Diamond rested behind it under the arching trees overhanging the beach. He began to spend time on the deck crowing. Crowing was always a good sign in a rooster who had been ill or badly injured. It was a declaration of power and possession of the surrounding territory. One day I found him fighting with Kubilai over the valued place in front of the house where I put their food, and from which he usually retreated when Kubilai came. Days when he was tired, standing on one leg with closed eyes, he placed himself in front of the door. If Kubilai approached, I opened it for him and he came inside. Then he crowed.

By then, Genghis' last chicks were the size of robins. They were never far from Diamond, and looking after them gave him a new purpose. The little group was often joined by Tricklet's son and his friend, who were both flowering into fully decked cockerels, one fire red and blue with black accents, and the other green and purple with gold accents.

One afternoon, I was lying flat on my back, after arranging myself with difficulty in a painless position. No sooner had I accomplished this when I heard the little mannikins giving their "hawk" warning, a high-pitched note reserved uniquely for the visits of swamp harriers. I decided that I would ignore it this time, but the roosters also began to give warning calls. When the cries of the mannikins intensified to panic mode, I rolled off the bench, landing on hands and knees, and leaped up in time to see a swamp harrier coming in low over the trees. By the time I reached the deck, it had whirled around and descended. As I raced down the stairs, it was winging up the lawn at ground level, straight in under the bushes, to where the flock was sheltering in front of the house. I rushed into its path and it dodged, and went up over the house.

Swamp harriers were a constant threat. Since they had taken Angel and the other fragile and affectionate fairy tern I had saved, I was attuned to all the words the garden birds used to announce their coming. Even the dogs eventually began to bark to let me know that one was threatening our family. The barking of the dogs was treated as an alarm call by all of the birds in the garden, including the alpha male cocks, an illustration of Arthur's concept of *interception*.

Gradually the tropicbird became more alert, and positioned her one foot beneath her instead of flinging it out behind. Her convulsions faded and stopped. So once more I took her swimming upon the sea—it was important that she drink sea water, and bathe, but I stood on the seaward side of her to make sure that she stayed in the shallows. I still couldn't expand my ribcage to do more than gasp in little breaths.

The first time, she just floated in the wavelets lapping upon the beach, but the next day she was more active, bathed, and swam a little. Later in the day, I found that her foot had puffed up with water like a little balloon. I stuck a needle in it and a tiny jet of clear water shot out—it was possible to squeeze the foot flat again, and in the next few days the last of the puffiness disappeared. No cause was ever discovered.

Using a piece of heavy plastic about three by four metres (10 by 13 ft) in dimension, I made a pool for her by laying it over a natural depression in the earth and filling it with water. A wire fence a metre high surrounded it, and kept her in and the other animals out. After such a serious and prolonged illness, being outside on water so she could begin to live, see the sea, and feel the wind and sunshine, seemed essential for her well-being.

I took her in the sea each day when it was calm, so that she could drink and gain a sense of greater freedom. She stayed close to me as I drifted in the water beside her, one hand on the shallow bottom so that I didn't have to use my muscles to stay afloat. She often touched my face with her beak. She bathed, waved her wings, and preened, flipping easily due to a tendency to overbalance, with just one leg.

While she was ill, I had tried to reassure and comfort her during her shocking convulsions, holding and soothing her as much as possible. As a result, a strong bond had formed. She was extremely affectionate with me, cuddling vigorously against my hand when I caressed her chin, face, and throat, and with her whole body when I held her, in an expressive display of love. She was always eager to be held. She was strong, and her feathers were pristine, water droplets flying in rainbow showers from dry feathers when she played upon the sea.

On January 24, the weather calmed. Gazing across the hazy sea and trying to sense the atmosphere, I decided to try to get out to the sharks. If the trade winds increased, as they had done erratically since my last trip, I would let them blow me homeward, and not fight them. Only the south wind could blow me out to sea.

I was increasingly worried because I had assumed that as my broken ribs healed, the pain would subside, and I would be able to breathe normally. But there was no change. It was as impossible to expand my ribcage to breathe as it had been right after the fall off the deck, and any movement resulted in sharp pains in several areas. It seemed like a terrible risk to entrust myself to the sea, but the time was passing, and I went.

Once again, concentrating just on pulling upon the sea enough to glide onward, I slowly approached the lagoon, then crossed it to Site One. The current was extremely bad there, and the sea surged through the lagoon in long, smooth waves with alarming power. I didn't dare to anchor—even if I had my strength, it would have been impossible to swim in such a torrent. The weight of me and the boat would drag the anchor till it caught on something, and it would be impossible to pull it up. I didn't dare to risk a situation in which I would have to go underwater. I was already frightened to be in such pain and weakness in a surge of overpowering waves and current.

A multitude of riffles surrounded me, as my sharks swirled around the boat, undulated against it, and descended. After each caress, the shark's tail appeared briefly above the surface around me, and I tried to avoid them as I dipped the paddle to keep the boat aligned.

As I positioned myself to throw in the food, one shark slammed the boat hard. Punishment for being late! I glimpsed Carrellina at the surface. A high percentage of the resident female sharks remaining were her gang of bad sharks! But no one else slammed the boat, and I managed to get the food into the site by repeatedly paddling to the reef side of it, and tossing food as the current swept me back across it. In the soft afternoon, with huge swells crossing the lagoon, lifting and lowering me as the sharks cavorted and splashed, my worries dissolved away like crystal salt in water. I drifted homeward with the breeze.

By then, the tropicbird was spending her days on her pool. One day I went out to check her, and found her resting quietly on the grass at its verge. Moments later, as I passed on down the garden, shrieks of alarm filled the air. Every bird in the garden was warning of danger. When I rushed back, the pool was empty. I gazed for some time at this incomprehensible scene, turning slowly, looking around, and out across the water, before I saw a crumpled white heap, with skewed wings, lying in front of the house.

It appeared that a swamp harrier had descended into the enclosure just after I had passed it. He had lifted the tropicbird, but had apparently had difficulty clearing the wire fence with such a burden, and had dropped her several metres farther on. She began to scream as I scooped her up and carried her into the house. Her expressions of terror rivalled those I had heard Diamond make, an intense torrent of emotion. For many minutes she screamed, in spite of my efforts to reassure her. But amazingly she was unhurt, except for a few ruffled feathers where the hawk had clutched her. Seabirds have thick plumage.

I kept her in the house beside me for the next days, since she was so frightened by her brush with death that she protested if I tried to take her outside, and was only calm when with me, resting on a cushion beside my computer, where I tried to fortify her with reassurance and solace.

Two other sad cases who were also hospitalized at the time, including another hawk victim, cuddled up to the tropicbird, and she responded, caressing them and fervently touching them with her beak. The birds

drowsed beside me cuddled together with their beaks crossing.

Do birds have feelings? It was impossible to watch, and deny it. If birds, why not fish? Why not sharks?

One afternoon, I was hurrying home in our pick-up truck with some supplies, when an enveloping dark shape appeared just above eye-level. A swamp harrier! I had just enough time to register it, surrounded by a cloud of white birds, as the great raptor reacted right at the windshield by releasing a white flutter, which vanished as I shot past. I stopped and reversed along the roadside until I found the fairy tern in the ditch, while the big hawk circled above the tree-tops overhead. As I slowly approached, the fairy tern flitted up, and, gaining in conviction, it mounted skyward to join its flock, wheeling above.

The swamp harrier must have been so low because it had just snatched the fairy tern on a steep dive, and the fortuitous appearance of my truck just at the right moment, caused it to release its prey as it made a desperate evasive move. The harrier circled for a long time before leaving the area. It seemed to be one of those coincidences that turns the tides. After so many losses, unexpectedly I had won a lottery with death and saved a fairy tern.

The next time I took the tropicbird out, she stayed very close while she floated on the waves. Soon, she paddled to the beach and grounded there, wriggling a little to get out of the water. I took her inside.

The following morning, we went out to the water again. She floated quietly while I rested, enjoying her silken, silver beauty in the golden light of the newly risen sun, the shining bay, and that blue abyss of sky. Then, quite without warning, the mysterious bird of the trackless ocean raised her wings, and took a tentative flight, just a few metres. She floated momentarily, then rose up from the still surface, her graceful wings lifting her easily. Soaring higher, she flew down the bay toward the ocean, fading quickly as she went, to vanish into the blue air above the sea.

I went again to see the sharks when fish scraps became available. As usual at that time of year, there were never very many, and sometimes weeks passed when none were to be had at all. Either the fishermen were on holiday, the sea was too hot, the boats were in dry dock, or some other type of fish were eating all the tuna. The truth of these stories could never be confirmed, and in the uncertainty of the future availability of shark food, I tried never to waste any. The weather had been unsettled with storms passing, so I was lucky to have a day in which once more, the atmosphere seemed to pause just long enough for me to travel, pulling softly on the sea, out to the site. There was a slight wind this time, but I managed to go without increasing my breathing rate.

I travelled so slowly that Cochita and Windy swam beside me as I crossed the lagoon, sometimes moving away, then returning. Unwilling again to anchor, I was trying to manoeuvre the boat in order to throw the food into the site, when I saw the beginning of their rush upward from below to caress the kayak. As they all began to undulate against it, Brandy wriggled against the paddle, and then together the sharks' tails flicked above the surface all around me. Droplets flew lightly as they went straight back down. Cochita spent some time at the surface, drifting slowly underneath the boat, while I stroked her.

It was extraordinary—another example of sharks all doing the same thing at the same time, apparently spontaneously. And many of these were the same sharks who had slammed my boat in shark-anger a year and a half before. They hadn't followed Carrellina's lead this time. They had feelings and their feelings had changed. I pushed their food in and watched for awhile, then drifted home before the sun dipped below the horizon. I was sad to leave them, but at least they knew that I had come and given them something.

A question in cognition is whether an animal knows that something continues to exist when he or she can no longer see it. An object apparently ceases to exist for dogs, for example, when it goes out of sight. So few people would agree that sharks could understand that I was in the boat, even when I had just left their company and climbed into it—Arthur refused to believe that it was possible. Yet they were aware.

Could they see me through the surface? It often appeared that they could, and when they raised their heads from the water, they raised them straight towards my face as if they could see it from beneath. Once their faces were in the air, they probably could see me in the boat, there above them in the mysterious volume above the surface. Great white sharks are known to look above the surface.

Their electro-sense works at close range, and possibly continued to inform them that my living body was just beyond the plastic hull when I vanished, too.

Then there was the likelihood that they could hear my movements in the hollow craft with their lateral line sense and sense of hearing. This way of perceiving the environment seemed to me to be dominant in sharks. If the sight of me underwater was replaced by the sound of my movements in the hollow plastic kayak as I got in, these perceptions could well continue to inform them that I was still present, but their view of me was blocked, just as it was blocked whenever they listened to me from beyond visual range.

Indeed, the many ways that sharks took advantage of the opportunity to hide behind the veiling light, and to approach when they were not visible, such as when a person's face was above the surface, strongly suggests that they are quite comfortable with the idea that something continues to exist, in spite of being out of sight.

One of my pet mynah bird's favourite games had been searching for toys that I had hidden for his amusement, especially inside containers, such as envelopes or match-boxes, so that he could enjoy opening them with his beak. He had no trouble with the idea that his treasures continued to exist though he could not see them. If birds, why not sharks?

I did not hear from the BBC, and by the beginning of February I was confident that they had decided not to come. As a general rule, it seemed to me, if things seemed to be too hard to accomplish at first, it was more likely that the more the matter was investigated, the more it would seem not feasible, rather than the contrary. If the decision was to be made soon after my talk on the phone with Stephen, it had been made long ago, and since I had not been contacted, it seemed obvious that the BBC had decided not to come.

Confident in my new assumption, I relaxed considerably, and went back to work on this story of my sharks. But one day, I was surprised by a message from Stephen saying that the BBC film crew would be coming during the last week in April. The documentary had been named *Jaws Junior*. Jaws? I dismissed it as TV hype, to draw in viewers prior to telling them what sharks were really like, but it bothered me, presenting an opposing image of sharks to the one we planned to portray.

Confronted again with the reality that I would have to take a film crew to the sharks, I began to tremble. Soon I felt so ill that I went to see my doctor, and poured out my story. She was incredulous that I had waited so long to get medical help due to frantic worries about sharks, treated me for a stomach ulcer, concluded that one or more ribs might be displaced, and sent me to an osteopath. The osteopath determined that the place on which I had fallen had caved in slightly, causing three vertebrae and the associated ribs to move out of place. As a result, my rib-cage was blocked and could not expand. As she slipped the bones back into their proper position, suddenly, after six weeks of gasping for oxygen, I could breathe!

I hitch-hiked home that day since Franck was using the pick-up, and while waiting for a ride I noticed a nearly naked bird standing in the sun, obviously at the point of death. She appeared to be a large domestic laying hen, but since someone had pulled out all her feathers behind her shoulders, it was hard to tell. As she had begun to die, her body had cooled, and she had moved into the sun; otherwise, it was unlikely that I would have seen her. Here was another despicable example of chickens and junglefowl being the most abused animals in Polynesia.

I scooped her up and she turned a face full of character to me. Murmuring soothingly to her, I sheltered her in my arms, trying to protect her from the sun and wind, as I continued to try to get a ride. Soon a man stopped for me, and as I thankfully got in, he asked about the conversation piece in my arms. When I told him about her, he was glad to turn off the air conditioning, which was blasting onto us—the bird was shaking with cold. We talked about the tragedy of animals in Polynesia all the way to my house, where I thanked him and rushed inside with her.

I tubed juice, water, yoghurt, and a rehydration liquid into her as I passed through the kitchen, grabbed a cushion and a towel as I went down the hall, and settled her down on the bed I made up for her in the quietest, and darkest room in the house. The bird looked at me, looked around, and softly purred, just like a cat. Then she sank on her breast and slept. Later I fed her more, and treated her for the dreadful burns she had received from standing in the sun. She remained semi-conscious for the first week, and I fed her by tube in her darkened corner.

When she began to take an interest in her surroundings, I carried her outside so she could see the garden, and meet the others. After a few days, I put her down near me when I was working outside, but Diamond was the only bird who didn't chase her.

She had become very attached to me and affectionate, feeling safest when she was in my arms. One evening, I was gathering flowers with her beside me, the other birds foraging peacefully around us, when one of the three cockerels began strutting around her, and seemed to be thinking of chasing her. But before he made his move, Diamond came over and walked beside him, tugging on grass stems and chittering softly. The youngster quieted and wandered away.

Diamond had warned him to leave her alone. And that was the beginning of her acceptance into our garden flock.

CHAPTER FIFTY-TWO

Preparations

The day I could breathe again, I went straight out to the sharks. Everything was getting better. For the first time since the rainy season had begun, the weather had settled, the trades were rising as the sun rose, and dying as it set as they should, and the sky was a deep, flawless blue, overlaid with reddening gold as the sun plunged seaward. Even the lagoon lay tranquil, its currents imperceptible.

I threw in the bit of food I had saved for the occasion, and slid underwater to find Christobel at my side and the site alive with sharks. Two of the big nurse sharks were circling. Carrellina was there with the residents, feeding, swirling, roaming around in search mode, and coming to see me. Sparkle had lost her hook! She had only a long, black line along her jaw where it had been. But it had taken six months for her to be rid of it! Fleet too, was hook free. I drifted among them for a while, but soon felt overwhelmed with weakness and the pain that still flamed in my side with the slightest movement. But having spent some time underwater with my sharks, I felt greatly reassured.

However, that night I was awakened by a searing pain between my shoulder-blades, at the place where my spine had been injured. It was impossible to get comfortable. The longer I lay there, the farther from that central focus the pain radiated out, and nothing relieved it. Finally I got up, and with my back vertical, it began to ease. But after only four hours of sleep, I went through the day in a daze. From then on, whenever I tried to rest, the weight of my body on my ribs or back caused the same unbearable pain to soon arise, and I moved through each day drowning in exhaustion.

Each day there were e-mail exchanges as the plans for the filming began to materialize. I tried to ease the way by helping with the practical arrangements involved in the transport and care of eight people and their twenty cases of equipment, totalling four hundred kilograms (882 lbs) Determining which would be the best hotel to stay in when the film crew arrived in Tahiti, how to transport the equipment from the airport to the hotel, the size of the planes flying from Tahiti to our island, whether they could carry that much weight, what sort of vehicle could be rented to transport it, where could it be rented, which ferry would be best to use, whether the times listed on the Internet for the various planes and ferries were accurate, and countless other details needed to be investigated.

The problems of coordinating such an endeavour were impressive, and endless phone calls and worries boggled my exhausted mind. But the mental picture that dominated was how exactly, to get the eight people and all of their things to the sharks. How was I going to get them there?

There was uncertainty as to whether the crew should stay at a hotel, or on a live-aboard boat. Renting a large motorboat to take the crew to the site was the obvious solution. But this was easier said than done in Polynesia, where outings were controlled by tour companies. They rented tour boats, with the captain included, for large amounts of money, and the outings took place on their schedules. Sometimes, the boat and captain just didn't come. I became very uneasy about entrusting my sharks' story to the reliability of a local captain. Further, such a boat would only be able to get us to the border. They would still need outriggers to cross the lagoon to the site. It was just not possible for a motorboat to travel through the lagoon, due to the coral formations that reached to the surface. I had worried and puzzled about the problem for some time before deciding on the solution of outrigger canoes.

There was a *pension* not far from the big luxury hotel in the centre of the lagoon, the one opposite Site

Three. If they could stay there, all my problems would be solved. It even had extra outriggers for the use of the guests. We would be free of any need to coordinate with outsiders, could go to see the sharks whenever we wanted to, could anchor and leave the outriggers while swimming and filming, then get back in and move on, or trail them as we wished. The equipment would be with us, and we could go back, a trip of just about ten minutes, to the *pension* at will. An added bonus was that we would be free of the worry of paddling against the wind, which would be with us on the way to the site. On the way back, if the wind was strong, we could minimize it by going along the shore.

If the crew stayed elsewhere, we would have to find someone living on the shore across from Site One, who would give us permission to keep the outriggers there. This would complicate things, since there was the danger that they could be stolen, we would not have the same level of privacy, and we would have to go there by vehicle. All the gear would have to be packed into the vehicle, then packed again into the outriggers, which would make twice as much work. I was particularly worried about being seen, and then accompanied out there by curious Polynesians who would ruin the filming by frightening the sharks.

The dimensions of the outriggers I had arranged to rent had come into question because some of the cameras were large, and the producer was concerned that they wouldn't fit in an outrigger canoe. I went back to the man who rented them to measure one, but he was away for two weeks. A friend of Franck's had a zodiac, and I asked him if it would be possible to make it available for some people who were coming to make a film, if they needed another boat for their equipment. He readily agreed, so there was another problem solved.

It was also important that the outriggers have shades to protect the cameras from the sun. There, I was stumped. But when the man who was renting them to us came back, and we discussed the proposed venture, he saw no problem with using a tarpaulin! We looked at the outriggers together and they were four metres (13 ft) long and sixty centimetres (2 ft) wide and deep. They seemed to me to hold quite a volume, but I still didn't know the dimensions of the equipment that the cameramen needed to put in them.

I went to see the *pension*, and found that there were two large bungalows available for the duration of the BBC's visit. It was in a peaceful and private setting and very beautiful, surrounded by well-managed gardens full of flowers. A white sand beach lined the shore, from which it would be easy to launch the outriggers. The bungalows were Polynesian in style, with hardwood floors, a heavy wooden table in the centre, and traditional cotton prints on the beds and in the doorways. They were clean and attractive, and the beds were protected with mosquito netting. Each had a kitchenette, with a refrigerator and stove. The lady told me that it would be fine for us to keep extra outriggers there for the days in question, and that there was a small safe for the use of the guests.

There was little extra space so I suggested to Sue, the BBC coordinator, that they could keep some of their things at our house if necessary. I told her not to worry, since my one interest while they were there would be making sure that they were all fine, and had everything they needed.

If everything went well for them, it would be more likely to go well for the sharks and me.

Sue wrote back right away to say that the *pension* and the outriggers sounded just right, so I returned to make the reservation.

While I waited for the manager, her husband, who had been working in the garden, wandered over and began to talk to me. I chatted with him, and was surprised when he brought up the subject of sharks in the lagoon. He informed me, to my astonishment, that there had been no sharks in the lagoon until the new hotel was built and began holding shark feedings out near the reef. He told me that he fished at night, which was illegal since fish are sleeping at night and are vulnerable, and that the sharks were dangerous!

Most of the Polynesians I had talked to about sharks admired them—some had assured me that I was protected by the shark spirit, even when I was not with them, because I admired and loved them too. Even spear fishermen who dove to twenty metres (65 ft) free diving, and were followed by sharks at that depth, were unafraid of them and surprised I even asked.

I studied this man and finally pointed out that female sharks used the lagoon, and always had, to keep their babies away from the dangers of the ocean. Did he not know, I asked, that long ago there were many more sharks, and that they came right up to the beach when people cleaned fish, and were given the scraps? I had been told that the children had played with them in the shallows. This was offered cautiously as information I had learned from chatting with my neighbours.

If sharks were dangerous, I asked him, would not shark feeding dives for tourists be impossible? If sharks came around as he fished, didn't the shark just want the fish? What did that have to do with tourists?

It was troubling because he seemed so angry. I was sure he was repeating things that others were saying. This attitude seemed to be originating and spreading among fishermen, in response to the outrage against shark finning, which was bringing them extra money. Before sharks had become the subject of controversy due to the Singapore company's finning operations, shark dives for tourists had been accepted as part of the tourist business. On the island, most jobs were related to tourism, so looking after tourists, speaking other languages, and involvement in their activities—diving and shark dives had always been prominent—was an accepted part of life and the economy. I was most relieved when his wife appeared, and with all the noble grace of the Polynesian style, accepted the reservation of the film crew from the BBC.

Concerned by the misinformation and vehemence toward sharks expressed at the *pension*, I mentioned the incident to Stephen, and suggested that they substitute the word 'fish' for 'sharks' when talking about their project locally. Filming small fish, as opposed to small sharks, was just the sort of silly thing that tourists would do, and would result in a complete lack of interest in our project by the local people.

Back at home, I sent an e-mail to the governmental departments to ask again when sharks would be protected from finning. There was no response. I was still trying to find out whether the law protecting the sharks had been passed, but no one knew anything about it. The Ministry of Fisheries said I should call the Ministry of Justice. The Ministry of Justice passed me on to the Department of the Sea. No one admitted knowing anything about the matter, nor where to find out. But in general, people seemed to believe that the law had been passed; it was just the details that they lacked. Even my media contacts didn't know.

One day I called around to find out where I could get some shark fins for my restaurant. While most governmental agents were helpful, trying to guide me to the proper place to get shark fins, one man became angry, and said that I certainly wouldn't be able to get shark fins much longer because sharks were going to be protected! Going to be. So why had the law not yet been passed? It was still hard, very hard at times, to resist inexorably slipping into immobilizing depression.

So when Franck came home one day with the mail, and handed me an envelope from the journal *Marine Biology*, I tried to avoid opening it by hiding it in a drawer. I couldn't face any more bad news. But Franck, gently admonishing me, went and got it, and waved it back and forth in front of my eyes.

"I think you should open it," he said. When I continued to refuse, he opened it himself and, reading and smiling, he told me that my paper on the gestation period of my sharks had been accepted for publication!

One of the reviewers, bless his heart, had written that as important as the discovery of the gestation period of the species, was the fact that I had found a way to study sharks without killing them, which was of significance given the threatened status of many species. He called my method a new and innovative one, which went a long way toward cheering my spirits. The other reviewer wrote that my English required "considerable attention," but also stated that my paper was suitable for publication.

When I wrote to tell Arthur that he had succeeded in pushing me to publish something, he wrote back with hearty congratulations, saying that *Marine Biology* was a fine journal. (He was a reviewer for not only *Marine Biology*, but for several other journals that published marine research.) He told me that he had finished the manuscript on cognition in sharks, and had named me as the second author! He had attached it to the e-mail for my approval. Arthur finished:

Thank you for all your aid over so long a period. I'm sure that someday, we'll meet and I'll be able to personally thank you for your all of your kindness, your thoughts, your insights, and your innumerable observations. And regardless if our manuscript is accepted for publication or not, I salute you as one, authen-

tic shark-ethologist—there's only a couple!

It was an amazing compliment, and helped me find the strength to go on.

The storms of the rainy season continued to howl across the sea, but I told myself to be patient; there were still two months for the weather to calm down enough for the filming, and for me to devote myself to spending quality time with the sharks.

In the meantime, I put myself on a self-improvement program, and swam as far and fast as I could along the drop-off at the edge of the fringe reef from our beach each evening. Though I was feeling better due to the medication I had been given and having regained the ability to breathe, I was still in very poor shape, and a lot of pain. I dreaded failing to play my role well in the up-coming filming, which would consist of two or three visits to the site daily, as well as a night-time session.

But my strict exercise regime came under threat when I discovered the home of a whitetip shark a few hundred metres along the shore. In the dusk, the animal was usually out hunting the drop-off, and instead of exercising, I followed him around. The problem was that the shark often descended too deep for me to follow, so I had to return to the surface and wait for him to come back up. It provided a welcome distraction from my neurotic worries and a much needed incentive to go on such purposeless excursions. At least I was swimming!

I also worked on getting the birds as independent as possible, so that they began to go into their respective shelters to sleep on their own. Thus I was freed to go to see the sharks, or swim at dusk without worrying about them. I just checked them and closed their doors when I came home. Those who had been too frightened to sleep in their house because they had been stolen from it, were simply permitted to sleep in the trees where they were safe. I burglar-proofed their building, where Diamond still slept, so that in the event of a break-in, a real effort would have to be made to actually take the building apart, and that would awaken the dogs. Kubilai and his flock had their own little building which was close to the house, and was always locked at night.

I had raised Diamond by hand since just after he hatched, due to the chicks having had fowl pox, and had always touched him with exaggerated affection. Yet he still went into a state of panic when he saw that I was going to pick him up. At the last minute he stilled, as if forcing himself back under control as I lifted him into my arms. Whatever his thieves and their families had done to him must have been traumatic in the extreme, that he still reacted with such irrational terror. I was determined to protect him and all of my birds from anything like it happening again.

He still walked painfully, and was on an anti-inflammatory drug for the swelling of his joints. But Dr. Jaime Samour felt that it was unlikely that his condition would improve any more. Still, I hoped that as his nightmare slipped into the past, his general condition and vitality would slowly improve. His other symptoms had faded. The skin on his face looked rosy and healthy, and he had grown a radiant set of brand new feathers.

He refused all affection from me, though prior to my desertion of him and the resulting nightmare, he had come often to sit near me to be stroked when I sat in my favourite place looking out across the ocean, which was in the heart of his territory. On the other hand, the three cockerels who had become so attached to me that they trailed after me wherever I went, lost no opportunity for a cuddle. One afternoon I was sitting looking out to sea on the deck, caressing one who had climbed onto my lap, when Diamond suddenly jumped up on the bench right in front of us, and seemed moved to jealousy when he unwittingly intruded upon our intimate moment. When the younger bird went away, he came close, and for the very first time since his rescue, he let me put my arms around him and affectionately hold him. Birds have feelings!

During the long, dark, stormy days, Diamond often came inside to rest on a cushion in the bird corner. The domestic hen I had rescued, and named Josephine, was resting there too. He ignored her, lying solemnly on one cushion while she lay on the other, as the rain beat against the house, the wind howled, and I worked. One storm lasted for three days. And once, when I looked up, Josephine was gently grooming Diamond's mantle, while he stood, head bowed, eyes closed. Each seemed to recognize that the other was suffering.

When the weather cleared and he went outside, she went with him, and the two were usually within a few centimetres of each other. Diamond refused to go to his house until I carried her there with him, and made up a bed for her. She was much bigger than Diamond, had gained weight, and grown soft fluffy feathers. Her face was smoothed out and much younger looking than she had been when I had found her. What an unexpected pleasure it was to see him in love again!

Josephine soon began to lay eggs, and I let her put them in her nest in the corner, deep in the house, where she had come back to life. She actually found and placed one of Diamond's decorative feathers in her nest. Not wanting to violate her trust, I did not touch her nest, and as the pile of eggs grew, we decided that after what Diamond and Josephine had suffered, we should let her hatch them. However, nearly twenty of them had accumulated before she became broody, and then, as if irritated with the enormous pile, she kicked most of them off the nest before arranging herself upon a sensible number. But her eggs did not develop. She was so much bigger than Diamond that they may not have been fertilized. She accidentally broke them one by one, and finally those that were left, came apart in layers. So I bathed the gooey mess off her frilly feathers, and put her back outside.

On February 26, 2004, there was another article in *La Dépêche* about the massacre of sharks in Rangiroa, the most popular island for diving in the Tuamotu archipelago. The newspaper reported that two fishermen had arrived at the beach in their outboard and dumped the carcasses of two large lemon sharks in the shallow waters, where they proceeded to fin them and cut out their jaws in full view of the tourists and divers going back to their *pensions* in the late afternoon. As the turquoise waters turned to the colour of blood under their machetes, they attracted a group of passers by.

They explained that they had caught the sharks while fishing at the pass—a shark had eaten the bait and the fish they had caught.

"Two sharks attacked the same fish?" asked an incredulous tourist.

"Stop asking dumb questions—you are really stupid," the fisherman replied as he worked on tearing the shark's jaws out of its mouth.

Others gathered, and the butchery continued beneath the eyes and cameras of tourists and divers. The fishermen became increasingly irritated by their presence as the sunset turned the skies as red as the sea in which they were hacking at the sharks.

One tourist who came upon the scene happened to be involved in a marine preserve devoted to shark protection and was well aware of the threat posed to sharks by the shark finning trade. However, she made no comment, but stood watching closely, her arms crossed.

"Get out," one of the fishermen snapped at her.

When she did not move, he leaped at her and grabbed her by the throat while waving his machete, and thus he roughly forced her backward, yelling that he would hit her if she did not back off.

Another person present used his cell phone to call the *gendarmes,* and half an hour later, a woman *gendarme* arrived. The tourist explained what had happened in excellent French, according to witnesses.

The *gendarme* called to the fisherman, who was by then leaving in his boat. The two discussed the matter in the Polynesian language in front of the victim, who couldn't understand a word, and then the *gendarme* turned to the tourist and accused her of provoking the attack.

"You stayed there when he told you to leave," the *gendarme* explained.

"But this is a public place," the victim said.

The *gendarme* responded, "Maybe, but here, you are in Polynesia, and that is how the Polynesians are. They don't like to feel provoked, and in staying there, you provoked him. You don't have the right."

The woman who had been choked began to cry, faced with this failure of legal protection from assault.

The article went on to describe how the fishermen left the mutilated, finned bodies of the sharks lying there in front of the marina, where they could attract tiger sharks inshore. It concluded by observing that the entire scene indicated how common-place shark finning was, even though divers came from all over the world to dive with sharks in Rangiroa, and the island's income was dependant on its sharks staying alive.

The story was a clear illustration of the racial bias against sharks and tourists in the country, to the point that the law would take the side of the Polynesians, no matter what they did. The same excuse had also been used to excuse rape by the Polynesians, who believed themselves above the laws that applied to other races. With the popularity of the Independentists, racism had become official.

I had been increasingly afraid of being accosted by fishermen on my trips to see the sharks due to the rising antagonism against white people, and the awareness of the local people that I was involved with sharks. Whenever I passed fishermen out on the water I was smiling and friendly if it was impossible to avoid them, and anyone who brought up the subject of shark feeding when I picked up the fish scraps was told that I used them to study, photograph and paint the fish. I even gave a painting of coral fish to the partners at the fish store. Still, I was regarded with suspicion, and the incidents with cock fighters and thieves had not improved my reputation.

So I contacted my friends overseas with the news, asking them to mobilize their networks to write again to the government about the extreme need to protect the sharks. The key points were that the coral reef had evolved in harmony with sharks, and depended upon them for its health, that the country's sharks had continued to be massacred since the law to protect sharks had been promised the previous September, that the foreign market for shark fins had nothing to do with the Polynesians, who had never traditionally fished sharks, and that Polynesia should be declared a sanctuary for sharks as Palau had been.

Franck called the Minister of the Environment and got through to Bruno Sandros. He raised a scandal over the issue. With his eloquent French, he presented a polished argument that the minister just couldn't respond to. Minister Sandros could not explain why he had said that sharks would be protected the previous September!

Then Franck wrote another open letter to the French government as well as to his local contacts, and his many international connections about the shark finning crisis in French Polynesia. The French government kept an eye on the political power plays of the Territory, particularly those likely to cause its income to fall.

The article in *La Dépêche* caused such a scandal that enormous pressure was placed again on the government not only by dive clubs, but also by the tourist industry. Tourism was the country's only source of income apart from hand-outs from France, and a high percentage of the income from tourism was generated by the incomparable dives in which several species of sharks could be seen. These were spoiled for tourists when, instead of being encircled by majestic and mysterious sharks as they descended into the blue, they found instead, their bodies lying finned on the coral floor. The rage of the dive clubs reached such a peak that I was told by a government official that the law protecting sharks had been passed in the Assembly, that very day!

By then I had become worried too, about my difficulties renewing my permission to stay in the country. I had applied as usual, well in advance, and permission should have been quickly and automatically granted, since I was married to a Frenchman and my circumstances had not changed. However, though by then I had lived in the country for ten years, the response to my request was not a new card, but more and more questions to answer. If permission was not granted, I would have to leave the country three days before the arrival of the BBC! I was frightened that I had been labelled a trouble maker due to my pro-shark activities, and that the government officials had decided to rid themselves of me.

Sue and I continued to exchange about the arrangements for getting the film crew and their four hundred kilos (882 lbs) properly set up. The latest news was that only seven people would be coming, making housing at the *pension* a little less strained. She had every detail visualized, including the arrangement of all the rooms in each bungalow, the beds and their location, and what was on the mezzanine. This bright woman thought of everything—the BBC was lucky to have her!

Her next concern was security. This, I had not thought of, so I went back to the *pension* to affirm that the bungalows could be properly locked up.

But there were no locks on the doors! I tried to convince the manager that the equipment of the visiting film crew really did need to be locked up, while she kept assuring me that it did not. She said that the crew could put some of their equipment in their house. I pointed out that that would not be convenient—the crew would need to access their equipment all the time.

I asked if I could speak to the owner, but he was not there, so I returned the following day. He was a kindly Frenchman named Denis, who assured me that there was no danger of theft. We pursued another long and rambling discussion about the matter and the necessity of being able to lock up the equipment in the bungalows. He repeated that there was a safe for valuables, and that some of the equipment could be put in their house. Someone was always there. After pointing out why these solutions were not appropriate for a film crew of this level of professionalism, I explained in different ways that the insurance for this expensive gear, including television cameras of different sorts, would not apply if it was not properly locked up.

Finally, I asked directly that the two bungalows have locks put on them, something which would no doubt be appreciated by subsequent guests. Locks were not expensive, and the shark-hating gardener would have plenty of time to get it done before the arrival of the BBC.

So finally Denis agreed that he would see that the two bungalows had locks installed. He assured me too, that he would reserve the outriggers at the *pension* for the use of the film crew during their stay, making a total of four available, including the ones I had arranged to rent. Further, he told me that there was a motorboat for the use of the clients that carried five people plus the driver. We wandered down to the beach so he could show it to me and indeed, it was a large and good quality boat. That might well be useful.

Then I went looking for a source of oxygen that could be made available to the crew, and investigated the problem of getting their gear from the airport to the hotel. The question of vehicle rentals continued to be a problem, which was just typical of Polynesia—now there would be all that they needed, and now there would be nothing available. It was as if they were playing games with tourists. Finally, I suggested to Sue that she go to the Internet sites of the two big hotels closest to the airport and describe the situation to them, emphasizing their status as a film crew from the BBC. With the harsh competition for tourists' money, I was sure that they would find at least one of the hotels willing to pick them up in a suitable vehicle at night and take them to the ferry on Sunday morning. Better deals often seemed to be available over the Internet, and in that way, they could be in contact with someone much higher placed than I would be able to telephone locally.

Sue told me that they might need to hire my friend's boat, so I contacted him about that again. The weight of the gear had grown to eight hundred kilograms (1764 lbs), so using the little planes that flew between the islands was no longer feasible. Sue asked me to book space for them on the most accommodating ferry, and to warn the staff about all the gear they would be taking. The staff at the ferry assured me that there would be no problem, and that no reservation was necessary.

She offered to send money so that I could get everything paid for but I reassured her on that question too, since it had been possible to arrange that all would be paid for when they actually arrived. The only thing left was to find a way to get them from the ferry to the *pension,* so I offered to come to meet them with our truck, to help to transport their equipment.

And the rain continued water-falling from the sky.

CHAPTER FIFTY-THREE

Food for the Filming

Ten days passed before the weather calmed enough to allow me to reach Site One. The sea swept so powerfully through the lagoon that after anchoring, the kayak aligned with the current instead of the wind. It trailed restlessly back and forth in the centre of the site, instead of on the north edge behind my coral lookout. I had wanted to try to photograph the sharks caressing the boat at the surface, but the waters were too ruffled, and only Brandy came around as I prepared.

I intended to return to my original routine in which I met the sharks underwater, and then put in their food. But having fed them from the surface for so long, I felt uncomfortable about getting in before feeding them. Apart from the previous brief time, I hadn't appeared underwater for two months, and I had abandoned them completely for two months prior to that. They would be expecting the food, not me, and the problem of how I was going to return to the former routine was one I had so far failed to solve. The last place I wanted to get in was the centre of the site. So I sat on the boat, frozen with uncertainty, watching. Finally, without thinking about it, I just slid in.

When the bubbles accompanying my descent cleared I found Hurricane from Section Three passing before my eyes. The site was full of visiting female sharks and Dapples, their herald, was to my left. The residents were with them, and all were soaring majestically around as if they had been waiting for me.

So I took the kayak to my lookout, climbed on, and pushed the food out of the back of the boat as I had done for years there, at my original sessions. I even fell off the edge of my perch once, when I lost my balance in the current, but had no qualms about it at all, so familiar and serene did everything appear.

It was a beautiful session, and I was able to remain underwater with the sharks for over an hour. The only concern was that I had not seen Trillium, whom I always looked for to make sure she was still all right.

Stephen phoned me again to discuss the plans for the filming and the practical problems of getting the crew to the sharks. He suggested I go out and buy "a nice new mirror" for the occasion, for which the BBC would reimburse me, and reminded me that I had promised to feed the sharks in just one preferred site, so that a large group would be present for the filming. He asked if I could do that in preparation for their arrival on April 24.

I replied that I had been feeding them as agreed, and was sure that we would have lots of sharks for the filming. My concern was that they may be shy at first on finding a group of strange people with unfamiliar equipment, when they were used to only me. We would have to be patient, remain quiet in the water, and give them time to look over the new situation. For this reason, I told him it was vital for the sharks to meet the film crew at the evening feeding session we had planned. Could we reserve that first Sunday evening for that purpose? In a group, in the course of their usual familiar routine, they would be far more accepting of the presence of the film crew. Then, when we held the mirror experiment, which Stephen was anxious to get on film, the sharks would be less concerned about their new visitors, because they had already met them.

Stephen agreed to this, which lifted a great weight from my mind. I knew my sharks, and if a crew of seven dropped into the site with the mirror, without them having a chance to look over the newcomers first, we would be lucky to see a shark, never mind get a spontaneous mirror reaction. I described the site to him, the way the scent from the food flowed toward the border, and how the sharks used swim-ways through the coral, which kept them organized.

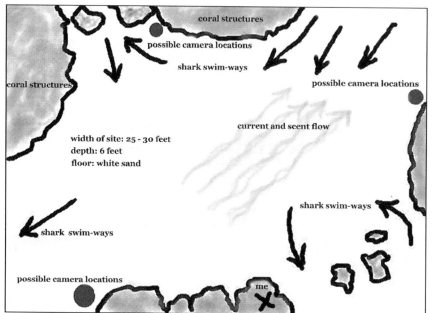

I told him I would send him a diagram of it, to help plan the filming in advance, and allow the cameramen to see the layout before they got there. Unfortunately I couldn't find the diagram software I thought I had, and so with trembling hands, I made a very rough diagram with paint software (reproduced here). But it showed the main features of the site. Knowing the swim-ways of the sharks, the crew could plan where to position themselves in advance, to get close-ups of them passing, while avoiding the scent flow.

A tropical depression circled through the Pacific during the first week of March. On Saturday night I watched and waited, and as the sun sank, the wind died down enough to allow me to get to the lagoon. But I would never have gone had I not been so worried about keeping up my feeding schedule for the BBC.

Windy followed the kayak across the lagoon, and passed slowly back and forth beneath it with Cochita. I stroked them from time to time as I prepared. Underwater, the current was almost overwhelming. The usual residents and the same group of visitors, including Hurricane and more of her companions, were there again. Jessica appeared for the first time in months. I was delighted. I had put her on my list of lost sharks. But there was an odd roughness of her skin on her head between her first dorsal fin and her nose, as if an extensive rash of eczema was about to break out there. It was nearly invisible, but the strong light coming from the side highlighted it sometimes. She had mated about two weeks before. Vixen was there, too, and had mated about the same time.

Later, even Twilight came, from far up the lagoon. She looked much older and the snowflake-like marks she had, where Kimberley's had been, were whiter and more clearly outlined than I had ever seen them. I was becoming convinced that the development and clarifying of these "snowflakes" were signs of ageing. I was so pleased that she was still alive.

The water was crystalline, and there was a huge gathering of fish, blackfins, and nurse sharks, including the pair of very large ones. My carpet of nurse sharks was beginning to reform, and the fish were again waiting for me to feed them, in spite of the long years of silent days since the wild parties we had held there before. As night fell, I watched two of the big nurse sharks munching on a scrap. One of them kept raising its tail above it to splash through the surface, back and forth, like an enormous heavy whip, lashing. The school of yellow perch was swirling around it, feeding on the scattering bits, and around this strange centrepiece, orbited the blackfins. They were all excited and moving fast. With the sinking sun casting a deep orange light upon the scene from the west, it was startling, and went on and on as the sun sank. If the BBC had a setting like this to film, all would be well. This was the sort of alien beauty I hoped they would be able to record there. As darkness fell, I put the anchor back in the boat, and held it as the wind and current swept it away. Flying underwater through the coral, I slowly overtook and passed a three-metre sharptooth lemon shark, who was leisurely carrying a fish spine away in his mouth!

Madonna had not appeared again, and I had no doubt by then that she had been finned. Trillium had also disappeared for too long. The visitors might surprise me, but I knew the residents. In the darkness of the storms raging over the island, I wrote an *Ode to Madonna* and circulated it on the Internet. Many people sent comments about it, and it was posted on a few websites devoted to the protection of sharks.

I grew positive about the coming filming, in spite of how infrequently I had been able to get food out to the site. Stephen and Sue sounded so nice that I was sure they would help me get through it, knowing I was inexperienced. One by one, the details of the visit and filming were being planned, arranged and organized until it had entered into the range of possible and foreseeable events.

And how exciting it would be for me after working alone for five years, to witness the filming of my beloved sharks by the BBC! It would be deeply satisfying to be with them, with people who appreciated them, who understood their plight, and who were there to record their story, after such overwhelming worry and grief over them.

These wild animals had accepted me into their society, had proved to be intelligent and interesting companions, and had provided some of the first examples of cognition in sharks, only to be caught, have their fins sliced off and be thrown away like garbage to supply a distant culture with soup. At least they would be memorialised. Their story, communicated by the BBC into the civilized world, would help to convince people that sharks must be protected by law.

Yet it had been impossible to go out during the day to spend time with Carrellina, Shilly, Flora, Cochita, or any of the others, as I had done with Martha. In order to make the point that sharks could distinguish one person from another, it would be crucial to have a companion shark like Martha. But to become her companion, I had to spend time with her.

Daily, I studied the sea and sky for signs of settling, and my concern mounted as storms continued to pass. When was the season of storms going to end? Mid-march arrived. I kept telling myself that I still had a lot of time, but then another week passed, and nothing had changed except that I had one week less. Just five weeks from the arrival of the film crew, my level of anxiety steadily mounted as day after day, wild winds swept the sea and blasted rain over the island. I imagined the film crew with all of their equipment slipping into those torrents, and being blown away, off through the coral like tumbleweeds in the wind. A storm would destroy the project and my hopes that my sharks' story would get out of Polynesia.

I had this one amazing opportunity to publicly make my point—that sharks were not the dangerously stupid killing machines that a horror-loving media had made them out to be, but rather that they were intelligent, ordinary animals in desperate need of protection. If there was a setback, it was unlikely I would ever have another chance.

I still awakened after just three or four hours of sleep, in such pain that it took all the courage I had just to manoeuvre off the bed and onto my feet. I would make coffee, and sit in the lotus position on the couch, remembering a recurring dream that haunted me those nights. In the dream I always trudged around, caring for suffering animals that had been abandoned. There was a dark green light on everything as I moved from chore to chore, much as I did in my daily life, but exaggerated, as dreams can be. Then the scene changed. A woman was standing within a wheel at the top of a building in a city. She turned and set the building alight. I could see the mounting flames and hear the shouts of people as panic spread through the city. Just then, my father appeared and led me down a trail through trees, away from the conflagration. We boarded a boat, and as I looked from the railing toward the glow from the burning city, he lifted the anchor, and turned it so that instead of looking back at the city, I was looking out to sea. A full moon partway up the sky cast a path of light across the water, and the scene was perfectly tranquil and beautiful. I looked appreciatively for a moment, then rushed across the deck to see if I could see what was going on in the city, now filling the sky with light from the flames. My father grabbed me by the arm.

"No," he said, "there are lots of people in the city, but you. . ." He pointed me back to the moon. "You, you look at the moon." I stared across the water, and took in the beauty and peace again, and that was always the moment in which I woke up. The dream was telling me that I had to stop being distracted by emotion and focus on the goal of my project. The people in the city were just parts of me who were wasting energy over emotional issues, symbolized by the woman in the wheel of life and her emotion, the fire. This message from the basement of my mind helped me to focus and allow daily life to happen, without excessive attention on my part.

But the physical anguish of having so much to do in such a state of pain and exhaustion drove me to seek more medical help. My sessions with the osteopath had ended, without relief. I telephoned a doctor I had known for years in Papeete, and he made an appointment for me with the top bone specialist in the country.

I made the long trip to the capital to see him. He was Chinese, and very polite and kind. He examined me closely and said that I reminded him of a French actress. He had read an article in the newspaper about my work with sharks when Arthur's work on cognition was being presented, with reference to the intelligence of Polynesian sharks—it had lead to my hope that they would soon be protected. So he joked about how good sharks were, to eat in shark fin soup, as he made arrangements for me to get an MRI scan of my back.

When I returned the following week, images in hand, he gave me a brilliant smile and told me he ate four sharks for breakfast. I gazed into his cool eyes, and said, "Please, help me—I got up this morning at 2:30," as I valiantly resisted the desire to shake him. He looked at my MRI, and told me that my back was fine, but that I had a worn disc in my neck, and I must stop going to see the sharks.

"I'm serious!" he assured me when I laughed and shook my finger at him.

He told me that it was important that I stop holding my head back in the way that one does while snorkelling, lying horizontally on the surface. Then he prescribed an anti-inflammatory drug, and more visits to an osteopath. But nothing had any effect on the excruciating, radiating pain that awakened me each night.

With the passing of time, there were periods when Diamond seemed almost normal, though he rested on his breast in the shadows too much, his face was too white, and walking was painful. He was at the centre of his own little flock by then, with Josephine always at his side. Whenever he found something good to eat, he called to the others and gave it away. And when his son's more aggressive companion attacked him in the course of a quarrel, Diamond chased him away, though his legs were so painful that he could scarcely walk afterwards.

His periods of health alternated with illness. Each time I cured something, he developed another set of differing symptoms, and any sudden movement near him still caused him to shriek.

If Diamond was threatened by another rooster, even before the fight broke out, Josephine would come calling anxiously at the door, and chatter to me as I ran with her to find him. Once, while I hid behind a tree, waiting for an opportunity to grab the other bird from behind, she became so agitated about my failure to act, that she leaped onto my shoulder, squawking in impatience. The idea that hens ally with the strongest rooster is erroneous. They love one, and want him to win.

Josephine continued to lay a big egg each day, though strangely she never became broody and settled on them. I eventually learned on the Internet that broodiness had been almost completely bred out of Rhode Island Red laying hens.

When Diamond was too ill to go outside, he rested with Josephine in the house, and I became aware of the patience she was exercising to stay quietly with him, hour after hour. As a large laying hen, Josephine needed to spend her time foraging. Once, with an exasperated squawk, she jumped on my lap when she saw me settle down for a close examination of a pile of papers. She wanted to go outside, and seemed to understand that when I began concentrated work, I wouldn't be getting up for a while, and there would be no chance of the door opening for her either.

Each evening, Diamond and Josephine stood together on the beach, close to where the waves lapped gently, looking out over the shimmering water. Diamond usually just stood looking, while Josephine foraged for sand hoppers. I came to believe that he was watching the tropical sunset and its reflection of rainbow colours melting across the glowing bay.

In mid-March, I was able to get a few scraps of shark food, and set off to the lagoon. Luckily the wind died down just enough to make it easier than I expected to get there. The lagoon was a river and the visibility was poor. That would be another thing that would kill the project—poor visibility. A big shark, probably Windy, followed the boat, joined by many smaller ones as I approached the site. I anchored, and since Carrellina was there, threw in the food. Exhausted, and in awful pain from the battle out there, I just couldn't face any further trouble.

As I busily extracted handfuls of the scraps and put them in the water, a big shark's head lifted to take something. Windy's jaws snapped shut so close to my hand I couldn't see the distance in between. How easily my hand could have been snapped off! I continued more carefully, and having already put on my gear, slid in right after the food. I drifted, unmoving, as the bubbles from my descent rose around me, clearing the view. The first thing that became visible was Windy passing slowly beside me from behind. As I continued to drift, looking, many of the sharks came up to my face, just as Martha and Madonna had faithfully done in the beginning. As the sun sank, a few rare visitors cruised through, including a group of males I rarely saw.

Since I had only been able to get a very small amount of fish scraps, Tamarack kept hopefully coming up to me at the surface, at the head of long, ragged lines of sharks, just like in the good old days at my original sessions. It was a relief to see that the few feeding sessions over the past few months had succeeded in renewing the phenomenon. Most of the sharks who arrived were young females, the older ones having disappeared, but there was a good congregation of them, as I had promised for the BBC.

The following week when I wandered into the fish store to say hello and find out how things were going, the younger partner looked up with a smile. "I was going to call you," he said. "Do you want some scraps?"

I smiled a jubilant affirmation. We walked into the back of the building, and he started hauling huge sacks out of the freezer, full of long fillets. Each sack weighed almost more than I could manage.

"Just a second," I hesitated. "Could I get the rest on Saturday? I only have a little boat."

He agreed, but said that I must come in the morning. I left with five of the huge sacks, less than half of them. By the next morning, the long cuts of meat from some unknown fish were thawed out enough to cut up. I was able to load the contents of two of the sacks onto the kayak.

It was 7:30 a.m. when I arrived at the lagoon. I planned to go early with the BBC, to take advantage of the calm conditions before the rising sun stirred the wind, so needed to be familiar with who was around and where, at that time. I hoped to locate Shilly and Carrellina.

It was a pleasure to see that Tamarack and Windy followed the boat from the border. They were responsive to me, even at unexpected times. I could count on my sharks. Increasingly it was becoming apparent that of all the factors involved, they were the most reliable.

And they were in for a treat after months of meeting me for nothing more than a few spines and scraps of skin. I anchored some distance west of the site, slid underwater, and waited while Windy and Tam glided up to my face. Windy turned just before she touched, and Tam passed over my shoulder. Merry, a new juvenile in the region, Ruffles, and Bratworst came behind them. I threw them a few pieces of the good food, and swam toward the site, trailing the kayak and jotting down the names of the sharks arriving. It was a delight to see the excited animals snatching up the chunks and chasing through their coral garden in the flickering morning light.

I anchored at the site, and watched the sharks zoom toward me. "Would you like something to eat? Have I got a surprise for you!" I threw in the food and they pounced on it. Sharks were stopped dead in the middle of the site, stuffed to the teeth, waiting for the big bites they had taken to descend. I had cut the long fillets up for them, but with teeth like theirs I hadn't worked too hard at it.

I left to swim for a while with Eden, trying to memorize the fins of a fleet of shark pups who passed us looking curiously, as we went. Eden circled through the area in a fashion similar to that of Martha, and returned to the site. There we found a large, dark male, strangely reminiscent of Lillith. He circled us. It was a rare visit from Blaze, from farther up the lagoon. Indeed, there were sharks around, even if they were not expecting food. I threw him a few chunks of the fillets, and a group of juveniles, just identified that year, accelerated to get some. They were named Sonata, Picatrix, Blue, Odetta, and Slinky.

One of the pieces of food drifted off in mid-water, and Fleur de Lis swept up and swiped it. I found it perfectly bitten through, showing the size and shape of her mouth. Her bite was about fifteen centimetres (6 in) across. She had matured since I had first identified her three years before; she had lost her slender juvenile look and gained the heavy build and greater length of an adult.

For a while there was no one at the site, so I shredded some food for the fish—the squirrel fish in their grotto, peeking out from their holes, the rainbow wrasses darting back and forth in front of me, the flashing silver needles at the surface, looking so solemn with those huge gazing eyes, and the last bit for the groupers peeking up from below. It was fun. I got another big chunk—what a treat it was to have lots for once—and did it all over again, as the space around me filled with the familiar multi-species cloud I had so loved to feed at the original sessions.

Suddenly, everything disappeared. A shark passed my eyes, snapping up the bits leaving my hands—Madeline. She whipped a hundred and eighty degrees around as I grabbed pieces of fish from the boat, and threw them in for her. She snatched them one after another as they sank. Twilight glided in with The Concorde. Junelight was with them; it was only the second time I had seen her.

Shilly came and I left the others to swim with her. We went winding through the supernatural surroundings nearly as far as the barrens before returning to the site. By then Storm, Pip, Gwendolyn, Jem, Shimmy, and Christobel, *once again the perfect shark*, I smiled to myself, were cruising around curiously, smelling food, and seeing the kayak, but not me, and not food! The sharks who had fed had already left.

I threw more pieces into the site, and after watching a while longer, took the kayak, and drifted on, tossing bits of food to the accompanying sharks from time to time.

By 9:30 a.m. I arrived at the barrens. Jem and Shimmy were still with me. Ondine approached in slow-motion, turning her head just slightly as she cruised past my face, which gave her a wondering air. I hadn't seen her for a very long time, so it was a relief to find her again. I anchored and threw in the rest of the pieces of food one at a time so that none would be lost, then watched and swam around with the sharks until 10:25. But Carrellina did not come.

As I was leaving, a male swam slowly by with a hook in the side of his mouth.

I took the rest of the fillets to Site One the following evening. The kayak was loaded to overflowing, and sunk to the gunwales. It was extremely heavy to paddle, and I was grateful that the evening was calm, so that I could enjoy it after so many difficult sessions. Never had I brought such a treat for the sharks!

Brandy followed the kayak to the site, pressing her body briefly against the paddle as I anchored. Few others seemed to be present, probably because they had stuffed themselves the day before. I tossed in a handful of pieces and slid in. Avogadro, Brandy, Simmaron, Marten, Fawn, Gwendolyn, Odetta, and Hurricane were shooting around snatching and dropping the falling bits, then whipping back to seize them.

I kept the kayak by me, and as they ate, threw in more pieces. Sharks were steadily arriving. Even Amelia came, on her only visit from her home at the far end of the lagoon in Section Five! They snatched the food, gulping and slowing as they swallowed.

Keeta soared into the site and gobbled everything she could find till she was rectangular from jaws to stomach. For a very long time she remained like that, barely moving on the white sand floor, apparently focused on the slow movement of her food down into her stomach. Eventually she swam slowly away. I became a little concerned. I had not wanted to cause them digestive problems, just give them a special treat.

Cochita glided out of the coral and fed, soon joined by an eager Tamarack. He had stuffed himself the day before, and stuffed himself again. When he left he looked as if he had swallowed a balloon. Yet he kept roaming around in search mode, though this time he did not come continually to me.

Shimmy and Melody flew through the site, whipped back, found the food, neatly bit up a piece, and inhaled it. There were many sharks who had not been there the previous morning. A group of rare visitors were delighted with the find. One was Capri, a young male who had come infrequently as a juvenile, and very rarely since he had grown up. The presence of the occasionally available food had not affected his

Sharks were stopped dead in the site . . . .

roaming. Many juveniles attended, so when the big sharks left, I alternated between sketching their fins and feeding the fish. The scene, with the sharks shooting in, and circling through the exotic environment, was beginning to resemble the original sessions when my fish parties had been at the peak of their popularity.

Finally, as the soft purple twilight descended on us again, the sharks dissipated. I drifted down-current, and Capri came to me, the closest he had ever come. At that moment Anne Boleyn arrived and started to charge me from different directions as I moved above the largest coral structure. It was awkward for her but she didn't let that stand in her way. All the food had been gone for some time, but she must have been able to detect the last traces of odour. Swiftly she was joined by Glamorgan, Vixen, Tuvalu and Cotlet, all sharks from Section Two. It was their turn to be disappointed to have missed out on the food after having had first choice when I held the sessions in their range. They glided off into the darkening waters toward the border.

Flora did not come. I had not seen her since I had been able to go underwater again, and felt that familiar crawl of anxiety as I climbed back into the kayak, counting the missing. Trillium had not appeared for so long that she must have died of her injuries or been successfully caught by the shark finners. I had assumed that she would be filmed by the BBC, as proof of what harm a hook can do, when the subject of the fishing and finning of the sharks was broached.

It was my only hope—to know that the BBC were coming to document this tragic situation before it was too late. Even the sharks I had counted on for the filming were vanishing.

On Saturday, I still had some of the huge fillets left. When I went back to the fish shop, I brought a six-pack of beer to facilitate negotiations, handed it to the younger partner with my most winning smile, and asked if it would be possible to save the rest of the fillets for me until the end of April, because some people were coming to make a film about the fish I had been studying.

"No problem," he said. The way he agreed right away without question was almost too easy. He said that he would tell his partner, and assured me that the scraps were old and would otherwise have been thrown out—it was better for me to give them to the fish. The big main freezer was always turned on, so it would not be a problem. I could scarcely believe my good luck. The fillets that appeared neat and clean would be excellent for the filming. For anyone not used to them, my usual fish scraps looked revolting, and I didn't think that they would be suitable for a film.

He told me to come the following Saturday for the regular scraps, and his partner would be there then. He said he would let him know that I would pick up the frozen fillets at the end of April.

I bought a second six-pack of beer for the senior partner, just to make sure that he too, was aware of my appreciation, and when I returned on Saturday, handed it over with a radiant smile. He smiled back. He told me that he unfortunately did not have any fish scraps. I mentioned the arrangement I had made with his partner to pick up the sacks of fillets in three weeks, for the film that was to be made. He explained that the fillets weren't his, and anyway were for sale for thirty francs a kilo.

He paused a moment, then continued, "Or maybe fifty francs per kilo."

I considered this. "But they're old," I said.

"Yes," he agreed.

I asked if I could buy them, and then pick them up in three weeks time.

"No," he said. We would wait until Saturday, April 24, and then we would see if he had anything available. If he didn't have any left, I could buy a box of sardines. No further efforts to sway him made any difference, even to the patient tone of his voice. Yet five huge bags had been handed over to me by his partner, who would have given them all to me!

My heart just fell. I had thought it was all arranged—at least the sharks' food, if not the weather. Why would he not let me buy the fillets?

CHAPTER FIFTY-FOUR

The Final Days

Whenever my mind was free to wander, I mused over the filming and what was needed most. I suspected that there would actually be time in the documentary for only the core of what I presented in the interview, so I put the main ideas into key phrases, which I pondered, reconsidered, and refined, until the essence of my message was unmistakeable, yet stated in just a few words.

That the sharks recognized me as an individual, seemed to be a main point of interest, so I considered how I could make that evident for a film crew of whom the sharks would be suspicious and nervous. How could I demonstrate that they singled me out? When I swam and then stopped, the sharks normally circled me. If I waited in the site for my thirty sharks to circle me, would that demonstrate our connection?

How could I show, visually for television, that the sharks were responsive, intelligent animals? It wasn't as if they were dogs. They were wild animals, and they lived underwater. There were limitations in terms of what I could do with them, and how long I could spend with them. Hopefully the mirror experiment would provide interesting behaviour, which then could be discussed and explained in terms of cognition in sharks. Their behaviour in front of a mirror had changed my own ideas about their potential intelligence.

Depending on what happened and what was filmed, I could weave the context of it into my interview. How I regretted the loss of the film that Bou Bou had taken, which had recorded Martha leaving the site, then returning fast to swim beside me when I dove to gather up the scraps that had been pushed under the coral. If she had still been alive, no matter what else happened, her behaviour toward me was so companionable that the film crew would have no trouble getting sequences of a shark interacting with me.

Another idea was having them swim with me as a group. Most of the time they followed me while out of sight. This was a pattern which was not suitable for filming. While I was sure that they would swim with me if I came to the site with food in the kayak, then swam around the area before giving it to them, it was not at all certain they would do that with other people present.

I had stopped calling to the sharks after I changed sites, but had begun again when I slid underwater in the hope that it could become a useful signal. Several of the sharks had been quite responsive to my call when I had done it systematically.

Each time the clouds blew apart, and sunbeams bejewelled the richly robed peaks, turning the sea to a glory of light, I was sure that the storms had finally passed and the weather was settling. Giving the sea a day to calm down, I would plan to leave the following morning to look for Carrellina, to swim with Shilly, and see who was around. And each time, before tranquillity fell upon the sea, the horizon darkened, breezes fanned across the waters, and another storm hit the island.

At the end of March, when I went to the smaller fish shop to pick up the usual load of scraps, I was told that there would be none for the next several weeks. The captain had fallen ill, and there was no one to replace him. At the other one, the partners had been selling the few fish they had been able to get whole, so no cutting had been done. Another storm hit the island anyway that weekend, so I was stuck again at home, frantic about the failure of the weather to settle and the loss of any source of shark food right at the time I was supposed to be convincing all the sharks in the lagoon to head for Site One.

Finally I bought a box of sardines when a calm day dawned, just to keep up some semblance of a schedule for the sharks. Ten days had passed since my last session. We were between storms, and the sky was low and brooding. There was no wind, but the current was strong—I anchored and was swept into the site as Carrellina came to the surface for a sniff at the boat. When I was ready, I threw in a handful of the sardines to distract the sharks, and slid underwater.

A multitude were swimming straight toward me; they were excited and moving fast. Flora shot past. She was alive! Deeply relieved, I tossed some of the little silvery fish in her direction, and watched the results. She and Sparkle, Teardrop, Cinnamon, Brandy, Tamarack, Bratworst, Breezy, Merry, Carrellina, Cotlet, Eden, Storm, Avogadro, Christobel, Blade, Ruffles, Odetta, Shimmy, Simmaron, Tristan, Cochita, Keeta, Tawny, and the little pup Slinky, swirled through the site, snapping up the fish that I threw to them. In a few minutes it was all over. There was nothing for the fish, and nothing for the nurse sharks.

The sharks glided around for a while, their excitement fading, and then began to disperse. I followed Flora, calling to her, but she sped up and in the current it was impossible to stay with her. Franck was away, and I had to get back early to make sure that the birds were settled for the night, so I returned to the kayak. As soon as I threw in the anchor and prepared to hoist myself up, the sharks noticed and reacted. Flora, Tamarack, Sparkle, Cochita, and Eden flew over and swiftly circled me as if to say, "You're not going yet?"

I hung on the kayak, concerned and watching. The sharks and lagoon fish formed an opaque, swirling cloud around me. Finally, since I had no choice, I climbed into the boat, and took off my gear. The current carried me swiftly away and the whole group moved with me. Their dorsal fins passed me, one after the other, slicing through the glassy surface.

The boat went on and on, while I sat unmoving, watching the little dorsal fins. Flora, Tamarack, Carrellina, Cochita, Eden, Merry, Ruffles, Brandy, Sparkle, and Bratworst passed closely around me over and over again. I patted my hand on the surface, and Tamarack and others rose up to it, not quite close enough to touch. At the border I moved over the indigo waters of the bay, and they began to stream away toward the lagoon then circle back around me again. Then again they streamed back toward their lagoon, only to turn and come to me again. Their circles became larger and larger as I paddled on down the bay.

Never had they done that before. At this late stage, again I was seeing a new behaviour pattern.

Finally, only Tamarack kept coming along behind, then circling back to the lagoon. Each time he left, I thought he was gone for good, but a few moments later when I looked back again his little fin was coming along behind the boat. He would drift up beside me, pass under my seat, and then circle far back towards the pale waters of the lagoon, now fading into the distance as I paddled steadily down the bay. Each time his circle was longer than the time before, and each time I was sure that he was finally going home. Yet each time he returned and resumed following the kayak.

I would have given a lot to know what he was thinking. Did he just hope I might have another fish for him? Did he want to know where I disappeared to each time? Had my visits been so sporadic for the past years that he was unsure if I would ever come back? None of the above? Finally I was sure that he had departed for good. The slight surface ripple of his passing had vanished into the distance; he must have returned to the lagoon. My thoughts wandered for a while before I thought of him again, and looked back to check. And there was his little fin just visible above the surface right behind the boat! I was nearly halfway home and well down the bay. As I paused, he drew alongside me, passing right beneath me once again.

I began to wonder what I would do if he accompanied me home. I could imagine myself getting out as the boat slid against the beach, the shark circling my feet. I would have to tell him, "Well, I have something in the freezer, but you'll have to wait while I thaw it out. Do you like. . . aahhh. . . *cooked* fish???" But after that, mercifully, he really did return to the lagoon.

Why had the group followed me so far? The sharks had all received something to eat. There had been a lot of sardines in the box I had brought, and the entire volume had been edible, whereas the scraps that they were used to provided little real food. Though it was easy to attribute fear and aggression to them, it seemed so much more difficult to ascribe the positive feelings of attachment or fondness. Yet they appeared to experience what passed for happiness in sharks, to see me, and here they seemed to be expressing what passed for regret in sharks, to see me go.

It was an affirmation that it was important to continue to go to them, because it was important to them that I come. I could not ascribe their reaction to me as due solely to the desire for an easy meal, especially when they could so easily snap up a fish whenever they wanted one. Perhaps the association with an enjoyable social situation had something to do with it.

The next two weeks were wildly stormy, and going out to the sharks was impossible.

One morning Franck called me from town. He had picked up the mail and there was a letter from the government branch that was responsible for renewing my permission to stay in the country. The letter said that since I lacked an income—I didn't have the right to work as a foreigner, and the art market was non-existent—I no longer had the right to stay in Polynesia! My permission to stay would run out on April 21, and the BBC were coming on the twenty-fourth!

Franck called the official handling my case, who mentioned some new law that applied, in spite of me being married to a French citizen, having lived in Polynesia for ten years, and having a record of good conduct. But under French law, a man's wife is considered to have half of what he has and vice versa, so accordingly I should be considered, legally, to have half of the family income rather than being treated as if I were on my own. Franck sent him the records of his own income, saying that I continued to be "in his charge" as I had been since we had come to Polynesia. The official then informed us that the matter would be decided at a Counsel of Ministers.

I was sure that this was due to my efforts to protect the sharks. Surely a Council of Ministers didn't decide every case. If I had been agitated before, now my panic went sky high. Boxed in by the storms, feeling like the proverbial cat on the hot roof, I began poking around on the Internet to see what I could see. When I typed the name of the president of the country into Google.com, while looking for his e-mail address, and those of others in the government, I was stunned to read, in big letters, "The Work of Ila France Porcher." One of the websites that had posted my writings mentioned the president in connection with the government's promise to protect the sharks. I was appalled. Could they just throw me out of the country on the twenty-first of April?

I was still writing articles and letters of different sorts and sending them out over the Internet to encourage people to write to the government to plead for them to act to protect the sharks. Now I added, "Please, don't mention my name."

To distract myself, I revised my paper on the gestation period of the sharks, and sent it to Arthur, who had graciously offered to go over it for me before I sent it back to *Marine Biology*.

Stephen wrote that the team was preparing to leave for Australia, and he enquired how I was doing with my preparation of a new mirror for the mirror experiment. All was fine, I reassured him. I planned a trip to Papeete to buy one of just the right size and thickness, which I had already researched. All I had to do was collect it, spread the back with a paste of silicone, then smooth a piece of plexiglass onto it to protect it not only from the salt water, but from accidentally breaking. One never knew what sharks would do!

Stephen was concerned that if the mirror was too large, the sharks would be able to see their reflection and react while still far away, which would be a problem for filming.

It was a good point. A large mirror would change the familiar landscape and might frighten the sharks. Further, it would be impossible to swim with underwater. I described how I arranged the situation, by placing the mirror parallel to the sharks' swim-way, with pieces of food in front of it, so that the sharks saw their reflection when they reached the mirror, as they came up the scent flow at their usual distance above the sand, of less than thirty centimetres (1 ft).

I assured Stephen that when I had visited the lagoon recently in the early morning, groups of sharks were passing. It was a time when they were on the move. Between 7:30 and 9:00 a.m., a steady stream of visitors would pass through, never too many at once. If we were there with the mirror, the reactions of twenty to thirty sharks could be filmed in a two hour period. At 7:30 the sun was high enough to be bright, with a very attractive, warm light, increasing in brightness and sending slanting rays of light through the scene in the lagoon. We would have until after 9:00 a.m. to film, which was when the wind became strong enough to be annoying.

There were still arrangements to be made about acquiring air and oxygen, and Sue was having endless problems hiring a van and eight seater minibus to transport everyone and everything. I promised to meet them at the ferry with our pick-up truck just in case things went wrong at the last minute. I was very anxious that the arrangements for their stay fall into place perfectly. If things unfolded smoothly for them, they would be freed to concentrate on the filming of the sharks, so things would turn out better for us, too.

As the law promised for the protection of sharks was once again delayed, on April 5, Polynesia's diving centres and clubs organized a day to inform, educate, and create more concern for the problem of the Polynesian sharks being slaughtered for Asian soup. Instead of following their usual diving routine, they did all they could to emphasize the desperate need for shark protection. Letters poured in. *La Dépêche* reported on this, as well as people who wrote and copied their letters to me.

A letter my closest friend, Janine Perlman Ph.D., an American biologist and wildlife nutritionist wrote, is a good example:

Dear Monsieurs and Madames:

We are writing to beseech you to establish a shark sanctuary in French Polynesia, in which sharks could not be fished, hunted, finned, or otherwise harmed.

The Polynesian species of sharks that are currently being fished and finned are on the IUCN Red List of endangered species. It is a legal and ethical necessity that these wondrous creatures be protected immediately. In addition to the ethical and legal considerations, diving to view these wonderful animals constitutes a large and growing contribution to the Polynesian economy. Because of the barbarism of finning in particular, these animals are disappearing very rapidly. Soon, divers won't be interested, and you (and the world) will have lost an irreplaceable resource.

Please show leadership that will be admired the world over; protect these majestic animals now.

Sincerely,
Janine Perlman, Ph.D.
James Fuscoe, Ph.D.
Arkansas, USA

The schedule of the BBC's filming tour arrived. It was a long document listing every plan in every place by each person involved. I was in awe of the work involved in preparing and coordinating not only the plans themselves, but the document, too. Indeed, Sue was a magician—I was becoming very excited about meeting these people with whom I had been in such intensive contact for so long.

I went through the schedule carefully. There was my own name, even the mention that I would meet them at the ferry dock, and the precise time. Reading on however, I was taken aback. The schedule for the night of Sunday, April 25 was not the feeding session Stephen had agreed on the phone that we would do, but filming the interview at our house. How could they film the interview before we knew what we would have on film? I had planned to talk about the sharks the viewers had met. Knowing the story of each individual was the key to the deeper level of understanding I had gained about their lives, and I couldn't know in advance whom we would film, doing what.

Then there was the practical problem of their food. It was only available on Friday or Saturday, and I would have to keep it cold until Sunday night. In the tropics I would be unable to keep it fresh enough longer than thirty-six hours at the most, so I had to be able to plan in advance how to proceed. The fishing boats brought in the fish for the weekend, and the shops expected me to clear out the scraps on Saturday morning when they had finished cutting them up.

I had gathered a large group of sharks to film at the BBC's request, but to have the phenomenon unfold as planned, sharks and cameramen had to synchronize. It was vitally important for the success of the project that the sharks and film crew meet at their familiar evening feeding session.

If the sharks met the film crew at the mirror experiment instead, they would be more interested in the crew and their equipment than in the mirror. Then there would not be much of an evening feeding if they were fed at the mirror. All my efforts to get a big group together for a dramatic filming of lots of sharks in the beautiful coral surroundings in evening light would have been in vain.

It could be possible to persuade the partners at the fish shop to keep the scraps for me until Monday, just that one time, but usually they turned off the refrigerated room over the weekend, due to the cost of the electricity to keep it running. So this change was a serious problem for me. A further complication was that Monday morning's filming was to start at 7:00 a.m., and the shops didn't open until 8:00. If I had to pick up the scraps on Monday, or buy a box of sardines for the filming, I would have to wait until then, and it would be frozen solid. The morning's filming would be delayed, the wind would rise, and precious time would be wasted. I asked Stephen if the Sunday night planning was flexible or not, and outlined my concerns about the change of plans. He replied that he would discuss the matter with the crew and send me a message back.

It was April 7.

Then the clouds began to thin, the sky lightened, and one morning the sun rose upon a clear, bright sea. The trade winds, blowing ever from the east, resumed their pattern of rising in the morning, and dying at the end of the day. Ten days remained before the arrival of the BBC, and given the signs that the weather was finally settling, I bought a box of sardines for the following day—the sea could have a day to settle down.

But that evening the wind rose and howled all night long. I awakened at my usual hour, at 2:30 a.m. and lay listening to it, frozen with pain and afraid to move. Whatever was I going to do? Just getting out of bed required more courage than I often felt I could summon. The pain radiated like light from a star, from the place in my spine where the vertebrae had moved, and spread intermittently around my torso and down my right arm. The only way to relieve it was to get up, but I longed to sleep.

I made coffee and sat on the couch in the lotus position waiting, tensing one muscle after another around the damaged area. Once my spine was vertical, the pain began to ease. I had expected that the problem would go away in time, but night after night, I awakened at the same time, in the same condition, and still dragged myself through each day so tired that all I wanted to do was lie down. Napping during the day was impossible, given the need to watch over the birds and the frequent attempts of cock thieves to steal them. I was always on guard. If I did lie down during the day I had to get up every few minutes because the dogs barked or the birds gave alarm calls, and visions of swamp harriers and cock thieves forced me back on my feet. I listened to the wind. I had spent the equivalent of more than twenty dollars on the sardines, so had to get out there. No daughter of my mother would waste a franc, never mind the lives of all of those fish!

I reached an intolerable peak of anxiety as the day dawned and the wind screamed down the bay, tossing up white-caps, whickering through the jungle, and flinging the coconut trees around as if they were flowers. Every few minutes, I rushed out onto the deck to check the speed of the clouds, the look of the ocean, and the height of the waves careening down the bay, forming theories about the weather that were dashed a few minutes later as conditions changed. It was a unique day, but analysing it critically, I predicted that at the very end of the afternoon, there could be a calm period, which I would use to go to the sharks.

But when my moment of calm softly descended, just as foreseen, I was deeply frightened. Exhausted, in pain, and worried sick, it seemed mad to entrust myself to the sea. However, there were no scraps to sort out and pack, just the box to throw in the back of the kayak along with my gear, so it took only a few minutes to get ready. If I just threw the food to the sharks and hurried home again, I could be back in an hour.

The winds were turbulent, coming from different directions, and a storm to the northwest kept sending blasts toward me as I set out. At the border, waves clashing into peaks from three directions met the river pouring from the lagoon, and broke in chaos. Luckily I had years of experience by then, and knew the precise place I could cross, and how to time it. There was always the danger of my light craft being overturned at this treacherous place when the sea was wild.

At the site, many sharks arrived swirling, thrilled, around the boat. They were moving faster than usual, and their childish excitement was so touching that it made everything worthwhile. Cochita especially, but others too, swam up to me very fast, raising their heads toward me. With the gunwale of my boat centimetres above the surface, I prepared for a shark in my lap. Reaching behind, I started tossing them the fish.

Watching from above the surface was fine. They often roamed around me, waiting, raising their heads up toward me, and then going down to search on the sand. Since their fins were out of the water for much of the time I could easily identify them. They were all my usual sharks, the ones who had followed me past the border on the previous occasion, and many were present who did not come to me at the boat, and whom I was unable to identify. So once again, my sharks and more, were impressed with the idea that regularly checking Site One for food was a profitable enterprise! I hurried home afterwards, feeling triumphant, and was safely back in the house when the wind's wild wailing arose again.

Three days later, the weather calmed, and after waiting during the long hours until the sky began to lighten, fortifying myself with pain killers and coffee, I rushed out to the site as soon as it was light enough to see. The lagoon lay innocent in the sunrise, shimmering with the light blue and gold colours of a clearing sky. Even its waters were quiet.

I slid underwater just over the border, and swam slowly toward the site, which was my usual pattern when I went early in the morning. The water was crystal clear; the lagoon had been washed clean by the ocean waves during the storms. I had the last pieces of the fillets I had saved in the freezer with me.

Brandy and Breezy were the first to join me, then Sparkle, Ruffles, Bratworst, and Tamarack appeared. At the site I dropped the anchor and threw in some snacks. Only Tamarack got excited and came to circle around me. I continued eastward looking for Shilly or Carrellina, not trailing scent. I still had a few days, if the weather remained calm, to spend with them. . .

Dapples swam languidly up to me, circled, and moved on. At the barrens, Brandy, Breezy, Ruffles, Merry, and Bratworst appeared. I tossed in a few scraps to engage them, anchored, and watched while Tamarack repeatedly passed, and the usual sharks gathered. I swam around the area, first with one, then with another, but they stayed nearby in hopes I would produce more treats. Eden swam up to my nose, then coiled down to the barren floor and went into search mode. I waited until she swam in my direction, and then dove down and swam up to her nose. She lifted it just enough to sail over my head. Once I was back at the surface, she came up to my nose again, then I swam down to her nose again. It felt as if we were playing, and I had a wonderful time with the sharks that day, swimming up to them face to face, and enjoying the way they would often come straight back up to nearly touch my nose in response, when I surfaced.

When Cochita arrived, I threw one of the two pieces of food that remained for her, but she missed it. She turned, but couldn't locate it, while Eden came in from left field and swiped it. For a moment, I was reminded of Madonna and Martha together. I waited until Eden was at the farthest point of her circle, and threw in the last piece for Cochita. She reacted instantly, zooming in to it and snatching it, then moving on slowly, while gulping it down. I swam around the area, sometimes with them, and sometimes alone, searching for Shilly and Carrellina. But they did not appear.

By the time the third week of April approached, I had extricated the most important points to convey in the BBC interview from out of the haze in my mind, and written them down. Then I worked on how to word these key ideas in a crystal clear way. I memorized these phrases, and repeated them to myself as I did the housework and cared for the birds. On the phone, Stephen had told me not to worry about that part, since they would "come with the story," but it seemed important to make sure that the core concepts were conveyed, so that these key facts would come across no matter how much my interview was edited.

Just a week before the film crew was due to arrive, on April eighteenth, I received a message from Stephen to let me know that he had discussed the problem of the Sunday night filming with the others, and that he was sorry, but the crew would not go out to the lagoon that night. The preparation required for the equipment would be too much and everyone would be tired. He suggested that I get the sharks' food and plan the full day of filming the next day on Monday, April 26. I had no option but to let events unfold as they would.

That day an envelope arrived from the government containing a brand new visa card, giving me permission to stay in Polynesia for another ten years. Now I could focus my worries exclusively on the weather.

In the middle of the following week I unexpectedly received a small load of fish scraps and took it out to the sharks when evening came. The island was shadowed with cloud, and a restless south wind blew, which was worrying—it was within days of the arrival of the film crew.

Rain began falling as I hurried out past the island's shelter, and in the wind and the disturbed sea, I couldn't find the site. Mordred was the first shark who came to the boat as I roamed back and forth looking for it, fighting to control the kayak in the wind. More sharks joined us including Windy, who stayed so close beside me that I paused my frantic paddling to stroke her. Brandy, Bratworst, Simmaron, Shilly, and Jem appeared at the surface, and finally there were so many sharks around the boat that I could no longer manoeuvre for fear of hurting one of them with the paddle, so I threw in the anchor over a clear area. Cochita circled tightly beneath me, and I intermittently caressed her while I prepared.

A large group of sharks were circling through the darkened surroundings, and approaching me in their usual way. I held onto the kayak to push their food in, and drifted around the periphery as they fed. Sparkle, Christobel, Pip and Storm were the other adult females present, and Cotlet swam into the site from the east with a big remora, who switched to me. I dove to extract a bit of meat from a spine, and fed him. He ate eagerly, then, stuffed with food, began roaming around a nurse shark who shook a scrap in slow-motion, and raised clouds of sand in the dusk. In the stillness beneath the storm which was wildly flinging the surface, the sharks soared, the time passed and the scene darkened. The first group began dispersing, though they still came into view as they roamed through the area. Tuvalu, with Glamorgan and Cheyenne, glided in from Section Two, and behind them, Clementine and Odyssey swept in with Hurricane!

It was the dark of the moon corresponding to the one in which they had appeared with Cheyenne the previous year. For the fourth year in a row they had visited during that lunar phase, and at no other time.

What an ideal additional detail for the BBC's story of intelligent sharks with individual patterns! It provided proof that sharks know each other as individuals, have a schedule of their own, and preferred travelling companions—companions they kept over very long periods of time. This single event inferred much more complexity in the lives of sharks than was usually acknowledged.

With the arrival of the visitors, some of the female residents returned, and they cruised through the area until the light faded and they were the colour of water, just motions in the gloom. But still I stayed and watched them, feeling real happiness for the first time since we had returned from our voyage to France. I was ready for the BBC. The sharks would come, as long as we could get out there, and though the weather had been unsettled, I was sure that there would be periods of calm in which we would be able to film them.

All my life I had heard about the eminence of the *British Broadcasting Corporation*. I had read of their supreme role played in the Second World War—in which both my parents had been involved—in countless books. My Scottish mother had listened to their broadcasts each evening on the radio when I was a child, so I could scarcely have felt more honoured if the Queen had decided to come.

On the evening before the BBC film crew was due to arrive, I went to pick up the outrigger canoes and take them to the *pension,* far along the shore of the lagoon. When I had called to confirm these arrangements, I was assured that it would be an easy job—the trip would take me about half an hour, paddling one boat and towing the other. I timed the excursion using this estimate, to give the wind time to die down, and thus I left late. The next night the interview was to be filmed at our house, and I still had last-minute clean-up and preparations, as well as the evening chores to complete.

That evening was fair, and the wind was fading away. But when I started to paddle along the shore, I found that the way the two outriggers had been attached together resulted in the second one being pulled sideways. They were big and heavy compared to my kayak, and pulling the trailing boat sideways made it almost impossible to control my direction, or to make any headway at all. They were designed to be paddled by two people, I soon realized—one person in the middle could control the craft but poorly, and pulling a second outrigger was an almost impossible job.

I tried different ways of managing the two boats. The problem was, that when I paddled forward, the rope jerked on the one I was trailing. Since it pointed in another direction, the heavy craft moved away at an angle and yanked mine off course. It was impossible to reach the place where the two boats were attached from the outrigger I was in, and I regretted that I had not gone back to shore to fix it right away. But it was so hard to move at all, that negotiating a one-hundred-and-eighty-degree turnabout in the shallows of the fringe reef, and losing ground, had seemed out of the question. I tried to rearrange the tow-rope in various ways but there was no way to pull it straight ahead with it tied at the side! I yanked the second boat ahead with my hand, then paddled swiftly until I had taken up the slack with the first, and went on to try many other increasingly desperate tactics.

Nothing I tried worked for more than a few strokes of the paddle; by then the heavy boats drove out of line again. For a short distance I was able to tie the second one beside mine, and move along holding the two boats parallel with one hand. However, this arrangement was impossible to maintain, and the farther I went, the more I was affected by the wind as I came around the mountain and started along the exposed northern shore. Whenever the wind gusted, I lost ground. I was afraid that I wouldn't succeed in getting to the *pension* at all. I was out in the bay and being blown across it, frantically trying to get the two heavy boats to face into the wind, when it died away, and very, very slowly, using all of my power, I began to close the distance between myself and the shore where the wind wouldn't be so strong. Had the wind not faded away at that moment, I wouldn't have been able to reverse the outriggers' flight to the other side of the bay. I couldn't understand the boat owner giving me such advice. Had he ever tried, I wondered, to paddle one of his outriggers alone, towing a second one broadside against the wind? It was he who had prepared them for me.

When I got to the shore I was too tired to paddle another stroke. I felt it would take me a week to get over the ordeal; the pain between my shoulder-blades where my spine had been injured, had lit up hot and spread like wildfire. I climbed out of the outrigger, rearranged the ropes so that I could pull the first, with the second following straight behind, and began to walk along the shore toward the *pension*, bent painfully against the drag of the heavy boats. There was much laughter by the Polynesians at the white woman who was pulling, instead of paddling her boats, but I scarcely heard them, so focused was I on my progress toward my goal. Two hours had already passed. At the *pension* I hauled the outriggers up the beach, tied them to a tree, and went looking for the owner to let him know that they were there, before hitch-hiking home.

My birds had all gone to bed by themselves. I carried Diamond in for his nightly treatment, put him to bed close beside Josephine, closed their doors, showered, and made dinner.

During all this time I had been at such a peak of anxiety that I had been very careful about mentioning the situation to Franck at all, because I was afraid of upsetting him with my level of tension. When he had learned that the BBC crew was coming to film my sharks, he had tipped his glass to me and smiled, but agitation in the house was an irritating distraction for a person in the intense concentration required to write complex computer programs, so I tended to avoid disturbing him.

That night as we ate, Franck suddenly started questioning me about the BBC's visit. What exactly, had I planned? What did I intend to do? I just stared at him, too tired to think, while trying to formulate a logical reply in my hazy mind. Before I could think of something to say, he demanded, "When are you going to wake up? They'll be here in twelve hours!"

My skin turned hot and then cold as I observed the wall. A fleet of sharks passed, and I said to myself, *You. You look at the moon.*

Franck continued to talk about what I should be doing, and I listened. It soon became clear that from his perspective, I should long ago have produced a five star press release in four colours.

"It's a film, not a press conference," I pointed out.

"The BBC *is the press*," he declared, and told me I should be treating them as such.

After that, Franck took on the role of my press agent, and his first project was the preparation of the press release that I had so thoughtlessly neglected. After my struggle with the two outriggers, I could think of nothing but how beautiful it was to sleep.

CHAPTER FIFTY-FIVE

The British Broadcasting Corporation

I awoke at 3:00 a.m. on Sunday morning and sat lotus-style on the couch, my back as straight as it would go, stretching and relaxing the muscles on either side of the spine to ease the radiating pain. Before dawn, as the world slept, nothing could happen, and I could tranquilly consider and review my position with respect to the filming, and my key phrases. As the light gradually rose in the east, I did the chores. Then Franck and I ate quickly, and, trembling with anticipation after the months of preparation, I drove to the ferry dock for the scheduled meeting with the film crew from the BBC.

I parked and waited as two ferries approached, from two different companies. Sue had not told me which one they had decided to utilise in the end, but the quay was so small, I wasn't concerned. I had made a sign reading "*Jaws Junior*," and stood at the roadside holding it up as the cars descended from the ferry I had recommended to Sue. They were not on it. No minibus or van appeared among the passing cars. The other ferry dock was quite far away—I ran at top speed and arrived breathless and hot, just in time. Once again I held up my sign and watched as the cars came down the ramp. There were only five. It was obvious that there were no rented vans or minibuses on either ferry!

Transfixed with disbelief, I went into each ferry to double-check, verified the absence of large vehicles having been on either ferry with the respective personnel, searched the parking lot, and drove home in a turbulence of frantic emotions. Franck told me that two minutes after I left, Sue had called to say that they would arrive at 4:15 p.m., due to the plane having been late the night before.

All of my expectations having been tossed in the air, I got through the day with difficulty. Settled weather appeared to have arrived for the filming—the final cog had clicked into place. But as I went about my work, a familiar sound riveted my attention. The wind had begun to blow! The strength of the wind increased dramatically as the day passed, and I was in panic mode when Sue called unexpectedly in the mid-afternoon. She told me that they were waiting for the ferry, and that they would like to film me meeting Mike.

This was a surprise. For me, there were the sharks, and there was the interview about the sharks. Apart from that, I neither expected nor wanted to be filmed. I was rarely even photographed. I told Sue that would be fine! Then I tried to calm myself, and remember who Mike was. I recalled vaguely that Stephen had mentioned him—a man from California. Only sharks smaller than him were permitted on the documentary. Thus I had been asked to provide the size of the sharks, which I had never measured. But since they were about the same size as me, except for a very few that were larger, I felt sure that they would be smaller than a man, unless he was a particularly little one. That had puzzled me at the time since it was so arbitrary and unscientific, but again I had dismissed it as television hype, not unlike the title, *Jaws Junior*.

Sue sounded very kind, and apologised for what had happened that morning. Reassured, I put on my favourite dress, and drove tensely back to the ferry dock. Standing on its very edge, I watched the ferry approach through the blue, in the dramatic light of the sinking sun, off-set by flying cloud shadows.

This time, I could see a group of people on the bow, and recognized the silhouettes of TV cameras. They waved and I waved back. We recognized each other spontaneously.

As the vehicles descended, there was no missing Stephen grinning at me, driving an Econoline van off the ferry. It looked very heavily loaded. The minibus followed behind, and I trotted along with the vehicles as we crossed the road and parked at the store there.

Stephen, who was the scientific researcher, Sue, the coordinator, and Andy, the producer, jumped out, and we joyfully smiled and greeted each other. I was thrilled to learn that Andy was Scottish. He was very young and tall, with a gorgeous deep voice. Stephen too, was very young, very good looking, and exuded an aura of friendliness, confidence and strength, which had the immediate effect of relaxing me. Sue was pretty, radiant, and vivacious, all warmth and smiles. I told them how honoured I was to be in their presence, and she laughed delightedly, saying, "You won't say that when we leave!"

I turned and a man with soft eyes, and an intense face walked up to me, accompanied by a TV camera aimed straight at me. I was introduced to Mike. He asked me questions and I answered, growing increasingly relaxed with him as his eyes looked deep into mine.

The first of my key phrases rolled out as if it had never been rehearsed: "The sharks you will be meeting are not the ones I got to know closely as companions. They all disappeared when the shark finning began here. These are sharks who have known me since they were little."

"Finning?" he asked. "I didn't know that."

"Yes," I said, our connection deepening through eye contact.

And he said just the right thing: "That must have been awfully hard for you."

I thought of so many ways to continue, realized it would spoil the moment, and just repeated, "Yes."

Mike was pleased because he said that he had wanted to get into the issue of shark finning in Australia, and hadn't been able to. Now it had come up naturally.

We got back into the vehicles and Sue rode with me. Just before we got to the *pension*, we stopped at the hotel to ask when the restaurant would open. The team had not had a chance to eat, due to a lost credit card, so were looking forward to a good meal.

The big vehicles rolled heavily into the *pension's* tiny parking lot and stopped. Transported into a flowery haven alive with bird song and the warm afternoon light, we walked through the honeyed air toward the quaint little buildings. Denis met us, and accompanied us as we looked over the bungalows I had reserved.

He had not bothered to put locks on the doors and the four cameramen and assistants expressed deepening concern about the matter. I could see the place through their eyes. It was so basic, small, and unsophisticated. Finally, the senior cameraman Paul, with Mike, who was the presenter as well as a cameraman, and the assistants, Beau and Mike, left with Andy to get a civilized room at the luxury hotel next door, so they could unpack and prepare their equipment. Sue and Stephen decided to take one of the bungalows, and while they were making the arrangements, I walked down to the beach to check on the outriggers I had brought there with such difficulty the night before.

They were gone! I rushed back to ask Denis what had happened to them, and he, typically, had no idea. Back at the beach, feeling another panic attack rapidly arising, I saw two people some distance away, walking in the shallows, trailing two outriggers. As they neared, I recognized them as ours. What had happened to them—why they had disappeared and then were returned later by neighbours—was never explained.

We decided that I would return at 7:00 p.m. to take the film crew to our house for drinks, then we would all have dinner at the hotel and discuss how to proceed for the filming.

Back home I found Franck, ever the conscientious French host, busy preparing a fascinating variety of delicious appetizers. Gallons of punch were cooling in the fridge, in our various small pitchers. I helped as well as I could, while making sure that the birds were taken care of. No matter what was happening, Diamond had to be treated, held, consoled, and put gently to bed, and Josephine had to be lifted onto the perch beside him. The sharks had suffered a terrible tragedy, but so had they.

At 7:00 p.m. I returned to the *pension*. The team had so much equipment to unpack, check, assemble, and polish, that they were still busy, but Sue, Andy, and Stephen came with me to our house so that we could get acquainted. After introductions, we excitedly talked and restored ourselves with Franck's hors d'oeuvres while he refilled our glasses. At least our spirits were high, even if the wind was too, I thought to myself. I was too tense to eat, so sipped the punch and watched our guests relax. At least we had managed this far!

The animated discussion continued at the hotel, where we joined the cameramen and technicians at the table. The senior cameraman, Paul, was concerned about getting his largest camera to the site, and declared the need for a motor-boat. The outriggers were simply not wide enough for the camera. Thinking of Franck's friend with the zodiac, who had promised that it would be available if needed for the filming, I suggested that he ask Franck about it—they were seated next to each other. Franck said that his friend had moved away and taken his boat with him.

A lively discussion followed about how to proceed. It was already late, the wind was howling, and I suggested that we wait and go out to the site in the late afternoon the next day. Thus we would have the morning to solve the problem of how to get the big camera to the site, and for everyone to rest. My secret motive was to regain my original strategy of introducing the sharks to the film crew during their sunset rendezvous.

The idea was bounced around and Paul remained firm in his wish to go first thing in the morning as planned, to "rec the site," even if we couldn't take the big camera. So that was what was decided. I did a rapid sketch of the location of the site relative to the *pension* so they would set off in the right direction from the beach in the morning, and indicated where I would be and at what point we would be in visual contact. Slowly the group dispersed and we left. Sue came with us to the house for some supplies for breakfast so she wouldn't have to go out to buy something before they ate in the morning, and to collect weights, anchors, and some accessories I had bought to shade the equipment in the outriggers.

That week the fish shop had promised me two sacks of the frozen fillets, to be picked up on Saturday morning. But instead, they had provided only one small bag of the precious fillets, and two bags of ordinary fish scraps. I was stunned beyond belief, having explained carefully and repeatedly to both partners that I had to keep them until Monday morning and it would be impossible for me to keep the ordinary scraps cold that long. The sharks wouldn't want to eat rancid food, and the film crew would be thoroughly repulsed by the smell, and also, doubtless, by me and my rotten fish scraps. So it was very, very important to have the frozen fillets. They had said that would be no problem.

Having checked on these arrangements many times, not forgetting to bring copious supplies of beer, I was convinced that the two were playing games with me, on seeing how concerned I was over being properly prepared for the project.

During one of the storms, waves had swept up across the property and deposited some of the bay's contents on our lawn, and one of the things it had brought us was an old mattress. It was missing its cover, but nevertheless, it was a large, thick piece of foam. It had been stored ever since in one of the property's decaying outbuildings. So I laid out the foam, carefully arranged the scraps upon it with the frozen fillets at the core, and enveloped the whole mass in the foam, hoping that this exceptional amount of insulation would keep the scraps cold enough so that they would not rot before they were needed.

On Monday morning, at dawn, I dragged the bags to the shore, cut up the fillets, and carefully packed the boat. The scraps were partly rotten, but not as badly as I had feared. I was late by the time I had finished, but the wind had died down a bit. Franck was in contact with Sue by telephone as he watched my preparations from the window, and the crew left the *pension* at the same time that I left home.

I hurried out along the familiar course, through a dark grey, windy morning. But I was able to slowly advance, and it was a consolation to know that from then on I would be paddling out from the *pension* with the others, which would make subsequent trips far shorter and easier. Eventually, I was able to discern the two white outriggers proceeding slowly along the reef far away. We met a bit east of the site. Mike and Paul, who were doing the filming, and their assistants and sound experts, Beau and Mike, had come. I took them to the site, where waves ruffled the grey surface, making it impenetrable to the eye.

Everyone was frustrated with the outriggers, which they found irritating to paddle and hard to control. No wonder I had had such trouble with them, if they were hard to handle even with two men in each one!

It was 7:45 a.m. when we all went underwater. With extreme difficulty, I placed the mirror across the site from my usual perch. It was so heavy with the protective backing I had added, that as I held it, it planed downward, and I could scarcely guide, never mind control its descent. I thought it was going to drown me, since I dared not let go of it, and had to struggle wildly to get it into an approximate position without dropping it, before going back to the surface to take a breath.

Once the mirror was in place, it kept slowly falling over, and time after time I shot to the surface, gasped in a breath, and shot back down to grab it before it was too late. Due to my need for oxygen, I was unable to stay down long enough to expend the energy to pick it up and actually move it into the proper position. Finally Paul set it at a slightly increased angle, so that it stayed there. I grabbed the handful of fillet pieces ready in the back of my kayak, and placed them in front of the mirror.

By the time I caught my breath and looked back underwater, sharks were flashing past it going in both directions. But instead of waiting and watching, Mike was moving along the bottom toward the mirror in his bright-coloured outfit, with his big camera, until he was right beside it.

After I had stressed that the sharks would be nervous, and that it was crucial for them to have some time to get used to the presence of such unusual visitors, I was appalled. Apparently, he had no experience filming wild animals. I wanted to say something as diplomatically as possible, but we were underwater.

Paul remained across the site, and easily filmed from this more discreet distance.

I had placed the mirror so that sharks coming up the scent flow, as shown in the diagram I had sent to Stephen, would pass close by it. But Mike was creating an obstacle to any approach to the mirror. They needed space to find the food, see the mirror, and react—there would never be a spontaneous reaction to the mirror with him there. Couldn't he realize by himself that what we would be seeing was not the reaction of the sharks to the mirror, but their reaction to him? Even sharks had no trouble with the concept that they were observable to others.

I got more food, and tried to entice the sharks to come past the mirror by scattering particles from up-current, so that they would see themselves reflected as they swam up the scent flow. But, overcome with suspicion, Ruffles, Windy, Sparkle, Eden, Merry, Ondine and Jem cruised the site over and over again. They appeared frightened of the mirror and Mike, and looked askance at the crew and bright equipment, while ignoring the mirror and the food.

I did everything I could think of to get them to approach the mirror, but with Mike with his shiny paraphernalia in the middle of their circle, they behaved as they had done the first time I had taken food out to the sharks. They circled warily and kept their distance.

One thing in our favour was that the water was perfectly clear for once, and when Madeline sailed in front of the cameras in mid-water, I was pleased. That would make a beautiful sequence. Madeline was magnificent and had a long and interesting history. Cochita didn't appear until 9:02, an hour and a quarter after our arrival. I put more pieces of food in front of the mirror, hoping that this shark would provide the footage for which we had prepared so hard and done so much. Christobel was now passing at a distance in the coral, and she was normally confident too.

But I was waiting for Carrellina.

Time and time again, Cochita passed the mirror, each time shying away from the strange object edged by a bulky creature equipped with bizarre-looking things. How could Mike possibly be unaware, given the repeated fearful reaction of the sharks to his presence, that by staying right there beside the mirror, he was ruining the experiment, which was the highlight of the documentary? I was frantic, witnessing the failure of all I had promised Stephen, due to the ignorance of one member of the film crew!

Typically, when Mike turned his attention to have an exchange with Paul, Cochita zoomed in, dipped down, grabbed a piece of fish, missed, snatched again, and accelerated away. Neither of the cameramen saw her. This was actually the self-awareness demonstration—she was aware of the moment they were not watching her—but how could one explain that under the circumstances? Mike was there being observed by over a dozen sharks, but was completely unaware of it!

Eden

Cochita's action showed that if Mike had remained at a reasonable distance from the beginning, the sharks would have felt comfortable enough to approach, and we would have had the appropriate reactions to the mirror. They wanted the food. What worried them was Mike lying beside it.

Slinky, who was still a small pup, had been hovering around on the periphery, passing and disappearing and never approaching. But finally the little shark came to me and circled at arm's length. Paul approached smoothly, and Slinky passed between me and the camera, just sixty centimetres (2 ft) away from it. At least they would have one sequence of me with a shark. Having a very small juvenile on film could lead naturally into a brief description of the pups born in the nurseries, growing up in the lagoon, and possibly remaining friends with their siblings. Sharks having preferred companions was another one of my key points.

The magnificent Glamorgan joined the suspiciously patrolling sharks, with Breezy, Sparling, Pip, and Marco. Marco widely circled the whole area counter clockwise as was his habitual pattern. This was the shark who had appeared at the moment that the sun touched the horizon, on each of his first widely-spaced visits, four days before the dark of the moon. I tried unsuccessfully to point him out to the cameramen.

Once I drifted down-current, hoping that some of the sharks would gain confidence if they spent some time alone with me, and would return with me to the site. As soon as I was a slight distance away, several sharks rushed up to me from different directions and surrounded me. Mike and Paul swam toward us, but the sharks whipped away as fast as they had come, so the moment was not caught on film. But both cameramen saw that the sharks would approach if they were given more space, and in the clear waters, they could have been filmed from a greater distance.

Returning to the site, Cochita approached me tentatively, but when I threw her a piece of food, she didn't come to take it. She was skittish, and very different from her usual calm self.

The mirror experiment was to be the core of the documentary, and for the two hours that we were there, I tried every ploy I could think of to encourage the sharks to approach the mirror. However, with the crew draped across the site, and Mike blocking the swim-way at the mirror, they shied away from even tentative efforts to take the food. Increasing the scent flow did nothing to entice them closer. At ten o'clock we left.

The sea was shrouded in a brooding, heavy greyness as we paddled back to the *pension* against strong winds. Franck was there, and after a few pleasantries, and a brief discussion of the sharks' behaviour, we went home.

I had scarcely finished showering and cleaning up the house, when Sue and Stephen drove into the garden in their big van. I walked out to greet them and we talked for a while. Sue asked if I couldn't take my things for sketching the sharks with me. "I did," I said, and showed her my slate, covered with the names of the sharks who had attended that morning. Clearly, she and Stephen had come for a reason, and I waited to learn what it was, painfully aware that my slate and scribbled names must look absurdly primitive to people who travelled the world with eight hundred kilograms (1764 lbs) of high tech equipment. How could one explain that it was what was in the mind that mattered, and that the little slate with its quick sketches and names, scribbled while being tossed about in waves, was just a prompt? I hurried to show her my book with the finished drawings of the fins of the hundreds of sharks I had known, the name of each shark inscribed between the two sides in calligraphy.

They expressed extreme doubt about the session that morning, and I reminded Stephen that I had told him that the sharks would be shy. We discussed the matter for a while. He was surprised to learn that all the sharks present had been familiar to me, saying that the crew had said that there was one that I had not known—the one Paul had filmed circling around me.

"That was Slinky," I said, "little Slinky," and showed them the drawing of his fin in my book. I realized that they had already critically watched the film taken that morning, and that this was the reaction. After only a little while, they left.

When we gathered at the *pension* that afternoon before the evening session, it was obvious that the film crew had concluded that the mirror experiment was a flop. They didn't even want to take it back to the site. I gathered that *Discovery Channel*, the corporation that was funding the documentary, had cut back the time allotted for filming by so much that everyone was upset about it, and felt that it made their job practically impossible. They told me that they could spend the short time that they had hoping that the sharks would settle down and cooperate, or change the story.

"Ila's the story here," Paul said.

I explained that the presence of the cameramen in the space that the sharks considered to be theirs, had alarmed them. They were used to that being their place—perhaps I had not been clear enough about that in the diagram that I had sent, but I had indicated the optimal places for the cameramen to be, on the periphery of the site, in the best places to view the sharks passing, using their habitual swim-ways. But the crew was adamant. They found all the information I had provided about the individual sharks and the visitors remarkable, particularly the information about Clementine and Odyssey who were currently in the area.

So they had decided to film me with the sharks, drawing them and talking about them.

It wasn't until late in the day that I went to the fish shop to get more of the fillets that the partners had said I could have then. I had been anxious to get them as early as possible so that they would have the day to thaw out, at least enough so that I could cut them up. Once they were small, they would soon thaw completely in the warm sea water. But Franck didn't want to waste the gas for the seven minute trip, and insisted that we pick up the sharks' food before we went to the *pension* in the afternoon.

The younger partner was there and told me that there were, quite suddenly, no fillets!

I bought two boxes of sardines at over twenty dollars each, remembering how the sharks had eaten one box and then been dismayed that I was leaving. One never knew what would happen. If interesting sharks came later, it would be imperative to have food available after the regular group had been satisfied and drifted away. If Odyssey and Clementine came, it would be at nightfall. Filming them was my priority, and the only way I had to tempt the timid old ladies to approach the cameras was with some good food.

When we got to the *pension*, I left the sardines in the back of the truck temporarily, while greeting the team. Whereas they always came well-packaged in a box, enveloped in heavy plastic, this time they were in a sack with holes in it. Fluid was already leaking from the block of frozen fish by the time we arrived, and had run all over the back of the truck.

The crew had noticed Denis' motor boat, and had decided to use it to transport the big camera to the site. The problem was that Denis refused to let us take it out on our own, in spite of the team's level of expertise. He insisted on driving it himself. I was adamant that he must not be involved in the project due to the local attitude to shark-loving white people, and the intensely anti-shark sentiments expressed by the husband of the *pension's* manager. Further, Denis could decide at the last minute that he didn't want to take the crew into the lagoon, and just stop on the border, leaving us stuck.

I carefully explained to the crew that the attitude of modern Polynesians toward sharks had changed in the last generation. This was partly due to the influence of American television, which dominated local programs and consistently promoted shark attack mania. On top of that was the profitability of shark fins, and the growing animosity toward whites, tourists, and shark dives, which they saw as the source of the threat to their shark finning profits. With the rise of the Independentists, the attitude had become official.

Andy, who was the only member of the team who spoke French, had talked to Denis that morning, and had finally gained permission for the team to use the motorboat unescorted. But as we sat talking at the tables on the grass by the beach, while the cameramen and technicians prepared the cameras and other equipment around us, Denis approached and stood hesitantly on the periphery. Andy went to speak to him, then came and told me that he wanted to drive the motorboat. So I agreed. The crew assured me that they would be able to get the craft to the site if one person stood watch in the bow and gave directions for guiding it around the coral formations.

I asked Mike and Paul how I could signal to them if an interesting shark such as Clementine or Odyssey appeared, that they should concentrate on filming. Any elderly shark, such as Twilight or Cheyenne should be filmed—there were so few of them left. They said to indicate the shark to either one of them.

There didn't seem to be anything else that they wanted to discuss about the filming, so I went back to the truck to get the frozen block of sardines and to try to thaw it out. In that short time, not only had the oily fish fluids flooded across the truck bed, but a cat had bitten through the plastic and was chewing at the little fish on the surface that were beginning to defrost. I took the block of sardines, rock hard at minus eighteen degrees, to the lagoon, and put them in. This was the fastest way to thaw the fish, as sea water is five hundred times more dense than air, and in Polynesia, the temperature of the sea is always warmer than the atmosphere. As the fish on the surface of the icy block softened and separated, I put them in the back of the boat, and loosened the next ones that were softening.

When he came down to the beach, Mike looked at what I was doing so urgently, and offered the opinion that I had brought too much food. How would he know? I pointed out that it was better to be safe than sorry—the most interesting sharks often came latest. He wandered away. That was the only time he mentioned the subject of feeding the sharks.

By the time everyone was ready to leave, only about half the block had thawed, and many of the delicate little fish had broken. I put the now slushy, frozen core into the boat, where it bled an oily fish broth into the well of my kayak. The benefit of having sardines was that they resembled clean, shiny leaves in the water, and the sharks swept up to take them, providing a dramatic spectacle of sharks looping in from all directions up to the surface and back down, instead of roaming boringly over the sand in search mode as they did with the fish scraps. However, if they were broken, the effect was spoiled. With my heart set on this evening session at which I would present the large group of sharks I had promised to Stephen, I was very concerned that the little fish remain clean, and rinsed them continually with sea water—this was the reason I had been so anxious to get them from the fish store in time for them to thaw out. I hoped that since the cameramen were aware of the direction of the current and scent flow, indicated in the diagram I had given them, they would be filming on the up-current side, so the sharks would appear in perfectly transparent water. There would be multitudes of sharks soaring through the coral in all directions, so in the crystalline water we had already seen that morning, it would be easy to film them moving through their coral surroundings.

This was what I had waited for, what I had worked so hard for, and it was about to become reality! I had the food, the film crew was there, the conditions were passable, and the sharks would certainly arrive at their evening feeding, so everything should be all right!

I sat in my kayak near the shore watching the crew prepare, while Franck talked to Sue, and photographed the scene. Paul was filming Mike introducing the outriggers, in a hilarious monologue about their construction, paint job, the *Discovery Channel*, and other details. While I didn't understand the references to the *Discovery Channel*, I knew it was of primary importance to them, and I was awed by Mike's natural talent and flare for humour. I was watching a television professional for the first time in my life, and felt deep appreciation for the privilege to which I was being treated, witnessing him and a team of this calibre functioning together.

Finally we headed out to the site. I was with Mike, and Paul and Stephen were in the motorboat with the assistant and sound specialist, Beau and Mike. The motorboat followed the channel close to the shore, and by the time we got to the site, it was far away in the bay. Mike grew increasingly impatient about this, and wondered aloud what they were doing. But I could see that Denis knew the topography of the lagoon very well. He stood at the wheel, the dominant figure in the boat, with his white hair blowing about in the wind, inching his precious craft forward, directly toward the hidden channel, the only place where it was possible to cross the border. I could understand why he did not want anyone else to drive his boat. He knew the lagoon while the visitors did not.

It took a very long time for him to manoeuvre through the coral to a position about thirty metres (100 ft) away from us in the direction of the shore, luckily out of the scent flow, which angled toward the border. I had my fingers crossed that Denis would not notice the sharks.

They were close around my boat, but not coming up to the surface with Mike's big outrigger parked nearby, and the presence of yet another odd thing—the motorboat buzzing in the water. I patted my hand on the surface when one of them rose up, and occasionally tossed them a small treat. I was watching for Carrellina, but couldn't see her. There was still a lot of wind, so the view through the surface was too distorted to see details. The clouds seemed lighter, but not a ray of sunlight penetrated the sombre sky.

Mike did not speak to me.

Finally, the motor boat was in place, and Paul approached underwater, trying to film the sharks around my boat. But when he neared, they vanished. We tried several times, but they remained out of his visual range. Mike and I slid in, and as I had foreseen, there was not a shark in sight. I followed my normal procedure, pulling the kayak to my lookout perch, steadying myself there, and trying to maintain the boat against the wind while throwing handfuls of fish into the site. In between, I dipped my face underwater to look. The sharks were nervous, but approached timidly to eat. Teardrop, my once-black shark, was the first! She would be shown on television! Usually she came late, since her home range was far to the east.

The wind was so strong that it kept yanking the kayak from my grasp, and overbalancing me. I often had to struggle frantically to keep my grip on both my lookout and the boat, as I tossed handfuls to provide a steady supply of food for the sharks. Determined not to make any mistakes during this one chance to present a perfect scene of my sharks' gathering, in desperation, I finally lifted one foot to place it against the boat on the leeward side to hold it steady. In this contortionist position, I was horrified to lay eyes on Paul, to my right, with his television camera pointed straight at me! What was he filming me for? I was going through this struggle so he could film the sharks, and would not even have been there in such wind had it not been for our filming schedule!

Mike began to cross the site for close-ups. Trinket came circling around the swirl of sharks in the centre of the site. He passed in front of Paul, and I excitedly pointed him out. The shark was actually turning right in front of Paul at that instant. But again, that huge lens was pointed straight at me! I pointed to the shark again, more urgently, yet the camera didn't move, and he didn't look where I was pointing!

By then the site was full of sharks—why wasn't he filming them? I couldn't understand it. He could film me any time, but the sharks were only there then! Paul did eventually film Tamarack turning in front of him, but missed Trinket completely. Trinket was the only shark I had kept track of since he was a baby, right up to his move into his new range in the ocean. It was two years since I had seen Trinket in the site!

The grand and beautiful Keeta was cruising around. Flora, Gwendolyn, Windy, Ruffles, Vixen, Eden, Sparkle, Storm, Brandy, Tamarack, Simmaron, Jem, Fawn, Tristan, and Odetta were there, and many more I did not note. But Carrellina was not. I did not know it then, but I had seen my favourite bad shark for the last time. Had my star shark been finned just days before the arrival of the BBC? I would never know.

With a motorboat down-current, and so many people in the water, there were no nurse sharks, and fewer sharks than usual. Many who had come in the morning—Cochita, Ondine, Christobel, Pip, Merry, and Slinky—were absent. The shy ones may not have approached, but listened, instead, in shark amazement, to the session from beyond their curtain of blue. Or should I say, in this case, their curtain of grey.

But Eclipse soared through the site from right to left, just one time. Months had passed since I had seen her. It was the first time she had ever approached from my right. Her hit-and-run visits always involved a left to right pass. No one filmed her.

Amazingly, The Concorde joined the circling sharks as the water darkened—it was just my fourth sighting of her in five years. I hoped she heralded the approach of Odyssey and Clementine, but she herself was a majestic old shark. My efforts to point her out to the film-makers failed again.

As the light faded, Bratworst and Tuvalu drifted through. I was still throwing a few fish in from time to time, to keep up the scent flow, but had only brought one box of fish, so was making sure they lasted.

When it became too dark to film, the crew turned on the lights, and the scene became mystical. I was rapt by the sight of Keeta turning and turning in the light of an underwater moon. Many others were still circling through the area and passing through the light. Beau and Mike later said they were very impressed by Keeta's behaviour in front of the lights. She was very close to them there, and they delighted in watching her. Tamarack too, seemed to like the light, and stayed within its radiance.

Suddenly there was a commotion down-current and Stephen, dressed in shorts and a mask, appeared between the coral structures. The sharks vanished with the noise he was making, upright in the water. It was his only visit to the site. I think that he just wanted to see it after our two-hour talks about the sharks and the site on the phone, and having waited patiently for so long in the motorboat.

Then we packed everything back into the boats, and paddled back to the *pension* in the prevailing wind.

Sue had food and drinks on the table as soon as we started coming across the grass. Someone handed me a beer, and I joined the group in excited discussion around the table. Beau, who was very kind, asked me how I was—I was still in my wet skins and was shaking with cold.

I smiled delightedly, and told him, "I'm happy! If you're happy it doesn't matter if you're tired and cold!" Everyone seemed so excited, talking about this shark and that one. Even the very serious and conscientious Paul was pleased. He agreed things had gone very well, and he had taken a lot of film of the sharks. I considered him the thermometer for the response of the team—if Paul was happy, well! Everything must be all right!

But how different it would have been, had Martha been at my side, had Madonna still been alive, had Carrellina come. . .

The next morning dawned to a shadowy light with rain in the wind, and darkness rimming the northeastern horizon. Every time I looked up, it seemed to have moved toward us.

Again we sat at the tables talking, while preparations were completed. Sue always had snacks and refreshments laid out for everyone, and it was the place we naturally gathered before the sessions. But when the conversation moved to politics, a human construction that has always just sailed over my head, I noticed the two cameramen standing talking on the beach, and went over to join them. They started when I appeared, and I knew I was unwelcome.

"Well," I said apologetically, "since it's you I'm working with, I thought I'd see if there was something we should discuss about today's filming." There was no response, so I went to my kayak and checked that everything was properly arranged.

Once more we paddled out, and the motorboat moved slowly with us, closer to shore. Paul stood in it shouting directions to Mike and me, while he filmed us moving together and talking.

We moved in one direction paddling and talking, and then Paul yelled at us to change direction and do it all over again. As a result of these manoeuvres, we went far off course, and had much farther to paddle when we were finally free to go to the site.

We sat waiting in the gloom, the grey water rhythmically, restlessly slapping the boats. Mike was sitting head in hand, and looked very glum. He didn't speak. I touched the water gently to see if any of my sharks would rise to me, but no one did. Finally, the silence became so oppressive that I said to him, "You don't look like you're having a good time."

Mike looked up, indicated the distant motorboat easing itself over the border, and made a gesture indicating extreme displeasure with them. "What are they doing now?" he asked impatiently.

"Do you not usually work with them?" I asked.

"No!" he responded with such vehemence that it was obvious that he was not happy about doing so this time. We continued to sit in silence. A small shark was investigating my boat, and I patted the surface. Another appeared. I distracted myself with them as the motorboat wended its slow way through the coral and anchored in place. Paul entered the water with his big camera and moved slowly toward the site. I slid underwater, and he ordered me back in my boat saying, "Are we in a hurry?"

"No," I said. "I didn't know what you wanted me to do." I was often unsure, since little was communicated about what the precise plan was. This session was to fill in the gaps in the story, but I wasn't sure which gaps, exactly, those were. There was never any time to talk about it. We were always preparing, going, or leaving.

Paul had seen no sharks underwater; at his approach those around my kayak must have fled. We waited. I dropped the anchor again so that he could film me doing it. After another wait for the sharks to appear, I suggested that they must be around, but not approaching.

Paul filmed me underwater, descending from the kayak, and I spent the next hour feeding the fish while we waited for sharks. Very few came. With the motorboat down-current, and the feeding the night before, they were not interested in the food, and didn't come to see me in their usual way due to their suspicious attitude toward the film crew.

Windy swept past me with Marco as if she didn't know me. She had doubtless been listening beyond visual range, and this was her one pass through, to take a closer look. A new juvenile resident, Tawny, swam slowly toward us up the scent flow several times, only to turn at the edge of the site, and disappear in the coral.

Paul filmed me drawing Marco's fin as he cruised through the site. Marco was so familiar I could have drawn it from memory, so I knew that it was perfectly accurate in defining the line of his fin tip. But it wasn't the finished sort of drawing most people are used to seeing, (which are not done in seconds underwater with a heavy pencil, while bobbing about in waves). I held up the drawing after for the camera to focus on. Then Paul filmed my face as I drew.

Amazingly, the beautiful, rare visitor Samphire glided by—an enormous shark. To my regret, she was not filmed, though she was one of the most interesting visitors who came.

We had been in the water for nearly two hours when we decided to leave. The day had lightened somewhat by then, and hints of sunlight began to glisten on the disturbed surface of the sea. But not one ray of sunshine had fallen upon my sharks during the filming. No rushing golden light and turquoise shadows had transformed the scenes to magic, and the gloom had dulled the velvety sheen of my beautiful sharks.

But, the water had been clear as crystal, the visibility perfect, and my promise of a large group of sharks had been kept. Thirty-five different individuals had appeared, some at more than one session. The world was going to find out about my sharks and what had happened to them. I sat thoroughly relaxed in the warmth in my boat, and contentedly took off my gear.

Mike was still sitting glumly on his outrigger, and I wondered if he had been upset because Paul had not wanted him to come underwater. Maybe the veteran cameraman had noticed that the novice failed to comprehend what was required for filming sharks, and had told him to stay out.

Paul informed me that we were now going to do the "top-side stuff," and he explained that we were going to pretend that we were now on the way to see the sharks. Unfortunately, my hair was soaking wet, but the wind was like a hair dryer, and soon took care of that problem. I was sure that the marks left on my face by the mask, after being underwater so long, would be obvious. But this was not the moment for vanity. Mike, the sound specialist, gave me a dry yellow T-shirt to wear, and attached a tiny microphone underneath it. While we prepared for this new phase, the sky lightened even more, and sunshine actually began to break through.

Mike and I paddled out onto the bay, to a place where the view of the encircling volcanic ridge was optimum, and then began paddling back toward the site, while the motorboat cruised beside us, and filmed us talking. We repeated this many times, with the motorboat at different distances. Each time Mike asked me the same questions, and each time I answered, trying to sound authentic each time. These questions and answers established the framework for the scenes with the sharks that were to follow, and addressed the fact that the population I had originally studied had been decimated by shark finning.

As we went back and forth, and up and down the bay, the wind strengthened, and I began to have trouble. The waves on the bay rapidly increased in size, and I found myself at the end of my strength. I had scarcely eaten nor slept since the film crew had come, and my reserves had run out. The ever-present pain required extra courage, and had become overwhelming. I had not dared to take any pain killers, since they could affect my level of alertness with cameramen and sharks.

Finally, something broke on Mike's outrigger. Mike, the sound specialist, fixed it temporarily for him, but it continued to be a problem. Everyone seemed to be feeling the effects of the midday sun and rolling waves, and finally Denis suggested that we go into the sheltered bay to continue. At the last minute, I managed to hand Mike my anchor, which he held, so that I could fly over the waves behind the motorboat, which was towing him. I didn't think I would have made it otherwise. When we had finished, we were deep in the bay, and I managed with difficulty to get Paul's attention to ask if we were going back out that night. He assured me that we were definitely not.

So I suggested that we leave the outrigger where it came from, adjacent to us on the shore, and I could go straight home. Our house was visible close by.

Without further discussion, Mike the presenter got into the motorboat, and Mike the sound specialist took charge of the outrigger. As the motorboat zoomed away and vanished, I accompanied Mike to the home of the man who had rented us the outriggers, and tied the irritating craft to a tree. The owner wasn't there, but I left a message for him with his son, while Mike started walking to our house. I got back in my kayak, and rushed it home. Pulling it up on the beach, I noticed that one of my fins was missing. It must have blown out of the well of my boat in the winds on the bay! Mike was just arriving when I got out to the road to meet him.

Franck had gone to pick me up at the *pension* at ten o'clock, and by then it was noon.

I gave Mike a glass of pineapple juice on our deck, then rushed out to salvage his wet hat from my kayak, and change out of the yellow T-shirt, while he waited for someone to come and pick him up.

Franck drove in. The motorboat had swiftly arrived at the *pension*, and Mike and Paul had told him that I had gone home by kayak. I was pleased and relieved to hear that in the interim he had enjoyed talking to Sue, Stephen, and Andy, and felt deeply grateful to Sue for her beautiful way of making people feel fine, and looking after everything and everyone.

Mike had told him that I had been "just great," Franck said proudly. The minivan pulled in right after he did. The efficiency of these people was stunning! There was never a moment wasted, as they cheerfully and flexibly worked together around all sorts of changes and obstacles.

Stephen said that everyone would come back around three o'clock to film the interview at our house. To my enormous relief, he seemed pleased.

I showered, and got busy washing the dishes and doing the usual cleaning up and chores. We just had time to have something to eat when the crew arrived for the filming.

Franck brought out a delectable selection of tropical juices. I had just made coffee for myself, since I felt I needed a jump-start before the interview, but I didn't want to appear drinking it in case someone else wanted some. So I asked if anyone would like some coffee, and Paul was intensely appreciative. We sat together sipping it and smiling.

It was obvious that the team were all tired from the whirlwind documentary filming, the travelling, and the jet lag. Now they busied themselves setting up to film me talking about the sharks, in the light of the setting sun, which shone through the trees overhanging the beach. The place they chose for me to sit was the precise spot I had sailed over the bench and nearly broken my back on the bowl of the leaning coconut tree. Irony and coincidence seemed to haunt my history with the sharks.

While they worked, I rushed around getting ready and putting out grain for the birds.

"What are you doing, Ila?" Paul asked. I explained that tiny chestnut breasted mannikins came in a flock at certain times, demanding food. The little tyrants would cling to my hanging plants until they resembled beads on multiple strands of a necklace, peering in the windows and shouting at me with tiny voices. If that didn't work, they would throw themselves against the glass to make me produce their food. Given such a performance, it was best that I look after the little things before we did the filming.

Everyone was standing around looking at the bird table, as hundreds of the tiny birds descended like golden snowflakes to eat. But there were too many people too close so they rose up again. Finally Mike, the sound specialist, put the camera on the bird feeder and left it there. Birds don't like suspicious looking objects any more than sharks do, but nevertheless, the tiny birds were once again descending when Franck came out the door, and they all flew up again.

Mike drew me to the other end of the deck, and told me that he would ask me a series of questions about the sharks, including why I fed them. He said I should be prepared for that. I was surprised he brought it up, when he knew I was using the feeding to facilitate the filming. This had been planned in concert with the BBC, and we both knew that the primary subject of concern was the killing of the sharks for shark fin soup.

So I acknowledged him, but felt it was an unnecessary distraction. He himself had seen that even with food, the sharks had been reluctant to come near. Their behaviour had been totally changed by the presence of the film crew, and the food had been there to compensate and bring the sharks close anyway, so that we could get other information about them across. In the case of the mirror experiment, with him lying beside the mirror, even food had not compensated for his intrusive behaviour.

But I was keeping my key phrases, properly ordered, in the front of my mind, so considered the feeding question to be minor—there was no indication that it would not be easily passed over with a few words.

We took our assigned places on the bench around the deck where it was angled so that Mike and I were slightly turned toward each other. We watched the others filming the birds while the TV camera was set up, along with accessory equipment, and a veil set on a frame to soften and reflect light on us. Mike, the sound specialist, clipped a tiny microphone to my dress, which connected to a box attached to a belt at my back.

Time passed, and quickly the evening light began to fall. Franck was standing watching, holding his camera. Finally we began.

Courtesy of Franck Porcher

I used my most important key phrase, spoken in a natural way so that it wouldn't sound rehearsed. Indeed, I felt completely relaxed there with these people of whom I had grown so fond, Franck watching with a pleased look on his face, my birds around me, the soft and fragrant airs, the warm light of the setting sun glowing in the fluttering leaves, and my ambivalent love, the sea, lapping gently behind me upon the sand.

It emerged as part of my introduction, that it was by observing sharks as individuals that I had been able to witness a deeper level of their daily lives and activities. I went on to describe sharks having friends, and how they travelled in loose contact, not always together, but frequently meeting due to their pattern of swimming in great circles, which joined in the centre. This pattern permitted the sharks to intercept not only the scent trail of food sources, but each other's scent too. The case of the long-term friendship of Clementine and Odyssey led naturally to tell how the shark pups were born in nurseries and left them in little groups in contact with each other—possibly these companionships endured.

The conversation was interrupted with each question as Paul filmed Mike asking the question, then changed position to film me responding. I answered guided by my plan for best presenting the most interesting material about the sharks.

Each time Paul changed the camera's position, there was a pause, and I was aware that the rest of the team were sometimes there, sometimes not. Once I was startled to see Mike and Beau very slowly closing in with a television camera on Diamond, who was foraging upon the grass, ablaze with his new fire-coloured frills in the sunset light. How remarkable that they would be interested in my birds!

Mike asked me why I fed the sharks. I smiled as Tamarack sailed by in my mind, globs of sardines scattering from his stuffed mouth, and said, "Well they were certainly happy. I brought them something special because you were here. But usually they don't get that." He repeated the question in different ways. I couldn't understand why he kept on about it and finally became unsure, and hesitatingly said that it was the only way to win their trust, to get them to accept me among them so that I could watch them, yet they would behave as if no one was looking. My phrase, "gesture of benevolence" failed to occur to me.

My actual research method, of providing feeding sessions once a week to identify the sharks in the area, then swimming with them without food during the week to learn what I could of their daily lives, had not been discussed. Such an explanatory answer seemed to me to be too long, and not part of the planned story.

Isolated as I had been for so many years in Polynesia, I had no idea that elsewhere in the world, there was a controversy among some circles on the subject of shark feeding dives. In any event, such controversies didn't concern me, since I wasn't trying to profit from eco-tourism. I was just trying to get to know a very interesting set of species, not even aware that they had never been studied underwater as individuals before. I had certainly had my share of problems with feeding the sharks, but lacked the time to explain all that then. Why use the short time available in the documentary to discuss something that was irrelevant to the planned story?

I talked about the things that I felt would be most interesting and surprising to viewers: the lack of aggression among sharks, and toward me, how Martha and Madonna related to me, and behaviour that demonstrated cognition. I had chosen the sharks' social lives, their long-term friendships, and their intelligence as subjects with the greatest potential to evoke positive feelings for sharks in viewers.

Finally, I described how nearly every resident female I had been close to had fallen victim to the massacre for shark fins.

We moved inside as the sun went behind the mountain, so I could show Mike my book of drawings of the fins of the sharks. This took a while since all the equipment had to be moved inside, set up again, and the lighting had to be properly adjusted.

We looked through my shark fin book, and talked about it and my records. I pointed out the drawing of Trinket's fin on my slate, that I had made at the session the night before, then showed the matching drawing in my book, on the very first page. Then I described how I had met him when he was a pup, watched him grow up, move outside the reef as males do, and that two years had passed since he had appeared in my study area.

Then Mike returned to the subject of shark finning. I answered that sharks must be protected because otherwise there would come a day when we would still be here, and the sea would still be here, but they would be gone. Forever. This was another key phrase, one I hoped would be lyrical enough to be sure to be included, and a fitting ending to this short glimpse of my sunset rendezvous.

There was a short scene of me working on my computer, and then the filming of some close-ups from my book of drawings of the fins of my sharks. When I was asked if there was anything else I wanted to say,

I rushed to the bedroom where I had left out my notes for handy reference.

But as I passed him, Franck handed me the telephone, explaining that my son, Peter, had been repeatedly calling and insisted on talking to me immediately. I took the phone and received the news that my father was dying. It couldn't have come at a worse time. There was nothing I could do, ten thousand miles away, and it laid a dark and awful shadow over the scene. I forgot all about checking my notes, and wandered back into the living room.

By then the last details had been filmed, the equipment was being put away, and drinks were coming from the kitchen, served by Sue and Franck.

I joined Franck and Andy who were in an intensive discussion about whiskey. Franck was a connoisseur, and we had learned a little about this unusual industry when we had visited Scotland several years before. Franck got his favourite whiskeys from the cabinet, and served some to everyone.

I went looking for chairs to arrange, as slowly everything was done, and the film crew came to join us. As we all congregated in the living-room, they said that they had filmed everything—the mannikins eating, the garden, and the junglefowl. I sat on the floor beside Paul to be sure to hear his opinions, as Franck handed me a glass of punch. Paul had lost his serious air and seemed just delighted, to my intense relief. He said that my interview was great, and that I had "really lit up" talking about the sharks. Looking at his watch, he said that it had taken twelve hours of filming. I sat drinking my punch rapidly due to an intense need for nourishment, and critically observed the scene, keeping a close eye on Paul. Sue was sitting on the couch with Stephen, and the others were grouped around, all making humorous comments and bursting into laughter. I had rarely in all my life seen such a high level of happy emotion in a group.

As the glasses of whiskey that Franck brought around were consumed, this joyful atmosphere reached a peak which transported me, and made up for all of the anguish I had felt preparing and worrying over such an ambitious venture. Everyone seemed to be thrilled with the filming, the sharks, me, and the interview. Paul was just radiant. No negative threads were detectable anywhere. Everything must be all right, I assured myself.

It was a memorable interlude, but did not last long, because they had so much to do, that they soon returned to their hotels. But they left me reassured that everything had turned out all right. They would not have been so unreservedly pleased had they not been truly content with what they had accomplished with me.

Franck too, seemed delighted. He thought my interview had been very interesting, and that I had spoken well. Though a few things I had wanted to include had been left out, I was confident that my message had been delivered. Franck went on to talk about how powerful the scene had seemed as he photographed it, moving around the periphery, watching. He said that everyone there seemed to feel it, that what was happening on that deck was right in the middle of a vortex that was going to spread out, and affect the whole world.

I was exhausted and famished after the continual strenuous activity, with not enough food or sleep, so Franck took me out to dinner. The hotel restaurant was partly outdoors, and surrounded by tropical gardens and lighted coconut trees. We sat in the luminous air beneath the stars, with the breeze wafting the sweetness of nocturnal flowers over us, sipping wine, and reminiscing over all the amazing things that had happened in the past few days.

After I had imbibed sufficient wine, I reminded Franck that if there had been one event that had turned my attention to sharks, it was that dream of the two sharks, one black, one white, swimming to me in the supernatural lagoon. When I had leaned down, the black shark swam into my arms and spoke to me, and I realized that all of its fins had been cut off. It had been such an overpowering experience that I had awakened riveted with anguish, and desperate to remember what the shark had been saying. It had felt like a call for help.

"Then," I said to him, "You remember, just two days later, that shipload of shark fins came into Papeete harbour, and we went down to the dock and argued with the crew."

Franck remembered my disturbance about the dream so many years before, and the subsequent trip to the rusty ship after the article appeared in the paper. Franck had been especially impressed by the number of sharks estimated to have been killed to fill such a ship with fins. Maybe they had been painfully dying not far away, at the time I had the dream.

In the surreal surroundings, in which the southern stars glowed so close, the supernatural seemed to be just beyond the curtains of awareness, and Franck recognized my meaning. He agreed that indeed, it was as though the shark spirits had sent a call for help, and as if my way, ever since, had been cleared to put me into a position to help them. Franck urged me to tell the BBC about this facet of my story, but I didn't think that they would understand.

The next morning we went to the hotel where the technicians had set up the equipment, to look at the film that had been taken. Everything except the sharks was fast forwarded. I gave the names of the different individuals on film, and Stephen wrote them down. Later, I would be able to send him information about the ones that they were going to include in the sequence, and Mike, the presenter, could give some details about them. I was anxious that Trinket be included, but did not see him on Paul's film. I drew his dorsal fin for Stephen, in case they found him on Mike's film when it was developed, stressing that this shark provided a good example of the life of a pup growing up. Five years of his story could be summarized.

Franck had called a press conference for eleven o'clock, and Andy and Mike came, obviously tired after the pressured sessions. Mike took me by the shoulders, and told me he had heard that I made just wonderful coffee. Would it be possible for him to get some?

"Well of course!" I said, and rushed off to make some for us.

Jeannot Rey, the island's representative for *La Dépêche*, arrived, and Mike and Andy were interviewed. Mike's natural talent for expressing himself shone, as he said that his approach was to show people the beauty and fascinating things about nature, to encourage them to treasure and preserve it. He and Andy gave an awe-inspiring interview. They had both worked on the renowned film, *The Blue Planet*, which amazed everyone. When Jeannot asked Mike if there were any personalities in the documentary, Mike said, "Ila is the only personality!" Jeannot was astonished, not knowing that I went underwater with sharks. All of his interviews with me had concerned my sea turtles, seabirds, and art work.

After Jeannot left, I drove to the *pension* to see how things were going, and if there was anything I could do to help. But Stephen and Sue were tired, and overwhelmed with work, so I didn't stay long, but picked up the heavy mirror. It was an emotional good-bye, particularly with Sue, of whom I had become very fond. Stephen said that if I could get a film of the sharks in front of the mirror by myself, he would add it to the documentary, and I said that I would try, though without access to a TV camera, I had no means to do so.

I took my leave of Denis as well, thanking him for all he had done for our project, for his hospitality and help. He was very kind as he smiled an acknowledgement, and I breathed a sigh of relief as I left, that he had not said anything about sharks. He had probably not noticed them from the boat. One more facet of the project had turned out all right.

Jeannot wrote an article about how this "little woman," all by herself, had learned everything there was to know about the lives of sharks—what they did, where they went, their gestation period, their social lives, who was friends with who, everything, and the BBC had come in secret to make a documentary about it.

Franck sent out his five star press release, and for the rest of the day the phone continuously rang. Franck refused to let anyone talk to me, preferring that I remain the mysterious figure in the background due to the prevailing hostility against white people and sharks. If the television stations had wanted to interview the film crew from the BBC and me, they could have attended the press conference that morning. I was not available.

This was typical of the English articles that were published. In spite of certain inaccuracies, such as the mention that I was scuba diving, it reflected the information given by Mike and Andy at the press conference.

BBC Films *Jaws Junior* Shark Documentary

A BBC television crew recently visited Tahiti's sister island to film sharks' social intelligence as part of a feature documentary movie entitled Jaws Junior for the Discovery Channel.

Special correspondents of the BBC's Natural History Unit were part of an eight-person team that spent three days filming the research work of Ila France Porcher, who has lived in French Polynesia since 1995.

The filming for the documentary began in Australia, continued [in Polynesia] and then continued in Hawaii. The correspondents have reportedly decided to use the images filmed in Polynesia as the main part of the documentary.

The project began when a researcher for the BBC Natural History unit was given the task of finding subjects and preparing them for filming. That was when he discovered the research work of Ms. Porcher, who has spent the past eight years observing shark behavior in their natural surroundings. Her goal was to try and uncover the mysteries of the sharks' intelligence.

"Her work is unique and recognized unanimously by the scientific world," said Franck Porcher, her husband. "She has succeeded in proving that sharks are totally harmless, that they know how to react when faced with unexpected elements and were capable of being friendly.

"Ila has been scuba diving with two sharks for such a long time that she has managed to establish a real relationship with them," Porcher said.

The BBC correspondents said that after three days of filming they had become astounded by what they had seen, Porcher continued.

The Discovery Channel plans to air the documentary in August, after which it will be sold for rebroadcast throughout the world. Porcher said the BBC correspondents described the film as "the best they had ever done".

This article is from GeoChannel "*Global Nature News, Research, Education and Media Network.*"

Chapter Fifty-Six

Peace

Two weeks of storms followed the departure of the BBC, during which my father sank deeper into a coma and died. Exhausted after the long preparation, the hours of filming, and this devastating news, I collapsed in the silence that followed, reading in the greyness, and listening to the wind and waves pouring past the house. I had communicated my message for all my dead sharks, and those still living, via a major television channel, which was more than I expected ever to accomplish in my life. Finally, it must be all right for me to rest. Franck agreed and told me, "You can die now."

Arthur sent me a few suggestions for the revision of my manuscript on the gestation period of the blackfins. It was the first time he had read it, and he said it was a fine article and a wonderful study. Pleased, I reworked it, and sent it back to *Marine Biology*.

As May began, the second fish shop reopened, and called to ask if I could pick up the scraps. So, in a brief period of calm, I loaded them on the kayak, and went back to the sharks.

It was a pleasure to go out that evening as the sun broke through scattering clouds to glimmer upon a sea that had fallen quiet. No sharks followed across the lagoon, but they soon gathered as I reached the vicinity of Site One. Cochita glided slowly beneath me, and I stroked her. So many sharks congregated that I threw in their food before sliding in myself, and called it learning from experience. If sharks could do it, surely I could.

Underwater I found myself in a bright, golden fog. There were many sharks, among whom I circled, looking at each as they fed and socialized.

Ali cruised through the golden glow toward me. She looked like Mephistopheles—part of her nose was cut off. It was hard to see in the glowing water, and since she tended to avoid close passes, I had to cruise around her discreetly for some time, to discern what was wrong.

It appeared that someone had hacked straight into her nose with a machete, the favoured tool in Polynesia. There was a wide V-shape of flesh missing from her rounded nose, and the cut extended up the middle of her face where her skull was exposed. There were also slashes at the back of her head. The injuries were hard to assess without being able to examine her, and black lines showed that they had begun to heal, so she had been attacked more than five days before, probably in a desperate battle for her life.

Shilly was with her. I threw in the last of the scraps for the friends. Ali had always been a shy shark, and it was difficult to get near her. How had she been in a position to have been slashed so badly? Caught by shark finners and escaped? How else could she have had her head near enough to a machete? It would be impossible to hack up a shark with a machete underwater without being able to hold her head still. Someone must have been trying to haul her into a boat by a hook in her mouth, while she thrashed wildly.

More sharks drifted through while I watched, wrote down their names, and sketched the fins of the new juveniles. A group of large females from the eastern end of the lagoon began passing down-current, and Mordred cruised into the site.

Jessica appeared, and in the strange lighting, the barely perceptible rash I had noted before was highlighted. It now covered the region on her back between her nose and her first dorsal fin, and resembled hives, or eczema before it breaks out into a rash of itchy scabs. Once, she began twisting and jerking her body in the strange way I described in the unknown male at the unusual Site Two session before the great shark disappearance. I lost sight of her as she moved behind a coral formation, and when she reappeared, she was gliding normally away through the coral. Later she passed me much closer, and the rash, a myriad of raised bumps close together, seemed to be all over her; they were just more developed in front of her dorsal fin.

More and more sharks gathered. I put the last of the food out for them, and threw in the fish heads. Twilight picked one up and carried it some distance up-current, where one of the three-metre nurse sharks swept down to it and began to tear it apart, scattering particles to which the fish flocked. Other fish heads were dispersed throughout the vicinity by blackfins, causing several scent flows as the nurse sharks began their methodical munching and shaking, and the mêlée that resulted attracted everyone back. Reddened sunbeams flickered through the glowing clouds rising around the nurse sharks, through which the sharks soared and chased in all directions, spread far across the landscape.

I watched Ali trying to eat. It seemed hard for her to pick up the food from the sand—that sliced nose was painful! I was still watching her when Hurricane and Anne Boleyn swept in close together and circled me at arm's length several times. What special attention from these rare visitors! They were true companions. I had identified them together for the first time in January, 2001, and they had often reappeared together, yet had also been in widely separated locations for weeks at a time.

The glow from the sun reddened, and finally went out, leaving a deep violet gloom. I drifted down-current to make sure I had identified everyone. A large, dark shark appeared and slowly circled me clockwise. Carmeline. It was the third time she had come, and I had made a note to check her dorsal fin, so I carefully drew both sides and double checked to make sure they were accurate. She rarely showed her left side, but I was sure I had it drawn accurately by the time I left. She had an unusual face. Her mouth appeared dark on both sides behind the hinge of her jaws, as if she had been hooked and hurt in the past on both sides of her mouth. She cruised around me as I swam, as she had done when I had first met her at Site Two. She even came over to look at me when I dove, and held onto the bottom, to try to inspect her injured mouth.

Back at the site, the big sharks Twilight, Glamorgan, Cheyenne, Hurricane, and Anne Boleyn were in search mode on the sand, with the usual males. My fish shot and fluttered around me. And there, passing, was a large grey female, whose fin was the one I had mistakenly drawn as Carmeline's. I waited to see the other side of her fin, and indeed, it was not unlike that of Carmeline, from a distance. By then it was too dark to see more, and though I waited, and food remained on the sand, I did not see Carmeline again, or the similar shark. Or was she just another view, at a different distance, in different light, of Carmeline?

I checked my records and found that I had seen her in June and December in 2003, each time when Carmeline had also been present. It was surely not possible that two different sharks would have been present at the same time and at no other. Now it was May, 2004. Perhaps with the passing time I had become confused. I was sure I knew Carmeline—not only was her look unique but so was her behaviour. Each time she arrived as darkness fell, circled me clockwise, then left, to return a few minutes later and circle counterclockwise. In my mind I could see her on both sides. Once again I made a note to check her identity.

In spite of being well fed, the sharks stayed until I left, circling and often approaching me.

A very large nurse shark came as darkness descended. It roamed constantly, like the one who had followed me at the barrens sessions that I had held before leaving for France eight months before, but I could not be sure if it was the same one or another. I regretted that my efforts to know and understand this unusual animal had been interrupted by our voyage to France, the disaster on our return, and then by the need to prepare for the visit of the BBC's film crew. I hoped I would be able to find him again.

It had been a beautiful session apart from Ali's hideous injuries—if only the BBC had been able to film her. But the golden fog would have been a poor environment for filming.

That week, Arthur wrote that he had heard from the editor of *Animal Cognition* about our paper on shark cognition. He told me that she wanted several aspects of it changed. She didn't think the title was appropriate since it was too negative, and had asserted to Arthur that readers didn't doubt that cognition was a fact in animals. She felt that the title suggested that cognition was just a possibility in sharks, whereas he should make the point that cognition was a fact and its acceptance was only a matter of time.

He wrote:

Although she may believe that readers accept cognition in animals, that belief may be extremely optimistic. I informed her that I had written to every known expert on sharks, having anything to do with their behavior, ecology, and natural history. I contacted over fifteen throughout the world.

All, but one, doubted that sharks have cognition in the sense that they recognized its meaning. They were very nice in their replies, but only one ventured his brief observations, which were not too relevant to our story. I told her that we wanted to show evidence that sharks likely have cognition since no one can prove such and that thousands of people likely believe, contrary to her belief, that such animals do not posses cognition.

And so, it must be shown, as difficult as it is to show, evidence that cognition may well be present rather than to disregard any consideration.

The editor kept the Figures, so she'll likely accept our manuscript, so long as I do her bidding, which I shall. . . to a point. But the point of the paper will not change. It's directed at people who consider that sharks likely do not possess cognition and its our job to provide what little evidence there is to counter that consideration.

He told me he would let the paper sit for a while as he considered how to revise it, while working on his important project of *Interception*.

My research on the Internet suggested that people certainly did not consider that sharks were capable of cognition—the public attitude to them seemed severely skewed by 'shark attack' hysteria. The general belief that sharks were killer robots, excluded the idea that they were ordinary animals with ordinary responses who naturally used cognition in the pursuit of their daily lives.

People are intrigued by the gruesome, as the popularity of horror films and books attest. It seemed to me that since modern science had discovered no sea monsters of mythological stature, sharks had been called upon to play that role. This false and ugly stereotype had crippled the search for a true understanding of them, causing people who could accept cognition in their dogs, cats, and parrots to categorize sharks as cold, mindless killers, instead of normal animals. Since it benefited fishermen, particularly sports fishermen, to adopt and broadcast this attitude, it had become deeply ingrained in people's minds. Fishermen met sharks more often than other people, and thus their stories had largely shaped people's attitudes. I feared that few people were likely to be able to discover what they were really like before they were decimated.

Thus, Arthur's initiative in presenting evidence that sharks were capable of cognition seemed to me an important step to help pave the way for positive changes in the prevailing public attitude toward them. If people learned what sharks were really like, perhaps they would insist upon their protection before it was too late. It had already happened with whales and wolves. Why not sharks?

Concern among nature lovers in Polynesia heightened as the government continued to allow the massacre of the sharks. Peter Schneider, a film-maker based in Rangiroa, told me that he believed the sharks would not be protected before October, depending on when the elections were held—the president wouldn't do anything to alienate the fishermen before then. He had found data published by the Ministry of Fisheries indicating that Polynesia exported seven tons of shark fins per year, mostly to Hong Kong, and that the amount was rising. Peter also mentioned the plastic bottles everywhere, along the roadsides, decorating the beaches, and blowing across the sea before the wind. How incomprehensible it was, he said, that the government would let companies produce so much plastic which was not recycled, on tiny islands in the middle of the ocean, when the country's future depended upon it remaining beautiful.

After such a long period of distraction, I began to put the sharks first again, and went to see them at least twice a week. As well as keeping a close watch on my well known residents in Sections One and Two, I returned randomly to the different locations where I had identified sharks in the past, including the ocean across the reef. It seemed important to continue to be their witness.

With the advancing season, the weather calmed and my submarine outings became easier, though the water kept its golden glow until well into June. I returned regularly to the site of my original sunset rendezvous due to the behaviour of the sharks. Cochita always came to be stroked, and many of them spontaneously undulated against the boat and paddle as I arrived. The apparent affection of the group there was a mystery, since no similar behaviour had been expressed in any other situation by the sharks.

Ali slowly healed, but for weeks she had a sharp black line up the centre of her face from the V-shaped cut out of her nose, high into her head. The lines at the back of her head became very black suggesting that they had been deep. They looked more and more like stab wounds as the healing process defined them. Stabbing a fish behind the head is a method commonly used to kill it. It was miraculous that she had escaped from the situation; probably the line had broken, or the hook had torn out of her mouth as a result of her mortal battle for life.

She and Shilly were usually together, often with Ondine; the three used overlapping home ranges centred at one end of the barrens. I kept trying to take photographs of Ali's face but she was so shy that I never did succeed. The only time I managed to get close enough, the shark in the photograph was Shilly!

During the day, the best place to find the sharks was the barrens, and I spent much time there after checking the border area and passing through the site. Many sharks, including some of the males, such as Jem, Tamarack, and Mordred, were willing to swim with me, but their habit was to come with me for a time, then circle away and return. In this way, I was nearly always accompanied by a group of sharks, whenever I roamed through that large area between the two sites of Sections One and Two.

Once when I swam out to where the waves broke over the reef to get a special photograph for a painting, a large remora accompanied me, often touching my legs, then remaining quietly attached to my stomach. The entire time I dove and manoeuvred in order to capture the image I wanted with the camera, the fish remained there, though I thought it had left. When I turned back toward the deeper parts of the barrens, I found Madeline and Ruffles following me out of sight. They stayed for a few minutes, before circling away through the rushing light of the shallows.

I returned to the boat, and watched underwater, while reaching for my bottle of diluted juice to fortify myself. Finding it sticky, I shook it underwater, and several sharks came over to see what had caused the noise. After that I kept a bottle partly filled with water in the boat to shake to attract nearby sharks closer for identification.

When Bratworst drifted by, the remora left me and joined her. I went on with her and Ruffles, and the remora left Bratworst and returned to me. But when she headed for the barrens, I went to the boat. Gwendolyn glided by. The remora stuck close, showing no signs of going to find a shark, so I was obliged to abandon it when I left. I realized how long it was since a remora had accompanied me. At the beginning of my study they had seemed to be much more numerous, and I couldn't remember the last time I had seen a spangled emperor, or any of the other big fish I had once seen regularly.

Toward the end of June, Stephen contacted me to say that they had been working hard on the documentary, and my sequence was finished. He was very happy with it, and said he would send me a copy as soon as possible. He asked me if I had a photograph of a blackfin for a graphic. All the sharks who were subjects of the documentary had to be smaller than Mike, and those in charge of the production were concerned that my sharks were bigger than he was. They wanted to put the photograph of the shark beside a photograph of Mike, to prove that the shark was smaller.

For most of my study, I had been able to get only a few plastic disposable cameras, and had saved the precious shots for photographs of the rare visitors, who nearly always came at night-fall. Thus my photographs were dark and of a poor quality. When I had been able to get an underwater camera in a housing, it had been the cheapest available, and the quality of the photographs were no better. After years of having to rely on my memory and rough sketches, even for paintings of submarine scenes, it was miraculous to have photographs at all. But the idea of offering one to a mega-corporation with the superior technology of the

BBC was embarrassing. My best example was the one of Willow soaring past me at the surface during one of my visits to Site Three. It was the one Arthur had used at the conference. The problem was that the end of her tail was cut off. I went through all my photos, only a few of which had been put on discs, and finally sent a small selection, apologizing profusely for the quality, and offering to put the end of Willow's tail on by artistic means if they chose that one.

To my astonishment, the photograph was considered acceptable! Stephen told me that it was not something they could pay for, and asked that I send permission for them to use it. Of course, that was fine with me! I wrote to Stephen with my congratulations on the successful completion of his project, adding, "If you are happy with it, then I think that's all that matters."

One day I met Denis while shopping at the local grocery store. Always the French gentleman, he greeted me formally. Then he looked over my loaded shopping cart, piled high with food for dogs and birds, and topped with leafy green vegetables, and he asked me, "Do you think you have enough there for the sharks?"

Taken by surprise, all I could do was laugh. Then I told him why I had felt it was important to keep the whole affair secret until the filming was done. He laughed too, and said he quite understood. I was most grateful for his intuitive discretion, and after that we were friends.

I usually avoided the area of Site One when I went to the lagoon during the day, so I could spend time with Ali and Shilly without interference from Windy. This worked. Shilly met and accompanied me. It always seemed such a miracle to be able to find the very shark I had gone to look for as soon as I arrived underwater, as though through some alchemy we were linked. I often saved a piece of food from the feeding sessions for her. On one such occasion, when she joined me, she swam slowly up toward my face as had always been her habit. As she came close, clear in the sunshine, I saw a long gouge under her nose, as if it had been badly cut and healed some time before. There was even a groove on her nose in the same place that Ali's was injured! She had not escaped the mayhem of the fishermen after all. There was an explanation as to why, in poor conditions, on approaching her face, I had thought she was Ali. She too had been cut.

I floated her treat toward her, but she missed it and it sank beneath her as she circled. Then she went to the boat and checked it, swimming on to orbit, watching, while I dove down to retrieve her bit of food. This time I threw it to her, and she took it immediately, apparently cued by the splash it made. She continued to circle, slowly spiralling toward me until she was at arm's length. She had filled out and matured—she would surely mate during the next season of reproduction.

Just at that moment, as we drifted together, another female shark passed us slowly, looking. I had seen her only once before on September 6, 2001, also while roaming through the lagoon without food. She looked just the same! Surely some sharks did roam very far, for them to appear so rarely—this could have been one who had come from another island. Shilly and I swam two large circles through her home range, and when we returned to the kayak, I let her go on alone.

Other sharks were cruising on the periphery, but not until Sparkle rocketed in and whipped around me, did anyone seem to be doing much more than sleeping. I drifted away underwater, using the kayak as a sail, disappointed not to have found Ali. Farther on, a large group of flute fish, the long and slender, white-patterned fish with snake-like tails, hovered, all facing into the current. Beyond them was a large crocodile needlefish like the one I had befriended, but I could not get close to her. Sparkle bulleted through the blue, shot around me, and blasted back in the manner in which she had come. I flew away with the wind.

Sometimes I took food to the western side of the barrens, then went looking for Shilly in her home range. She nearly always came back with me to the feeding site.

Once, I was watching Ruffles, undeviating in his aim for the food, when Breezy and Marten nearly collided close by. Breezy, who was by then an adult male, turned to avoid slamming Marten's shoulder. As he did so, his mouth opened and closed. But he made no effort to bite Marten. I never witnessed the sharks bite spontaneously as the mammals we know more intimately do. The more I watched them, the more I was convinced that they lacked the instinct that has always served us and our other mammalian relatives, so well—to open the mouth and bite in moments of provocation or defence. I wondered if the unquestioned assumption that sharks share our instinct to bite, was the deep core of 'shark attack' hysteria, given their dentition.

In mid-July, on a dark and windy day, the big fish shop called to ask if I could come and pick up some fish scraps. There was far more than I could comfortably handle. Since the load came already packed in sugar sacks, and I, as usual, was pressed for time, I just laid the sacks on the kayak, rationalizing that things would be simplified when I got to the site. I would dump the sacks out for the sharks instead of handling the spiny scraps.

Out in the exposed region near the reef, the wind was so strong when I anchored that it nearly capsized the kayak as the anchor rushed overboard, and I tried unsuccessfully to free the rope from under the heavy sacks. Paddle in one hand and anchor rope in the other, slammed broadside by the wind, with sharks raising their heads from the water all around the kayak, I tried to shake out the sack which was blocking the anchor rope, and preventing the kayak from lining up freely with the wind. The spiny scraps caught in the sack, which was heavier than I anticipated. It took all my strength to hold it up high enough to shake with one hand.

The sharks, with their heads out of the water, could see that their food was supposed to be coming down, but was not falling as it should, so they took hold of the sack and shook it too, from underneath. As many sharks as would fit on it!

The effect of them hanging onto the sack and shaking it so unbalanced me that I had to let go of the anchor rope and went speeding down the site in the wind, dragging the anchor till it caught on something. With difficulty, I freed the spines caught in the sack, and the food fell free, so the sharks let go.

I paddled back up the site, still surrounded by sharks, and dumped out the next sack. Cochita appeared at my right hand, lifting her head toward me as she came up to the boat. I glimpsed Windy, Sparkle, and Bratworst, but the ruffled, heavy waves concealed the countless sharks just discernible frolicking beneath. Finally, the anchor line was freed and the kayak stabilized in its usual place. I let the contents of the last sack fall, and slid underwater.

Sharks were flying everywhere through clouds of sand particles. I swam around writing down names, checking to make sure I had seen everyone, and searching down-current before the first arrivals had time to leave the area. Ali swam up to me, and I clearly saw her face for the first time since it had been injured. It had basically healed over, but there was still a V-shaped notch in her nose, and a deep line up her face where she had been cut to her skull. Shilly was with her.

The darkness was deepening again and one of the bigger nurse sharks began to sweep through the region. I threw in the fish heads. Trinket took them one at a time for a shake, which the blackfins rarely did. How self-possessed he had seemed when he had been a pup, always doing something different from the others. And out in the ocean, he had been the only shark to rise up from the ocean floor to touch the anchor with his nose. Perhaps my little Trinket was a smart, if clumsy, shark, and not mentally handicapped after all!

Dapples arrived, heralding the appearance of Cotlet, Cheyenne, and more sharks from the east. Then Carmeline circled. Her face seemed to glower due to the large chunk of missing flesh at the side of her mouth. I had to wait a long time before I was able to double-check the other side of her dorsal fin.

When she turned and displayed it, it seemed similar enough to my drawing of it that it was possible, drawing in poor conditions without double-checking for accuracy, that I could have exaggerated the details of her fin. With the poor drawing differing so much from the accurate one, I must have concluded erroneously that there were two different sharks. It was impossible that there were two sharks, because that would mean that neither had ever come alone, and I didn't think that shark companionships could be so faithful. It was more likely that I had mistaken the left side of her dorsal fin by drawing it inaccurately, then viewed it later under different conditions, from different distances, and thought she was a different shark.

Yes! That must be the explanation. Pleased with myself to have resolved the inexplicable quandary of having identified two sharks when there was only one, I turned my head to see the second shark sail through the site. Carmeline was still visible to my right. The second shark had arrived about a minute after Carmeline. The left sides of both sharks' dorsal fins were similar, but the features were less prominent in Carmeline than in this shark—both my drawings were accurate.

It was another unexpected truth, that two sharks could be such close friends that not only had they always travelled to Site One together, but they had arrived in the same order, following the same pattern every time. Carmeline always came first, circling me clockwise, and her companion glided in a minute later, always circling me counter-clockwise.

I gazed at the companion, who was circling me as she habitually had done. As she drifted on, I followed amazed, trying to take her in, to meet her up close. She was so surprised to find me following, that she turned, rose in the water column, and circled my head slowly at eye level, looking. It was another one of those moments when time pauses, and in all the world, there was only us, so near yet so far removed, gazing at each other.

Diamond's condition had seemed to be improving, but worsened again. His vitality lowered, his face and eyelids became swollen, and his skin broke out in scabs in some places and peeled in others. His hock joints developed large oedema-like swellings and the skin broke down, bled and finally scabbed over. Then the areas healed and the condition broke out elsewhere.

I treated him for every possible known cause, including the fungus, *Microsporum gallinae*, and certain parasites. There was even some evidence that his symptoms could be a reaction to his anti-inflammatory drug, but stopping it did not help. Instead, the swelling quickly worsened. I added greater variety to his diet, vitamins, and natural tonics such as the juice of the nono fruit, commercially known as *Noni*, and aloe vera. But as the weeks passed, when one symptom improved with intensive attention, another presented. He was always under intensive care for something, and spent so much of his time resting on his breast beside the house, that the sore on his keel worsened, and I couldn't heal it.

Then one of his hocks swelled dramatically, and massive scabs developed across the area. His face worsened—lesions spread across it and there were large areas of discolouration. Yellowish plaques lined parts of his mouth, particularly beneath his tongue. For the first time, he stood around with fluffed feathers, his wings limp and hanging, showing the typical 'sick bird look.' He rarely foraged for food, much to Josephine's distress, and ate little. Soon he developed a serious eye infection. Though it filled with thick yellow pus, I treated it locally, since I was convinced that his problems had been amplified by treatments with antibiotics, facilitating invasions by various sorts of fungus, including *candidiasis* and *trichophyton*. I was determined that he not go back on antibiotics.

Josephine stuck by him with admirable faithfulness. Laying a large egg each day, she was always hungry and would normally have foraged continuously. Instead, she had to stand around with Diamond for hours at a time. Often her distress over her quandary was evident. The hen loved him. Whenever another rooster threatened him, she ran to me calling loudly. Once when he fought and was losing, Josephine came running and flying to get me from another part of the garden, absolutely frantic.

Chapter Fifty-Seven

Sharks: Size Matters

At the end of July, *Discovery Channel's* popular *Shark Week* was announced. Suddenly I received an e-mail from the BBC, saying that the name of our documentary had been changed from *Jaws Junior*, to *Sharks: Size Matters*, and would be shown within a few hours! I e-mailed friends in America to let them know, and some were able to watch it. My friends from the shark discussion lists regularly watched *Shark Week*, too, so I waited in suspense for the reaction to my sequence.

The first report of the documentary was from a trusted American friend and scientist. She informed me that it was so bad that she hadn't been able to watch it to the end, and she sincerely hoped I never saw it.

My blood ran cold. I asked, "Did they show me as an eccentric woman?"

"No," she said, and told me that the story that had been planned had not been shown. None of my findings had been included—nor was the vital information that these intelligent animals were being finned. There was no conservation message. My sequence was a criticism of shark feeding, and had been edited to make me look defensive about doing what the BBC had asked me to do.

Another friend reported that she was disappointed not to have seen me swimming with the sharks, and said that Mike had sounded negative about my work. Others, who were unaware of the story that had been planned, said that I had been fine, but the sharks were scary. I heard many times about the nice pictures of my book of shark fin drawings. But any artist with too little to do could draw five hundred shark fins! What was important was the information I had provided as a result of observing them as individuals!

From the many reports I received over the next days I found that what people remembered was that I had been criticized for feeding the sharks, and that Mike had cast doubt on my findings. Not one person said that they had been surprised by the interesting things I had revealed while talking about the sharks, nor sorry to hear that my companions had been finned.

It was incomprehensible. Stephen, who had asked me to feed the sharks, had told me that my segment was great. Feeding the sharks had been the key to the mirror experiment, the key to having a large group of sharks of a species that normally roams alone, in couples or small groups, and the key to having nervous wild animals come close enough to the cameras to be filmed. I had done my very best to fulfil my part of the bargain to produce the segment they had said they wanted, for all of my dead and threatened sharks.

I was stunned by the magnitude of the betrayal.

Those involved in making the documentary, had known I would be feeding the sharks. The scent flow was marked on the diagram I had sent them, specifically so that they could make allowances for it in their filming plans. They had discussed my wish to hold the evening feeding session that first Sunday night, so that the sharks could become used to the presence of the film crew before we held the mirror experiment. Everyone had known in advance why I was feeding the sharks.

I wrote to Stephen, summarizing what I had heard, that what people remembered about the sequence was Mike presenting what he called a "mob feeding frenzy," while criticizing me for feeding the sharks. (*Feeding frenzy* is actually an old-fashioned term which has been shown by scientific studies to be an inaccurate representation of sharks eating.) Friends reported that Mike had said that, contrary to what I claimed, he didn't see any individual differences in behaviour by the sharks, and that the "jury is still out" on that.

I reminded Stephen that I had told him on the phone that I had stopped holding the weekly feeding ses-

419

sions at that original site two years before, but I would start feeding the sharks there again to attract as many of the most familiar sharks as possible. The primary concern for him at that time had been that I have a big group of sharks to film.

Mike had entered the water twice, and both times was looking into a camera and paying no attention to the different individuals. I couldn't even get his attention to point them out. Given his lack of knowledge of my intensive study and the complexity of my findings, it was indefensible for him to cast doubt on my five years of carefully recorded observations in a widely seen television documentary. No matter what his personal belief was, he had no reason to say it on television when he didn't know, and in saying it he discounted my work. "That he would emphasize that I had been feeding the sharks when in fact they are being killed, defies reason," I wrote.

I asked what Mike would have said if I had not brought food, but assured him that there were thirty sharks just beyond visual range.

Finally, I reminded Stephen of our discussions about it being a documentary involving solid science, and that I had never doubted the integrity and quality of the BBC, but that what had been shown was not solid science, but a superficial and arbitrary opinion.

They had misrepresented their plans to me.

Stephen replied that while I had made a few points, I really should see the film before we could discuss it, and asked for my mailing address. So I waited for my copy of the documentary to arrive.

Later I received a message from Germany to say that the documentary had just been broadcast there. In that version, I had been shown in my kayak saying that my sharks had been finned. This was proof that the original BBC version had included my statement.

It had been the *Discovery Channel* that had taken the trouble to remove the vital message that nearly all the mature female sharks in my study had been finned. Why would they have deliberately suppressed this news, when they were reputed to present true, scientific subject matter? Why go to the trouble to remove the pivotal moments of a documentary? Could *Discovery Channel* be censuring conservation messages to such an extreme degree?

When I received the tape of the documentary, I saw for myself that the reports I had received were true—there was no mention of the deeper level of understanding I had gained about sharks' lives by keeping track of individuals. None of my key phrases, so carefully worked out to express key ideas with a minimum of words, had been included.

Without telling me, they had changed the story from a "solidly scientific sequence," to Mike expressing his personal disgust about feeding sharks. Yet he gave no reason, scientific or otherwise, for his opinion. My initial impression that Mike had not observed what was happening around him when he was in the site was confirmed. His reasoning had not extended to noticing that at the mirror experiment, the sharks had showed no interest in the food at all—indeed, no observable facts had influenced his personal prejudice.

On the screen glided Keeta, Tamarack, and even little Trinket, but no information about their histories was given. When my book of the drawings of my shark fins was shown, I mentioned that Trinket, whom we had "met the night before," had been identified as a pup, observed while growing up, and then had moved to the outer slope of the reef where the males live. I mentioned that he had not appeared in the study area for two years. This information in itself suggested the unimportance of the feeding sessions to the sharks. It also suggested that individual stories were there, invisibly, on the screen, just waiting to be told.

But as the sharks swooped around gracefully picking the sardines out of the water, Mike said, "Ila claims these sharks have individual behaviour patterns, but I'm not seeing any here!" Then he went on to apply the words *feeding frenzy* and *shark rodeo* to the scene.

In calling my carefully planned introduction of the sharks to the film crew, a "staged shark rodeo," Mike corrupted what the audience was really seeing. Further, he discounted my years of careful observations before more than twenty-two million viewers. Had the show been concerned at all with science, he could have asked to know my observations about the pros and cons of feeding sharks, since I had so much data on it.

He had not done a study to learn about the responses of sharks to the presence of shark feedings, and was just expressing a personal opinion.

I watched my sharks coming forward timidly on the screen to take the sardines from near the surface. It would have been laughable if it hadn't been so disappointing. To call that a "shark feeding frenzy" suggested further how complete his ignorance of sharks was.

Whatever would he have said if he had met Carrellina?

The simple, surreal beauty of the sharks flying through their coral garden had been ignored, as if it had been invisible to them all.

As a preamble to the interview, Mike, in frowning seriousness, said he had to ask me some "hard questions," about feeding sharks. Yet in an earlier sequence of the documentary, he had been delighted to have a researcher's permission to film a shark in a small aquarium! He hadn't asked what such a large animal was doing in such a small tank, for example. He could easily have treated my case in the same way, saying something to the effect that, "Ila kindly brought some sardines so we could meet all her sharks at once this evening," and then go on to the information about their lives which I had provided.

But the only part of my interview that was included was my self conscious defence of having fed the sharks, which Mike had pressured me into providing. My initial smiling response that I had brought them something special for the filming had been cut.

It troubled me too, that Mike had publicly said that I had claimed that the sharks had "personalities." In informally talking, I may have used the word carelessly, but in writing and scientific circles I had been careful to use the terms "individual differences," "individuals responding differently," "flexible behaviour, differing in different sharks," and so forth. Using the word "personalities" for sharks deliberately opened me up to the accusation of anthropomorphism. The root of the word is "person" and is reserved for humans, who are primates. Primates have a different set of characteristics than sharks, or elasmobranchs.

Genetic studies have established that all individuals are different, so the concept of individual differences was not novel. It was my documentation of what sort of individual variation might exist, in terms of behaviour in sharks, which was new.

The kayak conversation about the sharks being fished and finned for shark fin soup was included in the version supplied to me by the BBC, though it had been removed from the American *Shark Week* version. But since the companionships I had developed with these sharks, leading to a new level of understanding of their lives, had not been mentioned, the reason why their deaths were of such concern to me was missing. Nor was the wider framework of the serious threat posed to sharks by shark finning outlined. If a viewer didn't already know what finning was, and the scale and danger of it, he or she could completely miss the significance of the statement.

The sharks were literally being finned at night and filmed during the day. But that part hadn't mattered! The focus of my segment was public criticism of me, for doing what Stephen had asked me to do—what they all knew I had done, especially for them, to facilitate their filming.

Farther on in the documentary, the segment on cookie cutter sharks did not involve seeking them underwater, as Stephen and me had discussed doing, but by pursuing them on a long-lining fishing vessel! The bites of cookie cutter sharks were filmed, not on living wild dolphins such as would have been possible on my island, but in an enormous fish warehouse with carcasses wall to wall. My fish shops could have provided such photos in a more intimate, exotic, and non-industrial setting if the bites rather than the sharks, were what they were interested in.

The ugly scenes of people cutting up cadavers, all subliminally indicated that for Mike, long-lining is fine. He talked enthusiastically to the fishermen while searching through acres of dead tuna for the bites of cookie cutter sharks.

The crew went out on a long-line fishing vessel, and some of the tuna that were caught had been partly eaten by sharks, who were then cast as the villains! There was no attempt to put into perspective the fact that sharks and people are predators in competition for the same food. The message that came across was that long-lining, which is responsible for a large fraction of the shark finning conducted world wide, and which poses one of the major threats to shark populations, was perfectly acceptable, while my feeding of the sharks to facilitate knowing them as they really are, was not.

The scene underlined the peculiarity of the negative slant given to my segment, and mine alone. Regardless of the face he had presented to me, it was clear that Mike favoured fish and shark killing, and disfavoured my effort to understand sharks as they really are. The disquieting memory of the way he had blocked the sharks' approach to the mirror, and ruined my mirror experiment returned, and I wondered what his actions had really meant. Other memories, such as the way he had spent so much time alone with me waiting at the site, yet had never spoken to me, suggested that even then his attitude to me and my work must have been negative.

Andy wrote that they had worked to the commission of the *Discovery Channel*, but he said that he didn't think that the segment had harmed me. He did insist that though it had been difficult, he had managed to make sure that my statement about the finning of my sharks had been included. This confirmed that the *Discovery Channel* had, indeed, taken it upon themselves to cut my message from the sequence.

My friends were outraged at this evidence of how the two companies had conducted themselves, and urged me to contact *Discovery* and tell them how disgusted my scientist/elasmobranchologist colleagues were, and in what a bad light the whole affair had placed the *Discovery Network*.

Even the BBC suggested that I write to *Discovery* about removing the information that my sharks had been finned for shark fin soup, but at that point I just gave up. If the BBC really was concerned about *Discovery* altering its productions once they were finished, why didn't it confront *Discovery*? *Discovery* would put any letter I sent to them straight into the garbage. If they cared about sharks or ethics, this would not have happened. I was stunned into silence.

After the investment of months of my time, and an intense personal commitment, the story of my sharks had not been told—not even sketched.

Chapter Fifty-Eight

Life and Death Go On

As August drifted by, the sea rested. There was rain on the mountain, sun on the lagoon, and the water lay so still that the conditions below the surface were worse than those above. The lagoon became a soup of microscopic organisms that created a submarine fog as impenetrable to the eye as particles stirred by wind and surge.

The sharks stayed at home. There was little roaming, and the only visitor who strayed through was Samphire. The sharks who usually visited the ocean during daylight hours, such as Flora and Pip, returned between two and three in the afternoon. Each session my slate was full of sketches of the dorsal fins of young sharks who had drifted into the area, as they did each year at that time.

There were still many youngsters growing up in the lagoon. Would they be finned when they began roaming the ocean?

Windy sailed up to me expectantly whenever she saw me. Given her attentions I often tried to swim with her, but she was an energetic shark with a fast cruising speed, so my success depended on her willingness to stay with me. Usually the farther we went, the faster she swam.

Tamarack would sometimes speed over to me if I dove down to the sandy floor, so I often played with him. When he zoomed over, I tried to photograph him passing above me by flipping onto my back at the right moment, without startling him. We were quickly joined by several other young males, and their antics seemed like play, particularly since they kept it up for long periods.

Kilmeny and Marco appeared together frequently. It was a long-term companionship that had been difficult to detect because the two sharks roamed separately as often as they were together, and usually separately even when they were together!

The latest corrections to my manuscript on the sharks' gestation period arrived from the *Marine Biology* editor, and I was disturbed by the number of my statements that were questioned. Uncertain as to how to respond to such fierce criticism, I wrote to Arthur for help, and while waiting for a reply, sent a summary of my observations of the shark pups in and around the nurseries to the editor, who was concerned about my neglect of them in my article.

Arthur answered a few days later. He spent some time discussing my questions about the reviewer's comments, essentially telling me to go ahead and answer all the questions as clearly as I could. Such tedious attention to detail was necessary for the accuracy and quality of the leading scientific publications, he explained. It was a long letter, and I began to feel happy and amused by his many joking commentaries and anecdotes.

Until I neared the end.

He wrote that he was away from his office since he had been ill, and had gone there to check his messages. He had been amazed by how many hundreds there were, but, he said, he was looking for messages from certain people, me among them, and these he had forwarded to his home address to answer. He said that more than a month before, he had been diagnosed with cancer, and was undergoing treatment.

He had been unable to start revising the manuscript on cognition due to his illness, and the many other things he had to take care of. But, he told me, he would send a message to the editor of *Animal Cognition* informing her that he was now revising it.

He said:

I have not told the hosts of the previous Symposium of my predicament, as I don't want special treatment based on that. I'll somehow do some revising as I owe you for your great help. I've still got time before me. There is always hope and I'm a pretty good fighter. So, I hope that some day in the future, we'll meet and I'll be able to tell you, face to face, what a joy it has been to communicate with you for so long a period. . .

I was devastated. I focused on his ability to fight and desperately hoped that I would not lose this valuable friend, who, through the intimacy of writing, had grown real and very dear to me. Besides being one of the most intelligent and kind people I had ever contacted, Arthur was the only shark ethologist I knew and could discuss things with.

I answered each of the editor's questions, and sent the article back to *Marine Biology*.

Diamond had entered a period of relative health. Even the sore on his keel had healed. It was a confirmation of my hope that with time, peace, and good food, he could get better.

But in the middle of August his diarrhoea returned. Hoping it was caused by something he had eaten, and would pass, I treated him for a common intestinal parasite called *coccidiosis*. But the scabbing and swelling of his face and hocks returned too, so I resumed treatment for the fungal infection. That didn't work, and I tried one ointment after another. His nose began running again, and he developed a large dry scab on the sole of his foot. An odd swollen area on one toe appeared to be dying, as if it were no longer nourished by the blood stream. The keel sore returned and scabbed over; it appeared deeper than ever and actually stuck to the keel bone itself, which is sharp in sad, thin birds. Bad keel sores require surgery to heal, so in desperation I covered it and the sores on his foot with eosine and betadine poultices. There was no veterinarian capable of operating on birds in the country.

Then there came a night when he did not eat, and his call, unique to each rooster, lost its complexity and musicality. When I handled him to tube-feed him, he felt strangely weak.

The next day I was busy when Josephine came to the door, calling. When I was free to rush out, she was foraging on the periphery of Kubilai's flock. It could only mean that Diamond was dying.

He was lying on his sore breast under the house, and it wasn't easy to get him out. He was weak and cold. He had been manipulating his tongue in his mouth a lot in the past days, and I had looked inside often, but could find nothing wrong, apart from his usual poor colour. This time, his mouth was full of thick strands of saliva, suggesting dehydration.

I tubed rehydrating liquids with charcoal into him, alternating with tubes of liquefied potato, yoghurt and fruit, and put him back on the full range of medications for all of his digestive, respiratory and fungal problems. I added lots of love, keeping him beside me propped on his seabird bed, where I could talk to him and keep his crop full of good things.

The next morning he seemed better, and I began to hope that the treatments would work. He foraged with Josephine on the lawn, and had lost the feeling of weakness I associated with oncoming death. However, as the day passed, it was obvious that he was seriously ill, and I brought him inside. His face was white, while his crest was an unnatural scarlet red, so I gave him a shot of Vitamin B12 to help remedy his anaemia.

His level of vitality seemed simply insufficient. His terrible experiences at the hands of the cock thieves—the loss of his hens, his home, his security, and his freedom—had destroyed him. He had never recovered his health since. And nine months after his rescue, he still screamed! Whenever he knew I was going to pick him up, he shrieked. Even though he knew I was just going to gently lift him onto his perch, he screamed. He no longer moved when he wanted me to pick him up—he stood there waiting, but he couldn't help screaming at the same time. Yet this was a bird whom I had hand-raised and always treated with extreme gentleness. He was with the cock fighters for two weeks, and it had permanently affected his mind and destroyed his health.

I installed him in his own little building. It was nicely furnished for birds, with a variety of perches, a white sand floor, a large screened window looking west across the bay, nesting boxes, resting places, cushions, and a hospital bed on a table with the space under it partitioned off to create a separate, private back room. I had an injured and traumatized young hen at the time who needed rest and quiet, so put her with Diamond in this safe and protected house. The arrangement was a comfort to both birds. They spent much of their time quietly murmuring and resting together in their private back corner. Josephine was anxious to join them in the evening, and slept on her usual perch, pressed against Diamond. The frightened young hen pressed against him on his other side. But in the morning, Josephine was anxious to leave.

A week passed. Diamond and his young companion seemed content as long as they were left alone. Neither wanted to go outside, nor showed interest in anything. Diamond ate all the papaya I brought, but nothing else. I continued his tube-feedings, along with his medications and rehydrating fluids. He slowly digested what I squirted into him, but remained dehydrated, and seemed increasingly sleepy and withdrawn. One day he saw Kubilai and his own son fighting furiously, as I carried him back to his building after a treatment, and he showed no reaction. Each day, he seemed a little weaker, and spent longer lying alone on his bed. As the end of August approached, his foot became uniformly swollen like a little balloon.

It seemed that in treating everything that went wrong with him—giving him something to kill his bacteria, something to take care of any invading protozoa, fungi, or parasites that might have ended his suffering, by administering medications, fluids, food, tonics, ointments, poultices, and vitamins—I was making sure that the bird stayed alive to suffer until the bitter end.

In September I returned again to the site of my sunset rendezvous. Clouds covered the Pacific, and it had rained intermittently for days, so the surface was ruffled and restless. An interlude of soft southern wind and sun between two storms, was passing. Several sharks followed from near the border, though I was not paddling—just surfing out on the wind. The surface at the site was so wild that I couldn't see who they were. Below, a long, slow surge swept the scene, and I found my sharks with a few visitors.

The females were beginning to look rotund with pregnancy. Shilly had dramatically changed into a heavy mature female, and had lost the graceful, serpentine look of the adolescent. Columbine, Keeta, Storm, and Cheyenne were taking on that particularly powerful look they gained in pregnancy. Cochita was not there. A few other young adults had been missing recently, too.

After the first group fed, they drifted away, and for the first time in memory at a feeding session at that site, I was alone. I drifted for a while, and then Ali glided up the scent flow to me. A group following behind her darted toward us. Jessica was with them, and it was a shock to see that her extensive eczema had broken out into darkened splotches of different sizes. Intuitively, it seemed that such a serious and wide-spread skin condition pointed to a deep problem, as Diamond's condition did. In places, the epidermis had broken away to reveal the white dermis beneath. This was particularly evident on the dorsal ridge where the trailing edge of her soft dorsal fin waved continuously back and forth with her movement, and had swept the skin off. Jessica circled through the area as it got dark, which had been her pattern as long as I had known her. I drifted with her when I could, without disturbing her, and affirmed that her swimming rhythm, weight and strength appeared normal.

I waited until the last of the sharks had fed and drifted away into the gloom. Very few had come.

Diamond grew imperceptibly thinner, whiter, weaker and more tired each day. Finally, he could no longer balance on his perch for the night. I brought him into the house, and installed his seabird bed in the bird corner. The injured young hen was well enough to begin a more normal life outside, and when necessary she could retreat to the shelter of their little building.

Diamond's legs had become too painful to keep them bent fully all the time, so I propped him slightly to one side, then turned him every so often so he could extend them. A hole in the netting of the seabird bed permitted his legs to go through some of the time, and a piece of foam protected the big scab on his keel. He had become a shrivelled creature of skin and bone, so weak that he would lie flat on his side, not realizing that he had fallen over after trying to get up. His outflow ran out of him continuously, and he made feeble efforts to move away so that he wouldn't dirty his plumes, but the advantage of the seabird bed is that the liquid drips through the netting, leaving the bird's feathers clean. He had been inside for a week when I realized that he was no longer sinking each day as he had been. He was staying the same.

One day, Josephine came inside, loudly demanding a snack. Diamond was on a cushion on the floor at the time, and I put her dish of food near him so that the two could see each other. As she ate, he crept over to eat with her, and he kept picking at the food, eating for the first time in weeks, hunched over the dish long after she had gone outside.

The next morning, he crept on his hocks to look out the window. This was strange behaviour for a dying patient. In the following days there was marked improvement. He became more alert, more active, stronger, and hungry. His diarrhoea improved. I had speculated that parts of his digestive system were working only slightly and what remained was sufficient to keep him alive in this torpid state for a while, but in fact, healing seemed to be taking place. I had been very sad to see him dying after the long and desperate effort to win him back to life, and his unexpected return from the door of death filled me with hope and wonder, anew, over the miracle of life.

But as the days passed, Diamond became sleepier, and finally scarcely moved. In spite of the copious fluids I kept tubing into him, he seemed to be drying up, shrivelling before my eyes. The dried skin flaked off him, and he was increasingly unconscious.

One evening as the sun set, he astonished me by walking to the window. He stood looking westward to where the afterglow shimmered upon the sea, just as he had done, evening after evening with Josephine, when he had been better. When night fell, his head slumped as he began again to sleep, and I arranged him for the night on his couch of netting.

Overnight he did not move, and in the morning seemed flat and cool. His apparent improvement had been due to imminent death. I had often seen this phenomenon, wherein an obviously dying animal improved dramatically, sometimes for several days, and each time it fooled me. He roused and gave little cries as I adjusted him on his bedding, and gently tubed some sugar-water into him.

At about ten o'clock when I checked him, tears streamed from his eyes. His face was white and relaxed, his body had sunk into the pillow. I went outside briefly, and when I returned, he had moved. He rolled his head one way and then the other slowly, and gasped. His eyes were closed. I got the stethoscope and his wings jerked a bit as I positioned it to listen to his heart, which was humming strangely. Gradually, a crescendo of vibration like a rising wind started low, and seemed to approach his centre, as if the life forces were zooming in from all over his body to oscillate at a high frequency in his chest. When the wave passed, his heart was faintly clicking within, as if it had nothing to do with this sound. He had stopped breathing, but his wings jerked a few more times. His head lay on its side, and his pale face was covered with tears. His heart continued to beat for a while after his death, then gradually slowed, and became fainter until the sound of it faded away.

It was daunting to think of the suffering that the bird had gone through in his two years. Something in him had been irreparably damaged when he was stolen and tortured, and it had taken him nine months to succumb.

I felt my heart had broken.

Josephine insisted on sleeping always on the perch she had shared with Diamond, and whenever she came to the door of the house looking for something extra to eat, the first thing she did was look in the bird corner, at the place where she and Diamond met, and the place where they parted. Such an attachment was extremely rare, especially when the birds met late in life. Josephine never took another mate.

It had become a natural escape, when struck with grief, to leave for the lagoon. The sharks were always there, and it was a comfort to be with them for a while. It was impossible to be with them and be sad. Then, when I returned home, I was too tired.

So after burying Diamond and building a cairn for him, I threw my gear in the kayak, and set off for the lagoon. Utter calm had fallen upon the sea. It seemed bewitched into stillness, a mirror to that far, clean horizon that always encircled my life.

But on the way a nasty east wind arose, the sky darkened and lowered. Avoiding the region of the site, I battled along to the lower barrens to look for Shilly, giving a spear fisherman there a wide berth. Odetta, one of the new juveniles, was coolly circling when I slid underwater, and Shilly was approaching.

Odetta slid past my shoulder, while the now slow and newly rotund Shilly circled gravely. Beyond, the water was alive with sharks slowly passing—Madeline, Glamorgan, Kilmeny, Ruffles, Jem, and the juveniles.

I swam away with Shilly, but she was on her way somewhere and left me behind, so I joined Glamorgan. We swam slowly southward, accompanied by ever-changing members of the rest of the group. When I was left behind in the current, Ali passed, going in the other direction, and I swam with her as she turned the typical, enormous, counter-clockwise circles through a wide region of the lagoon, each generally in the opposite direction of the one before. Many sharks were resting in the barrens. When a pup came close to circle and look at me, I left Ali, to sketch his fin and play with him. The little thing was delightful, very curious, and not shy.

I had brought some small fish that had become too old for seabirds, and put them on the sand for him. The baby shark was swiftly joined by a second one, and when their tiny tails swept the little fish off the seafloor, they swiftly turned and picked them out of the water column. They then swept the fish again from the sand with their tails, and turned to eat them. I wondered if the first time was by chance, but whether subsequent instances of using their tails in this way to lift the fish off the sand to eat, were intentional. It was impossible to tell for sure by watching them. All I could do was put the idea on hold and watch for further evidence. No larger shark had shown such initiative in trying to eat the fish scraps.

Ondine appeared from farther inshore, and I joined her for another four counter-clockwise circles, south of the boat and around it. As we passed the kayak, she suddenly approached my face, then turned to sniff the boat. Once again, she seemed to be referring to a mental reference—a declarative memory—that there was sometimes food in it, which she could detect by sniffing at a specific place. Then she passed on into the coral east of the barrens, while the awful wind slammed waves and glitters of sunlight into my face. Gradually she drew away from me, then accelerated and was gone.

I had to keep checking for the location of the fisherman, who had come quite close, so swam away up the eastern edge of the barrens. A large female was resting on the sand at the place where I had once found Amaranth and Samaria sleeping, long before shark finning had ended their lives. She accelerated away while I was still in the distance. A small crocodile needlefish passed slowly in front of me, looking.

After an hour, I trailed the boat back toward Site One, through the place where the juveniles and pups lived, all the way to the border, and did not see a shark.

By the end of September, the latest revision of my paper on the sharks' gestation period had been deemed acceptable by the editor, and it was sent to Germany to be published.

I got out my original paper about the roaming habits of the sharks, including all the graphs and tables, and reviewed my calculations, which had taken months of work and thought. This was the next paper, but I had let it go once the shark finning had begun, because the ratio of females to juveniles and males was no longer natural. It was going to need some work to update it.

I was still waking up at about 3:00 a.m. due to the unbearable pain in my back, and trailing through each day in a haze of pain and exhaustion. So when I received a message from a friend that there was an American chiropractor in Papeete who could work miracles, I made an appointment and flew over. The chiropractor's examination revealed that part of my spine had been caved in by the fall, and he easily found the place where it curved the wrong way on my MRI scan, though this detail had been missed by the shark-eating bone specialist. Through a series of treatments, he gently manipulated the bones back into the right place, and for the first time in seven months, I was able to sleep and function painlessly.

My letter writing campaign continued to keep pressure on the government, who were the only people with the power to end the slaughter of the sharks. I wrote and distributed a petition, emphasizing that shark finning is in contravention of the United Nations' *World Charter for Nature*. It described the degree of the depletion of sharks and the folly of eliminating the top predators from the oceans, then quoted the applicable sections of the *World Charter for Nature*. I summed it up by saying that if there is one practice which best characterizes the wasteful, careless attitude of humanity toward nature and living things, it is shark finning, and asked that the United Nations take measures to see that sharks be protected on a global scale.

Though the petition was widely circulated on the Internet, posted on web-sites, and collected thousands of signatures, it had no effect. Shark finning was profitable, and it continued. But I hoped that in spreading the word as widely as possible, more and more people would become aware of the plight of sharks, and that the growing public concern might bring about their protection before it was too late.

Ultimately, if the consumers of shark fin soup could be reached, and convinced of the harm that their tradition was causing, the demand for shark fins would stop, and so would the slaughter. Could they not change the recipe? The situation defied reason.

The months passed, and the sharks' reproductive season proceeded. Due to the poor fishing, scraps for feeding sessions were only available about once a month, and one of the fish stores had closed. My daytime trips to the lagoon continued, and when I could get scraps I took them to my sharks at Site One.

One day I was unexpectedly given a large quantity of shark food. There had been a long period of cloud and rain, but as I prepared, the atmosphere and sea fell quiet. I paddled through hazy light and shifting cloud to a still lagoon. Due to the invisibility of the landmarks in the haze, I couldn't find the site, and wasted a lot of time looking for it while sharks accumulated and milled around me. There was rain on the ocean, and black cloud rimmed the horizon. In the unnatural silence it seemed that a storm could explode at any moment.

Windy had been with me intermittently during my confused search for the site, and once I anchored, attached the paddle, and had a drink of juice, she appeared remora-like beneath the kayak as Madonna once had, drifting just beneath the surface. She raised her head toward me and nosed along the gunwale of the kayak, looking for an accessible scrap.

But Cochita did not appear beneath my hand, as she had done so faithfully before. Nor was Flora orbiting the boat clockwise as she had habitually done while waiting for me over the years.

There was a lot of brushing against the boat by Windy and her juvenile companions. I began handing food to her, and she and the others fed from my hand. It was just like handing food to our pack of dogs.

Soon, so many joined in and were so excited, shooting around the boat so fast, that I tensed for the possibility that one would slide onto my lap. They bumped and brushed the kayak, but didn't slam it. I threw more food to them and slid in.

Underwater, the coral was alive with sharks, a multitude sailing excitedly through crystalline waters, in rushing golden light that suddenly poured through a gap in the skies from a sun setting, one more time, upon my sharks. The shark tornado whirled as more and more joined in while I wrote down the names of those who had come, and whether or not they were pregnant. Many whom I still thought of as little ones were growing large. Keeta was still the biggest of all, and she showed no signs of ageing. Being advanced in her pregnancy, she was magnificent. It was almost unbelievable how corpulent her head and cheeks were. Teardrop too, had become just grand in adulthood. Gwendolyn showed no signs of ageing. She was at least eleven years old, assuming that she had been a minimum of five or six as a mature shark, when I had met her. She too, was enormous in pregnancy.

Shilly was pregnant! My curious baby shark had grown up!

Hurricane and Anne Boleyn soared up to me from the east, and circled me several times where I drifted at the western end of the site. They were the same bronze colour this time, pregnant and lovely. For the second time, these infrequent visitors greeted me with special attention.

I was watching the two of them when the unknown male who had startled me twice at Site Two, did it again, darting into my face and away, showing just a glimpse of the same unknown fin pattern. He must have come with the others from the east. Cotlet was among them, and she swam with Tuvalu throughout the second half of the session. They were the same age, a male and female just maturing; Tuvalu had been there when I arrived, while Cotlet had come half an hour later. That they began roaming together provided one more bit of evidence that sharks from the same range knew each other and chose each other's companionship. Beautiful Glamorgan soared in with a big hook stabbed straight into the front of her mouth over her lower jaw.

As darkness fell, a large grey nurse shark began circling through the coral. Usually they were pink. There was always some new mystery.

Among the corals down-current were three shark pups cruising together, and I began trying to sketch each of their dorsal fins. They sped up when they saw that they had my attention, and began a series of manoeuvres around me that made it very hard to keep track of them to double-check both sides of each fin.

Ruffles appeared with them suddenly. His pectoral fin was sliced straight up the middle, obviously by a knife. The two edges of the cut were separated by his movement through the water; each side of the fin flowed in its own direction, so that the sides of the cut were far apart. How would it heal?

It was the last time I laid eyes on Ruffles.

The deepening darkness brought more visitors. One was a male so grand, so perfect, that I went drifting after him. He was large and well rounded and perfectly formed. I threw in the last of the food saved just for such late and interesting visitations, and he orbited close around me several times, which was unusual for a stranger. When I began sketching his fin I suddenly realized that this was my beloved Antigone! He had always been golden, so I hadn't recognized him clad in foxy red-brown. It was just the seventh time I had seen him, yet each time, I was struck by his natural elegance before recognizing his fin patterns.

By then there were sharks everywhere, socializing. It was a long time since I had seen so many sharks, streaming all together, passing scraps back and forth, soaring through the surroundings, and sometimes past me, as if I weren't there. They were partying. While all of them were briefly away, chasing through the distance, I found myself alone, and suddenly Shilly was there, circling me slowly at arm's length. I called to her and swam with her as she scouted through the area down-current.

Eden appeared unusually late. She was swimming slowly and off balance, and there was a hole in her right side, as if she had been unsuccessfully shot. After the slashing of Ali and Ruffles, and the disappearance of Cochita and Flora, this latest blow from the continually culminating tragedy of the shark finning racket seemed unendurable. I found a last scrap for her in the bottom of the kayak's well, and drifted, gazing sadly through the deep purple gloom. In the airy world above, it was raining.

Madeline soared through the site with total lack of interest on her way to the border from the barrens. She had grown enormous. Her naturally wide face looked twice its size with her advanced pregnancy.

The lagoon was a kaleidoscope of shadows beneath a wildly tossing surface. While drifting silently down-current, sketching the fin of a tiny shark who had come flitting up the scent flow, I sensed a movement on the periphery, and turned to see a lemon shark coming straight toward me, where I drifted in the centre of the sharks' swim-way. I waited for him to see me, amazed again at the sight of his face, and that smile, moving back and forth, back and forth, with each horizontal undulation. His mass darkened the space from surface to sand, and there was a bright white spot on the tip of his pointed nose.

Less than two metres away, he saw me, paused, and turned back. His path took him within centimetres of the little pup, and neither shark reacted to the close proximity of the other. I went back to my perch to wait. After a few minutes, he appeared briefly just within visual range, then disappeared again. He had come to look.

I got my disposable plastic camera—the other was once again being repaired—and waited. A few minutes later he came slowly into the site, picked up a fish head, turned, and faded again into the dimness beyond. It was very dark and I was shaking with cold. Waiting for him to return, and occasionally checking behind me, I wanted to leave, so finally went to the kayak. I was putting my camera and slate inside when I reflected that often it was just when my patience ran out that the animal returned, his patience lasting just that little bit longer. Sure enough, as I returned to my lookout, the lemon shark passed me from behind.

I waited, poised, for his next move, and when the water was so dim that the sharks had become more movement than substance, he drifted into the site closely followed by a second lemon shark. I didn't move a muscle. There was a swirl, as the first dipped down to pick up a fish head. Then the two flowed away to merge with the shadows beyond. As I left I was glad that some food remained for the mega-sharks.

Three days later, the day dawned with supernatural glory, and I glided out across a surface as transparent as glass. The coral garden was as clear as if I were looking into an aquarium, but what I was seeing, was not the vivid habitat, alive with fish and multi-coloured coral that I remembered gazing upon with awe when I began. It was mostly dead and algae-covered. Much of it was collapsing into grey pieces.

At the site a crowd of fish greeted me. They were nibbling on the spines still lying around from the recent feeding, so I picked them up. While I was busy, Eden passed. She had birthed since the session three nights before. The hole in her side was just beginning to turn grey around the edges. She had birthed right after being shot—because of being shot? I hadn't been able to pin-point the date that she had mated, so couldn't tell if she had given birth early, between November 14 and 17, 2004, or not.

She came as I swam on to cruise through the barrens. There was a strange cloudy light underwater, and I noticed again how seriously the coral had deteriorated. In many places it looked diseased.

I returned several times during that period of stillness, swimming from near the border to the barrens, noting who was around, and looking for Shilly, Ali, Ondine, and Jessica. Sometimes I brought a few pieces of food to get a better idea of who was in the region. Eden nearly always accompanied me.

Pipkin, a young female, was often present, and began to come to see me when she appeared. She and Melody, who was her age, were friends, and usually together. In spite of all the deaths, there was new life, a new community of young sharks pursuing their lives in protected waters before going out into the dangers of the ocean.

Tamarack had disappeared for a long time, but returned then. I suspected that the little animal was finally growing up and finding a home for himself outside the reef. The next time I saw him, he had a half-metre (20 inch) length of fishing net stuck through one of his gills. Mercifully, he had lost it by my next visit.

One day when I was with Eden, I put two pieces of food on a sandy place for her. But instead, the tiny sharks Oleander and Trafalgar came and took them. With her little mouth stuffed full, Oleander swam off past me streaming globs of food. Eden darted after her, then turned to dodge around the coral as the crumbs of food filled the water like snow. A very large and pregnant female passed briefly into visual range downcurrent, then disappeared into the veiling light.

I gave the sharks another piece. Gwendolyn targeted it as it drifted down, and coiled up to get it. Storm joined in. Since I was at the end of my swim for the day, I threw them the last two pieces I had.

Suddenly the huge pregnant female charged in. As she turned toward the place the others were feeding, I sketched her dorsal fin, trying to remember who she was. But suddenly she was shooting up to me. My hand shot out in self-protection, and she turned at the speed of light when she felt my touch upon her head.

I had met her once in the lagoon, several years before, and she had come to curiously circle me. Yet I had never seen her at a feeding session. I decided that it was time to leave.

The hole in Eden's side slowly healed, first edging in black, then filling in as December began. Two weeks after her injury, she was swimming strongly, and there was no trace of the vagueness in her movements that she had shown initially.

As the month passed, the winds rose, the ocean swell changed direction and began coming in from the north, and the current in the lagoon made swimming with the sharks very hard. I often had to content myself with dangling from the anchor rope, and calling them. Shilly usually came within a minute. She circled while I tried to watch her without the current breaching the seal of my mask. But it soon became too painful to maintain myself in the torrential flood, and I had to leave her.

When the current slowed enough to allow me to swim away from the boat, Shilly again appeared when I called. She approached at an angle from the ten o'clock direction, made an arc around me and departed at two o'clock. I fed the fish after she left, then threw her a piece as she came lazily back through the flickering shafts of light. She spiralled in to the falling scrap and took it placidly without accelerating.

Bratworst was drifting through the coral, watching, but there were very few sharks around, and though I fed the fish for a few minutes, those passing in the distance did not approach. I swam on eastward looking for Jessica. Shilly stayed with me briefly, left for ten minutes, and returned.

We were together when three males, Blaze and two companions, met us head on. I tossed in the last two pieces of food I had and began to sketch their dorsal fins, while Shilly circled and I called to her from time to time. She was uninterested in the food, though she had scarcely eaten, and the males did not eat either, but went on after only a brief pause.

I fragmented a tiny scrap in the water to feed a remora flitting around me. It swam to Shilly and attached itself to the tip of her pectoral fin, so that she looked as if she were flying a pennant, but let go just as I tried to photograph them. Thus, amused, it was a surprise to see for the first time in daylight, the bizarre face of Mephistopheles as he glided toward us. He must have been passing with the other males when he crossed the scent flow and came to see. It was the fifth year that he had visited in December! With the exception of just two visits during November and January, he had come at no other time. And still I had not found anyone else who had ever seen him, which was mysterious indeed.

He was about the same size and same colour, a deep warm grey. Alone, he was very shy, but he did come cautiously as long as I kept my distance, to take the two pieces of food still lying on the sand. He took one down-current to eat it, then came again minutes later for the second, and returned for a last check of the area before leaving.

December passed, and another year began. Carrellina had never reappeared, and Lillith and Trillium were presumed dead. Instead of Flora and Cochita who had once drifted beneath me for caresses when I arrived at the site, Windy anxiously poked out her head at me, then nosed around the back of the boat, looking for a bite. Twilight no longer came, nor did Cinnamon. Ali was missing. Madeline was bypassing the sessions altogether. I could see her pass the site down-current, barely within visual range, late in the evenings on her way to the border. Finally, she too, disappeared.

I went to the lagoon regularly, whenever I could get away, whenever the ocean was welcoming. Among the sharks there was loss, after loss, after loss, and those who remained continued to appear with hooks driven into their mouths, cheeks, and jaws. One was a tiny pup, trailing a long length of line, his little body leaning toward one side as he struggled to swim ahead. Needless to say, he never appeared again.

By the end of another season of storms, I had confirmed that Arcangela, Raschelle, Merrilee, Pippet, Danny, Faroe, Muffin, Darcy, Megan, Cochise, Dante, Grace, Lavoisier, Emerald, Revel, Papillon, Killy, Marten, Marco, Eclipse, Victoria, Capri, Muffet, Vixen, Storm, Fleet, Samurai, Coco, Blade, Hurricane, Kilmeny, my beautiful Antigone, The Concorde, Clementine, and Odyssey had failed to appear on schedule. None of them came again.

Compared with the natural death rate I had carefully recorded in the first years, this massacre was a loss of heart rending proportions. That the killing encircled not just my island, but the entire planet, and that my effort just to tell the sharks' story had met with derision on television, filled me with despair.

I intensified my efforts to have the sharks protected. Even though elections had been held, and the Independentists had won, no law to protect the sharks had been passed. The dive clubs were deeply concerned, and so were all who knew of the situation and cared about nature. The flow of letters to the government increased again. The best letter that I saw was that of Arthur's closest friend, Professor Samuel Gruber, who wrote to President Temaru:

Rosenstiel School of Marine and Atmospheric Sciences,
University of Miami,
4600 Rickenbacker Causeway,
Miami, Florida 33149-1098 USA

March 24, 2005

Oscar Temaru, President of French Polynesia
BP 2551,
98713 Papeete,
Tahiti

Dear President Temaru,

As founder of the American Elasmobranch Society and the IUCN Shark Specialist Group and on behalf of the Longitude 181 NATURE association, which speaks for over 20,000 scuba divers and sea enthusiasts who have already signed this letter, I respectfully ask you to consider the following:

Over the past 20 years shark populations world-wide have plummeted so that today some species such as the oceanic whitetip shark which were once among the most abundant large animals on planet earth have become scarce. Certain of their populations have been shown to be at 1% of pre-1985 levels. Likewise I am told that the abundance of sharks in French Polynesia has recently dropped by 90%. Many kinds of sharks once common in Polynesia are now on the IUCN's Red list of endangered species. In recognition of this immense problem, here in the United States several species such as the great white shark are entirely protected while 39 species are specifically protected by the US National Marine Fisheries Service.

In 1990 I co-authored a volume called SHARKS: A FRAGILE RESOURCE. At that time we in the scientific community begin to recognize the ecological damage being wrought by over-fishing of these important apex predators. I called them fragile because like other endangered sea creatures such as whales, sea turtles and many seabirds (i.e. albatrosses) they have slow growth, late sexual maturity and low fecundity. Thus they cannot withstand targeted, mechanized fishing which is now prevalent in all the world's oceans. Also in 1990 I estimated that 100,000,000 sharks and their relatives were caught and killed each year. This estimate was based on landings given in the FAO Fishing Yearbook of 1986—that is, on data collected nearly 30 years ago! What must the death toll be today in 2005?

Until recently, my sources in French Polynesia suggested that shark populations were relatively stable and healthy. But today, it seems that they are being decimated by foreign fishers seeking to cut off their fins for the Asian market to make a quick profit.

In a protein-starved world, today's fishing aimed at sharks, purely for their fins, is incredibly short-sighted; it prioritizes short-term profit for a few individuals to the detriment of a long-term ecological balance.

The substantial development of local long-line tuna fishing, whereby sharks end up being killed as by-catch, considerably reinforces the threat to all shark species.

My Polynesian colleagues feel that urgent action is needed to conserve the extraordinary biodiversity of Polynesian waters, which visitors from all over the world come to admire. They believe that the government must act with urgency to continue protecting sharks and other marine species that started with the whale sanctuary!

Therefore I respectfully request that you give priority to the protection and wise use of your unique marine resources, and avoid the route taken by other, less enlightened nations, by recommending emergency le-

gislation to immediately stop the wasteful and cruel practice of finning; and to disallow the taking of sharks by foreign fishers. The model for this legislation is available in the EU, US and Australia. I believe your fishery regulations ought to include a ban on targeted fishing aimed at sharks, and on possessing, trading or transporting shark fins in the Polynesian waters. Furthermore, long-line fishers must be required to return accidentally-caught sharks to the sea.

Sincerely Yours,

Samuel H. Gruber
Professor

He was the only person I knew of who received a reply, signalling that someone in the government was concerned enough to pass the letter on to another department:

Dear Sir,

I understand your reaction and I thank you for letting me know about it.

We are actually aware of that very serious problem and the French Polynesian government is working on it. I have forwarded your e-mail to the Environment Department Head who is in charge of this issue.

Best regards,

Tourism Department Head
Clarisse Godefroy

In April, another crisis was reported. Patrice Poiry, of Aqua Tiki, a dive club operating in the Tuamotu archipelago, found dozens of reef sharks of all species finned on one of the wild islands he visited intermittently. While previously, sharks had been numerous on the island, he saw very few there and there were none in the pass where they had once been plentiful. The silky shark, a large oceanic species he had commonly sighted outside the pass, had disappeared.

At another remote atoll, he had seen a fishing boat enter the lagoon and fish out all the blackfins over a period of a few days of nocturnal fishing. At the same atoll, the local people began fishing the tiger sharks for their jaws and fins. He said that all fishing boats he encountered on his tours through the islands were decked with shark fins hanging up to dry.

At another atoll, a significant fraction of the blackfin sharks seen on dives were missing their dorsal fins only; the sharks had been able to survive the finning since they had been left their wings and tails.

Since in Polynesia there were hundreds of islands spread out in separate archipelagos across the ocean, many of them uninhabited and rarely visited, the implications for sharks were dire. The reefs and lagoons of the island nation were being systematically stripped of sharks, by countless Asian vessels moving freely through them, avoiding the usual shipping lanes, and voraciously finning the sharks.

Other dive clubs operating closer to Tahiti reported the disappearance of the oceanic species they had once encountered regularly. I remembered how Katoa, our gardener when we had first come to Tahiti, had colourfully described his encounters with the whitetip oceanic shark (*Carcharhinus longimanus),* and others, only seven years before. He had seen them often while fishing outside the reef on Tahiti.

Katoa had fled to Tahiti from the Cook Islands when they became independent, so all his life he had fished in traditional ways off the islands of the South Pacific. We often talked because we were both native English speakers, in a French culture. Loving the lagoon and its inhabitants, marine life was what we discussed, and we often shared encounters with sharks during the two years in which I knew him. He had urged me to look out for oceanic whitetip sharks beyond the little pass through the barrier reef opposite our first house on Tahiti, where he often saw them.

Sharks would harass him, trying to take possession of the fish he speared, and he would try to drive the shark off. A common tactic was to hold the fish above the surface and kick the shark, and he told me once that his brother-in-law had done that and the shark had bitten his fin off. "Lucky he did not lose his foot as well!" Katoa laughed. He respected sharks, and was not afraid of them.

Fishing for his family, he defended his fish, and refused to let the sharks have any. Blackfins were the worst, he told me, with their habit of darting up repeatedly at the speed of light. Oceanic whitetip sharks would leave the area and would not return once he warned them off, but the blackfins always circled back, and continued to follow and harass. In Tahitian, the species was named for the way it would dash up close to one's face, then soar away. When a shark made this gesture fast to a fisherman, it seemed likely that it was trying to intimidate him, in hopes of getting the fish he had.

So I had watched for whitetip oceanic sharks when I went outside the reef, in Katoa's assurance that they would not bother me and were wonderful to see. But I had not seen one, and now, common as they had been so recently, they were gone. Who knew how close to extinction they were? Tahiti and its far-flung flock of islets and atolls were the most isolated in the world, equidistant from the surrounding continents. If they were gone from Polynesia, what could be their status around the continents, where they were so much more accessible?

As the months passed, I thought often of Arthur. He had not written since he had told me he was ill. I had a sad feeling that in this case, no news was not good news. His friend, Dr. Samuel Gruber, to whom I wrote from time to time, told me that Arthur was aware of my concern for him, but was no longer writing many e-mail messages.

One day, an announcement from Dr. Gruber appeared on the Internet saying that Professor Arthur A. Myrberg Jr. had died. It was April 8, 2005. Dr. Gruber wrote to let me know he would read something for me at Arthur's memorial service, which was a comfort, and very kind of him. I wrote:

It was with deep sadness that I learned of Arthur's passing. Humanity lost a great and good man when he left, and I will always remember him fondly, as one of the most exceptional individuals I have known.

I never met Arthur. We corresponded by e-mail for the last few years, and through the immediacy of this medium, I gained an intense sense of his presence. I was awed by his visionary powers and the probing way he dissected and analysed each facet of shark behaviour, and ecological intelligence, from all angles at every depth. He tirelessly thought, analysed, and wrote. Not a year ago he mentioned that he was working all night on four different papers.

There was an honesty about his intellect too, for he did not jump to conclusions. His final comment on one question we discussed was that in some cases, until we and sharks evolve to the point that they can tell us why they do things, their behaviour must remain a mystery.

Besides having a brilliant and singular intelligence, Arthur was kind. When a catastrophe befell me, he wrote with great sympathetic concern and understanding of the consequences.

The last thing he wrote me was his wish that one day we could meet, and I still hope that one day we will, to pursue sharks together in a better world.

Arthur had been unable to complete the revision of our manuscript providing the first evidence of cognition in sharks, and it remained unpublished (please see the Appendix).

CHAPTER FIFTY-NINE

Jessica

When the storms passed, there was no wind, but heavy rain each morning. The water cascaded from the sky, evaporated during the day, then poured back down for a week, during which the island felt like a sauna. Finally it dried up enough for the wetness to stay airborne. I waited for the sea to clear before venturing out again.

Windy had not seen me for two and a half months, since I had been avoiding her on my way to see Shilly. But she came with me across the lagoon at the surface under the boat. I softly caressed her, got ready, and threw in a few pieces of fish as I slid underwater.

The water was fairly clear at the site, and I was swiftly joined by many of the usual sharks as I fed the fish, who as usual, seemed delighted to see me. Several excited juvenile sharks began to feed, but Windy swept around the food, and charged me, time after time. She had mating wounds that were about two weeks old. I tried to photograph her and she slammed into the camera. I threw in more food, but she continued to charge, and finally slammed me! I instinctively leaped back into the boat, whipping my legs in, in record time, and she slammed the kayak!

She was very upset about something. But it hadn't been food she was after, since the food was lying on the coral floor.

Her behaviour reminded me of my pet skunk's action, years before, of heaving over the garbage can when he became angry. Was Windy upset because I had abandoned her for so long? She had always been quite unpredictable in her behaviour, but never had she behaved so emotionally. I concluded, one more time, that her reasons must remain mysterious.

Disappointed to find myself back in the kayak so soon, I went further east looking for Shilly. But once underwater, I found that Windy and her entourage of juveniles had followed. The food had been left behind at the site—Windy had come with me instead of eating! But now she did not approach. In this new location away from her home range, she was just a passing female in the distance.

Ondine and Eden swam with me. When I concentrated on taking a picture of Eden, she came over to me looking, apparently responding to my focus upon her while waiting for the right moment to take the photograph. I swam with her for the next half hour. In strong current, she raised herself forty-five degrees in the water column, and hovered there on an angle. We were down-current from my boat, which was far beyond visual range. The young shark must have detected a few molecules of the traces of scent from the food I had brought to Site One. Then she slowly moved on.

Back at the kayak I found Tamarack, and swam with him for a while before leaving. One of the vast, flowing schools of convict tang appeared—the pale gold, palm-sized butterfly-like fish that filled any region in which they appeared with scattering, shifting light. We swam among them, and I experienced one of the most breathtaking visual spectacles of my life. The first time I had managed to swim with a shark, when we were still living on Tahiti, she had glided through such a school, and I thought that I had never seen anything so beautiful as the little shining fish, parting like a curtain, to allow her to pass. Here it was happening again, but with a shark who knew me, who was with me willingly, and with fish, too, who had accepted me into their community. No more was I an alien. I had become part of their world.

In May, Jessica suddenly appeared in the barrens, cruising slowly into view over the grey background.

She was in shocking condition.

Her back was covered with deep black pock-marks. The eczema-like condition had changed to a far deeper problem. Her flesh beneath her damaged skin appeared corrugated, as evenly as the inner layer of cardboard, but magnified. The pattern ran evenly lengthwise. Her tail appeared to be shredding. It also looked as if it could have been bitten, and torn through the teeth of a predator, a larger shark, but if so, it was perfectly fresh, and the laws of probability ruled strongly against me having found her within twenty-four hours of her injury. It was more probable that the condition of her tail was due to the same condition affecting the rest of her.

I had first noticed her skin problem fourteen months before in March, 2004, and had last seen her at the end of December, five months before. At that time her rash had developed into black dots, from which the skin had begun to fray. Now, the entire shark was emaciated, corrugated, blotched, and fraying.

I posted messages on the Internet lists for elasmobranchologists and avian veterinarians, many of whom were experienced in wild animal care, providing Jessica's history, and asking if anyone had identified the cause of this sort of condition in a shark.

No one had. Dr. Berend Westera, a veterinarian for *Kelly Tarltan* of Aukland, New Zealand, wrote to tell me that the most frequent skin lesions he saw on sharks were from abrasions and bite wounds.

He wrote:

"We have cultured the bacteria in some of these and found Vibrio carchariae. With others fungi have been present. It would be hard to know what it is unless you can do a scraping and find out what is growing there. I have seen protozoan parasites in scrapings from other fish where these appear to be living on the eroding tissue and probably causing the erosions."

Obtaining a skin scraping would be impossible because Jessica would never come around the boat at the surface. If I could manage to scrape her underwater, the precious cells would be washed off and contaminated by the sea water. Though it would be satisfying to learn what was wrong with her, whether her condition was due to parasites, bacteria, fungi, or a virus, I could see no way to medicate her regularly. Dr. Westera recommended that Povidone-iodine solution be spread on her skin; this would kill both bacteria and fungi. But I was unable to lift her from the water for treatments as I had done with Merlin, so could see no way to help her. All I could be was a witness.

I returned to the barrens with food as soon as possible, and anchored adjacent to Jessica's traditional range. Windy and some juveniles followed the boat, and she came and raised her head to me from the water. On subsequent passes, she nosed the back gunwale until she found an overhanging scrap at which she snapped, then pulled. The juveniles began darting upward, and as they left the water, they bent their faces toward the boat and tried to snatch a scrap from the back of the kayak. They were imitating Windy, and demonstrating social learning, and continued, apparently, until each had managed to get a piece!

Windy was gone by the time I slid underwater. The bit of skin she had taken from the back of the kayak had sufficed! Breezy, who had disappeared several weeks before, passed closely to look. Then he left for forty-five minutes. Glamorgan drifted in. Sparkle swept over to me before checking the food, and some sharks from Section Three joined them.

Shilly glided up to me at the surface, with Ondine behind her. This time both companions came to seek eye contact. The sharks I swam with responded, just as my long lost Martha had done.

There was a tiny shark with a deformed pectoral fin flitting around the food, whom I had named Blighty. Christobel soared in, looked over the scraps, and swept up to me. Her strongly directed motion affected the others, and several of them joined her. I had to push away Blue Blazes, who was a delightful youngster, with the hand which held the camera. Strangely, Blighty was in the photograph I took of this sudden approach, as if he had made a quantum leap. The shark following Christobel was Columbine; she had accelerated swiftly!

Cheyenne was lurking just within visual range down-current, but not approaching. Soon I discerned

Hurricane with her.

Bratworst deftly searched the coral close-by as Cheyenne glided toward the scraps, moving almost directly toward me. I raised the camera. She paused, coiled her body to change direction, and shot vertically upward to snatch a fish from above, a movement so fast that it was scarcely perceivable. I took the photograph, but it was already too late. She had caught a fish some distance above her head! Her aim was remarkable.

Upon the lightning act of Cheyenne, the many sharks in the region turned to her. Moving parallel to her, they cruised toward me. There must have been a scent in the water, released by the bitten, dying fish, or even a fish vocalization I could not hear. There was an atmosphere of checking, waiting and watching. The sharks flowed together, then suddenly whipped away with everyone following, like a crowd hearing rumours of events in different directions and rushing one way then another.

Tristan came swiftly up to me at the surface, as if to check whether I had anything to do with the event. Pip and Cotlet were close behind. Then everyone went into search mode at top speed, in the place where Cheyenne had eaten. Fleur de Lis soared past looking, as if she had just arrived and was surveying the cause of the commotion.

One of the best advantages of the barrens was the distance from which I could observe across the landscape from the surface in the deeper water, so I was able to watch the entire group of sharks, led by one after another, as they soared through the area. Many more arrived, and their excitement brought the nurse sharks out to see what was going on. I couldn't remember another time when large nurse sharks had circled, and come to eat in the middle of the day. Was it possible that the three-metre nurse sharks really did spend their days in the lagoon? Where could they hide to sleep?

I looked aghast at the horse-sized fish sweeping around me in midday, and wondered if anyone other than me knew about them. Indeed, I was lucky that the love of the men for marijuana and beer kept them out of the lagoon for most of the time, so that they did not know much about what was in there. The big nurse sharks possessed kilos of fins, long, wide, thick fins, each exaggerated to the extreme. They undulated them prettily in big graceful waves as they swept majestically by.

I was observing a fortune soaring by beautifully, as if this sunshine, this place, and this company was all they needed to be happy. . . As if they had not a care in the world. Deeply touched by their innocence and splendour, I promised them that I would not publish my hard won findings of the patterns behind the roaming of sharks. It would only be used by a science dedicated to serving corporate profits, to kill more sharks than ever.

The Polynesians shared my distrust of those in power. They were so disgusted with the government that pocketed the billions of francs sent from France—money that should have been used to develop the country—that they had lost all respect for it.

"The only profitable business is stealing," was the new philosophy they expressed as they discussed the politicians they had elected, and in their minds, there was no difference between stealing from other people and stealing from nature. The local people I had come to know over the many years we had lived there, through exchanging smiles and anecdotes about the bay we loved, told me that no one cared any more. People broke the coral, used nets with such tiny weave that they caught the baby fish, and indiscriminately killed the sharks.

I saw Jessica again at the end of June. She was in about the same condition, but the frayed part of her tail was edged in black, suggesting that it was an injury after all, and was in the process of healing. She was extremely emaciated, and seriously ill.

She reappeared one more time when the moon was full, but never again.

CHAPTER SIXTY

The Sharks are Protected

I had been in contact with one of the local television stations when they had come to film a seabird I was caring for, notable for his two-metre wingspan and heart condition. So, in early April, 2006 I called them to see if they would broadcast some information about the shark massacre. The response was immediate, and I was interviewed about my observations of the sharks in the lagoon. I only had a few minutes to talk, but described how I had become familiar with the sharks while trying to learn about what they were like, and that the shark finning had decimated nearly all of the adult residents. As a result, I stressed, no old sharks remained, and most of the sharks left in the lagoon were juveniles. When they matured, they went out into the ocean and were finned, which I had been noting for two years by then.

A few days later, on Wednesday, April 12, 2006, *La Dépêche* announced that the Counsel of Ministers had decided that the practice of shark finning would be banned in Polynesia for a period of two years, in which a study would be launched to learn the numbers of sharks and determine future policy. But the mako shark would not be protected.

Two years after the shark slaughter had begun, when many species had been fished to the verge of extinction, and eradicated completely in some areas, the government was to begin a study—they had given the fishermen free reign to kill all they wanted, and when the sharks were nearly gone, gave themselves credit for protecting them.

I left for the lagoon in a cloud of turmoil, of sorrow and relief. After several days of calm, a storm front was drifting across the sun, but the wind was light. No one was visible when I slid underwater, but within a minute, Bratworst appeared. She passed slowly back and forth, watching from just within the veiling light. I had brought two fish, acquired for my last large seabird, which had become too old—they were all I had to bring on this celebration visit. The number of seabirds had fallen to a tiny fraction of what I had known on my arrival on the island eight years before. Fewer and fewer were placed in my care.

The fish were wrapped in plastic, and lay above water level in the front of the kayak so that no scent could escape.

I swam away with Bratworst, but she cruised too fast, in wide circles around the area. So while keeping an eye on her, I drifted toward the lower barrens looking for Shilly. Soon Jem appeared, gliding back and forth over a wide arc parallel to the reef, then he circled widely for awhile before curving southward. I stayed with him until he entered the thick coral east of the site and I saw a large female ahead, so, like a remora, I switched sharks. But she moved away and I wasn't able to identify her. For a while I drifted, looking around, and when no more sharks came, I headed for the kayak.

Shilly, Odetta, and Teardrop suddenly appeared—they had been following several metres behind me. I drifted with them while they did their shark greeting gesture—a circle and a close approach—and left with Teardrop, who seemed to be going somewhere. She skirted the sandy circle where I had once found Madeline and Samaria sleeping, and went on in her original direction. Eventually, she turned toward the island, accelerating as she went. She appeared to be heading toward the deep pocket region on the border where the fish spawned, and drew farther ahead of me. Two other sharks closed in, one on each side. After looking at each of them, I lost sight of Teardrop and rested, watching.

Jazz came straight from out of the blue, circled me, and swept on. He was an older barrens juvenile, one of the many there, for whom the thick coral on either side had become too limiting. His skin shone a radiant bronze, and his long, very slender form was the image of grace as he passed, gazing into my face.

Back at the kayak, Ondine was gliding. Her tail scarcely moved horizontally; she was almost drifting. Young females were usually impulsive and fast moving, while she was so tranquil. It made her interesting as well as a pleasure to swim with. She floated toward the island in the flickering sunbeams, through steep and ancient coral formations, from which seaweed slowly waved. Then gradually she began to turn in a clockwise direction, the water deepened, the coral grew larger, and we traversed a series of sandy strands, as though flying up forking canyons. Finally the meandering shark curved back into the heart of the barrens and I was moving against driving waves and current, so it was hard to keep up with her. She passed a coral structure, then suddenly turned at right angles, and swam slowly past me on the other side of it looking into my face. Then she turned away.

Whether her move had been planned or spontaneous was impossible to tell, but she had appeared to choose the coral structure for the manoeuvre. After that she left me. More than an hour had passed since I arrived, and I was chilled to the bone and tired from swimming at my limit to keep up with the sharks.

There was no one at the kayak, and Shilly had not reappeared, so I took the fish, enveloped in plastic, and swam eastward, planning to give them to the next shark who came to me. The sun, sparkling on particles raised by the surge, laid a veil of shimmering white light over everything. Sunshine fired straight down into the flashing water before my eyes, while beneath, all was dim, and no sharks glided. Deep in the coral east of the barrens, I stopped and drifted for a moment.

Bratworst appeared, lazily undulating in my direction. On finding me stopped in her path, she cruised back and forth, with no change in her speed. As I turned toward her, away from the sun, I moved from blindness to clarity, and saw her there perfectly defined in the great light of midday, shining down on us from behind.

I put one of the fish on the sand, but Bratworst took so long to return that I began to wonder if she would. At last she came ambling through the coral, the sleepiest looking fish imaginable, to slowly glide all around the circle of sand. She looked at me and constantly glanced around, but I couldn't tell if she had seen or smelt the food or not. She spiralled in on her next one-hundred-and-eighty-degree turn, swept down like a jet plane aligning with a runway, picked up the little fish and inhaled it.

She glided away out of view, and I placed the other fish upon the sand.

Soon she returned, and slowly again, she spiralled in and picked it up, without a change in speed. Then she pursued her lazy sinuous ways through the coral, sometimes visible, sometimes not. She had kept track of me for two hours, yet had appeared within visual range only four times.

As I drifted back to the kayak, Gwendolyn passed from time to time, just within visual range, but never approaching.

She always was a strange one.

Epilogue

In 2007, Franck and I had to move back to Tahiti, and soon after that I was contacted by Johann Mourier, a shark scientist from the *University of Perpignan* in France, who had been intrigued by my *Marine Biology* article on the blackfins' gestation period. He had come to study the sharks of Polynesia, actually working through the marine research centre that had formerly been so close to my home, yet was bereft of true information about sharks. We met soon after that, at a conference on the establishment of marine protected areas. After my years of study in isolation, it was fascinating to find another person who was as avidly fascinated as I was, by every detail that could be learned about the sharks I loved so much.

Later, when I got my website on-line, Johann noticed, and sent me recent photographs of Shilly and Trinket, assuring me that they were still fine. I was so very touched and grateful.

Johann's approach was different than mine, so that our two studies revealed complimentary information, and he began to find answers to some of the important questions my observations had raised. The most intriguing for me was the revelation that, far from being as lazy as they seemed, my placid and sedentary sharks were actually travelling between islands!

I invited him to write something for this second edition of the story of my sharks, and he sent this summary of his findings about them:

In 2006, after the sharks were protected, the government decided to launch a study to gain more data on shark populations, and on their degree of vulnerability. This began with a study of sicklefin [sharptooth] *lemon sharks, by my colleague Nicolas Buray, and other collaborators from CRIOBE, using photo-identification. I added my contribution to this study through my Master thesis on the genetics of lemon sharks in 2007.*

Then I decided to start a Ph.D. on blacktip reef sharks, since this species is the most common one on French Polynesian islands, and there are more individuals to study than there are lemon sharks. This also permitted us to compare the two species. My role was to investigate the organization of a reef shark population at an island scale in order to better estimate their degree of vulnerability.

Sharks have usually been thought to be solitary marine predators, despite increasing evidence that they show seasonal aggregations. Investigations of the social behaviour of sharks remained challenging due to the difficulties of following animals at sea and seeing their interactions between each other. Ila started to challenge these difficulties and her study on blacktip reef sharks was one of the pioneer studies on long term identification of individual sharks in their environment.

Conversing with her at the beginning of my studies back in 2007, I remember her strong feeling that these beautiful, poorly known creatures were far more social than we previously thought. It was a feeling I quickly shared with her when I dove those crystalline waters and was also progressively able to identify individuals by their fins. I was surprised to see that some individuals were observed always together, however, never with others using the same site. So my question was: can I find evidence of preferred association between individual blacktip reef sharks? And more importantly, how could I investigate this question? After an extensive reading of the scientific literature on the behavioral ecology of animals such as dolphins, primates, elephants, birds, and bats, I came across an interesting tool that I could use to investigate associations between more than 200 free living sharks.

Using recent advances in the field of network theory and its application in ecology, I rose to this challenge and revealed for the first time that sharks may be more sociable than previously believed. Using longterm underwater surveys along a 10 km portion of the reef, I was able to identify about 200 individual adult sharks from their fins, and documented their associations at different sites. Using social network analysis, I demonstrated that sharks not only displayed preferred associations but also split into communities based on assortment patterns. Tagging also helped me to show that this species is highly attached to specific sites and usually displays restricted movement with some individuals spending up to 70% of their time within a circle of 300 m radius. Sharks also were found to change their social groups between mating and non-mating periods. While some females left their fidelity site [home range] *for parturition and males left the area, others arrived in the system accompanied by new males. Other individuals that stayed in the network changed groups. These changes in the social structure may partly explain Ila's observations in which she reported that some individuals may travel and leave the area for weeks.*

The next step I am focusing on is the investigation of the genetic basis of these social groups after having succeeded in sampling DNA from most studied individuals. So the question here is to reveal the family relationships between these sharks and try to answer the question: do sharks tend to associate preferentially with their relatives?

The use of DNA pushed me further in trying to reveal the reproductive biology of this species. Ila previously described the gestation period of blacktip reef sharks as well as mating and parturition periods from her underwater observations. Her next question was certainly: where do those pregnant females give birth? This was the primary question of my work. Using parentage analysis, I was able to assign juvenile sharks sampled in their nursery back to their parents (the sampled adults) and reveal the females' migrations for parturition. I found that females came back to the same nursery for each of their birthing events; this is also called reproductive philopatry. More surprisingly, some females conducted migrations to reach nurseries at another atoll located 50 km from our island. Adult females from the North coast migrated to give birth in the nurseries of the West coast, and one of them gave birth at this distant atoll, during two reproductive seasons. More interestingly, some sisters were found to use the same nursery, supporting the possibility of the natal philopatry hypothesis in which females return to the place where they were born to give birth.

After Ila and I compared some of our fin pictures, we found that some of her sharks visited my study sites (located on the outer reef) about 10 years later. This was the case for Carmeline (2003 -2008), Shilly (2000-2010), Trinket (1999-2010) and Gwendolyn (1999-2009). Shilly and Gwendolyn were only observed once at my site, and Carmeline and Trinket could have been considered residents. Unfortunately, Carmeline disappeared after 2008 and was never spotted again. She was old, and one of the biggest females visiting the site.

One of Johann's sites, as you can see from his page comparing our photographs, is across the reef from my study area. This is the place that I visited occasionally too, as described, and even found females from the lagoon prowling. We still hope to get together to compare our dorsal fin pictures and other data, which could provide very long term records of some of the individuals. Unfortunately, as you know, most of the sharks I knew were killed long before he came.

But he confirmed what I wondered—whether my funny little shark, Trinket, had succeeded as a shark, and had made the ocean outside my study area his home range. He had indeed!

Johann is revealing more and more about the population structure and social lives of the blackfins, while making sure that they remain protected, along with their nurseries and their habitat.

Work on shark ethology, behaviour, and cognition has not advanced much since the days when Arthur Myrberg told me that no one believed that sharks could think, so I published a detailed book on shark behaviour in July, 2017, entitled *The True Nature of Sharks*.

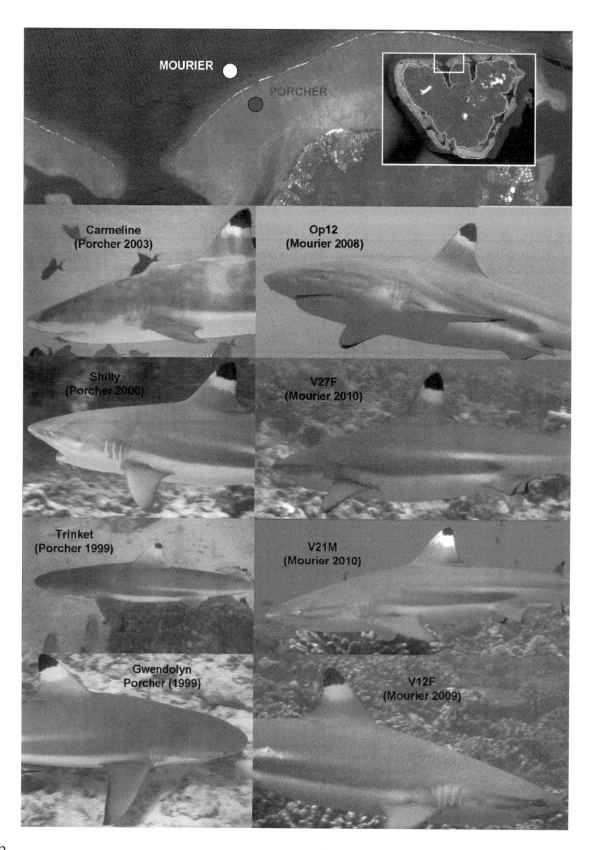

The vast numbers of seabirds I loved to see while zooming over the ocean on my dolphin tours, fell to a tiny fraction of what they had been; fewer and fewer birds came into my care. So with my work on sharks published, I will finish the book I began on the birds I knew, with respect to their intelligent behaviour.

I began writing down the story of my sharks overlooking the waters where they were being finned, then put it on hold for two years while waiting to see if they would be protected. When they were, I finished it, and added in the human framework—the first chapters to set the scene, the story of Merlin and the other case histories, the scientific aspect, including cognition, and my disastrous encounter with *Shark Week*. It was released at the end of 2010 under the title *My Sunset Rendezvous,* and for awhile I travelled to publicize it.

Returning to civilization after fifteen years in the simple life of Tahiti and her islands, I found that much had changed as I travelled slowly through Europe and then North America. Sharks were widely thought to resemble the monster in *JAWS*, and my public presentations about the community of lost sharks in their luxuriant coral garden, were welcomed with surprise and appreciation.

More and more divers were discovering the mesmerizing submarine presence of sharks, and from them, supported by many shark scientists, a powerful grass-roots movement was growing. Increasing numbers of shark protection NGOs were being formed by those who had understood the menace to them.

Among the shark advocates I had 'met' on the Internet while lobbying for the sharks' protection, was Jim Abernethy, owner and operator of *Scuba Adventures,* of West Palm Beach, Florida. He spent most of his time with wild sharks during dives from his live-aboard ship, *The Shear Water,* at remote sites in the vicinity of the Bahamas, and was on land for only forty days a year. He had targeted the sharks with the worst reputations—great hammerhead sharks, tiger sharks, bull sharks, oceanic whitetips, and lemon sharks—in order to show people their true nature, and he was the first eco-tourism guide operator to do so. For the first seven years of these interactions, the other operators who have now followed his lead, loudly voiced their opinions that people would soon be dying because of Jim's foolishness.

But he was right, and he was the one who showed all of the others, that even these larger sharks, "are peaceful animals who want nothing to do with humans as a food source."

After years of reading my pleas for my sharks over the Internet, Jim invited me to come to meet his sharks, and at last the day came when I stood on the deck of *The Shear Water,* gazing into sapphire waters where sharks circled at the surface, waiting.

My concerns about being carried away from my sunset rendezvous by a tiger shark had been based on local attitudes to them in Polynesia. Once more, the truth proved to be different from those fearful beliefs.

Underwater, Caribbean reef sharks (*Carcharhinus perezi*) approached us first, in a beautiful setting of sandy shoals alternating with reefs, decorated with soft corals and seaweeds. A friendly juvenile nurse shark was roaming among an assortment of gorgeous fish. The Atlantic species of lemon shark *(Negaprion brevirostris),* cruised through the vicinity in the same large, circling patterns I had seen in the blackfins. They were gold in colour and displayed an even more astonishing smile than the larger sharptooth lemons I had known. Finally, as the sun's rays flaring through the surface grew more and more slanted, a tiger shark drifted by, just at the limits of visibility. Another passed later, and after that a male tiger passed every few minutes, just within visual range.

The next day, at another location, we found ourselves at twenty-seven metres (90 ft), in a dim world of violets and greys, being investigated by bull sharks. In spite of their terrible reputation, these hefty brown sharks behaved very much like the blackfins, coming close and passing slowly looking at us, then circling away again, and not accelerating much. When they were given something to eat, they fed in the same relaxed way that my sharks had. Though often vilified as being among the most vicious of "killer" sharks, they too, behaved rationally and peacefully.

When again a tiger shark approached some seven metres (25 ft) above us, I breathed in and ascended without moving a muscle for a closer look.

Courtesy of Mary O'Malley

The tiger gazed back coolly, and showed no other reaction to my sudden appearance beside her. Later, my photographs revealed that this tiger shark came to each of our dives that week. She was following, watching, just like a blackfin!

Then we went to the shoal known as Tiger Beach, where tiger sharks gather in larger numbers.

I was enchanted by the days spent reposing on the ocean floor with Jim and the other divers, watching the sharks. There were crates of scraps to create a scent flow, and during the time we were there, more and more tiger sharks accumulated. They were curious animals who circled in and out of visual range, and approached in patterns much the same as those of the blackfins.

The tiger sharks seemed to be more interested in actually looking at the divers than the reef sharks I had known. Some who were comfortable with us, would glide very slowly by, close enough to stroke, looking intently at us one at a time. They had big black eyes that were more mobile than those of other species. As a result, their faces seemed more expressive, with their wide mouths open, and since they moved so very slowly, the eye gaze that one shared with them was long drawn out and quite spellbinding. A spirit of considerable power gazed back from those extraordinary faces.

Sometimes the shark suddenly turned around and came back.

The odd one did something completely different from the rest, showing the sort of individual differences in approach that the Polynesian sharks displayed. One was a large female tiger who sailed over us, then circled back into the blue. When she reappeared, she was high in the water, aiming straight for the crates of scraps hanging at the side of the ship, easily visible about nine metres (30 ft) above. Her body was silhouetted in the light of the sun glistening in the waves, as she snatched one crate after another, and shook it. The violent back and forth whipping of a tiger shark has to be seen to be appreciated—given her size she was the very image of power. After a few minutes, she let go of the crates, and soared back down to consider us again, before disappearing into the blue.

Two days later, she returned. With her nose to the sand as if she were following a scent trail, her back arched, and her eyes straight ahead, she passed me as she glided straight into the centre where the food was, and made another determined effort to get a scrap. After that, she reappeared only occasionally, while the other sharks tended to stay in the area, circling in and out of visual range, and intermittently approaching the crates. When one of them managed to find a scrap, they accelerated in excitement, sailing together through the vast blue surroundings, just as my sharks had done across the world in Polynesia.

One day I was watching a lemon shark who had managed to get a scrap, and was zooming along the sand, mouthing it, accompanied by a fleet of excited companions ready to snatch up any crumb that might fall. Suddenly the group changed direction, and knocked me over in a wild confusion of sharks. With the

The returning tiger.

heavy steel tank I was wearing, I could not regain my balance—when I tried to get up I was knocked over again and again by the sharks.

Later my photographs showed the moment I was knocked down—a lemon shark is blurred passing my hands, remoras flying in all directions, the horizon almost vertical. While trying to get up, I snapped photos of two tiger sharks who immediately materialized, and was knocked over again. One photo showed the blurred face of a tiger shark against the camera, while masses of lemons filled the surrounding waters. Another tiger was approaching behind that one, and the next photo showed her huge face bending down to the sand to look straight into mine as I lay on the ocean floor.

It was a strange, unreadable expression in the photograph I gazed at later from the safety of the ship—the face of a wild tiger shark who had not approached before, who had only passed twice in the distance in all of those days. Yet, the moment I had fallen down, this tiger had come. What a face! My painting of her expression is reproduced in the illustration.

The lemon shark had dropped his scrap under my body, and for more than two minutes, they all kept knocking me over in their effort to get it! Finally, I was able to get to my knees in time to see another tiger shark, twice my personal width, slowly approaching my middle. I was reaching out to take her head in my arms and push her away, when I was lifted up from behind. Matt, the dive master, had come to my rescue. He pushed away the tiger shark, and pulled me out of the mass of lemon sharks. Any of the tiger sharks could have picked me up and carried me away if they really behaved the way the shark attack mongers said they did. But they only came close to look, and did not bite me.

Jim talked to me a lot about his sharks, whom he loved as I loved mine.

Over the years, as he had become intimately familiar with them, he began caressing them gently on their faces when they approached him. As they got over their initial fear of his touch, he was able to massage their heads and remove their parasites. The sharks enjoyed these affectionate attentions, and responded by returning to him more often and more confidently, apparently considering him to be a type of cleaning station. When some of them appeared with hooks in their mouths, he was able to use these tactics to get them to swim up to him repeatedly, so that he could examine their hooks before removing them.

One of the large male lemon sharks was blind in his left eye. His name was Captain Ron, and Jim always gave him special attention, impressed that he had managed to survive in spite of this serious handicap. He knows how hard it is for an animal with any handicap to meet the challenges of living wild.

Jim had known Captain Ron for a decade by then, and described how he had recently appeared with a large 'J' hook piercing through the flesh of his nose and into his mouth. He had begun touching Captain Ron's face whenever he passed with gentle caresses, and as he relaxed, grasping his nose for short periods. The hook was close to his teeth, so Jim was concerned about being accidentally bitten, and waited for the right moment. When he sensed that the time was right, he held the shark's head still with his right hand, long enough to remove the hook with his left hand. Captain Ron remained relaxed and nearly motionless in the water, and when Jim released him, he circled and came back.

It was a week later that I was there with Jim, and Captain Ron swam straight up to him, allowing him to examine the wound. It had nearly healed. The hole made by the hook had filled in, and the redness around it had gone. Captain Ron was even more affectionate as a result of the incident, repeatedly returning to swim close beside Jim and let him touch his face.

Though most of the sharks drawn to Jim's dives remain distant and never do approach closely to interact with him, he found that no matter what the species, a natural bond would form between him and certain individuals over time, facilitated by his affectionate gestures. In the wide region known as Tiger Beach, there are approximately seventeen such tiger sharks, whom he calls "super models," who come to him on sight for the affection that he gives them on every encounter.

Jim's 'Martha' was a matronly four-meter (14 ft) tiger shark he had named Emma, after one of his guests. Over the years, the great shark had grown increasingly trusting and intimate with him, and their bond had steadily deepened. When Emma came to a dive with a fish hook driven through her lower jaw, Jim gently touched her until he had the opportunity to pull it out. Twice he was able to remove hooks piercing the outside of her mouth, by caressing her gently and waiting for the right moment to coax out the hook.

Then, one day Emma appeared at the beginning of a dive with fishing wire hanging from the side of her face, and a large hook driven into the muscle of the right hinge of her jaw. The hook was deep inside her huge mouth, and Jim observed her circling around him, wondering how he would be able to remove it, and especially about how he would protect his arm from her teeth when he reached inside. His first thought was to get a piece of PVC tubing to protect his arm, then open her mouth and remove the hook, but at that moment, Emma came straight toward him, and just before she reached him, she bit down on a large head of brain coral.

This was an incomprehensible move, since sharks never bite into the coral environment in such a way. Did she do it intentionally with some concept of stabilizing herself so that Jim could look at her painful mouth? We will never know, but her act remains unique and unexplainable by any stimulus / response theory of shark behaviour. Jim, hyper alert, moved swiftly to reach inside her mouth, take the hook in his fingers, and rip it out! Blood poured into the water. Emma took her jaw off the coral head, and soared around bleeding. She stayed in the area, and continued to approach Jim to be stroked as she always did.

Later that day, while Jim was stroking her head, he tried opening her mouth to get a look at the wound by sliding his right hand onto her nose, and using his left hand to open her lower jaw. Thus encouraged, Emma opened her mouth, and he was able to see that her wound was very swollen, and was between fifteen to twenty centimetres (6 to 8 in) long. A week later Jim returned to the area, and was again able to coax

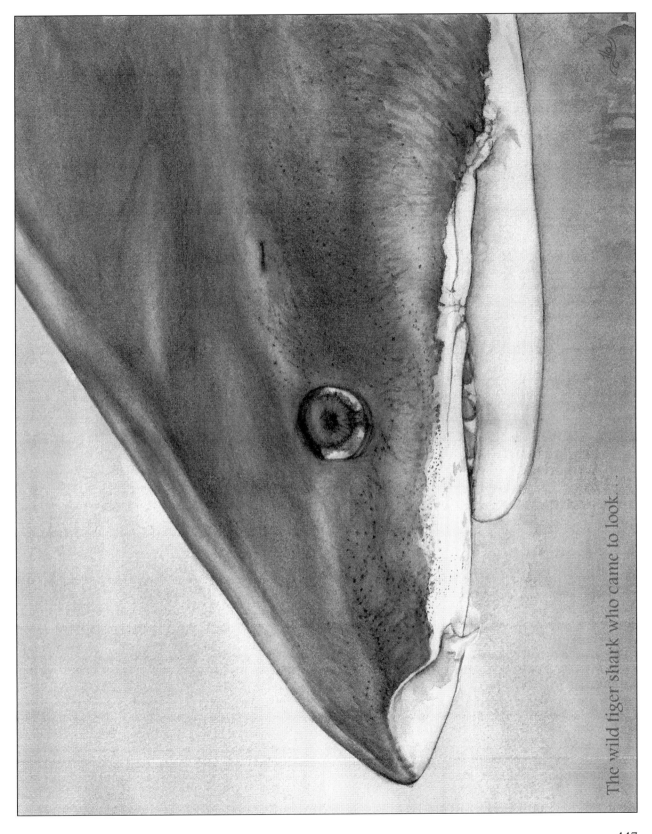

The wild tiger shark who came to look…

Emma to open her mouth for him. The open wound had closed, to his great relief—he was impressed by her ability to heal up quickly. For the next four or five weeks he opened Emma's mouth to see how her wound was healing whenever he saw her, which was about five times a week at that time. Eventually, she began swimming up to him and opening her mouth by herself!

I was intrigued by this story—it was the first case of love between shark and human that I had found since my own experiences with Martha, Madonna, Shilly, Flora, and the others. It was also relevant in terms of shark intelligence and cognition. But at the time of my visit to Tiger Beach, Emma had not been seen for six months, so I had little hope of meeting her. Yet one evening, an enormous tiger shark passed swiftly just at the visual limit. I thought of Emma—the shark was so much bigger than any of the others. She was pregnant, and the tip of her dorsal fin was missing. Soon she reappeared and glided straight into the centre where the divers were. Jim was on board at the time, so word was sent that Emma had come. She was roaming energetically through the area when Jim arrived on the seafloor, and she immediately swam to him. She recognized him from among all of the divers—there were about six of us present at the time.

I moved closer to see if she would open her mouth when she met him, unlikely as that seemed. Yet, in spite of the length of time that had passed since she had seen him, that is exactly what she did! Jim was still reaching out for her when she opened her mouth. He rested his right hand on her head, and looked inside. She had opened her mouth before he touched her—she was remembering him and their complicity over her hook wounds from six months before!

I was beside Jim, photographing them, when Emma unexpectedly turned her attention to me. I stroked and rubbed her huge head, which seemed to be what she wanted. She was drifting forward while I cuddled her, so that I was hard pressed, with the heavy steel tank, to rise up swiftly enough so that the enormous animal could pass beneath me!

Emma was excited, roaming around energetically, and approaching one diver after another to be affectionately stroked and rubbed, from her nose, all along her sides, to her tail—everyone present, all shark lovers including Stan Waterman, Mary O'Malley, and Douglas Seifert, knew her and gave her a lot of affection. She showed every sign of enjoying our visit and these attentions. Unlike my reef sharks, she really did like to be touched!

But most often she came to Jim. I saw her open her mouth when she approached him four times. Her momentum kept her moving forward so that she rose upwards, Jim moving with her. By the time he was able to take a good look inside her mouth, she was rising nearly vertically. The forward momentum of a four-metre tiger shark is daunting; Jim had to launch himself in synchrony with her to see.

Jim described how Emma had lost the tip of her dorsal fin. He and Emma had reunited soon after she had birthed the year before, and he noted that each week, she had another mating scar. (Females generally acquire significant mating wounds because the male holds her with his teeth to stabilize the pair during copulation.)

One day, a small male tiger

Emma opens her mouth for Jim.

shark of about three metres (10 ft) in length was with her, and kept trying to mate with her, biting her on the back of her head. Emma rejected him and eventually she swam away. Jim followed, trying to keep them in view, but he was left behind. When Emma reappeared, the upper part of her dorsal fin had been ripped off! Filaments of cartilage projected from the wound, and some of her fin was missing. As time passed, the fin rebuilt itself, though the tip of the fin has remained rounded off. I was reminded of Christobel's shark bite. Jim was able to document the healing of Emma's fin, which is another rare case of documented flesh-replacement in wound healing in sharks.

Jim looks in.

Jim described how a shark trophy hunter had visited the dive sites and fished some tiger sharks in the five and a half meter (18 ft) range, in hopes of setting a fishing record. As a result of the slaughter, the sharks disappeared from the area for a period of two months.

Far off across the world in another ocean, Polynesian sharks of a completely different species had also fled, as I and

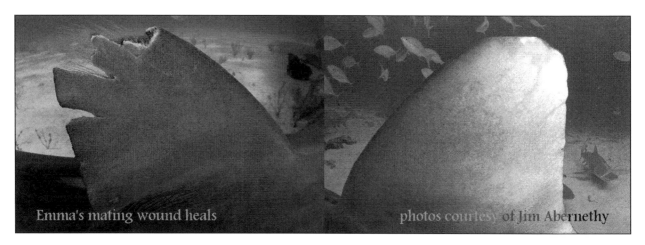

Emma's mating wound heals   photos courtesy of Jim Abernethy

others had documented, when some of their numbers were killed. In every way that I was able to learn about the behaviour of Jim's sharks, it was remarkably similar to the behaviour of the requiem sharks, far off, but not too distantly related, whom I had known in Polynesia.

This is to be expected since sharks have been evolving for approximately four hundred twenty million years, and many species travel widely and are found around the globe. The essential qualities that sharks evolved to be so successful would already have developed in the ancestral forms, before they evolved into modern species occupying the ecological niches we know today. These observations of tiger, bull, reef, and lemon sharks, support my position that my Polynesian sharks couldn't possibly be the only ones who possess the qualities I documented in *My Sunset Rendezvous,* as some have suggested. There is no reason to assume that my sharks were special or different. It is more probable that they were ordinary sharks, quite representative of their kind.

Sharks are Jim's life, and he is a powerful advocate for their protection. He has fought for them in one situation after another when some new law or project threatened them or the marine environment. When he was disparagingly referred to as a "loose cannon" by someone he had embarrassed who was damaging the environment for profit, he retorted that he was no cannon, but an F-22 Tactical Fighter for sharks!

While I was there, we were interviewed about shark finning by CBS Channel Twelve. The video is on youtube at: http://www.youtube.com/watch?v=EAhyuSI33p0

Jim's influence is growing and spreading. He is an award winning photographer, author,

Emma and Jim   Courtesy of Nigel Moyer

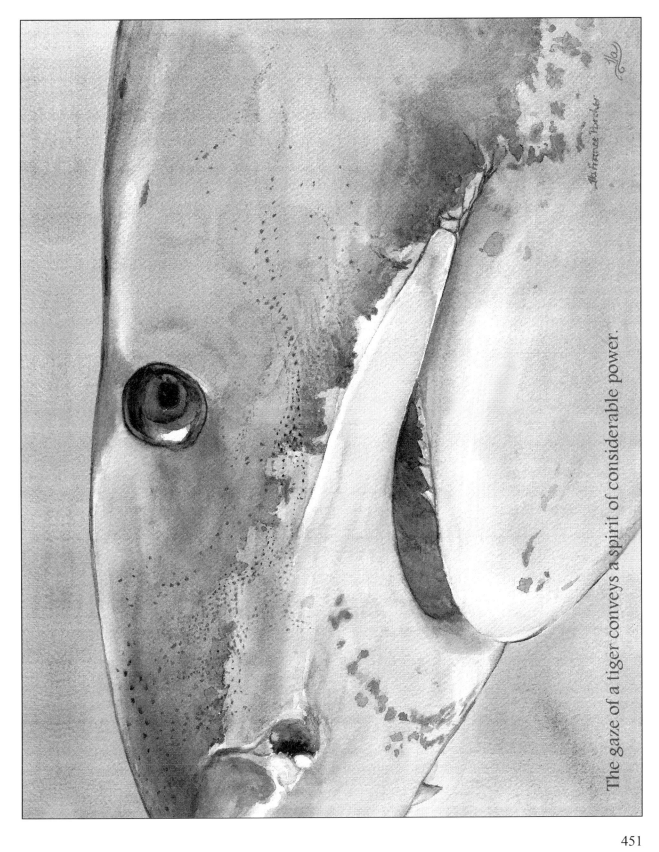

The gaze of a tiger conveys a spirit of considerable power.

and film-maker, and besides creating films of his own, he has hosted most of the world's top nature film-makers and magazines on his dive ship. His photographs have been widely published, and he has written books including *Sharks Up Close*. As well as fighting for sharks as an individual, he works through a variety of important NGOs, including *Operation Blue Pride*, which he founded in 2011.

Jim is a passionate man with a heart of gold, and is endowed with inexhaustible energy. I was astonished to see him flying among the condos along the beach in a little rubber raft suspended beneath a hang-gliding wing, on one rare day on shore, as he video-taped sharks migrating up the coast from the air. He had designed and built his flying machine himself!

After my exciting and touching experiences with Emma, I was angered to come across a video of her being abused on *Shark Week*. See http://www.youtube.com/watch?v=W7XLCo6MHH8

That shark is Jim's Emma, who comes to divers to be stroked, and wants him to help her when she is hooked. Note the way the film is sped up and slowed down to make her seem to accelerate and then pause dramatically, when in fact, her speed is steady. None of the comments about sharks hunting in this video are relevant. That is a shark who is accustomed to being greeted by divers, who is instead being fished. She is not hunting.

Discovery Channel and its dishonest portrayal of sharks on *Shark Week* has angered many scientists whose work has been twisted and falsified for dramatic effect. This week-long series of shows has been presented each summer since 1987, and has generated such hatred for sharks that many people believe that they should be exterminated. Thus, it has erected an impenetrable barrier to efforts to protect them from extinction. The shows have also generated an irrational fear of the sea in many people who grew up watching them. I was hoping that someone famous would write to *Discovery* and confront them about it, but when no one did, in 2007 I posted an outline for a possible letter on the Shark-L discussion list. We discussed and improved it, and when we agreed that it was finished, I sent it to the *Discovery Network*, to Mr. John Hendricks, Founder and Chairman, and Mr. David Zaslav, President and CEO. It was signed by three hundred fifty-two people who were involved with sharks and personally concerned about *Shark Week* and its dishonest, harmful message.

See: http://boycottsharkweek.blogspot.com/2009/06/our-original-letter-about-shark-week.html

As a result of this letter, four representatives, Maris Kazmers, Steve Fox, Julie Andersen, and David Ulloa, met with some of the *Shark Week* executives, including Executive Producer and Senior Science Editor Paul Gasek, and Senior VP of Development and Production, Paul Hassler, at *Discovery Communications*' headquarters in Maryland. The meeting lasted over three hours, and our reps felt satisfied that all of our points had been brought up.

They reported that the *Discovery* executives were proud of their shows, and were pleased with *Shark Week* as it is. They love the dramatic special effects that typifies them, the "killer shark" image, and the blood and teeth horror effects, which they call "shark pornography." They said that they are only giving the public what it wants, though the public's love of horror has nothing to do with *Discovery's* responsibility for having made sharks the subject of that horror, a marketing decision, incidentally, that has brought the company billions of dollars, we learned.

The programmers were uninterested in true shark behaviour, and were focused instead on how to make them look "scary and dangerous." They would not acknowledge that their shows are largely responsible for viewers thinking that sharks are "scary and dangerous" in the first place.

Paul Gasek had little scientific background. He had been a commercial fisherman for ten years, and thought that throwing turkeys to tiger sharks was good science, a recurring joke throughout the meeting, since he wanted to do it again. It seemed strange after the way they had treated my work. Apparently they have no problem at all with feeding sharks.

In the end, they said they would add some information about sharks being endangered by over-fishing, and try to show more real science, especially if we provided good stories about sharks for them. We did provide them with some interesting stories, but they ignored them. However, in subsequent years they have added more information on the threat of extinction sharks are facing. The problem remains that *Shark Week* has so effectively convinced its millions of viewers that sharks are vicious, and deserve to be hated, that many people think that the extermination of sharks is good.

That year, the titles included, "Top Five Eaten Alive," "Ocean of Fear : Worst Shark Attack Ever," "Deadly Stripes: Tiger Sharks," and "Shark Feeding Frenzy," so one of our specific requests was that *Discovery* tone down their use of such pejorative words, especially "man eaters," and "mindless killing machines." They have since reduced their use of these labels, but continue to use the words "monsters," "serial killers," and "jaws." The fact that *JAWS* was a horror film, and *Discovery* claims to be showing quality non-fiction, apparently escapes them!

A curious thing happened with my sequence, however. I purchased the CD of *Sharks : Size Matters* years later, and found that it had been changed. The sequence of me talking about the sharks being finned in the kayak was included, though no supplementary information about shark finning was given. The sequence in which Mike said, "Ila claims these sharks have personalities, but the jury's still out on that!" was omitted, and instead, at the end of the part in which he cross-examined me about why I had brought food for the sharks, he said that I had learned a lot about them through feeding them. So with those and other minor changes, the sequence was not so completely slanted against me as the original one had been.

Discovery never informed me that they had changed my sequence as a result of my outrage over their handling of the story of my sharks, who were being finned at night and filmed during the day when their crew visited, and I have no idea when the changes were made, nor whether the improved version was ever shown on television.

In a letter to them after their meeting with our representatives, I remarked facetiously that if they wanted to show-case horror, why not feature car accidents, which are much more dangerous than sharks? While writing this, I learned that they systematically film the worst of the car and truck accidents on a treacherous mountain highway in Canada for a horror show! What a commentary on the ethics of *Discovery Channel*, that they can treat people, their suffering, and their right to privacy, so callously, by televising shocking images of accidents, and the terrible things people suffer among the snowy wreckage.

In 2013, *Discovery* came into serious disrepute by claiming that the gigantic extinct shark *Carcharodon megalodon* might still be hunting the oceans and devouring hapless swimmers. This blatant falsehood was an inarguable public demonstration that *Discovery's* so called "quality non-fiction" is really nothing more than tabloid journalism, that does not reflect scientific knowledge.

But sharks have paid a terrible price for *Discovery's* multi-billion dollar fortune, generated by demonizing them. The hatred launched against them over the decades has continued to fuel their mass slaughter through sheer hate killing and monster shark tournaments. A good history of this phenomenon is given in the book *In the Slick of the Cricket* by Russell Drumm, which describes how *JAWS* changed the prevailing attitude of the public to these animals and resulted in monster tournaments being held all along the east coast of the United States, which filled the landfills in countless towns and cities with mountains of decaying sharks.

Though catch and release is described as the solution to this cruel decimation, Drumm reveals that the excited and malicious monster hunters fight more than eighty percent of the sharks gut-hooked, so that their fragile internal organs are sliced and torn apart during the fight, and upon their 'release' they simply sink.

Along America's east coast, the slaughter of sharks is obscene. According to the *National Oceanic and Atmospheric Administration of the U.S. Department of Commerce* (NOAA), two million, seven hundred thousand sharks were caught by sports fishermen in the U.S.A. in 2011, a travesty as shameful as shark finning. Since those were only the killings that were reported, this figure could be low compared with the true numbers killed if the toll from private boats that were not reported, were added in.

Between this mentality and the one that fins them, the way sharks are treated by our species is a disgusting illustration of the destructiveness and wastefulness of mankind.

While other countries have passed laws to protect fish from cruel treatment, in North America, the fishing industry has discredited the findings that fish are capable of feeling pain to the extent that most people continue to believe the old fisherman's tale about them—that you can't hurt them no matter what you do to them, because sharks and fish can't feel pain.

Indeed, fish and sharks alone, of all animals, are still seen by the public primarily through the eyes of those who enjoy killing them, and in this way a whole class of animals has been condemned to being considered lower and simpler and less feeling than all the others. Fishing has not received the cultural bias against it that hunting has, and fish have not been protected from needless suffering by law as reptiles, birds, and mammals have.

The numbers of sharks killed for scientific study in at least some regions of the east coast of America, are measured in metric tons, and an unholy marriage has developed between certain branches of science and fishing, to the degree that not only is fishing supporting science, but science is supporting fishing!

In the year 2012, fisherman James Rose wrote another review paper published in the same journal, *Fish and Fisheries*, asserting that fish cannot feel pain because they lack a human brain. His new paper included nothing new. Again he had done no study to back up his allegations, and he based his argument on the fact that fish, like all other animals besides humans and the great apes, do not have a human-like neocortex. When I originally wrote this book, I was unaware that mainstream scientists had found that his descriptions of the neuroanatomy required for suffering, eliminated all other mammals except the great apes.

Yet in all this time, he has presented no evidence proving that it is possible for an animal to survive if it cannot feel pain!

His argument condensed to the idea that fish pain is not human pain, so does not count. He dismissed it as a reflex reaction, and misrepresented the findings of Dr. Lynne Sneddon who originally established, through rigorous experimentation, that fish do feel pain and suffer. She sent a rebuttal to the journal—you can read it on her web-site—but it was not published.

It mentions that his article was written at the request of recreational fishing groups in the United States. This is important because this politically based, and not biologically justified, idea is being used to excuse shark finning, as well as the careless and cruel mistreatment of fish. Just to put this political affair in perspective, according to the NOAA, the fishing industry generated one hundred ninety-nine *billion* dollars in sales in the United States in 2011.

The idea that there is a huge gulf between humans and animals is a religious one, and its use to justify inflicting pain and suffering on other life forms, particularly for pleasure, should be fought wherever it rears its ugly head, in the interests of continuing to establish a moral society.

A study of the subtle differences between vertebrates as they grow increasingly complex, from fish to man, reveals the thread of evolutionary continuity, showing that no new ability or organ can suddenly appear in a class of animals without that feature having evolved in its ancestors. The implication is that the sensation of pain evolved along with all the other sensory systems.

Vertebrates all have the same basic plan—a head, spine, limbs, two eyes, two ears, one mouth at the anterior end of a tubular digestive system with analogous parts, a ribcage, pelvis, brain and spinal chord with its associated nervous system, a heart, respiratory system, vascular system, musculoskeletal system, blood, and so forth. This basic set-up is adapted in different animals, depending on whether they walk, fly or swim, and upon their niche in the ecology of the planet.

Is it not a leap of logic to think that though we are all similar to this extent, the pain system in fish, though wired like ours, doesn't work? Would it be as believable if it were the fishes' eyes which Rose arbitrarily stated did not work? If his argument is true, no animals other than humans and the great apes would be able to feel pain, but it has not been applied to dogs and cats, because the fishing industry is not concerned with them. And no one would believe that birds, cats, dogs, horses, and so forth, cannot feel pain, one reason being that pain has been used so effectively to train them!

My argument that fish feel pain was written in 2004 when I was isolated in Polynesia, trying to understand what fish may feel, and to defend my sharks, who were being finned. At that time, I had spontaneously used the findings of cognition and social intelligence in sharks and fish, as evidence that they couldn't possibly be as simple-minded as fishermen claimed. Now, preparing this second edition in 2014, it is gratifying to learn that my approach is considered valid by the mainstream researchers. They have referred to a growing pool of cognitive data as supplementary evidence, not only that fish feel pain, but that they are conscious.

Dr. Sneddon, of Liverpool, in the United Kingdom, made her initial discovery in 2002, and since then, research into the field has flowered, and been investigated from every conceivable angle by a wide variety of researchers, including Rebecca Dunlop, Victoria Braithewaite, Janicke Nordgreen, Temple Grandin, Joseph Garner, and Kristopher Paul Chandroo. All have verified her findings that fish show every sign of being capable of suffering, and Germany responded by passing a law that fish can no longer be fished for sport, but for food only.

Yet so many people have picked up on fishing stories that they can accept Rose's widely publicized idea that fish are so simple-minded that they don't feel pain. So here I will summarize the latest findings, and some interesting facts on the subject. Dr. Sneddon helped me considerably by sending me many of the relevant scientific articles.

As she has pointed out, it is not possible to know for sure what any creature except ourselves is actually feeling, though we, as people, can describe our feelings to each other, and empathize with each other. Since we cannot ask animals, and they cannot answer in a way we can understand, scientists have searched indirectly for evidence about their subjective states, in the studies of neuroanatomy, neurophysiology, and especially in their behaviour. They have developed strict criteria, all of which need to be met, before one can conclude that the animal can feel pain.

Firstly, there must be nociceptors, sensory neurons which respond to tissue damage by sending nerve signals to the spinal cord and brain. This process is called nociception, and causes the sensation of pain. There must be neural pathways from the nociceptor on the body and head to higher brain regions, and the signal from the nociceptor must be processed in the higher brain, not in the reflex centres in the hind brain or spinal chord. There must be opioid receptors within the nervous system, and opioid substances produced internally. Pain killing drugs should relieve the symptoms of pain that the animal displays, and it should be able to learn to avoid a painful stimulus. This should be so important to it that it avoids the threat of pain right away. The painful event should strongly interfere with normal behaviour, and this should not be an instantaneous withdrawal response, but long term.

Fish comply with all of these criteria, as shown in a wide variety of experiments. Their nociceptors are nearly identical to those found in mammals and humans. These are connected to the brain through neurons. There are also connections between the thalamus and the cortex as well as other regions of the brain, which are considered crucial to the experience of pain. The whole brain of the fish is active during painful events, not just the hind brain.

As well as neural activity, certain genes that are crucial to the experience of pain in humans are also found in fish. The activation of these genes is widespread throughout the fishes' brain during painful events. This activity of the brain at the molecular, as well as the physiological level, indicates that these are not reflex reactions, as Rose claims. If they were, such activity would not be seen in the higher brain.

Fish and humans share approximately fifty percent of our genes—the lion fish shares about eighty percent. They and puffer fish are considered so similar to humans genetically that they are being used as model organisms for some research.

Fish have displayed a variety of adverse changes in their behaviour after the infliction of pain. An extreme increase in their ventilation (respiratory) rate, may be followed by rubbing the damaged parts on the substrate, rocking on their pectoral fins, trying to stay upright, and no longer feeding. These, and other symptoms of distress, are relieved by the administration of morphine, which completes the circle, identifying pain as the cause of the change in behaviour. Like other animals tested, fish have been shown to self-administer painkillers if they can, even if that means going into a location that they do not like, to bathe in water that medicates them with a pain killer. This is another clue that the fish was experiencing discomfort, from which it found relief in the undesirable location.

Fish swiftly learn to avoid painful events, which researchers think indicates that they are conscious—they experience the pain so severely that they are strongly motivated to avoid feeling it again, even after just one exposure. Consciousness in fish is now often mentioned, whereas when I first discussed the subject with Arthur, the idea that fish could be conscious was generally unacceptable.

Though humans can over-ride pain at times in certain heightened mental states, and particularly when they are in danger, it seems that fish cannot do so. Studies have shown that after being hurt, fish become far less alert to danger, as if their pain is too overwhelming for them to ignore it to escape a predator. It is thought possible that due to their simpler neural design and mental states, they lack the ability to think about their pain, and put it in perspective as humans can. This suggests that pain for them is always an intense experience, and that fish may actually feel pain more intensely than humans.

I wrote down on my slate, time and again during my observations of sharks with mating wounds and other serious injuries, that the shark was swimming unnaturally slowly, and seemed interiorized—he or she appeared less alert than usual.

A state of basic consciousness is now thought to be present in animals whose nervous systems have evolved sufficient complexity, and it is believed that cognitive studies may be used to assess consciousness and sentience.

The concept of self recognition has advanced since I worried about what my sharks saw in their mirror. The mark test, in which an animal must recognize a mark placed on its face while it slept, and touch it when shown its reflection in a mirror, has given way to a wider view of ways that animals can indicate that they recognize themselves, and distinguish themselves from others. Fish live in an aquatic environment and are oriented to assessing smells. At least some fish recognize their own odour and prefer it over the odour of other fish, including their relatives. This is considered to indicate self-recognition.

Another indicator of self-recognition was found in fishes who watched fights between their conspecifics, and changed their behaviour when faced with either the victor or the loser. The behaviour of cleaner fish, in which hundreds of other individuals are recognized and treated differently, depending on their relationship to the cleaner, is also in this category. Fish have shown that they recognize themselves and discriminate themselves from others, an ability which was previously thought to be restricted to the higher mammals. These abilities are considered central to higher level cognitive skills and consciousness.

Fish have been shown to possess traits that distinguish one individual from another, just as my sharks showed individual differences. They exhibit clear preferences, avoid aversive events, use tools, can learn complex tasks, and have long term memory, showing that they have the ability to make decisions and to remember negative events. Given that they are conscious, and may suffer on an emotional level, fish welfare emerges as an important issue.

Research has been done to find ways to minimize the pain that fish suffer when fished. The physiological state resulting from being hooked and reeled in, and the types of damage caused by being fished and handled, have been closely studied. As a result, guidelines have been developed to minimize the suffering to which the fish is subjected. The length of time it is kept out of the water, and the hook damage are most important. Nets, especially knotted ones, cause abrasions, which can be avoided by using a soft net. The roughness with which the fish is handled, and the use of stringers, which are especially harmful, are also crucial to the degree of suffering inflicted, as well as the animal's survival. The fish should be gently handled, and unhooked with a minimum of damage, without removing it from the water at all, or for longer than is absolutely necessary. Stringers should never be used.

Fishing tournaments have been found to be so injurious to fish prior to spawning, that they have less offspring. Black bass on their nests who were fished, showed such difficulties afterwards, that they cared far less for their nests, and some abandoned them. The stress of being fished suppressed reproductive hormones in trout. Further, the mortality from fish who are predated while on the hook, appears to be as great as from catch and release, another destructive aspect of fishing which is overlooked.

Fishing should be subjected to close moral and ethical scrutiny, for the treatment of fish would be considered unacceptable in mammals.

Approximately seventy million tons of fish are produced through fish farming annually, and according to *The U.N. Food and Agriculture Organization* (FAO), ninety million tons of wild fish are caught globally. This amounts to more than one trillion fishes caught in fisheries and fishing each year. Given the way these fish are treated during and after capture, the degree of fish suffering far outweighs that of mammals caught up in agribusiness.

In fish farms, in spite of the information that is available about their requirements for stones and other materials for nest making, they are not provided with nesting materials, nor an enriched environment. Simply providing them with materials which would allow them to fulfil their natural inclinations would be an important step to reducing frustration and maladaptive behaviours when the fish are highly motivated to build nests and reproduce. By using knowledge of fish behaviour, important improvements could be made to their captive environment to allow them to express their natural repertoire of activities.

Knowledge about fishes' preferences can be applied to solve certain problems, too. For example, in one situation, low lights in crucial areas kept fish from entering power station intakes, after it was found that they would avoid such lighted areas.

Since the 1980's, over thirty species of fish have been altered by genetic modification, causing controversial impacts on behaviour and welfare. This has resulted in some worrying situations in which the genetically modified fish interacted to the detriment of the natural wild fish.

Fish are the most popular pet after cats and dogs, and the third most popular subject for vivisection after rats and mice. But the use of fish in scientific experimentation is highly regulated, to be sure that due care is given that live fish are only used if no other option is available, and that all measures are taken to ensure that they suffer as little as possible. This is opposite to the way they are treated by fishermen, by the food industry, and by the pet trade.

The consensus is that for an animal to enjoy a state of well-being, it should be free of illness, and kept in conditions that permit it to pursue its natural behaviour, free of pain or suffering. But these criteria are yet to be applied to fish in most of their uses by mankind.

More than a decade after Dr. Sneddon's discovery, there continues to be debate about fish being capable of suffering only because of opposition by the fishing industry, which stands to lose out financially if society begins to frown on fishing as it does on hunting and other forms of cruelty to animals. Yet, looking strictly at the facts established through rigorous scientific experimentation, there is no reason to continue to withhold the strict ethical considerations from fish that have been applied to reptiles, birds, and mammals.

Through my copious writing on the Internet while trying to get the sharks I loved protected, I gained many friends who were dedicated shark advocates. Eventually we established a discussion group, called *The Shark Group*, lead by shark advocate Alex Buttigieg, aka *The Sharkman*, of Malta.

First, we pursued the confrontation of *Discovery Channel* regarding *Shark Week* as described above, which we had begun as members of the Shark-L discussion list.

Then we declared the year 2009 to be *The International Year of the Shark*. The findings of Pew's *Global Shark Assessment* had revealed that the populations of heavily fished shark species was plummeting, and at current rates of decline, their extinction was forecast in ten to fifteen years. That was the trigger that decided us. I tried to persuade the United Nations to make it official, but got nowhere, so we just announced it ourselves, with the intention of making the unprecedented networking power of the Internet serve to our advantage.

I designed a logo for it, using a photograph of Carmeline.

Then we formed another group called *Let Sharks Live,* to act as a hub to coordinate and connect all of the people and groups who used the event.

It allowed us to share material and methods as we worked to draw attention to the plight of sharks, to distribute information, and to put a stop to shark fishing and the shark fin market in our local areas.

The main threats to sharks were identified as :

- Long-line fishing vessels looting the oceans with lines up to fifty miles long, and thousands of baited hooks
- For every ten pounds of fish killed, a hundred pounds of marine life is thrown away. This waste of an estimated fifty billion pounds of marine life yearly, including sharks, is casually referred to as 'by-catch'.
- The methodical killing for shark fin soup
- Smaller operations of a variety of sorts which have great impact taken together
- Habitat loss and destruction, particularly of nursery areas
- Pollution

The goal of *The International Year of the Shark* was to work on all of these problems. We created, shared, and distributed educational material in fifteen languages, including Chinese, and it was celebrated by dive clubs, conservation groups, and shark organizations all over the world. Our logo, which we offered freely to use for shark protection, was carried on a banner through the streets of Hong Kong, as well as in cities in North America and Europe. It was flown on flags, printed on tee-shirts, and displayed on posters and pamphlets around the world. We worked on raising public awareness of the need to stop the slaughter, and particularly on getting the message into China. I was sure that if the Asian people learned the harm that their shark fin soup was doing, they would change the recipe, and the market for shark fins would vanish. The shark fin racket is so wide-spread and so profitable, that it is largely in criminal hands, yet we had faith that those who eat shark fin soup could be convinced to stop supporting it, and if the market disappeared, the massacre around the world would stop.

Public education regarding the true nature of sharks was another project for countering the effects of shark attack hysteria dispensed by the media. Obtaining protection for threatened species, persuading fishermen to practice tag and release only, and countless more solutions to various local problems were taken up and promoted in the diverse international cultures that took part.

As a result of the efforts of shark advocates Wolfgang Leander and Brian W. Darvell, the huge Chinese Internet market, *Alibaba.com* (it corresponds to our *Amazon.com*), banned shark fins from its site as of January 1, 2009, instantly bringing our message to China. Restaurants began to stop serving the infamous soup in increasing numbers as our people around the world refused service in establishments that served it, and told the owners why. *Shark Safe Marinas* appeared in the United States, in resistance to the monster tournaments, and the shark slaughter on private boats.

The International Year of the Shark was essentially run by two people from their computers, one in the middle of the South Pacific, the other across the planet in the Mediterranean Sea, by effectively using the networking power of the Internet to reach like minded individuals and groups. And it took root and grew on its own. Many people told us that *The Year of the Shark* seemed so far reaching, appearing as it did all over the planet, that it was assumed to have an official source, rather than having been created to promote the idea that sharks must be protected, by a few people who felt the problem to be that urgent. Our plea became a movement which is picking up speed as more and more people understand the problem, join in lobbying, and support efforts to save sharks from extinction.

In the spring of 2012, the exhibition *SHARK* opened at the *Museum of Art* in Fort Lauderdale, Florida. It was curated by Richard Ellis, one of the foremost shark artists and authors in the world, who gathered together a history of sharks in art, and put it on display through the work of over seventy artists. He included three of my paintings, plus the art print I published from the illustration of me and Madonna. I had named this work *My Sunset Rendezvous*, since it is the only image that shows what the sunset rendezvous was actually about. No one ever photographed me with my sharks; I was always alone with them. This was the last illustration I painted, an afterthought because I was concerned that if I didn't put myself in one of the illustrations, readers might not be able to visualize how I really was with them. If I had had any idea of the interest it would generate, I would have spent more time on it! A copy of the book itself was included in the exhibition, along with an audio stop which described how I studied the sharks in Polynesia, and tried to save them when they were finned. My efforts to improve the image of the shark in society were mentioned, too. Finally, after all the anguish over it, the story of my sharks got out of Polynesia!

This epic exhibition was highly significant in its message, and had a powerful influence on the public consciousness. Richard researched the history of sharks in art and in our civilization in detail to create it, and wrote a book documenting it, entitled *SHARK : a Visual History,* released in tandem with the exhibition, which I highly recommend. I was greatly honoured that my story was included, and it did create a fine contrast to the traditional images of sharks that featured their teeth.

I have continued to paint them, to highlight their expressions and attitudes, and try to capture the essence of their gaze in portraits, as I first began to do following my dream of the call of the black shark for help, which set me on this long, long road. This watercolour is my favourite so far, of a tiger shark.

Some are posted on my online gallery, *Eternity's Children*, at : ilafranceporcher.wix.com/wildlifeart.

The battle for sharks is far from won. A recent study of shark fishing, "*Global catches, exploitation rates, and rebuilding options for sharks*," by Boris Worm, Brendal Davis, Lisa Kettemer, Christine A. Ward-Paige, Demian Chapman, Michael R. Heithaus, Steven T. Kessel, and my friend Samuel H. Gruber, was published in January, 2013. It revealed that the numbers of sharks killed each year are not falling, that no sharks have been saved, and that the ravenous market for the infamous party soup continues to climb, in spite of the increase in support for shark protection that has come in the last decade.

While many had rejected the previous figure, of seventy-three million sharks slaughtered per year, as being inflated, this in-depth study found that the actual figure is somewhere between sixty-three million and two hundred seventy-three million sharks killed per year, the larger figure being more likely if the sharks are smaller. And this could well be the case. When whole communities of animals are decimated, the average weight of those killed goes down, since the older, bigger ones are gone, and juveniles form a higher fraction of the remaining population.

I saw this effect among my sharks; the old ones quickly vanished, and after that, most of the adults were maturing sharks taking over the ranges of the dead. Most of these were soon finned as well, apparently because as adults they visited the ocean, leaving another set of maturing juveniles to take over their ranges in the lagoon.

In January 2014, the *International Union for Conservation of Nature*'s (IUCN) Shark Specialist Group (SSG), found that one-quarter of all known species of sharks and rays are threatened with extinction. Three hundred scientists systematically analysed the risk of extinction for one thousand forty-one species, comprising this entire lineage of exploited animals—sharks, rays, and chimeras. All available information was used in each evaluation, and modelling estimates were done for species for which data was lacking.

The study revealed the main causes of the depletion of oceanic biodiversity. Fisheries and globalized trade are swiftly expanding, and proving to be the most dire coastal and oceanic threat. Rampant over-fishing and habitat degradation have greatly altered the ecosystems of the ocean, but the researchers say that it is not clear whether the observed population declines are reversible, or a "chronic accumulation of global marine extinction risk."

The main factors contributing to the risk of extinction were found to be the size of the animal, and the depth at which it lives. Shallow water species are more accessible, and therefore at greater risk from fishing and extinction. The risk is highest in fresh waters—one third of fresh water sharks and rays are facing extinction due to their limited, often degraded habitat, and accessibility. The numbers of countries involved in the protection of a given species is another crucial factor, with the practice of finning for shark fin soup posing the greatest danger to sharks. The lucrative trade is mostly unregulated by the eighty-six countries and territories exporting fins to Hong Kong, and there exists no international agreement to monitor and protect the animals they are so heavily exploiting.

The status of sharks and their relatives is worse than any other major vertebrate lineage. Further, many species remain to be discovered, many of which are in high risk, high biodiversity regions of perilous waters, which are still relatively unexplored in terms of biodiversity.

While their findings are shocking, the researchers believe that their conservative, consensus-based approach tends to under-estimate the losses and risks, rather than over-estimate them. Catches are believed to be three or four times greater than reported. Most catches of sharks and rays are not regulated, are neither recorded nor reported, and are discarded at sea.

The researchers concluded that the level of threat that sharks, rays and chimeras are facing, shows a profound failure of conservation efforts, and that a dramatic change in the way they are managed is urgently needed to save them.

If you go underwater you find an intricate multi-species community in a state of continual communication of some sort. They are warning of danger, alarmed by passing ships, frightened by explosions, and disturbed by countless pernicious events, including destruction of their environment, and assaults through hooks in the food and plastics and poisons in the water. When you get to know them you see the stress they are under, and the caution with which they approach anything new.

I never got over the tragedy of my sharks being killed, for such an affront to their spirits as shark fin soup, and have continued to spread the word about them in whatever way I could find—to tell people what they are like, and that their slaughter must end. Given the ongoing massacre, diabolical in its extremity, I beseech you to help. Please tell others about what sharks are like, and how more than ninety percent of them have already been lost. Encourage them to read the story of my sharks, and raise their voices against the carnage before they are gone forever.

Much research is geared to finding out where sharks are, even some that is sponsored by NGOs claiming to be protecting them. This information is easy to get with modern tagging equipment. Little time or money is invested, and the scientist gets publicity for revealing the whereabouts of sharks whom he or she has neither the intention nor capability of protecting. The appalling truth is that this is exactly what shark finners want to know—where they are!

Given the value of their fins, and the threat of extinction, no one who is truly concerned for sharks would ever reveal such information. Revelations of their migration routes have resulted in the butchery of the vast numbers of animals using those routes, which become infested with shark finners.

If you dive and love sharks, I implore you to think twice about aiding and abetting any efforts to 'count' them. Few people other than divers appreciate them, and once the information is given, you have no control over how it will be used. Never reveal the location, species, or numbers of sharks you see while diving to any authority. Since sharks may either avoid or be attracted to divers, the numbers of sharks actually seen do not represent their true numbers on the reef, anyway, so such figures have no scientific value.

Some secrets of the sea should remain secret.

The numbers of millions of sharks being finned all over the world may seem unreal, but I assure you that if you find the sharks you loved to be with lying finned and alive—or dead—on the seafloor instead of circling majestically around you, you will see what I mean. Its one thing to hear about millions of sharks being finned, but its quite another to be truly faced with the reality.

Because when the sharks who were finned were the ones you knew, and loved to be with, you *reel*.

The End

Appendix

This is the article Arthur A. Myrberg Jr. wrote for the scientific journal *Animal Cognition*. It was based on his presentation at the international conference on cognition at the Max Planck Institute, founded by Konrad Lorenz, in Germany, in October, 2003, and provided the first evidence of cognition in sharks (elasmobranchs). As a result of our complicity in preparing the material, he named me as co-author.

Due to his untimely death, it was not published.

Cognition in elasmobranch fishes, a likely possibility.

Abstract

We provide a brief summary of the literature concerning learning in elasmobranch fishes as well as an example of the social organization shown by a colony of captive bonnethead sharks. A series of observations on free-ranging blackfin reef sharks, exemplifying their capability of rapid decision-making, and behavior suggesting the referencing of multiple mental representations, is provided. The evidence clearly suggests that elasmobranchs possess ecological intelligence and cognitive capability. Finally, the proof needed to demonstrate cognition in fishes, in general, is discussed.

Introduction

There is increasing recognition that when animals learn to perform new tasks, they think, consciously or unconsciously, about the problems they face and the solutions they attempt or achieve (Griffin 1992, 2001). It is commonly believed, however, that elasmobranch fishes (the sharks and their near relatives) are inferior to other vertebrates in the extent to which they modify behavior through learning processes. This is supposedly due to much smaller brains possessed by elasmobranchs. The direct comparison of relative brain sizes among vertebrates by Northcutt (1978), showed, however, that elasmobranchs possess brains equal in size or even larger than those possessed by many birds and mammals (Fig. 1).

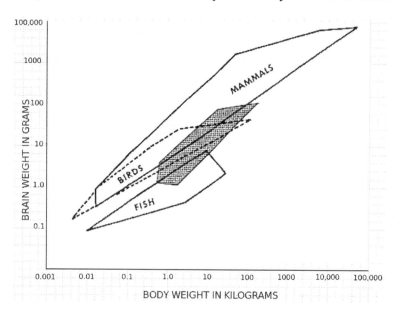

Fig. 1. Brain and body weights for four vertebrate classes expressed as minimum convex polygons, after Jerison (1973). Stippled polygon encloses elasmobranch brain-to-body ratios and overlaps polygons for bony fishes, birds, and mammals (Northcutt (1978).

This unfortunate, "small-brain" myth was the likely result of early workers, concentrating their studies only on readily available, small-brained species. There is also evidence that among other taxa, notably birds, the miniaturization of the brain has no evident effect upon the animal's cognitive faculties. (Barber 1993, Balda and Kamil, 1998, Pepperburg, 1999, Griffin 1992, 2001)

Learning

Excellent evidence exists that sharks not only readily learn operant responses, but readily learn, using the respondent conditioning paradigm as well. Clark (1959) trained an adult male and female lemon shark *Negaprion brevirostris*, to press a submerged target, connected to a submerged bell, five days a week for seven weeks. By the end of the seventh week, both sharks were readily pressing the target almost immediately after submergence. After ten weeks of non-training, testing was begun again and the operant response was repeated immediately by both sharks.

Brightness-discrimination by a female nurse shark *Ginglymostoma cirratum* was also amply demonstrated by differentially pressing small, submerged targets (light vs dark) after operant training (Aronson et al. 1967). The discrimination was learned by day 5 after one session/day, lasting 7-16 minutes and was maintained, without decrement, for one month of testing. The shark's learning curve of the discrimination was compared to the mean learning curve by lab-mice *Mus musculus* and also the cichlid teleost *Tilapia macrocephala*, trained in comparable apparati and similar programmatic techniques (Fig. 2).

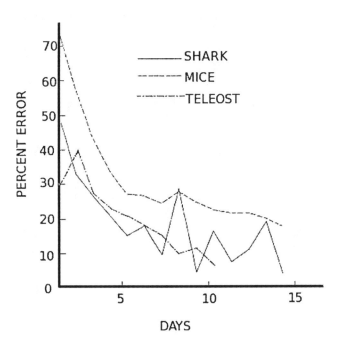

Fig. 2. Learning curve for shark in the light-dark discrimination problem. The average learning curve for 8 mice trained in a similar conditioning paradigm and the average curve for 5 teleosts *Tilapia* in a related problem, involving visual discrimination are provided (see text: from Aronson et al. 1967).

The great similarity among the curves does not equate learning by the members of the three species, but indicates that differences in learning must be examined by using more complex problems.

The next example demonstrated clear learning through use of the second major paradigm, i.e., the classical or respondent conditioning paradigm. Gruber and Schneiderman (1975) conditioned the nictitating eyelid-response (CR) of five immature lemon sharks *Negaprion brevirostris*, using a 1-3 v, electric shock (US) and a collimated, white light (CS) of 0.5 sec. duration with a CS-Us interval of 0.4 sec. Testing ended after 7 days with 100 acquisition-trials/day. Equal numbers of subjects (n=5), in three groups, controlled for spontaneous blinking, sensitization, pseudo-conditioning and backward conditioning. The learning curves of the progress of the CR are shown during the first three days of acquisition (Fig. 3).

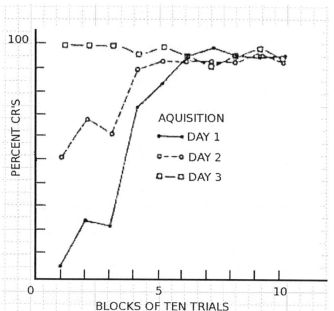

Fig. 3. Percentage of nictitating membrane CRs in the experimental group during successive blocks of trials in the first three acquisitions sessions (Gruber and Schneiderman 1975 @Psychonomic Soc., Inc.)

By Day 2, two of the 5 sharks showed 100% response during the second block of ten trials and by Day 3, essentially 100% response was noted during all trials and this remained throughout the remaining 4 days. Six days of extinction sessions followed the seventh day of testing (Fig. 4).

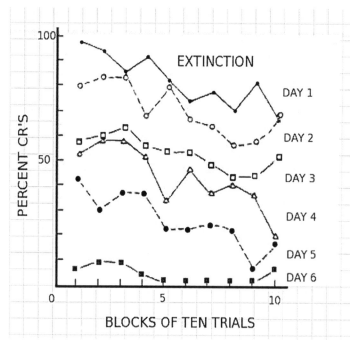

Fig. 4. Percentage of nictitating membrane CRs in the experimental group during successive blocks of trials in each extinction session. (Gruber and Schneiderman 1975 @Psychonomic Soc., Inc.).

One can readily see spontaneous recovery occurring during the second day. Extinction of the CR was essentially complete by the sixth day.

These three-mentioned cases are only a few examples of learning being readily shown by elasmobranchs (see also Wright and Jackson 1964; Banner 1967; Gruber 1967; Nelson 1967; Kalmijn 1971, 1978; Gruber and Cohen 1978; Berg and Schuljf 1983).

Free swimming blackfin reef sharks, *Carcharhinus melanopterus*, learned to recognize the second author during her long-term study (Porcher 2005) after 3 to 5 encounters. They showed their familiarity by approaching from in front, while unfamiliar sharks either remained at a distance or approached from behind, apparently to avoid being seen. As well as demonstrating learning, the sharks' behaviour indicates awareness of frontal images. They consistently responded to slight changes in movement or posture while attending to the researcher, as well. Other ethologists have noted comparable phenomena in studies of predation in terrestrial animals (Elgar 1989).

Social organization

Relatively complex social interactions are also evident among sharks. One such case involves a group of bonnethead sharks *Sphyrna tiburo* (Myrberg and Gruber 1974). Ten adult members of the species were closely observed for approximately two hours, almost daily, for seven months in a large (40 x 60 m) channel, about 7 m in width. Interactions seldom occurred.

However, one interaction invariably took place whenever two individuals approached one another head-on: one shark actively avoided the other by changing course. This resulted in what was termed "Give-way" (Fig. 5).

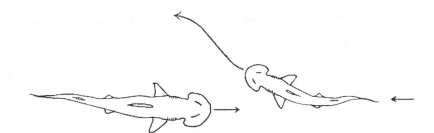

Fig. 5. Shark on the right Giving-way to the shark on the left (Myrberg and Gruber 1974).

The shark that Gave-way was operationally termed as subordinate and the other individual, dominant. Analysis of give-ways amongst all members of the group resulted in several noteworthy facts. First of all, a clear straight-line, size-dependent, social hierarchy existed among all members of the group. Additionally, it was apparent that females shied away from the three largest males (the latter were not the largest members of the group). But, why? Severe biting of females by males occurs during the reproductive season of elasmobranchs and even gender-segregation exists among numerous species. Results from the bonnetheads indicated that females are aware of a potential threat faced when near adult males. Also, since individuals approached one another head-on and at the same depth in the channel during Give-way, it is not unreasonable to consider that the size of a given individual was extremely difficult, if not impossible, to assess and yet the hierarchy showed clear, absolute size dependency. This suggested that bonnethead sharks individually recognize companions.

A similar social organization is also known in another species of shark, the smooth dogfish Mustelus canis (Allee and Dickinson 1954).

The second author repeatedly saw the same individuals visiting together from a distant home range. These companions sometimes appeared during the same month, year after year, yet at no other time, indicating that they knew each other as distinct from all other sharks. Recognition of other individuals has been demonstrated in many species of fish, notably at cleaning stations (Bshary, Wickler and Fricke, 2002).

Decision Making

Cognition is best demonstrated when an animal must make a decision between two alternatives. Evidence of decision making is a strong indicator that the animal referred to one or more mental representations before acting. Many examples of decision making have been found in fish, one of the most exceptional being the yellowhead jawfishes (*Opistognathus aurifrons*), of the western Atlantic, who choose rocks of the appropriate size and shape to build their nests (Colin 1972). Their ability is often given as evidence for cognition in teleost fishes (Bshary, Wickler and Fricke 2002). Blackfin and whitetip reef sharks (*Triaenodon obesus*), sometimes flip on their backs to wriggle in the sand, presumably to scratch or to free themselves of a remora. The substrate in the study area is made up of sand interspersed with reef flats, and the sharks were observed to choose only sandy places for such manoeuvres. They had made a decision with respect to the use of available parts of the environment to serve their purposes.

A more striking example of sharks making a decision occurred when two *C. melanopterus* left a "spawning dome" (Myrberg 1988) where they were hunting, to follow the second author to a feeding site they had not visited before, though they had not seen her for 8 months. Leaving the visible prey and scent-filled area to follow her required making reference to an old memory that she had sometimes supplied food, and granting the mental representation greater importance than the present-time stimulation of the spawning dome. In other words, the sharks thought about the situation and made a choice.

On the few occasions in which the second author brought another person to the study area, the sharks responded by swiftly approaching her, then vanishing into the veiling light. After several minutes, the same sharks reappeared in long files and went directly to the stranger. This initial disappearance never happened when the normal routine was followed. The sharks' actions provide further evidence that they recognize sudden changes in their environment and make consistent and rapid decisions to stay or leave, corresponding to events that are expected versus unexpected.

Eventual proof of cognition in sharks

Fig. 6 shows the hearing thresholds of lemon sharks Negaprion brevirostris that were obtained by two different means.

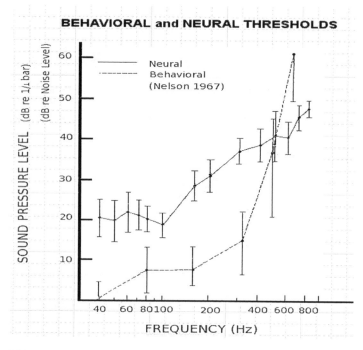

Fig. 6. Behavioral and neural frequency threshold curves for *Negaprion brevirostris*. The solid curve is a mean threshold for 7 lemon shark ears in dB re 1 μbar. Dashed curve is from Nelson's (1967 operant conditioning study of sound thresholds for the same species. Direct comparisons between the curves are nor possible since the values for the dashed curve are referenced to filtered noise level around each frequency. However, a low frequency plateau of best sensitivity below 200 Hz and a gradual threshold increase above that frequency are evident in both curves (Corwin (1981).

The lower graph is an audiogram, based on behavioral responses noted by Nelson (1967), while the upper graph shows compound action potentials, recorded at threshold, from seven lemon sharks, obtained from the CNS (8th cranial nerve) by Corwin (1981) at the same audio frequencies. The two curves show remarkable similarity to the same sensory stimuli by both neural and behavioral actions.

Proof of cognitive abilities in animals await precisely such correlates. But, rather than sensory events, such a correlate must occur between neural and mental events such as emotions and thoughts.

Such a correlate will not be easy to establish. But if it were, there would probably be little interest in doing it (Griffin 1992).

Until then, we must continue to seek justification for cognitive abilities by animals through an ever-increasing abundance of evidence for such abilities.

We consider this report on elasmobranch fishes as adding to that evidence.

Acknowledgements

We thank numerous colleagues, who study elasmobranch fishes, and provided their views on the subject of this report.

References

Allee WC, Dickinson JC (1954) Dominance and subordination in the smooth dogfish *Mustelus canis* (Mitchill). Physiol Zool 27: 356-364

Aronson LR, Aronson FR, Clark E (1967) Instrumental conditioning and light-dark discrimination in young nurse sharks. Bull Mar Sci 17: 249-256

Balda RP, Kamil AC (1998) The ecology and evolution of spatial memory in corvids of the southwestern USA: the perplexing pinion jay. San Diego: Academic Press.

Banner A (1967) Evidence of sensitivity to acoustic displacements in the lemon shark, *Negaprion brevirostris*. In: Cahn PH (ed) Lateral Line Detectors. Indiana University Press, Bloomington, pp. 265-273

Barber, XT 1993 The Human Nature of Birds. New York: St. Martin's Press

Berg AV van den, Schuijf A (1983) Discrimination of sounds based on phase difference between particle motion and acoustic pressure in the shark *Chiloscyllium griseum*. Proc R Soc Lond B 218:127-134

Bshary R, Wickler W, and Fricke H (2002) Fish cognition: a primate's eye view. Anim Cogn 5:1-13

Clark E (1959) Instrumental conditioning of lemon sharks. Science 130: 217-218

Colin PL (1972) Daily activity patterns and effects of environmental conditions on the behavior of the yellowhead jawfish *Opistognathus aurifrons* with notes on its ecology. Zoologica 57:137-169

Corwin JT (1981) Peripheral auditory physiology in the lemon shark: Evidence of parallel otolithic and non-otolithic sound detection. J Comp Physiol 141: 379-390

Elgar, MA (1989) Predator vigilance and group size in mammals and birds. A critical review of the empirical evidence. Biol. Rev. 64:13-33

Griffin DR (1992, 2001) Animal Minds. University of Chicago Press, Chicago

Gruber SH (1967) A behavioral measurement of dark adaptation in the lemon shark, *Negaprion brevirostris*. In: Gilbert PW, Mathewson RF, Rall DP (eds) Sharks, Skates and Rays. John Hopkins Press, Baltimore, pp. 479-490

Gruber SH, Cohen JL (1978) Visual system of the elasmobranchs: State of the art 1960-1975. In: Hodgson ES, Mathewson RF (eds) Sensory Biology of Sharks, Skates and Rays. Office of Naval Research, Arlington, pp. 11-10

Gruber SH, Schneiderman N (1975) Classical conditioning of the nictitating membrane response of the lemon shark *Negaprion brevirostris*. Beh Res Method Inst 7: 430-434

Jerison, HJ (1973) Evolution of the brain and intelligence. Academic Press, New York Kalmijn AJ (1971) The electric sense of sharks and rays. J Exp Biol 55: 371-383

Kalmijn AJ (1978) Experimental evidence of geomagnetic orientation in elasmobranch fishes. In: Schmidt-Koenig K, Keeton WT (eds) Animal Migration, Navigation, and Homing. Springer-Verlag, Berlin, pp. 347-353

Myrberg Jr. AA (1988) The reproductive behavior of *Acanthurus nigrofasciatus* (Forskal) and other surgeonfishes (Fam. *Acanthuridae*) off Eilat, Israel (Gulf of Aqaba, Red Sea). Ethology 79: 31-61 Myrberg Jr. AA, SH Gruber (1974) The behavior of the bonnethead shark, *Sphyrna tiburo* (L.). Copeia 1974: 358-374

Nelson DR (1967) Hearing thresholds, frequency discrimination, and acoustic orientation in the lemon shark, *Negaprion brevirostris* (Poey). Bull Mar Sci 17:741-768

Pepperburg, IM (1999) The Alex studies: Cognitive and communicative abilities of grey parrots. Cambridge: Harvard University Press.

Porcher IF (2005) On the gestation period of the blackfin reef shark, *Carcharhinus melanopterus*. Mar Bio 146: 1207-1211

Wright T, Jackson R (1964) Instrumental conditioning of young sharks. Copeia 1964: 409-412.

Bibliography

Alexander RD (1987) The biology of moral systems. Aldine de Gruiter, New York.

Alfieri, M.S. & Dugatkin, L.A. (2011) Cooperation and cognition in fishes, in Brown, C., Laland, K. & Krause, J. (eds.) Fish Cognition and Behaviour, pp.258–276, Oxford: Blackwell.

Arlinghaus, R. & Mehner, T. (2003) Socio-economic characterisation of specialised common carp (Cyprinus carpio L.) anglers in Germany, and implications for inland fisheries management and eutrophication control. Fisheries Research, 61, 19–33.

Arlinghaus, R., Schwab, A., Cooke, S.J. & Cowx, I.G. (2009) Contrasting pragmatic and suffering centred approaches to fish welfare in recreational angling. Journal of Fish Biology, 75, 2448–2463.

Aronson LR (1951) Orientation and jumping behavior in the goby fish Bathygobius soporator. Am Mus Novitates 1486

Aronson LR (1956) Further studies on orientation and jumping behaviour in the goby fish Bathygobius soporator. Anat Rec 125:606

Ashley, P.J. (2007) Fish welfare: Current issues in aquaculture, Applied Animal Behaviour Science, 104, pp. 199–235.

Ashley, P.J., Ringrose, S., Edwards, K.L., Wallington, E., Mccrohan, C.R. & Sneddon, L.U. (2009) Effect of noxious stimulation upon antipredator responses and dominance status in rainbow trout, Animal Behaviour, 77, pp. 403–410.

Ashley, P.J., Sneddon, L.U. & Mccrohan, C.R. (2007) Nociception in fish: Stimulus-response properties of receptors on the head of trout oncorhynchus mykiss, Brain Research, 1166, pp. 47–54.

Barber TX (1993) The Human Nature of Birds. Bookman Press, Melbourne

Bateson, P. (1991) Assessment of pain in animals, Animal Behaviour, 42, pp. 827–839.

Bekoff, M. (2007) Aquatic animals, cognitive ethology, and ethics: questions about sentience and other troubling issues that lurk in turbid water. Diseases of Aquatic Organisms, 75, 87–98.

Bekoff, M. & Sherman, P.W. (2004) Reflections on animal selves. Trends in Ecology and Evolution, 19, 176–180.

Bekoff, M. (2006) Animal Passions and Beastly Virtues: Reflections on Redecorating Nature. Temple University Press, Philadelphia.

Borsook, D., Sava, S. & Becerra, L. (2010) The pain imaging revolution: Advancing pain into the 21st century, Neuroscientist, 16, pp. 171–185.

Borucinska, J., Kohler, N., Natanson, L., Skomal, G., 2002. Pathology associated with retained fishing hooks in blue sharks, Prionace glauca (L.), with implications for their conservation. J. Fish. Dis. 25, 515–521.

Borucinska, J., Martin, J., Skomal, G., 2001. Peritonitis and pericarditis associated with gastric perforation by a retained fishing hook in a blue shark. J. Aquat. Anim. Health. 13, 347–354.

Broom, D.M. (1991a) Assessing welfare and suffering. Behavioral Processes, 25, 117–123.

Broom, D.M. (1991b) Animal welfare: concepts and measurement. Journal of Animal Science, 69, 4167–4175.

Broom, D.M. (2007) Cognitive ability and sentience: which aquatic animals should be protected? Diseases of Aquatic Organisms, 75, 99–108.

Brown, C. (2001) Familiarity with the test environment improves escape responses in the crimson spotted rainbowfish, Melanotaenia duboulayi. Animal Cognition, 4, 109–113.

Brown, C. & Day, R. (2002) The future of stock enhancements: bridging the gap between hatchery practice and conservation biology. Fish and Fisheries, 3, 79–94

Brown, C. & Laland, K.N. (2001) Social learning and life skills training for hatchery reared fish. Journal of Fish Biology, 59, 471–493.

Brown, C., Davidson, T. & Laland, K. (2003) Environmental enrichment and prior experience of live prey improve foraging behaviour in hatchery-reared Atlantic salmon. Journal of Fish Biology, 63, 187–196.

Brown, C., Jones, F. & Braithwaite, V. (2005) In situ examination of boldness–shyness traits in the tropical poeciliid, Brachyraphis episcopi. Animal Behaviour, 70, 1003–1009.

Brown, C., Laland, K. & Krause, J. (2011) Fish Cognition and Behaviour, Oxford: Blackwell.

Bshary R, Wickler W, Fricke H (2002) Fish cognition: a primate's eye view. Animal Cognition (2002) 5 : 1-13

Bshary, R. (2011) Machiavellian intelligence in fish, in Brown, C., Laland, K. & Krause, J. (eds.) Fish Cognition and Behaviour, pp. 277–297, Oxford: Blackwell.

Buhle, J. & Wager, T.D. (2010) Performance-dependent inhibition of pain by an executive working memory task, Pain, 149, pp. 19–26.

Chandroo KP, Yue S, and Mocci RD (2004) An evaluation of current perspectives on consciousness and pain in fishes. Fish and Fisheries 5. 281-295

Chandroo, K.P., Duncan, I.J.H., Moccia, R.D., 2004b. Can fish suffer? Perspectives on sentience, pain, fear and stress. Appl. Anim. Behav. Sci. 86, 225–250.

Chandroo, K.P., Yue, S., Moccia, R.D., 2004a. An evaluation of current perspectives on consciousness and pain in fishes. Fish. Fish. 5, 281–295.

Chapman, CR and Nakamura Y (1999) A passion of the soul: an introduction to pain for consciousness researchers. Consciousness and Cognition 8. 391-422

Clutten-Brock TH, Parker GA (1995) Punishment in animal societies. Nature 373:209-215

Colin PL (1973) Burrowing behaviour of the yellowhead jawfish, Opistognathus aurifrons. Copeia 1973:84-90

Colin PL (1972) Daily activity patterns and effects of environmental conditions on the behaviour of the yellowhead jawfish Opistognathus aurifrons, with notes on its ecology. Zoologica 57:137-169

Cooke, S.J. & Sneddon, L.U. (2007) Animal welfare perspectives on recreational angling, Applied Animal Behaviour Science, 104, pp. 176–198.

Cooke, S.J., Philipp, D.P., 2004. Behavior and mortality of caught-and-released bonefish (Albula spp) in Bahamian waters with implications for a sustainable recreational fishery. Biol. Conserv. 118, 599–607.

Cooke, S.J., Philipp, D.P., Dunmall, K.M., Schreer, J.F., 2001. The influence of terminal tackle on injury, handling time, and cardiac disturbance of rock bass. N. Am. J. Fish. Manage. 21, 333–342.

Cooke, S.J., Philipp, D.P., Schreer, J.F., McKinley, R.S., 2000. Locomotory impairment of nesting male largemouth bass following catch-and-release angling. N. Am. J. Fish. Manage. 20, 968–977.

Cooke, S.J., Schreer, J.F., Dunmall, K.M., Philipp, D.P., 2002a. Strategies for quantifying sublethal effects of marine catch-and-release angling—insights from novel freshwater applications. Am. Fish. Soc. Symp. 30, 121–134.

Cooke, S.J., Schreer, J.F.,Wahl, D.H., Philipp, D.P., 2002b. Physiological impacts of catch-and-release angling practices on largemouth bass and smallmouth bass. Am. Fish. Soc. Symp. 31, 489–512.

Cooke, S.J., Suski, C.D., 2004. Are circle hooks effective tools for conserving freshwater and marine recreational catchand-release fisheries? Aquat. Conserv. Mar. Freshwater Ecosyst. 14, 299–326.

Cooke, S.J., Suski, C.D., 2005. Do we need species-specific guidelines for catch-and-release recreational angling to conserve diverse fishery resources? Biodiver. Conserv. 14, 1195–1209.

Cooke, S.J., Suski, C.D., Barthel, B.L., Ostrand, K.G., Tufts, B.L., Philipp, D.P., 2003. Injury and mortality induced by four hook types on bluegill and pumpkinseed. N. Am. J. Fish. Manage. 23, 883–893.

Danylchuk, S.E., Danylchuk, A.J., Cooke, S.J., Goldberg, T.L., Koppelman, J. & Phillip, D.P. 2007. Effects of recreational angling on the post-release behaviour and predation of bonefish (Albula vulpes): the role of equilibrium status at the time of release. Journal of Experimental Marine Biology and Ecology, 346, 127–133.

Dawkins, M.S. (1998a) Evolution and animal welfare. Quarterly Review of Biology, 73, 1–21.

Dawkins, M.S. (1998b) Through Our Eyes Only? The Search for Animal Consciousness. Oxford University Press, Oxford.

Derbyshire, S.W.G. (2010) Foetal pain?, Best Practice & Research Clinical Obstetrics & Gynaecology, 24, pp. 647–655.

Dugatkin LA, Wilson DS (1993) Fish behaviour, partner choice, and cognitive ethology. Rev Fish Biol Fish 3:368-372

Duncan, I.J.H. (2002) Gordon memorial lecture. Poultry welfare: science or subjectivity? British Poultry Science, 43, 643–652.

Dunlop, R. & Laming, P. (2005) Mechanoreceptive and nociceptive responses in the central nervous system of goldfish (carassius auratus) and trout (oncorhynchus mykiss), Journal of Pain, 6, pp. 561–568.

Dunlop, R., Millsopp, S. & Laming, P. (2006) Avoidance learning in goldfish (carassius auratus) and trout (oncorhynchus mykiss) and implications for pain perception, Applied Animal Behaviour Science, 97, pp. 255–271.

Duzer EM van (1939) Observations on the breeding habits of the cut-lip minnow, Exoglossum maxillingua. Copeia 1939:65-75

Edelman GM and Tononi G (2000)A Universe of Consciousness. Basic Books, New York, NY

Eysenck, H.J. (1946) The measurement of personality. Proceedings of the Royal Society of Medicine, 40, 75–80.

Feldheim, K. A., Gruber, S. H., DiBattista, J. D., Babcock, E. A., Kessel, S. T., Hendry, A. P., Pikitch, E. K., Ashley, M. V. and Chapman, D. D. (2013), Two decades of genetic profiling yields first evidence of natal philopatry and long-term fidelity to parturition sites in sharks. Molecular Ecology. doi:10.1111/mec.12583

Flecknell, P., Gledhill, J. & Richardson, C. (2007) Assessing animal health and welfare and recognising pain and distress, Altex-Alternativen Zu Tierexperimenten, 24, pp. 82–83.

Fricke H (1971) Fische als Feinde tropischer Seeigel. Mar Biol 9:328-338

Galhardo, L., Almeida, O. & Oliveira, R.F. (2009) Preference for the presence of substrate in male cichlid fish: effects of social dominance and context. Applied Animal Behaviour Science, 120, 224–230.

Geiger W (1956) Quantitative Untersuchungen über das Gehirn der Knochenfische, mit besonderer Berücksichtigung seines relativen Wachstums, I Acta Anat 26: 121-163

Gentle, M.J. (1992) Pain in Birds. Animal Welfare, 1, 235–247.

Gentle, M.J. (2001) Attentional shifts alter pain perception in the chicken. Animal Welfare, 10, S187–194.

Germana, A., Catania, S., Cavallaro, M., Gonzalez-Martinez, T., Ciriaco, E., Hannestad, J. & Vega, J.A. (2002) Immunohistochemical localization of bdnf-, trkb- and trka-like proteins in the teleost lateral line system, Journal of Anatomy, 200, pp. 477–485.

Germana, A., Gonzalez-Martinez, T., Catania, S., Laura, R., Cobo, J., Ciriaco, E. & Vega, J.A. (2004) Neurotrophin receptors in taste buds of adult zebrafish (danio rerio), Neuroscience Letters, 354, pp. 189–192.

Glynn I. (1999) An Anatomy of Thought: The Origin and Machinery of the Mind. Wiedenfield and Nicholson

Grandin T. and Deesing M. American Board of Veterinary Practitioners - Symposium 2002 May 17, 2002, Special Session Pain, Stress, Distress and Fear. Emerging Concepts and Strategies in Veterinary Medicine

Griffin DR (1998) Animal Minds. The University of Chicago Press.

Grutter AS 1995 Relationships between cleaning rates and ectoparasite loads in coral reef fishes. Mar Ecol Prog Ser 118:51-58

Gustaveson, A.W., Wydowski, R.S., Wedemeyer, G.A., 1991. Physiological response of largemouth bass to angling stress. Trans. Am. Fish. Soc. 120, 629–636.

Harmes CA, Lewbart GA, Swanson CR, Kishimori JM, Boylan SM (2005) Behavioural and Clinical Pathology Changes in Koi Carp (Cyprinus carpio) Subjected to Anaesthesia and Surgery with and without Intra-Operative Analgesics. Comparative Medicine Vol. 55 No. 3 221-226

Hawkins, P. (2002) Recognizing and assessing pain, suffering and distress in laboratory animals: A survey of current practice in the UK with recommendations, Laboratory Animals, 36, pp. 378–395.

ILAR (2009) Recognition and Alleviation of Pain in Laboratory Animals, Washington, DC: National Academies Press.

Jarvis, E.D., Güntürkün, O., Bruce, L., Csillag, A., Karten, H., Kuenzel, W., Medina, L., Paxinos, G., Perkel, D.J., Shimizu, T., Striedter, G., Wild, J.M., Ball, G.F., Dugas-Ford, J., Durand, S.E., Hough, G.E., Husband, S., Kubikova, L., Lee, D.W., Mello, C.V., Powers, A., Siang, C., Smulders, T.V., Wada, K., White, S.A., Yamamoto, K., Yu, J., Reiner, A. & Butler, A.B. (2005) Avian brains and a new understanding of vertebrate brain evolution. Nature Reviews Neuroscience, 6, 151–159.

Kacher H (1963) Opithognathus aurifrons (Opisthognathidae) Graben einer Wohnhöhle, Wallbau Film E514, Encyclopaedia Cinematographica. Institut für den wissenschaftlichen Film, Göttingen

Kavaliers, M., Colwell, D.D., 1991. Sex differences in opioid and non-opioid mediated predator-induced analgesia in mice. Brain Res. 568, 173–177.

Kieffer, J.D., 2000. Limits to exhaustive exercise in fish. Comp. Biochem. Physiol. 126A, 161–179.

Kieffer, J.D., Kubacki, M.R., Phelan, F.J.S., Philipp, D.P., Tufts, B.L., 1995. Effects of catch and release angling on nesting male smallmouth bass. Trans. Am. Fish. Soc. 124, 70–76.

Kuhajda, M.C., Thorn, B.E., Kilinger, M.R., 1998. The effect of pain on memory for affective words. Ann. Behav. Med. 20, 1–5.

Lachner A (1952) Studies of the biology of the cyprinid fish of the chub genus Nocomis of northeastern United States. Am Midl Nat 48:433-466

Laureys S., Faymonville ME, Luxon A, Lamy M, Franck G, and Maquet P (2000) Restoration of thalamocortical connectivity after recovery from persistent

Lawson, P.W., Sampson, D.B., 1996. Gear-related mortality in selective fisheries for ocean salmon. N. Am. J. Fish. Manage. 16, 512–520.

Lee, S.T., Lee, J., Lee, M., Kim, J.W. & Ki, C.S. (2009) Clinical and genetic analysis of Korean patients with congenital insensitivity to pain with anhidrosis, Muscle & Nerve, 40, pp. 855–859.

Li X, Keith DE Jr, Evans CJ. 1996. Mu opioid receptor-like sequences are present throughout vertebrate evolution. J Molecular Evol 43:179-184.

Lovibond PF, Shanks DR (2002) The role of awareness in Pavlovian conditioning: empirical evidence and theoretical implication. Journal of Experimental Psychology 28 3-26

Luiten PGM. 1975. The central projections of the trigeminal, facial, and anterior lateral line nerves in the carp (Cyprinus carpio L.). J Comp Neurol 160:399-418.

Maes, J., Turnpenny, A.W.H., Lambert, D.R., Nedwell, J.R., Parmentier, A. & Ollevier, F. (2004) Field evaluation of a sound system to reduce estuarine fish intake rates at a power cooling water inlet. Journal of Fish Biology, 64, 938–946.

Maren S (2001) Neurobiology of Pavlovian fear conditioning. Annual Review of Neuroscience 24. 897-931

Marino L (2002) Convergence of complex cognitive abilities in cetaceans and primates. Brain, Behaviour and Evolution 59. 21-32

Matthews G, Wickelgren WO. 1978. Trigeminal sensory neurons of the sea lamprey. J Comp Physiol A 123:329-333.

Meunier, B., Yavno, S., Ahmed, S. & Corkum, L.D. (2009) First documentation of spawning and nest guarding in the laboratory by the invasive fish, the round goby (Neogobius melanostomus). Journal of Great Lakes Research, 35, 608–612.

Milinski M, Pfluger D, Külling D, Kettler R (1990b) Do sticklebacks cooperate repeatedly in reciprocal pairs? Behav Ecol Sociobiol 27:17-21

Millsopp S, Laming P. 2008. Trade-offs between feeding and shock avoidance in goldfish (Carassius auratus). Appl Anim Behav Sci 113:247-254.

Millsopp, S. & Laming, P. (2008) Trade-offs between feeding and shock avoidance in goldfish (Carassius auratus). Applied Animal Behaviour Science, 113, 247–254.

Molyneux, B. (2010) Why the neural correlates of consciousness cannot be found? Journal of Consciousness Studies, 17, 168–188.

Morris D (1958) The reproductive behaviour of the ten-spined stickleback (Pygosteus pungitius L.). Behaviour [Suppl] 6:1-154

Mourier, J., & Planes, S. (2013). Direct genetic evidence for reproductive philopatry and associated fine-scale migrations in female blacktip reef sharks (Carcharhinus melanopterus) in French Polynesia. Molecular ecology, 22(1), 201-214.

Mourier, J., Mills, S. C., & Planes, S. (2013). Population structure, spatial distribution and life-history traits of blacktip reef sharks Carcharhinus melanopterus. Journal of fish biology, 82(3), 979-993.

Mourier, J., Vercelloni, J., & Planes, S. (2012). Evidence of social communities in a spatially structured network of a free-ranging shark species. Animal Behaviour, 83(2), 389-401.

Myrberg AA, Montgomery WL, Fishelson L (1988) The reproductive behaviour of Acanthurus nigrofuscus (Forskal) and other surgeon fishes (Fam. Acanthuridae) of Eilat, Israel (Gulf of Aqaba, Red Sea). Ethology 79:31-61

Myrberg AA, Riggio RJ (1985) Acoustically mediated individual recognition by a coral reef fish (Pomacentrus portitus)

Neiffer DL, Stamper MA. 2009. Fish sedation, anesthesia, analgesia, and euthanasia: Considerations, methods, and types of drugs. ILAR J 50:343-360.

Newby, N.C., Wilkie, M.P. & Stevens, E.D. (2009) Morphine uptake, disposition, and analgesic efficacy in the common goldfish (Carassius auratus), Canadian Journal of Zoology, 87, pp. 388–399.

Nowak MA, Sigmund K (1998) Evolution of indirect reciprocity by image scoring. Nature 393:573-577

Oliveira, R.F., McGregor, P.K. & Latruffe, C. (1998) Know thine enemy: fighting fish gather information from observing conspecific interactions. Proceedings of the Royal Society of London Series B– Biological Sciences, 265, 1045–1049.

Ostrand, K.G., Cooke, S.J., Wahl, D.H., 2004. Effects of stress on largemouth bass reproduction. N. Am. J. Fish. Manage. 24, 1038–1045.

Overmier JB, Hollis KL (1990) Fish in the think tank: learning, memory, and integrated behaviour. Neurobiology of Comparative Cognition (eds Kesner RP, Olson DS), Lawrence Erlbaum, Hillsday, NJ. pp. 205-236

Pankhurst, N.W., Dedual, M., 1993. Effects of capture and recovery on plasma levels of cortisol, lactate and gonadal steroids in a natural population of rainbow trout. J. Fish Biol. 45, 1013–1025.

Pankhurst, N.W., Van Der Kraak, G., 1997. Effects of stress on reproduction and growth of fish. In: Iwama, G.K., Sumpter, J., Pickering, A.D., Schreck, C.B. (Eds.), Fish Stress and Health in Aquaculture. Society for Experimental Biology

Pasko, L. (2010) Tool-Like behavior in the sixbar wrasse, Thalassoma hardwicke (Bennett, 1830). Zoo Biology, 28, 1–7.

Pham, T.M., Hagman, B., Codita, A., Van Loo, P.L.P., Strommer, L. & Baumans, V. (2010) Housing environment influences the need for pain relief during post-operative recovery in mice, Physiology & Behavior, 99, pp. 663–

668.

Philipp, D.P., Toline, C.A., Kubacki, M.F., Philipp, D.B.F., Phelan, F.J.S., 1997. The impact of catch-and-release angling on the reproductive success of smallmouth bass and largemouth bass. N. Am. J. Fish. Manage. 17, 557–567.

Pitcher, T.J., Hollingworth, C., 2002a. Recreational Fisheries: Ecological, Economic and Social Evaluation. Blackwell Science, Oxford, UK, p. 271.

Portavella, M., Vargas, J.P., Torres, B., Salas, C., 2002. The effects of telencephalic pallial lesions on spatial, temporal, and emotional learning in goldfish. Brain Res. Bull. 57, 397–399.

Reilly, S.C., Quinn, J.P., Cossins, A.R. & Sneddon, L.U. (2008a) Behavioural analysis of a nociceptive event in fish: Comparisons between three species demonstrate specific responses, Applied Animal Behaviour Science, 114, pp. 248–259.

Reilly, S.C., Quinn, J.P., Cossins, A.R. & Sneddon, L.U. (2008b) Novel candidate genes identified in the brain during nociception in common carp (cyprinus carpio) and rainbow trout (oncorhynchus mykiss), Neuroscience Letters, 437, pp. 135–138.

Rink, E. & Wullimann, M.F. (2004) Connections of the ventral telencephalon (subpallium) in the zebrafish (danio rerio), Brain Research, 1011, pp. 206–220.

Roberts G (1998) Competitive altruism: from reciprocity to the handicap principle. Proc R Soc Lond B 265:427-431

Roques, J.A.C., Abbink, W., Geurds, F., Van De Vis, H. & Flik, G. (2010) Tailfin clipping, a painful procedure: Studies on Nile tilapia and common carp, Physiology & Behavior, 101, pp. 533–540.

Rose, JD 2002 The neurobehavioural nature of fishes, and the question of awareness and pain. Rev. Fish. Sci. 10:1-38

Rushbrook, B.J., Dingemanse, N.J. & Barber, I (2008) Repeatability in nest construction by male three-spined sticklebacks. Animal Behaviour, 75, 547-553.

Schmidt, M.B., Balk, H. & Gassner, H. (2009) Testing in situ avoidance reaction of vendace, Coregonus albula, in relation to continuous artificial light from stationary vertical split-beam echosounding. Fisheries Management and Ecology, 16, 376–385.

Seminar Series, 62. Cambridge University Press, Cambridge, UK, pp. 73–95.

Shettleworth, S.J. (2001) Animal cognition and animal behaviour. Animal Behaviour, 61, 277–286.

Sneddon, L.U. (2002) Anatomical and electrophysiological analysis of the trigeminal nerve in a teleost fish, oncorhynchus mykiss, Neuroscience Letters, 319, pp. 167–171.

Sneddon, L.U. (2003a) The evidence for pain in fish: The use of morphine as an analgesic, Applied Animal Behaviour Science, 83, pp. 153–162.

Sneddon, L.U. (2003b) Trigeminal somatosensory innervation of the head of a teleost fish with particular reference to nociception, Brain Research, 972, pp. 44–52.

Sneddon, L.U. (2004) Evolution of nociception in vertebrates: Comparative analysis of lower vertebrates, Brain Research Reviews, 46, pp. 123–130.

Sneddon, L.U. (2006) Ethics and welfare: Pain perception in fish, Bulletin of the European Association of Fish Pathologists, 26, pp. 7–11.

Sneddon, L.U. (2009) Pain perception in fish: Indicators and endpoints, ILAR Journal, 50, pp. 338–342.

Sneddon, L.U. (2012) Response to Rose et al. 2012 Can fish really feel pain? Fish and Fisheries, in press.

Sneddon, L.U., Braithwaite, V.A. & Gentle, M.J. (2003a) Do fishes have nociceptors? Evidence for the evolution of a vertebrate sensory system, Proceedings of the Royal Society B: Biological Sciences, 270, pp. 1115–1121.

Sneddon, L.U., Braithwaite, V.A.&Gentle,M.J. (2003b) Novel object test: Examining nociception and fear in the rainbow trout, Journal of Pain, 4, pp. 431–440.

Sneddon, LU 2003. The evidence for pain in fish: The use of morphine as an analgesic. Applied Animal Behavior Science. 83:153-162.

Snow PJ, Renshaw GMC, Hamlin KE. 1996. Localization of enkephalin immunoreactivity in the spinal cord of the long-tailed ray Himantura fai. J Comp Neurol 367:264-273.

Snow, P.J., Plenderleith, M.B. & Wright, L.L. (1993) Quantitative study of primary sensory neurone populations of three species of elasmobranch fish, Journal of Comparative Neurology, 334, pp. 97–103.

Thunken, T., Waltschyk, N., Bakker, T.C.M. & Kullmann, H. (2009) Olfactory self-recognition in cichlid fish, Animal Cognition, 12, pp. 717–724.

Valet, M., Sprenger, T. & Tolle, T.R. (2010) Studies on cerebral processing of pain using functional imaging — somatosensory, emotional, cognitive, autonomic and motor aspects, Schmerz, 24, p. 114.

Vecino, E., Caminos, E., Becker, E., Rudkin, B.B., Evan, G.I. & Martin-Zanca, D. (1998) Increased levels of trka in the regenerating retinal ganglion cells of fish, NeuroReport, 9, pp. 3409–3413.

Wells, R.M.G., McIntyre, R.H., Morgan, A.K., Davie, P.S., 1986. Physiological stress responses in big gamefish after capture: observations on plasma chemistry and blood factors. Comp. Biochem. Physiol. 84A, 565–571.

Wilkie, M.P., Brobbel, M.A., Davidson, K., Forsyth, L., Tufts, B.L., 1997. Influences of temperature upon the postexercise physiology of Atlantic salmon (Salmo salar). Can. J. Fish. Aquat. Sci. 54, 503–511.

Wilkie, M.P., Davidson, K., Brobbel, M.A., Kieffer, J.D., Booth, R.K., Bielak, A.T., Tufts, B.L., 1996. Physiology and survival of wild Atlantic salmon following angling in warm summer waters. Trans. Am. Fish. Soc. 125, 572–580.

Yue, S., Moccia, R.D., Duncan, I.J.H., 2004. Investigating fear in domestic rainbow trout, Oncorhynchus mykiss, using an avoidance learning task. Appl. Anim. Behav. Sci. 87, 343–354.

Zahavi A (1995) Altruism as a handicap—the limitations of kin selection and reciprocity. J Avian Biol 26:1-3

Zfin (2010) [Online], http://zfin.org/action/marker/view/ZDB-GENE-980526-118 [28 August 2011].

Zimmerman, M. (1986) Physiological mechanisms of pain and its treatment, Klinische Anäesthesiologie Intensivtherapie, 32, pp. 1-19

About the author:

Ila France Porcher is a published ethologist who has spent much of her life observing wildlife. She grew up in British Columbia, Canada, and at an early age became fascinated by watching and drawing wild animals. As a result, she naturally became a wildlife artist, and as time passed, began documenting the behavior of the animals she observed and painted, being especially intrigued by actions suggesting intelligence and cognition.

This approach was balanced by university studies in the fields of physics, chemistry, and mathematics, which has left her with a unique perspective on the biosphere of this world.

Other books she has written are *The True Nature of Sharks*, released on 17/07/17, and *MERLIN | The Mind of a Sea Turtle,* which presents the stories of the sea turtles included in this book as evidence of sea turtle sentience.

The True Nature of Sharks lays out her findings from these shark sessions, and other observations of wild sharks to present a full scale exploration of their behaviour. The True Nature of Sharks is a revolutionary book that does not flinch at denouncing the foundations of traditional shark science and how it has stood in the way of the true understanding of these maligned animals, through its unholy marriage with fisheries.